When Western missionaries introduced modern chemistry to China in the 1860s, they called the discipline *hua-hsüeh*, literally "the study of change" – an appropriate name for the branch of science that studies the nature of pure substances and the processes by which they are transformed into something new.

In this first full-length study of science in modern China, James Reardon-Anderson describes the introduction and development of chemistry in China in the late nineteenth and early twentieth centuries, and examines the impact of these events on Chinese language, education, industry, research, culture, society, and politics.

In a broader sense, *The Study of Change* also explores the relationships among science, state, and society in the modernization process. Professor Reardon-Anderson shows that science in China fared best when a balance was struck between political authority and free social development. Chinese science and scientists had difficulty when a too-powerful state restricted the pursuit of knowledge, or, conversely, when political chaos precluded the order needed to give direction and purpose to dispersed scientific endeavors.

Throughout the book Professor Reardon-Anderson sets the development of chemistry in the broader context of the history of science in China and the social and political changes of this era. The narrative moves from detailed descriptions of particular chemical processes and innovations to more general discussions of intellectual and social history. It is based on an extensive study of Chinese and English sources.

Scholars and advanced students of Chinese studies and the history of science will find this a fascinating account of an important episode in the story of modern China. For a list of recent books in this series, turn to page 435.

The Study of Change

STUDIES OF THE EAST ASIAN INSTITUTE, COLUMBIA UNIVERSITY

The East Asian Institute is Columbia University's center for research, education, and publication on modern East Asia. The Studies of the East Asian Institute were inaugurated in 1961 to bring to a wider public the results of significant new research on modern and contemporary East Asia.

THE STUDY OF CHANGE

Chemistry in China, 1840–1949

JAMES REARDON-ANDERSON

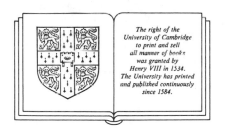

The right of the
University of Cambridge
to print and sell
all manner of books
was granted by
Henry VIII in 1534.
The University has printed
and published continuously
since 1584.

Cambridge University Press

Cambridge

New York   Port Chester   Melbourne   Sydney

Published by the Press Syndicate of the University of Cambridge
The Pitt Building,Trumpington Street, Cambridge CB2 1RP
40 West 20th Street, New York, NY 10011, USA
10 Stamford Road, Oakleigh, Melbourne 3166, Australia

First published 1991

Printed in the United States of America

Tables 11.1, 11.2, and 11.3 and portions of chapters 7, 11, and 12
appeared in another form in James Reardon-Anderson, "Chemical Industry
in China, 1860–1949," *OSIRIS* 2 (1986), reprinted by permission of The
History of Science Society.

Library of Congress Cataloging-in-Publication Data
Reardon-Anderson, James.
The study of change: chemistry in China, 1840–1949 / James Reardon-Anderson.
p.   cm. – (Studies of the East Asian Institute, Columbia University)
Includes bibliographical references.
1. Chemistry – China – History – 19th century.   2. Chemistry – China – History –
20th century.   3. Science and state – China – History – 19th century.   4. Science and
state – China – History – 20th century.   I. Title.   II. Series: Studies of the East Asian
Institute.
QD18.C5R43   1991   90–34975
540'.951'09034–dc20   CIP

British Library Cataloguing in Publication Data
Reardon-Anderson, James
The study of change: chemistry in China, 1840–1949.   –
(Studies of the East Asian Institute).
1. China, 1644–1949
I. Title   II. Series
951.03

ISBN 0–521–39150–4 hardback

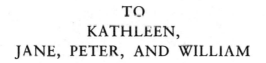

TO
KATHLEEN,
JANE, PETER, AND WILLIAM

Contents

Abbreviations   *page* XIII

Preface   XVII

Introduction   I

Part I *Science and self-strengthening, 1840–1895*

1   The advocates: chemical translators, John Fryer and
    Hsü Shou   17

2   Changing Chinese: chemical translations of the
    Kiangnan Arsenal   29

3   The limits of change: science, state and society in the
    nineteenth century   53

Part II *The interregnum, 1895–1927*

4   First-generation scientists: makers of China's New
    Culture   79

5   Learning about science   104

6   The beginning of chemical research   132

7   Chinese entrepreneurs and the rise of the chemical
    industry   153

Part III   The Nanking Decade, 1927–1937

8   Science and the state during the Nanking Decade   177

9   Scientific education: the balance achieved   208

10   Scientific research: the balance threatened   230

11   The chemical industry and the limits of growth   258

Part IV   The War, 1937–1945

12   Science in Nationalist China: the wartime experience   293

13   Science in Communist China I: innovations in industry   319

14   Science in Communist China II: scientists versus the
     state   339

15   Conclusion   365

     Appendix   376

     Glossary   391

     Bibliography   401

     Index   437

Tables

2.1	Billequin's characters	*page* 40
4.1	Chinese scientific publications, 1851–1936	83
4.2	Chinese terms for physical structures	85
4.3	Japanese and Chinese scientific terms	86
5.1	Semester hours taught in Chinese mission colleges, 1926	122
7.1	Chinese-owned factories that use chemical inputs, inaugurated 1895–1913	158
8.1	Membership of the Science Society of China by fields, 1930	183
8.2	Selected terms from the "Principles of Chemical Nomenclature" (1932)	193
8.3	Chinese scientific societies	197
8.4	China Foundation expenditures, 1926–1937	199
8.5	British Boxer indemnity grants, 1934–1937	201
8.6	Funds for education and research, 1934–1937	201
9.1	Student enrollments in higher education, by subject, 1928–36	216
9.2	Chemistry students, Yenching University, 1926–1934	217
9.3	Number of students in colleges and universities, by department and college, 1931	218
9.4	Summary of W. E. Tisdale's findings on scientific education in Chinese colleges and universities, 1933	220
10.1	Institutional affiliation of authors in the *Journal of the Chinese Chemical Society*, 1933–1937	231
11.1	Chinese manufacturers of sodas and acids	275
11.2	Chinese imports and production of soda	282
11.3	Chinese imports and production of acid	283
12.1	Production of synthetic gasoline from vegetable oil, Nationalist China, 1940–1944	299

12.2   Production of ethanol, Nationalist China, 1940–
        1944   302
12.3   Production of chemicals, Nationalist China,
        1940–1944   302
12.4   Enrollment in Chinese higher education, Spring 1944   307
13.1   Publicly owned factories in the Shen-Kan-Ning
        Border Region, December 1942   325
15.1   Higher education, enrollment by field, selected years,
        1936–1957   366
15.2   Chemical production in Manchuria and China
        proper, selected years   370
15.3   Leading chemists of pre-1949 China in the PRC   371

Abbreviations

CFJP   *Chieh-fang jih-pao* [Liberation daily]. Yenan, 1941–7. Chinese.

CHCYC   *Chung-hua chiao-yü chieh* [Chung-hua educational review]. Chung-hua chiao-yü-chich tsa-chih she [Chung-hua Educational Review Magazine Society]. Shanghai, 1912–37. Chinese.

CJP   *Chinese Journal of Physiology* [Chung-kuo sheng-li-hsüeh tsa-chih]. Chinese Society of Physiology [Chung-kuo sheng-li-hsüeh hui]. Peking, 1927– English.

CKKCSL   *Chung-kuo k'o-chi shih liao* [China historical materials of science and technology]. "Chung-kuo k'o-chi shih-liao" pien-chi-pu [Editorial Department of "China historical..."]. Peking, 1980– . Chinese.

CMB   The China Medical Board.

CMJ   *The China Medical Missionary Journal*, 1887–1909; *The China Medical Journal*, 1909–31. The China Medical Missionary Association. Peking, Shanghai. English.

CRMJ   *The Chinese Recorder and Missionary Journal.* Foochow, Shanghai, 1868–72, 1874–1941. English.

CYNC   *Chung-kuo chiao-yü nien-chien* [China education year book]. Ministry of Education. Shanghai, 1934, 1948. Chinese.

CYTC   *Chiao-yü tsa-chih* [Educational review]. Shang-hai shang-wu yin-shu-kuan [Shanghai Commercial Press]. Shanghai, Changsha, Hongkong, 1909–48. Chinese.

HCN   *Hsin ch'ing-nien* [New youth]. Shanghai, Peking, Canton, 1915–22. Chinese.

HHKC   *Hua-hsüeh kung-ch'eng* [Journal of chemical engin-

eering, China]. Chung-kuo hua-hsüeh kung-ch'eng
hsüeh-hui [Chinese Institute of Chemical Engineers].
Tientsin, 1934–49. Chinese and English.

HHKY   *Hua-hsüeh kung-yeh* [Chemical industry]. Chung-
hua hua-hsüeh kung-yeh hui [Chinese Society of
Chemical Industry]. Peking, Shanghai, Chungking,
1923–49. Title varies, vols. 1–3: *Chung-hua hua-
hsüeh kung-yeh-hui hui-chih* [The journal and proceedings of the China Society of Chemical Industry].
Chinese.

HHSC   *Hua-hsüeh shih-chieh* [Chemical world]. Shang-hai
shih hua-hsüeh kung-yeh chü [Shanghai Chemical
Industry Society]. Shanghai, 1946–66. Chinese.

HHTP   *Hua-hsüeh t'ung-pao* [Chemistry]. K'o-hsüeh ch'u-
pan-she [Science Publishers]. Peking, 1951– .
Chinese.

JACS   *Journal of the American Chemical Society.* The American Chemical Society. Washington, DC, 1979– .
English.

JBC   *Journal of Biological Chemistry.* Rockefeller Institute
for Medical Research. Baltimore, 1905– . English.

JCCS   *Journal of the Chinese Chemical Society* [Chung-kuo
hua-hsüeh-hui hui-chih]. Chinese Chemical Society
[Chung-kuo hua-hsüeh hui]. Nanking, Chungking,
1933–66. English.

JCP   *Journal of Chemical Physics.* American Institute of
Physics. Lancaster, PA, 1933– . English.

JSHS   *Japanese Studies in the History of Science.* Nippon
Kagakusi Gakkai [The history of science society of
Japan]. Tokyo, 1962– . English.

KCHP   *Ko-chih hui-pien* [The Chinese scientific magazine].
John Fryer. Shanghai, 1876–7, 1880–1, 1890–2.
Chinese.

KH   *K'o-hsüeh* [Science]. Chung-kuo k'o-hsüeh-she
[Science Society of China]. Shanghai, 1915–58.
Chinese.

KHYJSK   *K'o-hsüeh yü jen-sheng-kuan* [Science and the philosophy of life]. Ed. Ch'en Tu-hsiu and Hu Shih.
Shanghai: Ya-tung, 1923. Reprinted as *K'o-hsüeh
yü jen-sheng-kuan chih lun-chan* [Debate on science

and the philosophy of life]. Ed. Wang Meng-tsou. Hong Kong: Chinese University of Hong Kong, 1973. Chinese.

KJCC   *K'ang-Jih chan-cheng shih-ch'i chieh-fang-ch'ü k'o-hsüeh chi-shu fa-chan shih tsu-liao* [Historical materials on the development of science and technology in the liberated areas during the War of Resistance]. 5 vols. Ed. Wu Heng. Peking: Chung-kuo hsüeh-shu ch'u-pan-she, 1983−5. Chinese.

KYCH   *Kung-yeh chung-hsin* [Industrial center]. Chung-yang kung-yeh shih-yen-so [National Bureau for Industrial Research]. Nanking, Chungking, 1932−49. Chinese.

NCUSR   *National Central University Science Reports, Series A, Physical Sciences* [Kuo-li chung-yang ta-hsüeh k'o-hsüeh yen-chiu-lu, chia tsu, wu-chih k'o-hsüeh]. National Central University [Kuo-li chung-yang ta-hsüeh]. Nanking, 1930−3. English.

PCF   *Shen-Kan-Ning pien-ch'ü tzu-jan pien-cheng-fa yen-chiu tsu-liao* [Research materials on the dialectics of nature in the Shen-Kan-Ning Border Region]. Ed. Shen-hsi-sheng kao-teng yüan-hsiao tzu-jan pien-cheng-fa yen-chiu-hui, Yen-an ta-hsüeh fen-hui [Shensi Provincial Higher Schools Dialectics of Nature Research Society, Yenan University Branch]. Sian: Shen-hsi jen-min ch'u-pan-she, 1984. Chinese.

PSEBM   *Proceedings of the Society for Experimental Biology and Medicine.* Society for Experimental Biology and Medicine. New York, 1903− . English.

RAC   Rockefeller Archive Center, Pocantico Hills, New York.

SQNUP   *Science Quarterly of the National University of Peking* [Kuo-li Pei-ching ta-hsüeh, tzu-jan k'o-hsüeh chi-k'an]. National University of Peking, College of Science [Kuo-li Pei-ching ta-hsüeh li-hsüeh-yüan]. Peiping, 1929−35. English.

SRNTU   *Science Reports of National Tsing Hua University, Series A: Mathematical and Physical Science* [Kuo-li Ch'ing-hua ta-hsüeh li-k'o pao-kao, ti-i-chung]. National Tsing Hua University [Kuo-li Ch'ing-hua ta-hsüeh]. Peiping, 1931–3. English.

TFTC   *Tung-fang tsa-chih* [Eastern Miscellany]. Tung-fang tsa-chih she [Eastern Miscellany Society]. Shanghai, Changsha, Hongkong, Chungking, 1904–48. Chinese.

TJKHSYC   *Tzu-jan k'o-hsüeh shih yen-chiu* [Studies in the history of natural sciences]. "Tzu-jan k'o-hsüeh shih yen-chiu" pien-chi wei-yüan-hui [Editorial Committee for the *TJKHSYC*]. Peking, 1981– . Chinese.

TLPL   *Tu-li p'ing-lun* [Independent critic]. Tu-li p'ing-lun she [Independent Critic Society]. Peiping, 1932–7. Chinese.

Preface

In the spring of 1840, Lin Tse-hsü, governor-general of Kwangtung and Kwangsi, arrived in Canton with orders from the emperor to eradicate the opium trade. Lin issued a decree that all opium must be surrendered immediately, and eight days later the British consul delivered nearly 20,000 chests of the substance. Now that he had it, what was Lin to do with this great mound of "foreign mud"? Shipping it north to Peking would be both expensive and risky. Mixing it with tung oil and burning it, the usual means of destroying opium, might leave as much as a quarter of the drug undamaged, inviting thieves and tempting corrupt officials. Boiling down the opium and soaking it in brine and lime, a method favored by some because it left a residue that could not be recovered, was time-consuming, expensive, and left open the possibility that as the procedure dragged on, some of the opium might be stolen. Finally, after seeking the advice of experts, Lin adopted a new process, which proved remarkably effective:

[Lin] ordered two pits to be dug at a high point along the shore outside the gates of Canton. Each pit was a square, more than 15 feet on a side, surrounded by a fence to keep out thieves and connected to the sea by two ditches, one in front and one behind. At high tide, the rear ditch was opened, allowing sea water to flow into the pit. Once full, blocks of opium were thrown in and left to soak in the salt water for half a day, after which lime was added. Workmen, standing around the edge of the pit, stirred the boiling mass with iron shovels, causing the opium to react with the salt water and lime, forming little beads. As the tide began to flow out, they opened the ditch in front, allowing the mixture to exit to the sea. Once emptied, the pit was washed clean with new water and the process repeated. In this way, they destroyed 800 to 900 chests of opium a day, and the whole job was completed in a matter of weeks.[1]

It was so like Lin Tse-hsü, the first Chinese official to recognize the superiority of Western "ships and guns" and alert his country-

---

[1] Li Ch'iao-p'ing, *Chung-kuo hua-hsüeh shih* [History of Chinese Chemistry], 2 vols. (Taipei: T'ai-wan shang-wu yin-shu-kuan, 1976), 727–9.

men to the potential of modern techology, to try a new method for disposing of opium, which, although Lin could not have known it, relied on chemical reactions. Throughout the next century, a growing number of Lin's countrymen responded in similar fashion to a succession of foreign threats by resorting to technological innovation. The problems were not always so simple, nor the solutions so apt, but the results were to invest chemistry and other branches of science deeply and broadly in the fabric of modern China.

This book is about the first century of Chinese chemistry, and at times I feared it might take me that long to write it. I began about ten years ago, more out of personal curiosity than scholarly purpose, to look for information on science in China and discovered one of the ironies of China studies and, as I later learned, modernization studies in general: namely, that whereas almost everyone agrees that science is an essential feature of modernity, almost no one has written about the development of science in so-called "modernizing" non-Western societies. China, despite the attention it has attracted, is no exception. In fact, this irony is easy to explain. Scholars who have spent years learning esoteric languages and the intricacies of strange cultures have no time or energy left for the equally forbidding world of science, and vice versa. Some people study China, some study science, but few have the stomach for both. An important, interesting subject with few busy-bodies looking over my shoulder was all the invitation I needed to attempt the book that sits before you.

Now, I want to begin with a tip for my fellow sinologists: Science is not that hard! I say this with some trepidation, because I am not a scientist, and even this book is more sociology than history of science in the narrow sense of the term. Still, my "research" took me back to college to study chemistry, an experience I found entertaining, illuminating, and no more humbling than a similar introduction to Chinese. I want to thank Ronald Breslow, Clark Still, Kris Wynne-Jones, and other members of the Columbia University chemistry department, who helped enlighten me to their dark craft. And I urge other area specialists to enrich their studies with knowledge of a technical sort. It's fun, and it offers insights hitherto lacking in our study of the non-Western world.

Several other parties who helped support my work along the

way also deserve a word of thanks. The East Asian Institute of Columbia University, then under director James Morley, gave me an office and a pat on the back in the fall of 1982, while I was writing the sections on industry and education. I am proud to appear again in the Studies of the East Asian Institute, a series ably edited by Carol Gluck with the assistance of Madge Huntington. Much of the research and writing was done between and around my chores as Columbia's East Asian librarian, causing a schizophrenia that was made tolerable only by the understanding and support of the university librarian, Patricia Battin, and the marvelous staff of the C. V. Starr Library. The book was completed during my tenure as Sun Yat-sen Research Professor of Chinese Studies at Georgetown University, where dean of the School of Foreign Service, Peter Krogh, director of Asian Studies, Matthew Gardner, and my other colleagues have given kind counsel and generous assistance. Grants from the Joint Committee on Contemporary China of the American Council of Learned Societies and the Social Science Research Council and from the Father Walsh Fund of Georgetown University relieved me from other duties at crucial points along the way.

Many people have read all or parts of the manuscript and offered helpful advice. They include Raymond Chien, Robert Dernberger, Albert Feuerwerker, Thomas Fingar, Steven Levine, Andrew Nathan, Mary Rankin, Thomas Rawski, Laurence Schneider, and Nathan Sivin. Several research assistants – Dan Brown, Scott Davidson, Christie Hong, Margaret Huang, Margot Rogers, Carrie Schmitt, Jeffrey Schultz, Sun Chung-hsing, Zhang Jiahui, and Zheng Xuecheng – have contributed to the project. Dr. Shen C. Y. Fu, Senior Curator of Chinese Art, Freer Gallery of Art and Arthur M. Sackler Gallery, provided the calligraphy that appears in Table 2.1. Others who helped in less direct ways or whom I have simply forgotten should remind me the next time we meet. Finally, I dedicate this book to my wife, Kathleen, and our children, Jane, Peter, and William, who put up with it all and remain my chief collaborators in life.

Bethesda, Maryland, 1989   J. R.-A.

Introduction

When Western missionaries introduced the science of chemistry to China in the middle of the nineteenth century, they called it *hua-hsüeh*, which means "the study of change." The choice was a good one, and it has survived as the name for this discipline in China as well as Japan. "The study of change" captures the essence of chemistry, the branch of science that deals with the arrangement of atoms into molecules and their rearrangement to form new structures or compounds. Chemical reactions, which are all but invisible under the most powerful microscopes, may catch the eye of even the casual observer, as compounds are transformed with explosive effect or substances with unimaginable properties appear, leaving little doubt that some important change has occurred.

This book describes the study of change during its first century. We begin with the word, the problem of translating not only the name, chemistry, but a whole vocabulary connected to it, and ultimately the problem of replacing one language and associated mind-set with another. Since chemistry is central to the curriculum of modern education, we look next into China's schools, examine the students, teachers, textbooks, laboratories, and vocational workshops. In the early twentieth century, the Chinese undertook the application of chemical technologies to industry, which calls for visits to arsenals, factories, and industrial plants. As the quality and training of Chinese chemists improved, many turned to research, so we will consider the problems and sometimes stunning achievements of members of an international scientific community working in the backwaters of Asia. Finally, the expansion of these activities produced a growing body of practitioners who created societies, journals, conventions, and other symbols of a modern scientific profession. In all these ways, we will examine Chinese chemistry from the inside, to see what the chemists said and did to effect change in these corners of China.

The title has a second meaning, however, for this book also probes larger changes in China as they relate to the development of chemistry and of science in general. Foremost among these is the roller coaster profile of the Chinese state, whose descent, collapse, and resurgence help explain the inverse growth, flowering, and decay of science as an autonomous activity. Changes in Chinese society, away from the Confucian ideal of governance by an omnicompetent, classically trained elite, through the dispersion of power among diversified, technical experts, and back again to the dominance of political generalists, were both causes and effects of the parallel developments in the community of Chinese scientists. Science has had an impact on the elemental questions of knowledge, values, and understanding, and this in turn raises problems of cultural identity in a society that is trying to be both modern and Chinese. Finally, chemistry and other branches of science were never far removed from the quest for wealth and power, which has been a concern, often the chief concern, of Chinese leaders on all points of the political spectrum.

This is the study of change, from the narrow focus on one scientific discipline to the broad outlines of China's modern experience. For the benefit of those more familiar with Chinese history than with test tubes and pipettes, we will begin with a brief description of chemistry as the Chinese found it in the mid-nineteenth century. Students of science may be more interested in the following pages, which lay out major themes in the relationship between science and China's culture, state, and society.

Chemistry in the nineteenth century

When the Chinese came to this subject in the mid- to late-1800s, chemistry was already an established discipline, with its own models, methods, vocabulary, and community of practitioners, then confined largely to Western Europe. The foundation of modern chemistry dates from the 1780s, when Frenchman Antoine Lavoisier set forth a new paradigm to explain the composition and changes observed in material substances, which proved superior to the old Aristotelian and alchemical notion of the "four elements." Lavoisier also constructed a vocabulary to describe his system, bringing together the theory and language that have remained at the core of chemistry ever since. During the first half of the nineteenth century, chemists described the interactions among substances in gross proportions – the combining of "equivalent proportions" or "equivalent weights" of different materials to form new products. It was not until the 1860s that John Dalton's atomic theory gained general acceptance, bringing chemistry down to the level of the atom and the combination of atoms to form molecules. The full picture of modern molecular theory – with atoms arranged in the periodic table, valences, electrical charges, bonding, and so forth – emerged from discoveries made during the last decades of the century. In short, the roots of modern chemistry were firmly in the ground before the Chinese pulled up the plant to have a look.

Concurrent with developments in theory and laboratory practice, nineteenth-century Europe was also experiencing a revolution in the application of chemical knowledge to a wide range of practical uses. The boom in textile production that followed innovations in spinning and weaving created a corresponding demand for cleansing and bleaching agents, which was met with new techniques for the manufacture of sulfuric acid, bleaching powder, and soda. In surgery, chemical anesthetics introduced in the 1840s – nitrous oxide, ether, and chloroform – reduced pain, and antibiotics, first used by Lister in 1865, lowered the incidence of infection and death. Beginning around 1850, European agriculture profited from the introduction of chemical fertilizers: superphosphates, ammonium sulfate, sodium nitrate (Chile saltpeter), and potassium sulfate. Mining and warfare were advanced by the development of nitroglycerine, dynamite, and other high explosives. Toward the end of the century, a whole new chemical industry was constructed around the synthesis of organic drugs and dyes.

The most telling feature of nineteenth-century chemistry was not that a new scientific discipline had come of age or that chemical products were changing the economy of Europe, although both were important, but that the two sides of chemistry, the science and the technology, had been alloyed. On one hand, changes in technology led to breakthroughs in science. The rise of steam power and developments in mining, industry, and transportation called forth new responses from the laboratory. The manufacture of coal tar derivatives, chemical reagents, improved glass, metal alloys, and other products made the work of the scientist in some cases possible and in all cases easier. The opposite was also the case, for professional chemists whose main purpose was to investigate the working of nature produced knowledge and methods that were applied outside the lab. In France, physician Nicolas Leblanc invented a method for making industrial soda from common salt. Leblanc was followed by Clement, Desormes, and Gay-Lussac, other Frenchmen who helped improve methods for the large-scale manufacture of sulfuric acid. A young English chemist, William Perkin, in the course of experiments designed to make the drug quinine, produced the first synthetic dye. Perkin's breakthrough was more fully exploited in Germany, where during the last decades of the century, chemical industry, education, and research were first brought together to create the synergy that is so characteristic of modern science-based economies. The work of Boussingault and Liebig on plant nutrients, of Pasteur on the chemical basis of fermentation, disease, and other aspects of microbiology, and of Haber and Ostwald on the synthesis of ammonia and nitric acid all demonstrate the interdependence between chemical science and chemical technology.

The cutting edge of chemistry migrated in the course of the nineteenth century from France and Britain to Germany. Lavoisier, a Frenchman, based his theories on the experimental work of Englishmen Cavendish, Priestly, and Boyle and of Scotsman Black. Other names, French, British, Italian, and German, came to populate this Pantheon during the early nineteenth century. But by 1860, it was the Germans – Lieber, Wohler, Kolbe, Bunsen, and Kekule – who rose above the landscape of European chemistry. Britain remained the world's leading producer of the "heavy" industrial chemicals, acids and sodas, while Germans dominated the "fine" chemicals, dyes and drugs, and made almost all of the important breakthroughs in chemical technology during the decades before World War I. Americans played no part in the science of chemistry and only a minor role in the chemical industries of this period.

In sum, the chemistry that was introduced to China after 1850 came in the form of a well-organized, well-articulated body of ideas and instrumentalities. It had emerged from behind the dark veil of alchemy and the recesses of Europe's iron and coal mines to an honored place in the academy, and boasted all the theories, methods, journals, conferences, and other paraphernalia that mark the arrival of an independent discipline. Chemistry helped to create the wealth and power of the West and bring the undeniable might of modernity to the shores of East Asia. The Chinese would learn of chemistry through the translation of English-language sources, even while the leading edge of this discipline was moving to Germany, but this made little difference since China had no prior experience with these matters and needed the most basic information.

The choice of chemistry as the focus for this book was not a difficult one. One purpose of the study is to portray science in its interactions with modern China, and we will cast our glance from time to time across this broad landscape. But there is a danger that pictures taken with a wide-angle lens might miss too many details, so it seemed wise to pick one discipline, and chemistry had much to recommend it. Chemistry had the greatest impact of any branch of science on the world of the late-nineteenth and early-twentieth centuries. It transformed medicine, saving lives and ushering in the global population boom; agriculture and industry, making possible rapid, sustained economic growth; materials, changing the substance, color, and shape of our surroundings; and warfare, enabling us to destroy one another with greater effect. For these and other reasons, more Chinese chose to study chemistry than any other branch of science. There were more Chinese chemical teachers, researchers, institutes, publications, societies, and industries than in any other field. Among the sciences, chemistry penetrated most broadly and deeply into the fabric of China and did the most to change life in that country.

There is a second factor, less a reason for selecting this discipline than a benefit of a choice already made, namely, that chemistry is one of the physical sciences and is without precedent in China. During the period in question, the descriptive sciences – geology, biology, meteorology – relied, as the label implies, on taxonomy or classification, which could be performed with simple tools and related to elemental theories, readily accessible to novices and laymen. These are "local" sciences that rely on the accumulation of data, collected in the field or from historical records, that describe the flora, fauna, and topography of a particular place. These features made the descriptive sciences attractive to many Chinese scholars, who could quickly enter the game of world science, contribute to the nation at home, and gain recognition from the scientific community abroad, simply by reporting on what they found in their own backyards. Chinese who chose this route played at the low easy end of the scientific spectrum, often with early favorable results. The physical sciences – physics, chemistry, and astronomy – have all of the opposite characteristics: They depend on elaborate equipment, abstruse theories, complex experimental procedures, and data that are the same everywhere. Little in China's natural environment or library of traditional scholarship gave Chinese scholars a leg up. The challenge was frightening, but those Chinese who chose chemistry played in the big leagues of international science, and this gives their story added appeal.

Science in China and the West

Chemistry is one branch of "science," a difficult and elusive term that raises a number of interesting questions. As used here, science means that special set of models and methods for explaining the behavior of material phenomena, which issued from the great Scientific Revolution of seventeenth-century Europe. Science embodies a distinct approach to nature, a way of acquiring information, fitting this information together into theories, testing the theories against new data, and modifying or replacing these theories as the dialogue proceeds. It is a continuous tradition of ideas, methodologies, and social practices, all of which are devoted to capturing an ever more accurate picture of the universe. And it has won out against competing approaches through its ability to produce convincing, useful results.

This definition of science raises a number of questions, the first of which is the relationship between science and those explanations or models of nature developed in other times and places, including pre-modern (or pre-Western) China. It may be, as Joseph Needham has argued, that science is the outcome of a "grand titration" that has drawn together the contributions of many nations. Even so, the science that emerged from Europe in the nineteenth century was unrecognizable to the Chinese, who had to discover, adopt, and adapt it along with other strange new things from the West. There were, of course, links between traditional Chinese and modern Western nature studies: Chinese astronomers and mathematicians of the seventeenth century learned from Jesuit missionaries of recent European breakthroughs, which they adapted to Chinese astronomy, raising this discipline to new heights. Early recruits to science in the nineteenth century came to this subject through study of ancient Chinese texts and translations handed down from the Jesuits and their Chinese converts. Chinese scholars of the late Ch'ing dynasty (1644–1911) took great pains to square the new science with the classical concept of the "investigation of things." But connections of this type were tenuous and fleeting. With the passage of time, the gap between traditional Chinese and modern Western approaches to nature widened. Chinese science did not grow organically from the study of *yin* and *yang*, the "five elements," the principles of *li*, or the forces of *ch'i*. Rather, science entered China through a side door, and Chinese who studied chemistry and other disciplines took lessons directly from the West, often oblivious to the treasures of their own past.

This is not to say that the delivered culture played no role in this story. On the contrary, traditional approaches to learning, values, and habits of mind have had the greatest impact – for better and worse – on the study of science in China. Classical learning fostered a respect for scholarship, a care in organizing and preserving factual information, and training in memorization, all of which are virtues in the scientist. On the other hand, the bookishness, the preoccupation with philosophical and literary subjects, denigration of manual and technical skills, and undue respect for established authority often left the classically trained scholar unfit for the laboratory, field, or workbench. The point is that the interplay of tradition and modernity lay less between competing explanations of nature, than between Western science and the broader values and ideas that permeated Chinese society and culture.

A second question raised by this definition of science is its relationship to technology. Although the two should not be confused, they are closely connected and will be treated in tandem throughout this book. Science is a way of understanding or explaining nature. Technology is an instrument for manipulating or changing it. It is safe to say that there would be no science without the tools, starting with broken stones, through which man has interacted with his surroundings. Conversely, while their early development had little to do with a systematic view of nature, most modern technologies are "science-based." Both the conceptual distinction and practical connection between science and technology are amply demonstrated by the case of chemistry in China. Chinese scientists sometimes acted as though the two were separate, pursuing pure or basic research while ignoring problems close at hand. But the demands of China's economy and the pressure of government officials shifted attention back to practical applications. Science and technology are both mutually supportive and competitive, and at the end of China's first century of science, the tension between them remained unresolved.

A third question is the relationship between science and culture: whether science is limited to material objects of nature, or extends to human behavior and beyond? For many in China (and elsewhere) science has had a very broad meaning: in some cases a model or law, in others a set of methods and principles, and in either instance applicable not only to material objects, but to all things. At one extreme were the Marxists, whose law of dialectical materialism purports to explain everything from the big bang to the coming utopia. The Marxists enjoyed a large following in pre-1949 China, and philosophers in the Communist Party made ambitious claims for their ideology. Even more popular, particularly during the 1920s, was the devotion to scientism, the idea that the models and methods of science have universal applicability and should, therefore, replace the discredited Confucianism as the basis for China's "New Culture." In a society as unsure of itself as twentieth-century China, science could not be contained in the laboratory alone. But not all Chinese welcomed such radical ideas. In particular, conservative elements of the Kuomintang, who gained sway during the Nanking Decade (1927–37), wanted to use science and technology to enhance the power of the state, at the same time insulating traditional values against infection by foreign ideas, scientific or otherwise. There was, in the end, no agreement on China's identity or the place of science in it, but enormous interest in the question and a willingness on all sides to provide the answer.

Science, state, and society

Science is an idea, a way of understanding or explaining things, and can therefore be studied as intellectual history, the process by which one paradigm or picture of the universe displaces another. It is also a social activity, a set of values and beliefs, a means of relating and communicating ideas among people. Much of the history of science adheres to the former definition, and appropriately so, for its main concern has been to describe the evolution of man's image of nature, the way new facts are discovered and better concepts devised. In the case of China, however, the problem is not to trace the pattern of ideas, for it might be said that there was no pattern, at least in the period covered by this study, because scientists in China addressed piecemeal the questions posed by scholars in the West, and Chinese science developed without systematic coordination or agenda. This means that the more interesting question has been sociological, namely, What has been the relationship between Chinese science and the political and social context in which it emerged?

The answer presented in this book is that science rose and fell along with the rhythms of change in Chinese state and society. During most of the nineteenth century, the Manchu rulers and their Chinese allies shared an interest in limiting the penetration of new ideas and groups and succeeded in keeping both under control. Some small moves cracked the door to modern science and technology. Treaties signed with the Western powers admitted missionary doctors, teachers, and translators who came to China with knowledge of science and the desire to share its secrets. Chinese "self-strengtheners" established arsenals, shipyards, schools, and translation bureaus and sent students abroad, all for the ships, guns, and secrets of making them, needed to protect the empire against rebellion at home and invasion from abroad. But this is as far as the Ch'ing dynasty would go. The dominant, conservative forces in Peking rebuffed attempts to broaden the curriculum of the civil service exams, by which officials were chosen, or the scope of industrialization, which would have created greater demand for new knowledge and skills. Even the more progressive self-strengtheners in the provinces recruited and promoted men with classical backgrounds, while ignoring those with technical expertise. Finally, changes in local society – the "fusing" of merchant and gentry classes, their growing separation from the state, and the creation of an autonomous "public" managerial elite – failed to expand the market for technical skills.

Given this lack of hospitality, few people in nineteenth-century China took a serious interest in science or understood its message. The most persistent and effective proponents were Protestant missionaries, who led in creating a Chinese scientific language, translating and publishing texts, and establishing schools to teach the new ideas. A handful of Chinese, men who had failed the classical examinations and had an off-beat interest in nature studies, joined the missionaries as partners in the translation business. A somewhat larger number studied science in government and missionary schools or read the translations on their own. But this group was small and its horizons limited. So long as classical learning remained the favored credential, scholarly talent was drawn to this pole, students of science had difficulty finding work, and members of the literati who dabbled in science confused its message with traditional categories of thought. The vice of political and social authority was tight, the incentive to take up strange new studies weak, and the development of science correspondingly slow.

All this changed in 1895, when the defeat in the Sino–Japanese War shattered China's complacency and opened the country to revolutions of every kind. Like the peeling of an onion, the layers of the old order – the dynasty, the imperial system, central authority, and Confucian culture – were stripped away, leaving an emptiness that invited radical change. The Manchus made a last-ditch effort to save themselves by adopting modernizing reforms, before succumbing to the Revolution of 1911. Military strongman, Yuan Shih-kai, held the new Republic together for a time (1912–16), but after his death China was torn apart by warlords (1916–27), none of whom could unify the country or remain satisfied with only one piece of it. Meanwhile, the traditional elite and the Confucian ethic on which their authority rested were swept aside by a firestorm of social and cultural iconoclasm that left few charred remains.

The collapse of the old order had some positive effects on the development of Chinese science. Governments of the late Ch'ing and early Republic began to promote the study and application of science, either to increase the nation's wealth and power, or simply because they sensed the need for modern solutions. Even elements of the old Confucian elite recognized the necessity for change. But the chief point about this interregnum – the period from 1895, when the downward slide of the empire began, until 1927, when the Nationalists restored a measure of central authority – is that the absence of political and social form left the field open to a variety of forces, some of which established the first real foundation for the study and practice of science in China.

As the state lost control, three groups that favored the development of science, but had previously been held down, surged forward. First were the foreigners: governments, missionaries, and philanthropists. Western missionaries were the most important force behind science in China in the nineteenth century, but their influence had been contained. After 1900, the U.S. government, through remission of the Boxer indemnity, and the Rockefeller Foundation joined the churches in making scientific education and research available in greater quantity and quality to more Chinese at home and abroad. Second was the new urban, Western-oriented Chinese intelligentsia. These men, and now also women, created and promoted the New Culture that was to replace the Way of the Sage and connect China to the mainstream of the modern world. Few were scientists in the strict sense of the term, but many had been introduced to scientific ideas and applied them in shaping the society and culture of the 1910s and 1920s. Third were Chinese investors and entrepreneurs, who introduced new technologies based on a knowledge of science and called forth the services of a trained technical elite. The scope of economic change was still narrow, but the thin edge of the wedge was in the log.

Whereas the collapse of the old order cleared the field for new forces, the absence of effective political authority and accepted social norms left few structures to support and coordinate their scattered initiatives. There was in early twentieth-century China no escape from violence and chaos, no reliable source of funding and organization, no insulation of scholarship or industry against the winds of fate. It was good for Chinese science that the too-strong, too-conservative empire had been destroyed and the door to innovation opened, but bad that the door had come off its hinges. Some sort of order was needed, and missing.

If the problem of the nineteenth century was too much authority and the early twentieth century too little, then the Nanking Decade (1927–37) was just right for the development of Chinese science. During this brief period, the Nationalist or Kuomintang regime provided the unity and commitment needed to carry out a program of research, education, and the application of new knowledge to economic development and national defense. The Nationalists channeled money and people into the study of science in colleges, universities, academies, and institutes. They supported the introduction of new technologies in public and private enterprises. They established security, stability, and a sense of direction that attracted educated Chinese to return home and gave them confidence in the future. In all these ways, the Nationalists restored the central coordinative power of the state and placed it behind a program of scientific and technical development.

There was a temptation to overdo it, however, to place too tight restrictions on those forces that had emerged during the preceding decades and stifle the free flow of resources, people, and ideas that are essential to scientific and technical growth. The heightened sense of nationalism and the ability of Nanking to enforce nationalist demands, placed foreigners, particularly missionaries, under duress. Some elements of the new regime wanted to rein in the mission colleges, which had led the development of scientific education in the 1920s. Another target of the ultranationalists was the scientists themselves. Many members of China's new intelligentsia, men and women trained in the world's leading universities, had earned membership in the international scientific profession. This identity carried with it a faith in the importance of specialized knowledge, the right to establish an agenda in areas such as education and research that lay within their sphere of expertise, and to autonomy from the larger social and political environment. The Kuomintang right viewed the scientists with suspicion, pressured them to adopt a more utilitarian approach aimed at increasing the wealth and power of the state, and rejected the quest of knowledge "for its own sake." Finally, Chinese entrepreneurs, who led in the introduction of new technologies, faced an expanded army and bureaucracy that threatened to draw resources away from productive investments. Some recent scholarship suggests that far from being the tool of the capitalist class, the Nationalists exploited and undermined China's private sector, denying industry the opportunity to develop and grow.

The present study shows that Nanking, by choice, necessity, or inadvertence, resisted excesses of this kind. There were visible strains of antiforeignism in Kuomintang ideology, and the Nationalists limited foreign influence by regulating mission education and other means. But foreign funds were always welcome to support the conduct of science, and the flow of people and ideas into and out of China increased multifold during the Nanking Decade. Despite the pressures to direct Chinese science into a narrow utilitarian channel, the Nationalists continued to fund education and research that was favored by the scientists and promised few short-run returns. Even though the relationship between scholars and the state was often tense, Nanking accepted a measure of professional independence and pure learning. Finally, the record of the chemical industries shows that the government, far from exploiting or undermining the private sector, provided a combination of direct support and benign neglect that enabled China's chemical industries to expand to the limit of the home market. The failure to transcend this boundary was due to economic rather than political causes. In sum, the Nationalists achieved an effective balance between governmental authority and autonomous social forces that yielded substantial growth in all branches of science.

In the fairy tale, Goldilocks ate her lunch when she got to the "just right" porridge, but the Chinese were not so lucky. The invasion by Japan ended the Nanking experiment and the tenuous balance between state and society developed in this era. During the war (1937−45), much of modern China, including the people and institutions responsible for science and technology, retreated to the hinterland, either to the Nationalist strongholds of Chungking and Kunming or the Communist capital of Yenan, where they carried on the resistance to Japan. In both areas, poverty, backwardness, and the revival of state power undid the achievements of the prewar decade. In both, a community of scholars, teachers, and students tried amidst considerable hardship to carry forward an agenda that included broad basic learning, designed to keep Chinese in touch with the development of international science and enable them to continue the work of national reconstruction after the war. In both, the central authorities opposed this strategy on the grounds that it wasted scarce resources, and pressed the scientists to accept a narrower program of applied learning to serve the immediate needs of production and defense. In the Nationalist areas, the battle between scientists and the state was fought to a stalemate. But in Communist China, the authorities overwhelmed the opposition, destroying all vestiges of professional autonomy and a separate commitment to learning about nature. The resurgence of state power restored much of what the Chinese had lost in the last years of the Ch'ing, but it also brought back the constricted environment that had proven unhealthy for the development of science, then as now.

The study of change traces the episodic growth of the tendrils of chemistry into the nooks and crannies of modern China. Retreating a pace, it surveys the question, How did political and social changes affect the introduction and development of science in a non-Western setting? The answer, in this case at least, is that science fared best when there was a balance between constructive political authority and the free flow of autonomous social forces. This balance was achieved during the 1930s, but destroyed by the war. The imposition by the Communists of a harsh, narrow orthodoxy created a legacy that burdens Chinese science even today.

PART I

Science and self-strengthening 1840–1895

Self-strengthening, the program adopted during the latter half of the nineteenth century to defend and preserve China's imperial order, created a place for science and scientists, while limiting the scope of their development and the breadth of their appeal. Self-strengthening meant an accommodation with the Western barbarians, in order to build up China's defenses and put down internal rebellion, without compromising the integrity of traditional institutions and culture – a grafting on of Western "functions" to preserve the Chinese "essence." This opened the door to science, which was accepted by the Ch'ing dynasty and its allies as the handmaiden of technological change and promoted by Western missionaries as the cutting edge of an enlightened world order. At the same time, China's rulers effectively resisted those changes in political, social, and economic structures that alone would make room for the broad, deep development of science.

By the middle of nineteenth century, currents flowing from within late-traditional China produced a small number of men who were receptive to new explanations for the behavior of nature and new ways of dealing with it. Rapid population growth, restrictive quotas on examination degrees and official appointments, expansion of commercial opportunity, and the creation of new administrative and managerial roles: All these factors contributed to the emergence of a more diverse, dynamic Chinese elite. Intellectual and political trends invited thoughtful minds to seek new solutions for urgent social and technical problems. More than in any other recent time, China was ready for change. From among those scholars whose advancement through the civil service examinations had stalled and who were looking for new outlets for their talents came the early recruits to science.

It is unlikely that these men would have found their way to the main channel of science, however, had it not been for Western imperialism. The threat from abroad, coming on top of rebellion at home, prompted Chinese authorities to seek up-to-date military technologies and learn about the relevant branches of science. Protestant missionaries, following behind the gunboats, promoted a wide range of modern studies and helped reshape Chinese language and culture to accommodate the new ideas. Foreign patrons created a social and economic niche, within which Chinese interested in unorthodox subjects could pursue careers not provided by the traditional order. More than just passive sources of ready-made knowledge, foreigners played an active, even dominant role, pressing science deeply and broadly into the Chinese scheme of things.

The confluence of these trends created the context for the study and practice of science. Chinese with a flare for technical subjects went to work in the arsenals and shipyards that were set up to strengthen the nation's defense, studied in modern government or missionary schools, and in a few cases joined foreign partners in the translation of scientific texts. A majority of secularly minded British and American missionaries had their own agenda for enlightening and lifting up the Chinese. Chinese scholars and foreign churchmen, supported by the Chinese government, led China into the scientific age.

Until the very end of the century, the Ch'ing dynasty succeeded in controlling the pace and scope of these developments. Under the policy of self-strengthening, Peking accepted reforms deemed necessary to keep the dynasty in power, while rejecting proposals that would have exposed the empire to broader, deeper, more rapid change. The door was opened just enough to let in the men, machines, and ideas needed to strengthen China's defenses, while government institutions, social structures, and cultural standards remained intact. Peking refused to broaden the civil service exams to include modern Western subjects or sanction widespread industrial development, either of which would have threatened the position of the incumbent elite and the values that sustained them.

In the absence of major institutional change, Chinese society bent slowly, but did not break. Ownership of land, education in the classics, and access to officialdom remained the pillars of "gentry" status. Commerce was on the rise, and in some area merchants joined gentry to form a new stratum of entrepreneurial and managerial talent. Yet even in the most dynamic modern enterprises – the arsenals, shipyards, factories, and mines – and in the autonomous "public" institutions – schools, orphanages, relief and reconstruction bureaus – which were created to restore order after the mid-century rebellions, recruitment and promotion favored men with traditional training and administrative experience. Chinese society did not reward a knowledge of science or modern technical skill.

Finally, the introduction of science had little impact on the ideas, attitudes, and values of China's traditional culture. The emphasis during this period on the translation of Western scientific texts underscores the fact that "science" in nineteenth-century China was almost exclusively a literary affair, rich in reading and writing, but poor in observation and experimentation, and in this sense not scientific at all. Most translators shunned practical work, and in the case of chemistry, which found few applications in China at this time, even practical subjects, preferring the quiet, clean, noble library over the noisy, dirty, vulgar workshop. Nor was the audience for these works – mostly members of the literati with a curiosity for the strange or a flare for technical device – well prepared to receive them. Although the translations present a fair picture of science, its models, methods, and ideals, the evidence suggests that most readers, weighted down by the baggage of classical learning, folded these concepts into the fabric of traditional thought, and in the end failed to understand science at all.

In sum, the age of self-strengthening was a time of marginal change on the margins of a changing society. The protagonists, Chinese misfits and foreign missionaries, were given a narrow space to do their work and were opposed by powerful, stubborn institutions, an entrenched, confident society, a healthy, albeit static economy, and in the end by the traditional habits of their own minds. During the nineteenth century, advocates of science did the best they could, which was to turn the classics of European science into readable Chinese and place the record before their countrymen. Broader, deeper, more rapid development would have to await the collapse of the old order, which began in 1895.

1

The advocates: chemical translators, John Fryer and Hsü Shou

The leading advocates for the study of chemistry in nineteenth-century China were translators John Fryer and Hsü Shou, who worked together in the Kiangnan Arsenal during the 1870s and 1880s. Both men rode the major intellectual and social currents of their times, yet it was their peculiarities that brought them together in this historic enterprise. Hsü came from that flotsam of educated Chinese who failed to advance through the civil service examinations and bobbed about seeking to turn their learning to other ends. What set him apart was his interest in nature studies, or as he called it, "the investigation of things," a field that commanded little attention in China – until the decision was made to build ships and guns and recruit skilled craftsmen and knowledgeable scholars to help with this task. Fryer was also a misfit: failed missionary, disgruntled teacher, student of the forbiddingly strange Chinese, who finally found his calling as a translator of technical manuals. Spokesmen for a cause that found few adherents in nineteenth-century China, Fryer and Hsü did more to promote the study of science in that country than any other men of their time.

Hsü Shou and the "investigation of things," 1818–1860

The first Chinese to understand the essentials of modern chemistry and communicate this knowledge to his countrymen was an unsuccessful, somewhat off-beat scholar named Hsü Shou (1818–82). Hsü's failure to pass the civil service exam, the principal route forward for young men of his time, left him free to pursue his real interest, the "investigation of things." Chinese society of the early nineteenth century was changing in ways that welcomed the exploration of unworn paths. And when the crisis of the late Ch'ing called for men with scientific and technical skills, Hsü Shou was ready.

17

Hsü Shou. (Ko-chih hui-pien, 1877.)

Hsü Shou, who was also known by the sobriquet, Hsüeh-ts'un, hailed from a village near Wuhsi, in southern Kiangsu province, the region the Chinese call Kiangnan – "South of the River." Here was an ideal base from which to launch an inquiry into modern science in the middle of the nineteenth century. Kiangnan lay at the crossroads of the Yangtze River, which reached from the Szechwan basin in the west to the Pacific Ocean in the east, and the Grand Canal, which linked the rice bowl of the south to the northern capital, Peking. By the end of the Ming dynasty (1368–1644), this region, with its cotton textiles and silk brocades, its intricate network of waterways and thick urban settlement, was already the richest, most commercialized part of China. Wealth, traffic, and dense population contributed to a flowering of scholarship and culture that set a fast pace for the rest of the empire. After the middle of the nineteenth century, proximity to Shanghai and the other "treaty ports" opened the doors of Kiangnan to the world. Wuhsi, long at the cutting edge of commercial and intellectual changes that emerged from within China, became, in the course of Hsü Shou's lifetime, a frontier of the revolution introduced from without.

The Hsü family had produced no official degree-holder for several generations, removing them a step from the ranks of the gentry. But as "middle landlords," they were wealthy enough to support the unproductive interests of at least one son. Hsü Shou's grandfather had been a farmer and merchant, trades followed by Shou's father until his premature death, which left the son, age five, in his mother's care. Still, the family seems to have done reasonably well, for in Hsü Shou's time its holdings included several tens of *mou* (one *mou* equals 0.15 acres) of mulberry trees for the raising of silkworms and a filature for the reeling of silk. Hsü never earned an official degree or the government stipend that came with it and apparently found no gainful employment outside the household until well into middle age, so the family must have provided for his needs. Accounts written after his death describe a successful farmer, entrepreneur, and manager of the family estate who was recognized and respected in his home district. The picture may be generous, but is probably not far wrong.[1]

The early course of Hsü's education followed the well-worn path to the civil service examinations, success on which promised wealth, power, status, and the other rewards imperial China had to offer. Even in the best of times, the exams placed a premium on memorization, textual analysis, and the study of human affairs, siphoning intellectual energy away from the empirical, practical, and natural studies that underlie modern science. Moreover, many critics charged that the late Ch'ing was the worst of times, that the exams had declined to a base repetition of dogma, which was crammed into the head through hours of oral drill and onto the page in the rigid form of the "eight-legged essay" – a poor

[1] The primary sources on Hsü Shou's life are: Chao Erh-hsün et al., ed., *Ch'ing-shih kao* [A draft history of the Ch'ing dynasty] (Peking: Chung-hua shu-chü, 1977), 46:13,929–31; Min Erh-ch'ang, comp., *Pei chuan chi pu* [Supplement to the collection of epitaphs], 24 vols. (Taipei: Wen-hai ch'u-pan-she, 1973), chüan 43, 12b–18b; and Yang Mo, ed., *Hsi-Chin ssu-che shih-shih hui-ts'un* [Records of the four scholars of Wuhsi and Chin-k'uei] (Wuhsi, 1910). For a recent collection of biographical materials, see: Yang Ken, ed., *Hsü Shou ho Chung-kuo chin-tai hua-hsüeh shih* [Hsü Shou and the history of chemistry in modern China] (Peking: K'o-hsüeh chi-shu wen-hsien ch'u-pan-she, 1986). On Hsü family, see also: Yang Ken, "Wo-kuo chin-tai hua-hsüeh hsieh-ch'ü-che Hsü Shou ti sheng-p'ing chi chu-yao kung-hsien" [The life and important contributions of the vanguard of our country's modern chemistry, Hsü Shou], *HHTP* 4 (1984): 71; and Yüan Han-ch'ing, *Chung-kuo hua-hsüeh shih lun-wen-chi* [Collected essays on the history of Chinese chemistry] (Peking: Hsin chih san-lien shu-tien, 1956), 271. Mulberry trees: Liu Shu-k'ai, "Wo-kuo chin-tai hua-hsüeh ti ch'i-meng-che Hsü Shou" [Pioneer in our country's modern chemistry, Hsü Shou], *Hsü Shou ho Chung-kuo chin-tai hua-hsüeh shih*, 64.

standard on which to select China's leaders. When Hsü Shou failed the entry-level prefectural examination, he declared the whole exercise "impractical" and gave up further attempts to advance by this route.

After abandoning the exams, Hsü turned to the study of what he considered more useful knowledge, which today we would call science and technology, but was then referred to as *ko-wu* or the "investigation of things," a topic that lay outside the official curriculum. Hsü's initial readings included traditional geography texts, such as the *Yü-kung,* China's oldest book on this subject, and the *Shui-ching-chu,* a text from the Northern Wei dynasty (A.D. 424–532) that deals with "water management," as well as passages relating to geography in the *Book of Odes,* the *Spring and Autumn Annals,* and the *Histories of the Former and Latter Han Dynasties.* He combed the classics for information on mathematics, astronomy, calendrical sciences, acoustics (or musicology), mechanics, mining, medicine, and, according to some accounts, "chemistry," which we must assume meant alchemy, since the modern science of chemistry was without counterpart in traditional China. These readings led him to the maps of Matteo Ricci and the Jesuit translations, by then nearly two centuries old, but still the most recent material on modern science available in Chinese. All of Hsü's early studies relied on Chinese texts, for he had no knowledge of Western languages and probably never gained more than a passing acquaintance with English.[2]

Hsü's classical training inclined him to turn first to written sources, a literary bent that persisted throughout his career. But his interests and talents extended beyond textual studies to a wide range of practical skills. As a young man, he made compasses, quadrants, musical instruments, and other useful devices. In order to study optics, he polished down his quartz crystal chop (used for affixing one's signature to documents) to form a prism. And in one moment of avarice, he counterfeited silver dollars of such quality that merchants in the Wuhsi market could not distinguish them from the real thing.[3]

Nor was Hsü an aimless tinkerer, for his contemporaries credit him with introducing innovations that helped save the Wuhsi silk industry. The traditional method for manufacturing silk rested on the hand-reeling of thread from the silkworm cocoon. This process yielded a thread of uneven thickness and tensile strength, which had to be rereeled by mechanical methods in the filatures of Italy and France, in order to gain acceptance in the European market. In 1860, a Western-style filature was set up in Shanghai to produce silk for export, and purchasing agents were sent to the surrounding districts to buy raw cocoons, threatening local manfacturers. Hsü Shou developed two devices that proved a boon to

[2] *Pei chuan chi pu,* 43:14a, 16a.
[3] Compasses, quadrants, dollars: ibid., 43:15a. Prism: *Ch'ing shih kao,* 46:13,930.

himself and his neighbors: roasting ovens to kill the silkworm pupa without disturbing the cocoon and thus extending the period the cocoon could be stored, and mechanical reeling devices to produce a more even thread. One report credits these measures for sharply increasing the profits of silk producers of Wuhsi and neighboring Chin-k'uei counties.[4] Hsü Shou was remembered as a clear-eyed, hard-headed rationalist, who took a dim view of the religious and superstitious practices that engaged many of his contemporaries. According to one biographical sketch, he always warned others

not to talk wildly or spread rumors, not to speak of things of which you have no experience, of astrology, geomancy, sorcery or the occult. In dealing with ceremonies, marriages and funerals, he did not use the *yin* and *yang* to select the day. In offering sacrifices in the four seasons, he sacrificed only to his ancestors and not to outside spirits. In managing funerals, he did not use Buddhist monks to perform the rituals or musicians to beat [drums] and blow [horns]; he did not consult geomancers, but discussed the matter only with other people. No word of the five phases, of jinxes and omens, or of the superficialities of *li* [principle] and *ch'i* [vapors] ever passed his lips. He always guided his students with facts and proofs. One could see that his conscience was clear and his thoughts were free.[5]

Despite such eulogies, Hsü's early studies attracted little attention. Before the middle of the nineteenth century (and for the most part afterward as well), there was little call for men skilled in the investigation of things, and few scholars could afford to pursue research that promised such meager rewards. Hsü did find one confidant, the son of an official from neighboring Chin-k'uei county, named Hua Heng-fang (1833–1902), whose interest was mathematics. Although Hua was fifteen years Hsü's junior, the two men met regularly to talk, exchange reading materials, and perform the experiments described therein. Together, they established contact with a more famous contemporary, the mathematician Li Shan-lan, an employee of the Inkstone Book Store [*Mo-hai shu-kuan*] in Shanghai, where he joined members of the London Missionary Society to translate and publish works on Western science.[6]

[4] Western-style filature: Shannon R. Brown, "The Transfer of Technology to China in the Nineteenth Century: The Role of Direct Foreign Investment," *Journal of Economic History* 34, no. 1 (March 1979):190–1. Hsü Shou's role: *Ch'ing-shih kao* 46:13,930; *Hsi-chin ssu-che shih-shih hui-ts'un*, cited in Liu Shu-k'ai, "Wo-kuo chin-tai hua-hsüeh," 64–5; and Chiang Shu-yüan, "Chi-nien wo-kuo shih-chiu shih-chi cho-yüeh ti hua-hsüeh-chia Hsü Shou" [In memory of our country's outstanding chemist of the nineteenth century, Hsü Shou], *Hsü Shou ho Chung-kuo chin-tai hua-hsüeh shih*, 71.

[5] *Hsi-chin ssu-che shih-shih hui-ts'un*, cited in "Hsü Hsüeh-ts'un hsien-sheng hsü" [Introduction to Mr. Hsü Hsüeh-ts'un], *KCHP* 1, no. 2 (1877):1a. This document is reprinted in *Hsü Shou ho Chung-kuo chin-tai hua-hsüeh shih*, 352.

[6] Hua Heng-fang: *Pei chuan chi pu*, 43:15a. Biography of Li Shan-lan: Arthur Hummel, *Eminent Chinese of the Ch'ing Period* (Washington, DC: GPO, 1943), 1:479–80. *Mo-hai shu-kuan*: Tsuen-hsuin Tsien, "Western Impact on China through Translation," *Far Eastern Quarterly* 13 (1954):313.

When Hsü Shou and Hua Heng-fang went to Shanghai in the late 1850s to meet Li Shan-lan, they made an important and unexpected discovery. Up until that time, their reading had been limited to traditional Chinese texts and translations produced by the Jesuits some 200 years earlier. It is not clear how familiar they were with more recent works on science, but these, if any, would have focused on math and astronomy, the fields that received most attention from scientifically minded Chinese and missionary scholars. In Shanghai, Hsü and Hua came upon a new book, *Po-wu hsin-pien* [Natural Philosophy and Natural History], written by the English surgeon Benjamin Hobson and published by the Inkstone Press in 1855, which provided a general introduction to current knowledge on all the major branches of science. "This book," John Fryer later explained,

though of a very elementary character, was like the dawn of a new era upon their minds, enabling them to leap at one bound across the two centuries that had elapsed since the Jesuit fathers commenced the task of the intellectual enlightenment of China, and bringing them face to face with the results of some of the great modern discoveries.[7]

Hobson was the first writer to discuss in Chinese the concepts of modern chemistry, but it was his description of electricity that first interested the visitors from Wuhsi. Hobson laid out one of the great marvels of the modern world – that "lightning" [*tien*] that runs through the "forces" [*ch'i*] of the earth – describing its properties, methods of production, and practical applications, from the magnificent telegraph linking London to Paris to the use of electric shock as a cure for insanity. Hsü and Hua purchased several of the devices illustrated in this text and took them back to Wuhsi for study. During the next several years, they worked in an improvised laboratory, performing experiments and circulating papers reporting the results. Unfortunately, these unpublished manuscripts were destroyed when the Taiping rebels occupied Wuhsi in 1862. The excitement of their work was later recalled, however, by Hua's younger brother, who remembered the day Hsü Shou "folded paper into the form of a man that hung by its hands from a glass rod [apparently charged with static electricity]. The paper-man danced, and I howled with delight, for I could not understand the cause [of this strange behaviour]!" There is no evidence that these early experiments extended to chemistry, although Hobson's book described procedures for preparing various gases and other substances, and the recommended device for generating electricity was a simple wet cell made of copper and lead plates submerged in dilute sulfuric acid, so the would-be scientists must have gained some knowledge of this discipline.[8]

[7] John Fryer, "Science in China," *Nature* (May 1881):9.
[8] Quotation: Hua Shih-fang, *Pei chuan chi pu*, 43:15b.

Hsü Shou and "self-strengthening," 1860–1867

Prior to 1860, Hsu Shou's activities were a wholly private affair. He published no writings, held no office, and was unknown outside the small circle of men who shared his uncommon taste for nature studies. With the possible exception of his innovations in the manufacture of silk (and there is no record of when these were made), Hsü's early investigation of things served mainly to satisfy his personal curiosity. He probably knew of the scholarly currents that were most relevant to his interests – the "empiricist" [k'ao-cheng] movement, which contributed to a burst of creativity in the fields of astronomy and mathematics during the mid-Ch'ing, and "statecraft" [ching-shih] studies that directed attention back to the practical problems of reviving the dynasty in the nineteenth century. But we know nothing of Hsü's reading habits and have no evidence that he entered the great debates of his time. During his first forty years, Hsü was an amateur, whose work lacked social significance. All this changed after 1860, when the demand for military technology exploded, and Hsü Shou's skills, previously ignored, found their market.

The 1850s witnessed a burst of interest in new military technologies. Officials charged with defending south China against the Taiping rebels purchased foreign ships and guns and recruited shipwrights, gunnery experts, and experienced troop commanders. Around the same time, Hsü Shou also turned his attention to military affairs. He is said to have commanded a force of more than 100 ships in the defense of Wuhsi against the rebels and experimented with ballistics, firing guns up and down hill to determine how far the shot would carry. In 1862, as news of their technical achievements rippled outward, Hsü Shou, Hua Heng-fang, and Li Shan-lan were invited to join the staff of Viceroy Tseng Kuo-fan's new arsenal at Anking. Since their own work had been interrupted by the rebellion and Tseng offered a salary and the chance to work with facilities unavailable elsewhere, the scholars accepted. Hsü brought along his son, Chien-yin, who at seventeen already had experience in his father's laboratory, to assist in the new enterprise.[9]

Hsü's first assignment at Anking was to build a steamship. Although their previous introduction to the subject of steam engines was limited to the illustration and brief description in Hobson's book, the Wuhsi scholars widened their knowledge by inspecting foreign ships anchored on the Yangtze. In July 1862, they gave a demonstration of their model engine to Tseng Kuo-fan, whose diary entry reveals the great pleasure the viceroy took in China's newfound technical success:

[9] Command of water forces, hired by Tseng: *Pei chuan chi pu*, 43:12b–13a. Ballistics: *Ch'ing shih kao*, 46:13,929.

Hua Heng-fang and Hsü Shou brought here [Anking] the engine of the "fire-wheel boat" which had been made by them, for a demonstration. The method is to use fire to make steam. [Here follows a description of how the machine works.] The engine moves forward and backward as if it were flying. This demonstration lasted for one hour. I was so happy that we Chinese could do these wise and clever things like the foreigners. No longer can they take advantage of our ignorance.[10]

Three years later, after the arsenal had moved to Nanking, the Hsü's completed construction of China's first working steamship. Hsü senior personally directed all phases of design and construction, using imported iron and steel, but making all the components in China. No foreigner was engaged in any stage of construction. This ship, to which Tseng attached the classical name, *Yellow Swan*, weighed twenty-five tons and covered a distance of eighty-five miles in just eight hours on her maiden voyage.[11]

By the time Hsü and company began their work at Anking, the pace of change in China had quickened: the Second Opium War (1856–60) brought British and French troops into the capital and north to burn and pillage the Summer Palace, while the Taiping Rebellion threatened to topple the government in central China and unravel the whole fabric of Chinese society. Peking's response to this dual challenge, summed up in the slogan "self-strengthening," was to strike a deal with the foreigners that would assure peace abroad and help the dynasty crush the rebellion at home. Provincial officials hired foreign military advisers to train and command Chinese troops, purchased foreign weapons, and undertook to manufacture in China Western-style ships and guns under the supervision of foreign technicians. To support these efforts, the Tsungli Yamen, a foreign office, was established in the capital, along with a school for interpreters, the T'ung Wen Kuan, to prepare diplomats to handle China's expanded foreign relations. During the mid-1860s, arsenals were erected at Kiangnan (near Shanghai), Nanking, Tientsin, and other sites, and a shipyard built at Foochow on the southern coast. Attached to Kiangnan and Foochow were schools for training technicians in the skills and knowledge required to build and operate modern ships and guns. Later, more schools were opened to prepare gunnery experts, army officers, surgeons, telegraph operators, and engineers. These arsenals, shipyards, and schools provided official lodging for scholars and technicians who would introduce modern science to China.

During the 1860s, the Wuhsi scholars made several trips to Shanghai,

[10] Tseng Kuo-fan, *Jih-chi* [Diary], chüan 14, cited in Gideon Ch'en Ch'i-t'ien, *Tseng kuo-fan: Pioneer Promoter of the Steamship in China* (Peiping: Yenching University, 1935), reprint (New York: Paragon Book, 1961), 40–1.

[11] Pai Kuang-mei and Yang Ken, "Hsü Shou yü 'Huang Hu' hao lun-ch'uan" [Hsü Shou and the *Yellow Swan* steamship], *TJKHSYC* 3, no. 3 (1984):284–90; John Fryer, "Science in China," *North China Herald*, 29 January 1880, p. 77, reprinted in *Nature* (May 1881):10.

already the most westernized of China's "treaty ports." Hsü Shou and his colleagues had been urging Tseng Kuo-fan to build a larger factory, and when the Kiangnan Arsenal opened outside Shanghai, the lure of the city prompted them to request a transfer. Tseng, who intended to expand the Kiangnan operation to include shipbuilding, probably thought Hsü could help. The move was approved in 1867, after which the career of Hsü Shou took another significant turn.[12]

John Fryer and missionary science

John Fryer, an Englishman hired by the Chinese government to translate manuals and textbooks, was at the Kiangnan Arsenal when the Wuhsi scholars arrived at the end of 1867. In the course of the next thirty years, Fryer was the single most important force, Chinese or foreign, behind the introduction of science to China. He translated books, published journals, built a museum and school, and provided the rallying point for Chinese interested in the study of science. Hsü Shou, Hua Heng-fang, Li Shan-lan, and others, Chinese and foreign, did their bit, but Fryer provided the knowledge, vision, and drive that produced the greatest success.

John Fryer (1839–1928), the eldest son of a poor English clergyman and recent graduate of Highbury Training College, London, where he had been a scholarship student, arrived in Hong Kong in 1861, on the crest of the first wave of missionaries who had come East to test the rights granted them under the "unequal treaties." Fryer was typical of these early Protestants, most of whom issued from rural, middle-class, often clerical families in Britain or America. The Americans, mostly from farms and small towns in the north and midwest, had attended denominational colleges or seminaries near home. The core of their training was classical – Greek, Latin, philosophy, religion, mathematics – although by the 1850s, some of the better colleges had begun to introduce natural history and the experimental sciences, physics and chemistry. The British generally received training as physicians or nonconformist clergy; few had attended university. No missionary of this period was educated in a major center for the study of science. None can be called a scientist.[13]

Properly speaking, Fryer was not a missionary – he was never ordained, soft-pedaled the Bible in his teaching, and left the employ of

---

[12] Wang Erh-min, *Ch'ing-chi ping-kung-yeh ti hsing-ch'i* [The rise of the armament industry during the late Ch'ing] (Taipei: Institute of Modern History, Academia Sinica, 1963), 77–8.

[13] Biography of Fryer: Adrian Arthur Bennett, *John Fryer: The Introduction of Western Science and Technology into Nineteenth-Century China* (Cambridge: Harvard University Press, 1967), 4–28.

John Fryer. (Ko-chih hui-pien, *1881.*)

a mission school in Shanghai with no regrets on either side. Still, his commitment and career fit best into the progressive, secular wing of those Protestant churchmen with whom he lived and worked. Not all Protestants in nineteenth-century China favored the propagation of science. In the early years, the 1860s and 1870s, most were fundamentalists, who considered teaching and healing to be a distraction of their limited resources from the more important tasks of converting the heathen and saving their souls. But there was a minority committed to spreading the gospel by good works and introducing religion as part of a modern, enlightened world view. These "scientific" missionaries considered their religious and secular goals as two sides of a coin. As the American Presbyterian Calvin Mateer explained it, science was important not only for its practical applications, but also for its ability to dispel superstition and foster the "faculties of reasoning and analysis," and ultimately because the study of nature would provide potential converts with "an exposition of the unwritten laws of God."[14] Fryer was comfortable with such views. His objective, in his own words, was "to be named among those who are foremost in enlightening and Christianizing the Great Empire."[15]

What the missionaries brought to China was not a knowledge of science, but the ability to learn and a powerful will to put that learning to service. Fryer, like many of his contemporaries, proved this commitment first by mastering the Chinese language. He knew little about science and evinced no interest in this subject until the spring of 1868, when he ordered some materials from England for "the sake of showing experiments to the Chinese." After his appointment as a translator, he sent home for books and laboratory equipment, which he used to teach himself, confiding, "as I am only a half-educated man, I have to study pretty hard to keep pace with my duties."[16]

Missionaries were at the center of all those developments through which science entered China in the nineteenth century. Some worked in schools and translation bureaus set up by the Chinese government, others created their own schools, published journals, or pressed their concerns in unofficial circles. In both roles, they led the effort to reshape Chinese language and culture to accommodate the new learning. Outsiders by birth, they were for a time important actors in the remaking of China.

Conclusion

During the two decades between the opium wars, when the mind of official China was still closed to barbarian ways, the first small steps to introduce science to the Middle Kingdom were taken by a radical fringe of failed literati, with a curious taste for the investigation of things, and a handful of missionaries, out of sync with the fundamentalist spirit of their time. Early Chinese students of science took a notably traditional

---

[14] Mateer's view of science: Irwin T. Hyatt, Jr., *Our Ordered Lives Confess: Three Nineteenth-Century American Missionaries in East Shantung* (Cambridge: Harvard University Press, 1976), 186–7. For other details on Mateer, see: Daniel Fisher, *Calvin Wilson Mateer: Forty-Five Years a Missionary in Shantung Province, China* (Philadelphia: Westminster Press, 1911), passim.

[15] Fryer, letter of 15 March 1870, in Adrian Bennett, *John Fryer*, 26.

[16] Fryer, letters of 25 March 1868, and 1 November 1869, ibid., 22, 25.

approach to this nontraditional subject. They began by reading the classics, discovered among these texts a few that described nature and man's interaction with it, moved to Western works on astronomy and mathematics translated by Jesuit fathers centuries before, and finally found the contemporary books and articles produced by the missionary-translators of their own time. This gradual transition from traditional Chinese to modern Western studies of nature stands in sharp contrast to the experience of later Chinese scientists, who knew little or nothing of their own past and took their science directly from authorities in Europe and the United States. China's scientific pioneers were amateur scholars, who had broad interests and no thought of developing a particular expertise, very pillars of the Confucian ideal. This distinguishes them from the physicians, pharmacologists, and astronomers, who played the leading role in introducing Western science to Japan, but were notably absent in the case of China. Despite their traditionalism, however, Chinese who came to science in the nineteenth century had to leap out from the deep rut of established learning – which explains their peculiarities and the fact that they were so few.[17]

Without belittling their contribution, these avant-grade Chinese could have achieved little without the leadership and support of their foreign partners. The missionary advocates of science were equally strange: secular Christians in a fundamentalist age, willing participants in Chinese society at a time when most of their colleagues preferred to stand apart from and lecture to the heathens. More than just sources of information and advice, the missionaries were active agents, helping to create a new Chinese language and shaping Chinese culture to accommodate the modern world. For the next half century, they held a spot in center stage.

What drove these pioneers, Chinese and foreigners, was curiosity and spirit rather than utilitarianism. Among the missionaries, only the physicians had a practical use for science, whereas the others saw it as an expression of the Enlightenment, which they would learn themselves and pass on to the Chinese. John Fryer showed a great zest for learning language, but no particular penchant for science, until he gained an audience for this subject. Hsü Shou focused more consistently on objects of nature, but he too was driven primarily by the quest for knowledge rather than results. When their paths met at the Kiangnan Arsenal, it is not surprising that Fryer and Hsü should find in one another kindred spirits, or that they should borrow this office, which had been set up to make ships and guns, to begin their pursuit of the esoteric study of change.

[17] Japan: Nakayama Shigeru, *Academic and Scientific Traditions in China, Japan, and the West*, trans. Jerry Dusenbury (Tokyo: University of Tokyo Press, 1974), 195–202.

2

Changing Chinese:
chemical translations of the
Kiangnan Arsenal

Science entered nineteenth-century China through textbooks and other materials that were translated by teams of Chinese and Western scholars working in Peking, Shanghai, and Canton. In the case of chemistry, the most important studies were done by John Fryer and Hsü Shou, who served together in the translation bureau [fan-i-kuan] of the Kiangnan Arsenal. The Kiangnan translations are a monument, because they created the literature and language that established science in the culture of modern China. They are also a curiosity, because they placed the word of science before a society that was ill-prepared to understand and ill-disposed to accept it.

The Kiangnan Arsenal

Chemistry, which was well established in the science and technology of Europe, was virtually unknown in China when John Fryer and Hsü Shou arrived at the Kiangnan Arsenal in 1867. Novelty was just the thing to excite these men, however, for both had been drawn to their earlier studies more by curiosity than the prospect for gainful employment. They had been selected for the arsenal because it was thought they would be helpful in making more and better ships and guns. But this was not exactly what Fryer and Hsü had in mind. Instead, they borrowed on the reputation for utility to continue their private search, which they directed, for reasons known only to themselves, to the study of chemistry.

Hsü Shou and his colleagues requested transfers to the Kiangnan Arsenal, as John Fryer later explained, "for the convenience of carrying on their investigations and studies in the vicinity of foreigners. . . ." Tseng Kuo-fan's motive for moving them, on the other hand, was that he intended to extend Kiangnan's mandate to include shipbuilding and saw that Hsü's experience would be vital to this undertaking. When the Wuhsi scholars arrived in 1867, the arsenal had only one shop, filled with machinery of a general type – "machines to make machines" as the American-educated purchasing agent, Yung Wing, called them. Under Tseng's prodding, a shipyard was added and the first keels laid down in 1868. For several years thereafter, Kiangnan manufactured ships, guns, and ammunition of various types.[1]

Hsü Shou was not to be drawn to this side of the operation, however. Shortly after their arrival, Hsü and Hua Heng-fang persuaded the director that what the arsenal needed was more and better information, for which purpose they should be authorized to set up a translation bureau. The bureau was established in 1868, with Hsü Shou, Hsü Chien-yin, and Hua Heng-fang as its first Chinese staff, together with the English-speaking foreigners, John Fryer, Alexander Wylie, John MacGowan, and Carl Kryer. The bureau's assignment was to produce studies that would support the manufacture of arms and serve as textbooks for use in the arsenal school.

Since his earlier experience had been with the construction of steamships, it was natural that Hsü Shou's first translation, published in 1871, should be the "Origin of the Steam Engine" [Ch'i-chi fa-jen]. The translators were, however, free to select titles from among the hundreds of books Fryer had ordered from England, and Hsü, whose greatest asset was his capacity for growth, chose at the age of fifty-one to pursue a completely new subject, chemistry. Ignoring the fact that they knew little of this discipline and that it bore no connection to the military or any other industry in China at that time, Hsü and Fryer forged ahead.[2]

Chemistry in China before 1870

When Fryer and Hsü began their work at Kiangnan, chemistry was virtually unknown in China. During the seventeenth and early eighteenth centuries, Jesuit missionaries had introduced Western mathematics, astronomy, and cartography, but contributed little to the Chinese understanding of chemistry. There were two reasons for this. First, modern chemistry did not come of age until the end of the eighteenth century, *after* the suppression of the Jesuit order in 1773. Second, the Jesuits in China included few men with an interest in or understanding of this branch of science. In contrast to the demonstrable truths unveiled by

[1] Fryer: John Fryer, *North China Herald*, (29 January 1880), cited in Gideon Ch'en, *Tseng Kuo-fan*, 89. Other details: ibid., 43–52.

[2] For an overview of the work of the translation bureau, see: Adrian Bennett, *John Fryer*, 18–45. For list of books ordered by Fryer, see 73–81, and translations by Fryer, 82–109.

astronomy and mathematics, the study of earthly substances was still mired in the practical knowledge of craftsmen and the hocus-pocus of alchemy. The Jesuits were not particularly interested in these subjects and, judging by the reception, neither were the Chinese.

The early Jesuit translations that introduced to China European explanations for the composition and interaction of substances were depreciated or ignored. The *K'ung-chi ko-chih* [Treatise on the material composition of the universe] (1633), which outlined Aristotle's theory of the four elements, failed to impress the Chinese who already had "five elements" [*wu-hsing*] of their own. *K'un-yü ko-chih* [Investigation of the earth] (1640), a translation of *De Re Metallica,* the sixteenth-century classic on mining and minerology by the German author Agricola, was presented to the emperor, but apparently never published. If it had reached an audience of Chinese mining engineers, this book might have done some good, but it had little potential for advancing science, for as Joseph Needham notes, "Although Agricola belonged to the old assay tradition rather than the alchemical, there is not a word of the new chemistry in him." More popular was the *Huo-kung ch'ieh-yao* [Essentials of gunnery] (1643), which advanced the manufacture of gunpowder in China, by prescribing an explosive rich in nitrate and better adapted to the needs of seventeenth-century cannons than the familiar Chinese concoction. It contained nothing of significance, however, on the still undeveloped discipline of chemistry.[3]

Traditional Chinese approaches added little to the understanding of these matters. Alchemy, the principal mode of inquiry into the nature of substances before the coming of modern chemistry, flourished in China during the first millennium of the Christian era. After the eleventh century, however, Chinese alchemy declined, both in popularity and in the creativity of its models and methods. It may be, as Joseph Needham

---

[3] For details on Jesuit translations, see: Joseph Needham, *Science and Civilisation in China* (Cambridge, Eng.: The University Press, 1954), vol. 5, pt 3:220–41; Chang Tzu-kao and Yang Ken, "Ya-p'ien chan-cheng i-ch'ien hsi-fang hua-hsüeh ch'uan-ju Chung-kuo ti ch'ing-k'uang" [Introduction of Western chemistry into China before the Opium War], *Ch'ing-hua ta-hsüeh hsüeh-pao* 11, no. 2 (June 1964):1–14, also reprinted in Chang Tzu-kao, ed., *Chung-kuo ku-tai hua-hsüeh shih* [History of ancient Chinese chemistry] (Hong Kong: Shang-wu yin-shu-kuan, 1977), 311–31; P'an Chi-hsing, "A-ke-li-k'o-la ti *K'uang-yeh ch'üan-shu* chi ch'i tsai Ming-tai Chung-kuo ti liu-ch'uan" [Agricola's *De Re Metallica* and its transmission to China during Ming times], *TJKHSYC* 2, no. 1 (1983):32–44; P'an Chi-hsing, "Ming-Ch'ing shih-ch'i hua-hsüeh i-tso shu-mu k'ao" [Study of catalogs of chemical translations of the Ming-Ch'ing period], *CKKCSL* 5, no. 1 (1984):23–38; P'an Chi-hsing, "Wo-kuo Ming-Ch'ing shih-ch'i kuan-yü wu-chi-suan ti chi-tsai" [Chinese records of inorganic acids from the Ming and Ch'ing periods], *Ta tzu-jan t'an-so* [Explorations of nature] 3 (1983); and Li Ya-tung, "Hsü Kuang-ch'i ti hua-hsüeh ch'eng-chiu" [The chemical achievements of Hsü Kuang-ch'i], *Literature and History in China* (Shanghai) 3 (1984):29–39. Quotation: Joseph Needham, *Science and Civilisation in China,* vol. 5, pt 3:236.

argues, that the skills and instruments of the alchemist were passed on through Chinese metallurgists, distillers, and pharmacists. If so, they failed to enrich the practice of modern science. Some early Chinese recruits to chemistry, such as Hsü Shou, came to this field from a background in traditional nature studies, but none began with an interest in alchemy. Chinese scholars of recent times who have investigated the properties of native products have drawn their theoretical models and analytical techniques exclusively from the laboratory of modern chemistry. In contrast to the continuities exhibited in the fields of astronomy, mathematics, and medicine, no Chinese in the mainstream of chemistry has tried to combine traditional theories and practices with the models of modern scier.ce. (A similar disjunction has been noted in the case of geology and may be characteristic of other scientific disciplines in China as well.)[4]

Protestant missionaries, on the whole less scholarly and more evangelistic than their Jesuit predecessors, were slower to promote the spread of secular knowledge. Almost all of the nearly eight hundred Protestant works published in China between 1810 and 1867 were devoted to the propagation of Christianity, while only a handful dealt with medicine and science and only one of these with chemistry. Protestant translations from this period contained no word of Lavoisier or the revolution that had remade this discipline in the West.[5]

In the absence of scholarly leadership, what little chemical knowledge that found its way into China in the middle of the nineteenth century came through the practical experience of merchants, artisans, and workers. Clipper ships carried fire extinguishers that used sulfuric acid. The ships' pharmacy included tincture of iodine, boric acid, and Epsom salts (magnesium sulfate). Merchants assayed metals with the aid of nitric acid; welders used hydrochloric. Chinese commoners, who were the first to recognize these substances, assigned crude names, many of which remain in use today: *huang-ch'iang-shui* [sulfuric acid], *hsiao-ch'iang-shui* [nitric acid], *yen-ch'iang-shui* [hydrochloric acid], and *yang-ch'i* or *sheng-ch'i* [oxygen]. Later translators who sought the names of chemical substances often looked among "such Chinese merchants, manufacturers, mechanics, etc., as would be likely to have the term in current use."[6]

---

[4] Alchemy: Joseph Needham, *Science and Civilisation in China*, vol. 5, pt 3:208–20; Nathan Sivin, "Chinese Alchemy and the Manipulation of Time," *Science and Technology in East Asia*, ed. N. Sivin (New York: Science History Publications, 1977). Geology: Tsui-hua Yang Lee, "Geological Sciences in Republican China, 1912–1937" (Ph.D. diss., State University of New York at Buffalo, 1985), 43.

[5] Protestant publications: Tsuen-hsuin Tsien, "Western Impact on China through Translation," 311.

[6] Quotation: John Fryer, "Science in China," *Nature* (May 1881), 54. Other details: Yüan Han-ch'ing, *Chung-kuo hua-hsüeh-shih lun-wen chi*, 266.

One reason for the relatively low output of translations by Protestant missionaries, even after the middle of the nineteenth century, was that they could not agree on what language to use.[7] Although criticized by members of other Catholic orders for their worldly preoccupations, the Jesuits at least agreed among themselves that scientific and other secular knowledge should be spread through the medium of Chinese and that they should take the lead in making these works available. But the Protestants were divided. A substantial number of sincere and learned men doubted that Chinese with its peculiar writing and restrictive classical forms was capable of expressing modern ideas. Some argued that the language was a culture trap, which "can awaken only Chinese thoughts in Chinese minds." In their view, a true understanding of science and other modern knowledge could come only through the medium in which these ideas had developed. Without English or some other European language, Chinese students would view the world darkly, through few and faulty translations, and never come within reach of the rapidly advancing edge of knowledge. In any case, English was on the way to becoming a universal language, making Chinese expendable. As educators and foreign agents with congregations to answer to back home, these missionaries also knew that they could expand school enrollments by recruiting students from among the growing number of young people who saw English as the ticket to a lucrative career.

Protestant proponents of Chinese, on the other hand, included those most skilled in this language, which, they believed, had the same capacity for growth and development as any other. In practical terms, the Sinophiles argued, students educated in English generally left school as soon as they found well-paying jobs and before they had been much influenced by either secular or religious training. These young men and women departed with knowledge that made them "foreigners in their own country" and lacked the integration of language and substantive learning required to make good Chinese ministers, doctors, lawyers, and Christians. Use of the native tongue was the only way to reach the literati, who, Fryer pointed out, would "never consent to learn of the Barbarian whom they despise...in his own language."[8] In the opinion of many missionaries, the objective was not to educate a few individuals in knowledge of a sort separate from China, but to reform Chinese language and culture so that they would become part of the modern

---

[7] Debate on language: "The Advisability, or the Reverse, of Endeavouring to Convey Western Knowledge to the Chinese through the Medium of Their Own Language," *Journal of the North China Branch of the Royal Asiatic Society* 21 (1886):1–21; Liu Kwang-ching, "Early Christian Colleges in China," *Journal of Asian Studies* 20, no. 1 (1960):74–5.

[8] John Fryer, "Western Knowledge and the Chinese," *Journal of the North China Branch of the Royal Asiatic Society* 21, nos. 9–11 (1886):10, cited in Adrian Bennett, *John Fryer*, 28.

Christian scientific world. The circular of one mission college issued in 1893 summed up the case for Chinese as follows:

In spite of its manifest defects, every year sees the Chinese language become a more worthy medium for the teaching of even the most advanced lines of science, and to be content to teach the sciences from English textbooks is to admit defeat before the battle has well begun, and to retreat in advance from the efforts to build up such a system of education along the lines of national self-respect as in the end shall inevitably win the day.[9]

The first and for some time most influential Protestant writer on science in the Chinese language was Benjamin Hobson. Hobson, an Englishman, practiced and taught medicine in Canton and Shanghai during the 1850s and wrote six books on the medical sciences, which were widely regarded as the standard Chinese works in this field. One of these, *Po-wu hsin-pien*, which was published in Shanghai in 1855 and, as noted, had a great impact on Hsü Shou, provided a general introduction to meteorology, chemistry, physics, astronomy, geography, and zoology. The section on chemistry included descriptions of the properties and methods of preparation of the important gases (oxygen, hydrogen, nitrogen, carbon monoxide, and methane) and acids (sulfuric, nitric, and hydrochloric). Although the presentation was brief and the terms borrowed from the common dialect, the treatment was entirely modern, making this the first exposition of chemistry in the Chinese language. Hobson's books circulated during the 1850s and 1860s throughout the treaty ports of China and to Japan.[10]

A Chinese-language science textbook of even broader scope was produced by the American missionary-educator W. A. P. Martin and two Chinese assistants at the T'ung Wen Kuan in Peking in 1868. Martin's seven-volume work, entitled *Ko-wu ju-men* [Natural philosophy], covered hydraulics, meteorology, heat and light, electricity, mechanics, chemistry, and mathematics. The volume on chemistry included chapters on general principles, gases, metals, and organic materials, with laboratory experiments to demonstrate how to prepare and test for each substance. Like Hobson, Martin compiled this work on his own rather than translate an existing text, and described the new science using traditional terms that were unfortunately burdened with established

---

[9] The 1893 circular of the North China College, cited in Dwight W. Edwards, *Yenching University* (New York: United Board for Christian Higher Education, 1959), 29.

[10] Biography of Hobson: K. Chimin Wong, *Lancet and Cross: Biographical Sketches of Fifty Pioneer Medical Missionaries in China* (Shanghai: Council on Christian Medical Work, 1950), 14–16. Contemporary critique of Hobson's books: William Lockhart, *The Medical Missionary in China* (London: Hurst and Blackett, 1861), 154–8. *Po-wu hsin-pien*: Ho Hsin [Benjamin Hobson], *Po-wu hsin-pien* [Natural philosophy and natural history] (Shanghai: Mo-hai shu-kuan, 1855). For a review of Hobson's *Medical Vocabulary in English and Chinese* (Shanghai, 1858), see: *CMJ* (1887):115.

connotations. Martin was among the first, however, to use the term *hua-hsüeh*, the modern Chinese name for chemistry, which he explained as follows:

To investigate the substance of things and the forces by which they are blended, to separate them into pure elements and combine them into many forms, to study the changes in things and the principles [underlying these changes], in order to recover their most minute particles: this is called chemistry [*hua-hsüeh*].[11]

Both Hobson and Martin presented a broad range of scientific knowledge pitched to a general audience, using a vocabulary that most laymen could understand. They drew their information from various sources, borrowed Chinese terms from those in common use, and repackaged the whole in digestible textbook form.

By contrast, the first translation of a full-length chemistry text, using a formal scientific language developed expressly for this purpose, was the work of the missionary-physician John Glasgow Kerr. Kerr began in the 1860s to train Chinese assistants at his Canton hospital. Initially, his students read the works of Hobson, but Kerr recognized the need for better textbooks, and during a period of thirty years he translated and published thirty-four volumes on all aspects of medical science. The earliest of these, *Hua-hsüeh ch'u-chieh* [First step to chemistry], was based on a popular American text by David Ames Wells, translated by Kerr and one of his students, Ho Liao-jan, and published in four volumes between 1871 and 1875.[12]

At about the same time, the Frenchman Anatole Billequin, who taught chemistry at the T'ung Wen Kuan, translated the popular textbook by Faustino Malaguti under the title, *Hua-hsüeh chih-nan* [Guide to chemistry] (1873). Billequin was the only member of his generation in China who had been formally trained in this discipline. During his stay in China he performed a variety of chemical research, wrote on traditional Chinese chemical arts and translated a second text, *Hua-hsüeh shan-yüan* [Explanation of the principles of chemistry], a study of chemical analysis that appeared in 1882.[13]

[11] Martin and the T'ung Wen Kuan: Knight Biggerstaff, *The Earliest Modern Government Schools in China* (Ithaca: Cornell University Press, 1961), 108–28. *Ko-wu ju-men*: Ting Wei-liang [W. A. P. Martin], *Ko-wu ju-men* [Natural philosophy], 7 vols. (Peking: T'ung Wen Kuan, 1868). Definition of "hua-hsüeh": ibid., 6:1a.
[12] Biography of Kerr: K. Chimin Wong, *Lancet and Cross*, 23–8; K. Chimin Wong and Wu Lien-teh, *History of Chinese Medicine* (Shanghai: National Quarantine Service, 1936), 372–5, 391–4. *Hua-hsüeh ch'u-chieh*: *Hua-hsüeh ch'u-chieh* [First step to chemistry], translation of David A. Wells, *Wells's Principles and Applications of Chemistry*, trans. John Kerr [Chia Yüeh-han] and Ho Liao-jan (Canton: Po-chi i-yüan, 1871–1875).
[13] Biography of Billequin: *T'oung Pao* 5 (1894): 441–2. *Hua-hsüeh chih-nan*: *Hua-hsüeh chih-nan* [Guide to chemistry], translation of Faustino Malaguti, *Leçons élémentaires de chimie* (Paris, 1853–68), trans. Anatole Billequin [Pi-li-kan] and Lien Tzu-chen (Peking:

Hobson, Martin, Billequin, and Kerr began the process of introducing modern chemistry along with other sciences to the Chinese. Hobson's books, although crude when compared with those that replaced them, blazed the trail in China and gained wide circulation in Japan as well. Kerr's texts became the standard for use in training medical apprentices. Martin and Billequin were the most influential promoters of science in Peking, where students at the T'ung Wen Kuan read their works. But the contribution of these men was limited to these few volumes, and their effort remained a decidedly foreign affair. Hobson noted that his translations were prepared "by the aid of an intelligent native." The Chinese collaborators of Martin, Billequin, and Kerr are known to us only by name. By contrast, at the translation bureau of the Kiangnan Arsenal, a community of Chinese and foreign scholars, whose careers are fully documented, produced an entire library. Most important to the field of chemistry was the partnership of John Fryer and Hsü Shou.[14]

Translations of the Kiangnan Arsenal

During the four decades it remained in operation, from 1868 to 1908, the translation bureau of the Kiangnan Arsenal produced at least 178 books, primarily in natural science (66), military science (38), and engineering (35). Among the sciences, the largest numbers of works were on astronomy and mathematics (26), geology and mining (17), and physics (14), while only 9 dealt with chemistry. In most cases, the original sources were English-language textbooks that presented current knowledge in a given field. The method of translating was the same as that employed by the Jesuits: A foreigner dictated the text sentence-by-sentence in spoken Chinese, and his Chinese partner wrote it down in the literary style. The two men went over difficult passages together to determine their meaning and find the most appropriate way of expressing it. When the text was complete, it was copied, illustrations and diagrams were added by an experienced draftsman, and wooden blocks were cut to be printed by hand on the best-quality paper, which was bound with silk thread. "They present," Fryer noted proudly, "an appearance which satisfies the taste of the most fastidious native."[15]

T'ung Wen Kuan, 1873). *Hua-hsüeh shan-yüan: Hua-hsüeh shan-yüan* [Explanation of the principles of chemistry], translation of Karl R. Fresenius, *Chemical Analysis*, trans. Anatole Billequin [Pi-li-kan], Ch'eng Lin, and Wang Chung-hsiang (Peking: T'ung Wen Kuan, 1882). For details on *Shan-yüan*, see: Yüan Han-ch'ing, *Chung-kuo hua-hsüeh shih lun-wen chi*, 277; and Joseph Needham, *Science and Civilisation*, vol. 5, pt 3:252.
14 Hobson's books in Japan: K. Chimin Wong, *Lancet and Cross*, 16. Hobson quotation: William Lockhart, *Medical Missionary in China*, 155.
15 The most authoritative list of Kiangnan publications for the period 1870–1905 appears in Wei Yün-kung, ed., *Chiang-nan chih-tsao-chü chi* [Record of the Kiangnan Arsenal]

The first work by Fryer and Hsü, *Hua-hsüeh chien-yüan* [Authentic mirror of chemical science], was a translation of David Ames Wells's *Principles and Applications of Chemistry,* the same text chosen by John Kerr. The American edition, first published in 1858 and reprinted several times thereafter, was addressed to students in academies, seminaries, and colleges and meant "to furnish just that information which will prove most useful and practical in their future employments and relations of life." It contained a full account of basic chemical knowledge and theories, including the principles of chemical structure and laws of chemical combination, the atomic theory of John Dalton, the methods of chemical nomenclature and notation, a description of the sixty-two known elements, the most important compounds formed from them, and their applications to industry. The Fryer-Hsü translation covered the first half of Wells's text, including the items just described, but excluding sections on physics, which had been dealt with by W. A. P. Martin, and organic chemistry, which was reserved for a later volume. In most cases, the translators paraphrased, employing a simple classical style that conformed to the standards of the time, while avoiding the literary allusions and poetic license that make traditional Chinese texts ambiguous and sometimes opaque.[16]

Fryer and Hsü were not trained chemists; their understanding of this field was less than perfect. A mistranslation of one critical section, dealing with "chemical equivalents," illustrates the difficulty these novices faced in describing the strange new reality that confronted them. The discipline of chemistry in the nineteenth century rested (and still rests) on the observation that the various elements combine in fixed proportions *by weight* to form compounds. This concept was explained by Wells as follows: "The proportions, or quantities by weight, in which different substances unite to form definite chemical compounds, are called Chemical Equivalents. They are also sometimes designated as combining, or equivalent, weights. The numbers representing or expressing these proportions are termed equivalent numbers."[17] Although

---

(Shanghai: Shang-hai wen-pao shu-chu, 1905), 2:15–22. Number of books in each scientific discipline: Li K'un-hou, "Chiang-nan chih-tsao-chü yü Chung-kuo hsien-tai hua-hsüeh" [The Kiangnan Arsenal and contemporary Chinese chemistry], *Chung-kuo k'o-hsüeh-shih lun-chi* [Essays on the history of science in China], ed. Lin Chih-p'ing (Ling Chih-Bing) (Taipei: Chung-hua wen-hua ch'u-pan shih-yeh wei-yüan-hui, 1958), 47. Production of books: John Fryer, "Science in China," 55.

[16] In comparing the original text with the translation, I have used: David A. Wells, *Wells's Principles and Applications of Chemistry* (New York: Ivison, Phinney & Co., 1864), and Fu Lan-ya [John Fryer] and Hsü Shou, *Hua-hsüeh chien-yüan* [Mirror of chemical science], vols. 9 and 10 of *Hsi-hsüeh fu-ch'iang ts'ung-shu* [Collection on wealth and power through Western studies], (1902). The first and only edition of Wells's *Principles* was issued several times between 1858 and 1879. A favourable assessment of this translation is given by P'an Chi-hsing, "Ming-Ch'ing shih-ch'i (1640–1910) hua-hsüeh i-tso shu-mu k'ao," 24.

[17] David Wells, *Wells's Principles*, 165–6.

demonstrating a general understanding of this principle, Fryer and Hsü wrote that the elements combine by "fractions" [*fen*] or "proportions" [*pi-li*], omitting the key point that these shares are measured "by weight." They abbreviated Wells's definition of chemical equivalents, omitting reference to the weight of the substances involved and saying only: "The numbers used [to represent] the combination of elements are called equivalent numbers." Critics have pointed to this section of the translation as evidence that the work is inaccurate and muddled. A more generous view might be that the translators, who were starting from ground zero, succeeded in providing a reasonably clear description of a new science in a language that made it for the first time accessible to Chinese readers. Given the distance Chinese had to travel to reach an understanding of chemistry, I conclude that the contribution of this text was great and its flaws excusable.[18]

During his years at Kiangnan, Hsü Shou executed thirteen translations, all save one in collaboration with Fryer, ranging from metallurgy to law, but most dealing with chemistry. In addition, Hsü is generally credited with two vocabulary lists on chemistry and Western medicines. Fryer and Hsü's eldest son, Chien-yin, produced another volume on analytical chemistry. Some years later, Fryer and Wang Chen-sheng translated a three-volume work on the manufacture of acids and alkalis. Each of these books was a complete or nearly complete version of an important and authoritative European text. Despite the prejudices that may have lingered in the minds of some, the Kiangnan translations established beyond reasonable doubt that the Chinese language could be adapted to the needs of modern science.[19]

More important to the spread of science than transmitting factual information are the attitude, discipline, and spirit that must infuse this method of study. These values are best communicated by people, face-to-face in the laboratory or field, but textbooks can reinforce the point. Fryer and Hsü tried to explain the approach of the European masters, as for example in their *Hua-hsüeh ch'iu-shu,* a translation of

[18] Fryer-Hsu translation: *Hua-hsüeh chien-yüan,* 1:2b. Critics: Chang Tzu-kao and Yang Ken, "Ts'ung 'Hua-hsüeh ch'u-chieh' ho 'Hua-hsüeh chien-yüan' k'an wo-kuo tsao-ch'i fan-i ti hua-hsüeh shu-chi ho hua-hsüeh ming-ts'u" [Viewing our country's early translations of chemistry books and chemical terminologies through "Hua-hsüeh ch'u-chieh" and "Hua-hsüeh chien-yüan"], *TJKHSYC* 1, no. 4 (1982):349–55.

[19] A comprehensive Chinese-language list of translations by Fryer, Hsü, et al. appears in: Hsü Chen-ya and Juan Chen-k'ang, "Hsü Shou fu tzu tsu-sun i-chu chien-chieh" [Brief introduction to the translations of Hsü Shou, his sons, and descendants], *CKKCSL* 7, no. 1 (1986):48–55. An English-language list appears in Adrian Bennett, *John Fryer,* 84–106. Among the prominent works on chemistry translated at Kiangnan were the popular textbooks by John Bowman and Charles Bloxam, the classics on qualitative and quantitative analysis by Karl Fresenius, and the standard work on European industrial chemistry by Georg Lunge.

Karl Fresenius's *Quantitative Analytical Chemistry*. In their introduction to this work, the translators offer advice that will be recognized by any scientist as both comforting and sound. Students, who in the course of their experiments make a mistake and cannot find the cause, who because of careless technique have lost some of the material and cannot get the results, or who in their frustration begin to suspect that the balances are inaccurate, "must seek out the cause of the error and do it over again."

Take a look at the experiments of the ancients. After hundreds and thousands of times, they got it right, and can therefore infer [the correct answer]. Scholars must not place the saving of time and trouble first and the determination of quantity second. Much less can they make empty guesses and avoid experimentation. If they commit this error, they certainly cannot do the work of quantitative analysis.[20]

The Kiangnan translations were not without fault. One problem was the serendipitous manner in which books were selected. Reliance on the personal inclinations of the translators meant that some fields, like natural history, were ignored, but if the choice had been made on strictly utilitarian grounds, chemistry might also have been omitted. Hsü and Fryer made use of the most current European works: The gap between publication of the originals and their translations was only five to thirteen years. Later translators failed to keep pace, however, with the result that the most important chemical theories of the late nineteenth century – the law of periodicity (1869) and theory of ionization (1887), for example – did not appear in Chinese until after 1900. Despite these reservations, the work of the Kiangnan translators stands as the single greatest achievement of science in nineteenth-century China. Chinese interested in this subject did not want for accurate, informative material; even Japanese agents came to Kiangnan to procure the latest chemical texts.[21]

[20] Introduction to the *Hua-hsüeh ch'iu-shu*, cited in Yüan Han-ch'ing and Meng Nai-ch'ang, "Hsü i 'Hua-hsüeh k'ao-chih' ho 'Hua-hsüeh ch'iu-shu'" [Hsü's translations, "Qualitative analytical chemistry" and "Quantitative analytical chemistry"], *Hsü Shou ho Chung-kuo chin-tai hua-hsüeh shih*, 138.

[21] Time gap for Hsü's translations: Li Ya-tung, "Hsü Shou so-i hua-hsüeh chu-tso ti yüan-pen" [Original chemical works translated by Hsü Shou], *HHTP* 3 (1985):54. Li-K'un-hou, "Chiang-nan chih-tsao-chü," 62–3, points out that the last major Kiangnan chemical translation, which appeared in 1898, was based on the 1880 edition of Georg Lunge's study of chemical industry. The first description of the law of periodicity in Chinese appeared in the journal, *Ya-ch'üan tsa-chih*, in 1901. See: Chang Tzu-kao and Yang Ken, "Chieh-shao yu-kuan Chung-kuo chin-tai hua-hsüeh-shih ti i-hsiang ts'an-k'ao tsu-liao – 'Ya-ch'üan tsa-chih'" [Introducing reference material related to the history of modern chemistry in China – the "Ya-ch'üan Magazine"], *HHTP* 1 (1965):56–7. Japanese agent: *Hsi-Chin ssu-che shih-shih hui-ts'un*, cited in *Hsü Shou ho Chung-kuo chin-tai hua-hsüeh shih*, 363. The quantitative importance of the Kiangnan translations can be seen by measuring them against the 182 scientific

Table 2.1. *Billequin's characters*

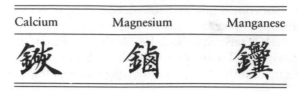

Calcium	Magnesium	Manganese

Chemical nomenclature

Whereas the principal task of the Kiangnan translators was to produce texts of immediate practical use, the lasting contribution of this enterprise lay in the establishment of a standard vocabulary to express the new ideas. There has been some disagreement over who deserves credit for inventing modern chemical Chinese. Benjamin Hobson and W. A. P. Martin, who were first to describe the science of chemistry in Chinese, relied on established terms familiar to their readers, but by the same token often clumsy and inappropriate for the ultimate purpose of expressing a distinct and precise reality. Martin called silicon *po-ching,* or "essence of glass," and named manganese after the traditional term for manganese dioxide, *wu-ming-i,* which means literally "nameless wonder," hardly a felicitous label for a substance the authors hoped Chinese scientists would come to understand. Anatole Billequin, following Martin, named the elements by inventing new characters that drew on their traditional names or described the nature of the element itself. In Billequin's system, the character for calcium, the chief component of lime, included the term for lime, *shih-hui;* magnesium, which was extracted from brine, contained the character for salt, *lu;* and manganese was represented by a monstrous character composed of the entire expression *wu-ming-i* (Table 2.1). There is no indication as to how any of these characters was to be pronounced.[22]

Ultimately, the debate over who invented the Chinese chemical nomenclature now in use is between supporters of John Kerr and Ho Liao-jan, who published the *Hua-hsüeh ch'u-chieh* in 1871, and John Fryer and Hsü Shou, whose *Hua-hsüeh chien-yüan* appeared one year later. The case for Kerr and Ho rests on the fact that their publication came out earlier and contained a large number of terms for the chemical elements that remain in use today. On the other hand, Kerr later

works from the period 1850–99, which are listed in Hsü I-sun, comp., *Tung hsi-hsüeh shu-lu* [Eastern bibliography of Western studies] (Shanghai: Chiang-nan chih-tsao-chü, 1899), 1–29.

[22] Billequin's characters: Liu Shu-k'ai, "Wo-kuo chin-tai hua-hsüeh ti ch'i-meng-che Hsü Shou," *Hsü Shou ho Chung-kuo chin-tai hua-hsüeh shih,* 63.

admitted that he borrowed several element names from a list provided to him by Fryer. More to the point, the contribution of Fryer and Hsü went beyond labeling the known elements. They also worked out systems for naming inorganic compounds and describing chemical reactions, explained the rules for inventing terms that could be applied in naming currently unknown substances, and made use of this new vocabulary in their several translations and other publications. Later translators recognized and built upon the foundation laid by them. For all this, Fryer and Hsü deserve credit as founders of the modern Chinese chemical nomenclature.[23]

The original intention of Fryer and Hsü was to append a glossary to one of their early chemical texts, but they delayed publication until they had refined the methods and broadened the scope. Their *Chinese–Western Glossary of Chemical Substances* [*Hua-hsüeh ts'ai-liao Chung-Hsi ming-mu-piao*] finally appeared in 1885. The five thousand terms in this list and the principles underlying their formation are the same as those used in previous Kiangnan texts and with appropriate modifications remain in use today.[24]

The first problem translators faced was to assign names to the elements, the building blocks of more complex chemical substances (Appendix, Table A.1). Long before Europeans set foot in the Middle Kingdom, the Chinese had names for the known elements, mostly metals.

[23] This priority debate has passed through several stages. The first historian to recognize Fryer and Hsü was Tseng Chao-lun, "Chiang-nan chih-tsao-chü shih-tai pien-i chih hua-hsüeh shu-chi chi ch'i so-yung chih hua-hsueh ming-ts'u" [Books on chemistry edited during the era of the Kiangnan Arsenal and the chemical nomenclature used in them], *Hua-hsüeh* [Chemistry] 3, no. 5 (1936):746–62. This view was later adopted by writers in both China and Taiwan: Yüan Han-ch'ing, *Chung-kuo hua-hsüeh shih lun-wen chi*, 277; Li K'un-hou, "Chiang-nan chih-tsao-chü yü Chung-kuo hsien-tai hua-hsüeh," 49–53; and Li Ch'iao-p'ing, *Chung-kuo hua-hsüeh shih*, 1:19–24. This orthodoxy was challenged by the view that Ho Liao-jan is the proper founder of Chinese chemical nomenclature: Chang Tzu-kao, "Ho Liao-jan ti 'Hua-hsüeh ch'u-chieh' tsai hua-hsüeh yüan-su i-ming shang ti li-shih i-i" [Historical significance of Ho Liao-jan's "Hua-hsüeh ch'u-chieh" for the translation of names of chemical elements], *Ch'ing-hua ta-hsüeh hsüeh-pao* [Tsinghua University journal] 9, no. 6 (December 1962):41–7; and Chang Tzu-kao and Yang Ken, "Ts'ung 'Hua-hsüeh ch'u-chieh' ho 'Hua-hsüeh chien-yüan' k'an wo-kuo tsao-ch'i fan-i ti hua-hsüeh shu-chi ho hua-hsüeh ming-ts'u," 349–55. A more recent study takes a balanced view: Chang Ch'ing-lien, "Hsü Shou yü 'Hua-hsüeh chien-yüan'" [Hsü Shou and the "Hua-hsüeh chien-yüan"], *CKKCSL* 6, no. 4 (1985):54–6. It is interesting to note that *no* Chinese historian gives substantial credit to Kerr, Fryer, or any other foreigner. Kerr borrowed terms from Fryer: *CMJ* 4 (1871–2):26.

[24] Glossary: *Hua-hsüeh ts'ai-liao Chung-Hsi ming-mu-piao* [Chinese-Western glossary of chemical substances], vol. 16 of *Hsi-hsüeh fu-ch'iang ts'ung-shu* [Collection on wealth and power through Western studies] (1902). A statement of the principles governing translations first appeared in the *Hua-hsüeh chien-yüan*, 1:5b–6a. For an analysis of the principles used in the "Glossary," see: Li K'un-hou, "Chiang-nan chih-tsao-chü yü Chung-kuo hsien-tai hua-hsüeh," 49–53. For other details on the assigning of names, see: John Fryer, "Science in China," 54–5.

Later, Chinese merchants and workers gave names to those gases –
oxygen, hydrogen, nitrogen, chlorine – that had more recently entered
into their experience. Now, it was necessary to label newly discovered
elements such as lithium, iridium, and neon. The Chinese language
lacked a single system to encompass all the known substances and make
room for those not yet discovered. But what should that system be?
Should one accept the existing Chinese names, borrow the characters
already adopted by translators in Japan, or invent new ones? How
should new names be formed: by translations that described properties of
the elements? or transliterations of the sounds of their Western names?
Should there be one or several characters for each element? Should these
be selected from among traditional characters, or new ones invented?
What about using Western names themselves or the chemical symbols
assigned by foreign scientists? There was a wide range of available
choices, yet it was in the interest of all Chinese that one system be
commonly accepted, and the sooner the better.

From the outset, all major translators agreed that they would use
Chinese characters only, excluding English, Latin, and other terms in the
Roman script, a decision that might have removed some ambiguities but
would have delayed the evolution of Chinese as a medium of modern
science. This prohibition against foreign terms and scripts also extended
to use of the standard symbols for each of the elements, which are based
on the first letters of the Latin names. Rejecting the character compounds
used by Martin, Kerr-Ho and Fryer-Hsü agreed to assign a single
character to each element, creating both a name and a brief label that
could be used in chemical formulae in the fashion of Western symbols.
Exceptions were allowed for common gases, such as oxygen – the
"nurturing gas" [yang-ch'i] – and hydrogen – the "light gas" [ch'ing-
ch'i] – which retained their popular two-character names when referred
to as elements, but dropped the second character to conform to the
single-character rule when used in naming compounds.

Having decided to name each element with one character, the next
question was: What should that character be? In keeping with the spirit
of sinification, Fryer and Hsü decided that if Chinese already had a
one-character name for the element, they would use it. They studied
native works and Jesuit translations and questioned Chinese merchants
and manufacturers in search of established names. Their findings
included several common metals – gold, silver, iron, copper, lead, tin,
and mercury – as well as the nonmetal sulfur. (The name for carbon was
also derived from the traditional character for coal, to which the "stone"
radical was later added.) If Chinese had an established name of two or
more characters such as pai-ch'ien [white lead], the term for zinc, or no
name at all, as was the case for most nonmetals, then the translators

would assign a new character. In these cases, selection of the new character followed two rules. First, the sound of the character should resemble the first sound of the Latin name for that element. (In this regard, Fryer and Hsü departed from previous practice, which had favored translating or describing the nature of the element.) Second, the character should contain a radical that reflected the chemistry of the element or the physical state in which it was normally found: namely, the metal [chin] radical for metals, the stone [shih] radical for solid nonmetals, or the water [shui] radical for elements commonly occurring as liquids. Later, this principle was extended to cover gases, which were assigned characters containing the gas [ch'i] radical.[25]

Where should new characters, when needed, come from? Again, the choice was to use traditional terms whenever possible. There are more than forty thousand characters in the storehouse of Chinese ideography, so the initial range of choice was quite broad, which was fortunate because the translators had one more requirement for selection: The character should have no other common use that might otherwise add to the confusion. In some cases, such as bismuth and uranium, it was easy to find a character that met the criteria of sound, radical, and uncommonness. But taken together, these restrictions so reduced the pool of available candidates that a loophole had to be introduced, permitting the translators to modify the sound of an existing character to make it conform to the Latin name and give it a unique status in the Chinese lexicon. Examples of this type include zinc [hsin], which was assigned a standard phonetic character [ch'ieh-yin-tzu] from the K'ang Hsi dictionary, and cadmium, which was represented by a character whose original sound, "li," was changed to "ke" for this purpose.

Despite the most exhaustive and imaginative search, Fryer and Hsü were left with some elements for which no existing character seemed appropriate. In these cases, which included calcium, magnesium, lithium, and a few others, they created new characters, again following tradition by combining an established radical (metal, stone, or water) with a particle selected for its sound. (Chinese chemists, more than scientists in other fields, favored invention, creating some 200 characters in the course of the next eighty years.) Applying these principles, the Kiangnan translators assigned one character to each of the sixty-four known elements.[26]

[25] Names assigned by Chinese merchants and manufacturers: Yüan Han-ch'ing, *Chung-kuo hua-hsüeh-shih lun-wen chi*, 266–7, 277. For a complete list of chemical substances mentioned in traditional Chinese texts with their modern Chinese and English equivalents, see: Joseph Needham, *Science and Civilisation*, vol. 5, pt 2:162–84.

[26] Wells's *Principles* lists the sixty-two elements thought to exist in 1858. In fact, two of these, ilmenium and pelopium, were found not to be elements, so the number of actual elements in Wells's list was sixty. Fryer and Hsü apparently had access to a more recent

The next task was to combine these units in a systematic fashion to provide names for the more complex substances, the compounds. The history of chemical terminology has been inextricably linked to that of chemical theory itself, since the name one applies to a given substance depends on what one takes its composition and structure to be. In the West this has meant that inventors of chemical language have had to defend the theories on which their new terms are based. Chinese translators had an easier task: They were not concerned with the theory (indeed, they were not equipped to question the theory), but were simply trying to match their own language with an existing nomenclature. By the time the Chinese entered this field, European chemists had already established two systems for naming inorganic compounds. The first of these, promulgated in France in 1787 and accepted throughout most of Europe during the subsequent generation, is the system of compound names (carbon tetrachloride, hydrogen peroxide, etc.) in which each word represents an element or radical, and suffixes and prefixes indicate the relative quantity of each. The virtue of such names is that they describe the actual makeup of the substances, while fitting easily into the common pattern of speech. The second system, developed somewhat later, includes symbols for the constituent elements and numbers to indicate the exact quantity of each. These labels, called molecular formulas, offer a more compact and precise description of chemical composition, but are somewhat less compatible with normal speech. Over the last century, the advance of knowledge has brought improvements in both systems, while the underlying principles have remained unchanged.[27]

The two systems were clearly described by Wells, so Fryer and Hsü had ample understanding of the available choices. Despite its virtues, they rejected the use of compound names on the grounds that in many cases the terms "must be very long and difficult to remember," and opted to follow the model of molecular formulas, substituting the appropriate Chinese character for each chemical symbol. In this way, readers of the Kiangnan translations were given precise descriptions of the various inorganic chemicals, although those who sought to use these names

table that included the four elements discovered between 1860 and 1863: rubidium, indium, cesium, and thallium. The list given in the *Hua-hsüeh chien-yüan* contained these sixty-four elements. One more, gallium, was discovered in 1875 and added to the list that appeared in Fryer and Hsü's *Pu-pien*, published in 1883. See: Chang Ch'ing-lien, "Hsü Shou yü 'Hua-hsüeh chien-yüan,'" 54–6. Subsequent changes in the Chinese names for these elements: Li K'un-hou, "Chiang-nan chih-tsao-chü," 52–3. Inventiveness of Chinese chemists: Tseng Chao-lun, "K'o-hsüeh ming-ts'u chung ti tsao-tzu wen-t'i" [Problem of the creation of characters in scientific terminology], *Chung-kuo yü-wen* [Chinese language] 14 (October 1953):3–4.

[27] History of chemical nomenclature: Maurice P. Crosland, *Historical Studies in the Language of Chemistry* (New York: Dover, 1978).

鈣
養
炭
養
二
上
硫
養
三
＝
鈣
養
硫
養
三
上
炭
養
二

in spoken conversation undoubtedly found them cumbersome. Similarly, equations, the shorthand record of chemical reactions, were written following current Western practice, save that the orientation was vertical in keeping with traditional Chinese writing, and the plus sign (+) was replaced by an inverted T. Thus the reaction,

$$CaO \cdot CO_2 + SO_3 = CaO \cdot SO_3 + CO_2$$

was written as shown to the side.[28]

The naming of organic compounds, whose identity rests not only on the type and number of atoms but also their spacial arrangement, was (and is) a much more difficult problem, and its solution defied Western chemists until the 1890s. Lacking guidance in this field, the Kiangnan translators chose terms that described the function of a given substance − chloroform, for example, was rendered *mi-meng-shui* [bewitching water] − or more commonly transliterated the English name − ammonia became *a-mo-ni-a*, methylene, *mi-i-t'o-li-ni*, ether, *i-t'o*, and so on. Such terms were unwieldy, susceptible to ambiguity, and, if accepted, would have imparted to this science the same indelible foreignness that Buddhism, burdened with transliterated Indian terms, has suffered in China. For all these reasons, they were replaced in 1932 by a more orderly Chinese nomenclature, modeled on the system that had, by this date, gained general acceptance in the West.

Popular science translations

Despite their many virtues, the Kiangnan publications were, as Fryer noted, "far too elaborate and profound, and consequently can only be understood by a few, while the masses can never master them." More digestible forms of communication were possible, and Fryer and Hsü made it their business to help widen the audience. Among their most successful projects were the translation and publication of popular books and articles in the *Ko-chih hui-pien* [Chinese scientific magazine] and "School and Textbook Series."[29]

Ko-chih hui-pien was not the first Chinese periodical in this field. *Chiao-hui hsin-pao* [The church news] and its successor, *Wan-kuo kung-pao* [The globe magazine], which were published weekly in Shanghai from 1862 to 1883, reached two thousand readers, primarily

---

[28] "Difficult to remember": *Hua-hsüeh chien-yüan*, 1:5a–6a. For a discussion of nomenclature used in this work, see: Li K'un-hou, "Chiang-nan chih-tsao-chü," 53–6.

[29] Quotation: John Fryer, "Science in China," 55.

scholars, merchants, and officials in and around the treaty ports, with information on science and technology aimed at helping them "contend successfully with many of the most potent superstitions." The Society for the Diffusion of Useful Knowledge [*Kuang hsüeh hui*], established by foreigners in Peking in 1872, published a journal, *Chung-hsi wen-chien-lu* [The Peking magazine], the object of which was to promote the spread of science and liberal thought. Several other missionary publications carried reports on various aspects of the Western experience. Many of these works were published by the Presbyterian Mission Press [*Mei-hua shu-kuan*], which succeeded the Inkstone Bookstore as the most important publisher of scientific books in Chinese at the end of the century. What distinguished Fryer's magazine from its predecessors was that it was devoted exclusively to science, technology, and the broad dissemination of this learning among the Chinese.[30]

Ko-chih hui-pien was published during those years when Fryer was in China: monthly from 1876 to 1877 and 1880 to 1881, and quarterly from 1890 to 1892. Each issue was forty to fifty pages in length and contained eight to ten articles, mostly translations of English and American texts, plus a few original essays by Chinese and missionary authors. The subject matter varied over time. In the early years, many articles described industrial machines and methods of manufacture, perhaps to satisfy the growing number of British exporters whose advertisements helped defray the cost of publication. After 1880, Fryer returned to basic scientific education, which had always been his principal concern. Among the most prominent series was a twenty-four-installment translation of a British handbook that explained the chemistry underlying everyday experience in terms laymen could understand. Another series illustrated chemical instruments and laboratory devices, presumably to help guide those whose intention was to teach and/or conduct experiments on their own. Although there were no reports on acids, sodas, or other industrial chemicals, several articles described the manufacture of secondary chemical products, including

---

[30] For details on *Chiao-hui hsin-pao* and *Wan-kuo kung-pao* and an index of their contents, see: *Research Guide to the Chiao-hui hsin-pao (The Church News) 1868–1874* and *Research Guide to the Wan-kuo kung-pao (The Globe Magazine), 1874–1883*, both compiled by Adrian Bennett (San Francisco: Chinese Materials Center, 1975, 1976, respectively). Also: Li San-pao, "Letters to the Editor in John Fryer's *Chinese Scientific Magazine*, 1876–1892: An Analysis," *Chin-tai-shih yen-chiu so chi-k'an* [Quarterly of the Research Institute on Modern History] 4 (1974):737–9. Society and Press: Tsuen-hsuin Tsien, "Western Impact on China through Translation," 313; Chou Ch'ang-shou, "I-k'an k'o-hsüeh shu-chi k'ao-lüeh" [Study of translation and publication of science books], *Chang Chü-sheng hsien-sheng ch'i-shih sheng-jih chi-nien lun-wen chi* [Essays in honor of the seventieth birthday of Mr. Chang Chü-sheng], ed. Hu Shih et al. (Shanghai: Shang-wu yin-shu-kuan, 1936), 428–9. History of *Mei-hua* Press: Gilbert McIntosh, *The Mission Press in China* (Shanghai: American Presbyterian Mission Press, 1895).

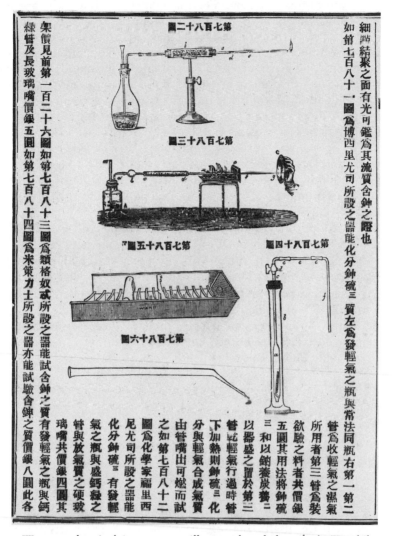

Western chemical instruments illustrated and described. (Ko-chih hui-pien, *1881*.)

glass, sugar, paper, leather, and matches, as well as the techniques of bleaching, dyeing, electroplating, and photography. The magazine was edited in an office of the Shanghai Polytechnic Institute, another undertaking of Fryer and Hsü that was designed to promote public education on modern science and technology.[31]

[31] For background on *Ko-chih hui-pien*, see: Adrian Bennett, *John Fryer*, 50–5, and Li San-pao, "Letters to the Editor," 733–44. Handbook: *KCHP* 3 (1880):1–12, 4

In addition to this magazine, Fryer sought to reach a wider audience through publication of introductory and intermediate science textbooks. This effort began with the establishment in 1877 by the General Missionary Conference of a School and Textbook Committee with Fryer as its secretary and general editor. The committee produced an "Outline Series" [hsü-chih] of primers or elementary texts and a more advanced "Handbook Series" [i-chih] that included texts and accompanying wall charts originally produced for British classrooms. During the first decade alone, more than 100 titles appeared, one-quarter of them translated by Fryer. The School and Textbook Committee later continued this work under the auspices of the Educational Association of China, which was established following the General Missionary Conference of 1890.[32]

In order to promote wider distribution of publications by the arsenal, the textbook committee, and other sources, in 1885 Fryer established the Chinese Scientific Book Depot. Within three years, this clearinghouse offered 650 separate titles on Western subjects, and sales reached 150 thousand volumes [chüan]. After 1895, when the defeat by Japan spurred greater interest in the outside world, sales at the depot skyrocketed.[33]

The partnership between John Fryer and Hsü Shou ended with Hsü's death in 1882. Fryer soldiered on, working with Hsü Chien-yin, among others, until 1896, when he left to become Professor of Oriental Languages and Literature at the University of Califor. , but he never lost touch with China. Upon his death in 1928, Fryer left a generous endowment to establish a school for the blind in Shanghai – a fitting tribute from the man who had done so much to open China's eyes to the world.[34]

Conclusion

Translation was the main channel through which knowledge of science entered nineteenth-century China. The results of this effort, particularly in chemistry where some of the best work was done, included a literature that explained the new reality and a language that would grow along with the science itself. But an assessment of the early scientific

(1881):1–12. Chemical apparatus: KCHP 3 (1880):5–12, 4 (1881):1–12. Institute: Knight Biggerstaff, "Shanghai Polytechnic Institution and Reading Room: An Attempt to Introduce Western Science and Technology to the Chinese," Pacific Historical Review 25 (1956):127–49.

[32] Textbooks: Adrian Bennett, John Fryer, 60–3. "Outline" and "Handbook" series: John Fryer, Catalogue of Chinese School and Text Books (Shanghai, 1894), 2–4, cited in ibid., 82–3. School and Textbook Committee: CRMJ 7–8 (1877–8):247–8; 10 (1879):468–71.

[33] Depot: Adrian Bennett, John Fryer, 63–6. Fryer: ibid., 37–8. For a complete list of books for sale by the depot, see: ibid., 112–35.

[34] Ibid., 68.

translations reveals in equal measure their strengths, shortcomings, and peculiarities.

The most telling fact about China's introduction to science is that it remained, almost without exception, a literary affair. Nineteenth-century Chinese "scientists" were preoccupied with book learning. Hsü Shou, who rose above his countrymen as a man of intellectual curiosity and technical skill, nonetheless devoted more time to reading and writing about science than to experimenting with or investigating the world around him. Nor was Hsü alone. I have been able to locate only one account of laboratory exercises in modern chemistry performed by Chinese scholars before 1900. This report, written around 1880 by one member of a scientific study group that met periodically at the Shanghai Polytechnic Institute, shows how dependent these men were on literary sources and how limited was their engagement with nature:

In recent years, Chinese have translated many books on chemistry. Yet most who engage in Western studies only dabble, assume a posture of great vanity, and gain no mastery of the subjects. Although one or two men have achieved considerable knowledge, even they have not performed experiments on their own, and in the end they have understood little. . . . At first, I too understood little of science, but seeing that several of my colleagues at the [Shanghai Polytechnic] Institute were interested, I decided not to be cowed by ignorance. After the Institute opened, we met one evening each week to study and discuss the *Hua-hsüeh chien-yüan*. The first section [*chüan*] of this book deals with the basic principles of chemistry, which are very mysterious. In the study of chemistry, however, one cannot skip over these princ' 'es, but must gradually come to understand them. . . .

Then we cam. .o the second section, which deals with oxygen, hydrogen and how to test for these substances. We found this section easy to understand and tried some of the experiments. First, we tested for oxygen. We heated potash, chlorine oxide and manganese oxide powders in a copper pot and collected the oxygen in a glass storage tube, then transferred it to a glass bottle. We put a lighted candle in the bottle to test for oxygen, and the candle burned brightly. . . .

In the fourth month, we tested for hydrogen. First we combined zinc shavings and sulfuric acid to produce hydrogen gas and collected it in a storage tube. We transferred the gas to an inverted bottle. When we inserted a lighted candle into the bottle, the flame went out, but the hydrogen at the mouth of the bottle burned on its own. Now we know that hydrogen burns and is different from oxygen. . . .

We were disappointed that our experiments were so elementary and that we were not able to do all that was necessary to satisfy the curiosity of the scholars. But the equipment at the Institute is incomplete and it is difficult to obtain that which is lacking.[35]

Since Hsü Shou and his colleagues failed to move their activities out of the library and into the real world, it is hardly surprising that they spent

---

[35] Luan Hsüeh-ch'ien, "Ko-chih shu-yüan chiao-yen hùa-hsüeh chi" [Record of studying chemistry at the Shanghai Polytechnic Institute], quoted in "San-shih-nien-ch'ien wu-kuo k'o-hsüeh chiao-yü chih i-pan" [Note on science education in our country thirty years ago], *KH* 8, no. 4 (1924):430–2.

so much time on chemistry, a discipline that found so few applications in China. Hsü won recognition because of his ability to make things. He rose through the patronage of the self-strengtheners, whose objectives were practical. And he defended the study of science on the grounds of its utility. Yet he could not ignore the fact that in backward, under-developed China, knowledge of chemistry was nearly useless. After arriving at Kiangnan, Hsü spent little or no time on the manufacture of ships and steam engines, the purpose for which he was called. He was happy among his books, and that is where he stayed, oblivious to the needs of his employers or of the larger world around him.[36]

This literary bias cannot be blamed entirely on the Chinese, however, for a second feature of these early years was the dominant role played by foreigners. In part, Western dominance reflected the current distribution of knowledge and skill. There were, during the nineteenth century, more foreigners adept in Chinese, knowledgeable about science, and able to produce useful translations than there were Chinese who could master the same information and express it in their native tongue. Equally important were organizational factors. The enormous size and poor internal communications of China made coordination of all activities difficult, particularly so when individual scholars separated by great distances each sought to invent an extensive yet integrated linguistic scheme. It was to the great advantage of the foreigners that in 1886 they established the China Medical Missionary Association, which set up committees to produce vocabularies in each of the medical sciences. By 1901, these committees had completed lists for anatomy, histology, physiology, and pharmacology, while work on chemistry continued. Without exception, Europeans and Americans, seconded by Chinese assistants, invented the terms, translated the texts, published the books, attended the conferences, presented the scholarly papers, and carried on the debates over whether and how Chinese could be made an instrument of modern scientific discourse.[37]

In this regard, the contrast with Japan is striking. Japanese scientific translators began work a full century before this activity was resumed in late Ch'ing China, and did so without the benefit of foreign collabora-tion. Working alone, the Japanese produced translations that were sometimes clumsy and inaccurate, but the experience prepared them to do high-quality independent work much earlier than might otherwise have been the case. By contrast, the first Chinese translations were done largely by foreigners, with results that were generally accurate and fluent,

---

[36] Hsü Shou defense of science: Hsü Shou, "Ko-chih hui-pien hsü" [Preface to Ko-chih hui-pien], KCHP 1, no. 1 (1876):1a–1b.

[37] CMMA and its nomenclature committee: K. Chimin Wong and Wu Lien-teh, History of Chinese Medicine, 554–5; CRMJ 32 (1901):305–6; CMJ 19 (1905):53–60.

but in the process the Chinese developed few independent skills, and the liberation of Chinese science occurred later, more slowly, and at greater cost.[38]

The third point to make is that translators in China completely ignored the work already done in Japan. By the middle of the nineteenth century, Japanese scholars of "Dutch learning," or *Rangaku*, had translated a wide range of texts dealing with Western medicine and natural history, including the twenty-one-volume work, *Seimi kaiso* [Foundation of chemistry] (1837–47), by Udagawa Yōan (1798–1846), based on his own research and reading of over twenty European texts. In labeling the elements Udagawa followed the system presented in Lavoisier's *Traité Élémentaire de Chimie*, selecting names that reflect the chemical functions of the elements – *sanso*, "principle of acid," for oxygen; *suiso*, "principle of water," for hydrogen, and so forth – rather than the sounds of their foreign names, as practiced at the Kiangnan Arsenal.[39]

Chinese translators paid no attention to the Japanese studies, which were written in Chinese ideographs and could have provided guidance for work in China. By contrast, the more advanced Japanese made great efforts to learn from their colleagues on the mainland. Benjamin Hobson's translations, published in the 1850s, circulated widely in Japan. Later, Japanese scholars came to the Kiangnan Arsenal to buy books, although nothing was imported in return. In fact, one of the earliest scientific influences to pass between these countries, translation of the term "chemistry" itself, moved from China to Japan. Udagawa called this discipline *seimi*, a transliteration of the Dutch term *chemie*. In China, the character *hua* had been used since the seventeenth century for that branch of the physical sciences concerned with "change." The term, *hua-hsüeh*, the "study of change," was coined by two missionaries, Alexander Williamson and Alexander Wylie, in 1857. These characters, pronounced *kagaku* in Japanese, first appeared in the title *Kagaku Shinsho* [New book on chemistry] in 1860, and have since been accepted as the standard in both countries.[40]

[38] This interpretation has been suggested by the Japanese historian Oya Shin'ichi, "Reflections on the History of Science in Japan," *Science and Society in Modern Japan: Selected Historical Sources*, ed. by Nakayama Shigeru, David L. Swain, and Yagi Eri (Cambridge: M.I.T. Press, 1974), 63.

[39] Tatsumasa Doke, "Yōan Udagawa: A Pioneer Scientist of Early Nineteenth Century Feudalistic Japan," *JSHS* 12 (1973):99–120; and Eikoh Shimao, "The Reception of Lavoisier's Chemistry in Japan," *ISIS* 63 (1972):309–20.

[40] Hua-hsüeh: P'an Chi-hsing, "T'an 'hua-hsüeh' i-ts'u tsai Chung-kuo ho Jih-pen ti yu-lai" [Discussing the origin of the term "*Hua-hsüeh*" in China and Japan], *Ch'ing-pao hsüeh-k'an* 1 (1981):62–5; and Kunika Sugawara, "Kagaku toiu yōgo no honpō de no shutsugen shiyō ni kansuru ikkōsatsu" [A historical study of the use of the term *kagaku* for chemistry in Japan], *Kagakushi* (Tokyo: Journal of the Japanese Society for the History of Chemistry, 1987), 29–40. *Kagaku Shinsho*: Minoru Tanaka, "A Note on the Development of Chemistry in Japan," *JSHS* 7 (1968):64.

Finally, it would be unseemly to leave the discussion of an academic enterprise without noting the pettiness and discord that flavored it. There was dissention among the translators and confusion among their readers. At the extreme, translators adopted totally incompatible vocabularies. Even those who shared basic premises, such as Fryer and Kerr, often disputed their application to specific cases, choosing different characters for the same element or combining the characters in different ways. Students of medicine normally adopted the vocabulary in Kerr's texts; those who studied general chemistry or its application to industry followed Fryer and Hsü. Many scholars probably agreed with one frustrated translator who found the existing systems "faulty, inexpressive, and incomplete," while the parties engaged in this business attacked one another in typical academic fashion, producing light and heat in roughly equal proportions.[41]

Despite the shortcomings, however, China's early chemical translators achieved some notable results. They invented a system of nomenclature that worked reasonably well at the time and has survived, with the same periodic reforms required in all languages, down to the present. And they applied this language to a small library of books that set forth in clear, accurate, and readable fashion the discipline of chemistry as it was known at the time. This record stands in contrast to the experience of geology, the field where the Chinese made their first notable contributions to science, but where the early translations were crude, and later development of the language had to follow terminology invented in Japan. It also contrasts with the record of the twentieth century, when many Chinese scientists came to rely almost exclusively on English, and the practice of translation declined. "Looking back," wrote Tseng Chao-lun, one of the leading Chinese chemists of the Republican period, in 1941, "the translations from the era of the Kiangnan Arsenal are better than those done in this country during the past half-century. It is very regrettable that this enterprise has not been continued."[42]

---

[41] Fryer and Kerr: *CRMJ* 26 (1895):187–90, 288–9. "Faulty": *CRMJ* 25 (1894):88. Attacks: Adrian Bennett, *John Fryer*, 30–3.
[42] Geology: Tsui-hua Yang Lee, "Geological Sciences in Republican China, 1912–1937," 39–47. Quotation: Tseng Chao-lun, "Chung-kuo hsüeh-shu ti chin-chan" [Development of Chinese scholarship], *TFTC* 38, no. 1 (1941):57.

3

The limits of change: science, state, and society in the nineteenth century

The scientific translations, books and journals, described in the preceding chapter were read by Chinese students and scholars of the late nineteenth century, but to little effect. China's traditional state and society, although weakened by a century of rebellion and forced to accept marginal changes required to keep themselves in place, were strong enough to push back the new ideas and social groups that threatened their hold on power. Science was taught in modern government and mission schools, but graduates of these programs found no market for their skills. The growth of industry was too slow to accommodate more than a handful of technicians. And even though a few members of the literati read scientific translations, their understanding of science was bent to fit the traditional contours of their own minds. However well the message of science was laid upon the page, it had little impact on a state that was unwilling and a society unable to receive it.

Science in China's early modern schools

One way to assess the impact of science on nineteenth-century China is to examine the record of the modern schools that were set up by the Chinese government and foreign missionaries to promote this study. The curricula, faculty, and facilities of the best programs were generally adequate to the task. But in most cases, the students were of poor quality and poorly prepared, and graduates found little market for their skills. Prior to 1900, technical education was a bad investment, and China's best and brightest were wise to reject it.

Modern government schools were created to train interpreters, diplomats, and technicians needed to carry out the policy of self-strengthening. The most important of these were the T'ung Wen Kuan, established in Peking in 1862 to prepare interpreters for the Tsungli

Yamen (foreign office), and two units set up alongside the Foochow Navy Yard in 1866 to teach shipbuilding and navigation. Later, more schools were created to train gunnery experts, army officers, surgeons, telegraph operators, and engineers.

The curricula of the technical schools began with Chinese and foreign languages, primarily English, then French, followed by mathematics, physics, and mechanics, the disciplines most relevant to the military tasks at hand. Only one of the industrial units, the Mining and Engineering College of the Hupeh Province Board of Mines, which was established in 1891 and took a special interest in the analysis of coal and iron ore, offered instruction in chemistry. The life sciences, unrelated to national defense, were ignored. The arsenals (as distinct from the shipyard) employed only low-level craftsmen and laborers, and training in these schools was limited to routine skills.[1]

The Foochow shipyard schools – the French school to train naval engineers and the English school for ships' captains and engine room officers – offered the most complete technical education available in nineteenth-century China. Prosper Giquel, director of the naval yard from 1867 to 1874, described the curriculum at the French school and the way this knowledge was to be used:

[I]n order to calculate the dimensions of a piece of machinery or of a hull, it is necessary to know arithmetic and geometry; in order to reproduce that object on a plan, it is necessary to understand the science of perspective, which is descriptive geometry; in order to explain the pressure exerted on engines and ships as well as on all bodies, by gravity, heat, and other phenomena of nature, it is necessary to understand the laws of physics. Next in order come the movements a body undergoes under the impulse of the forces to which it is subjected; the resistances which it will need to overcome, the strain which it is able or ought to bear, which is the science of statics and of mechanics; and for these the calculations of ordinary arithmetic and geometry no longer suffice; it is necessary also to possess the knowledge of trigonometry, analytical geometry, of the infinitesimal calculus, so as not to be any longer bound down to reason as to objects of determinate form and size, but be able to arrive at general formulae applicable to all details of construction.[2]

The English school taught navigation and, to support this craft, the sciences of astronomy and mathematics.[3]

From 1877 to 1900, seventy-three graduates of the Foochow schools spent three years each in England or France for advanced training in navigation and engineering. Several students took up more specialized

---

[1] Hupeh; modern government schools: Knight Biggerstaff, *Earliest Modern Government Schools in China*, 70. Other details: ibid., passim.

[2] Prosper Giquel, *The Foochow Arsenal and Its Results*, trans. H. Lang (Shanghai: *Shanghai Evening Courier*, 1874), 18.

[3] Foochow: Knight Biggerstaff, *Earliest Modern Government Schools*, 210–28.

studies, ranging from science and math to naval and international law. Only one, the English interpreter with the first delegation, Lo Feng-lu, studied chemistry at King's College, London. Lo went on to a diplomatic career, however, so his scientific training was at most a form of scholarly enrichment. The program of sending students to Europe was terminated in 1900, owing to a shortage of funds.[4]

Lacking first-hand descriptions of classroom and laboratory instruction, it is difficult to assess the quality of the Foochow schools. Contemporary sources give them mixed reviews, and a more recent study, based on interviews with the descendants of students and teachers from the French school, paints a dismal picture. The foreign teachers were said to be of low quality – unable to speak Chinese, poorly trained, overworked, lazy, and ineffective. The students were interested mainly in stipends, spent most of their time preparing for the civil service exams, and left school at the earliest opportunity. In one class, only 35 out of 130 completed the course. Some graduates admitted they "could not manufacture ships and had no true ability, but could only read books and do some translations."[5]

The overriding problem for Foochow graduates, and for those who sought to recruit students to this program, was the shortage of jobs in China that could make use of such advanced technical training. During its heyday, from 1866 to 1911, Foochow graduated 178 naval engineers, 241 deck officers, and 210 engine room officers. Job opportunities were scarce for the men trained in naval construction. Following the withdrawal of foreign technicians in 1874, only six engineers, all Foochow graduates, were employed at the Foochow shipyard. Three more graduates of the French school remained there as teachers and translators. Several others found jobs teaching in modern schools in Shanghai or elsewhere on the coast. A few worked as engineers or interpreters with foreign firms or consulates. Graduates of the English school fared somewhat better, many serving as deck and engine room officers in the Chinese navy. No graduate from either school achieved

---

[4] Foochow graduates in Europe, program terminated in 1900: Knight Biggerstaff, *Earliest Modern Government Schools*, 228–42. Lo Feng-lu was head of the first class of naval students at the Foochow Arsenal in 1871 and interpreter for the first group of students sent to Europe in the 1870s. From the late 1870s until 1896, he served as naval secretary and adviser to Li Hung-chang. He was later Chinese Minister to the United Kingdom, Spain, and Belgium, 1896–1901. See: ibid., 252; Kenneth Folsom, *Friends, Guests, and Colleagues: The Mu-fu System in the late Ch'ing Period* (Berkeley: University of California Press, 1968), 140.

[5] Contemporary sources: Knight Biggerstaff, *Earliest Modern Government Schools*, 225–6, 246–50. Recent study: Fang Ai-chi, "Wo-kuo tsui-tsao ti tsao-ch'uan chuan-k'o hsüeh-hsiao – Fu-chou ch'uan-cheng-chü ch'ien-hsüeh-t'ang" [Our country's earliest shipyard technical school – The Foochow Naval Yard First School], *CKKCSL* 6, no. 5 (1985):57–62. Quotation: ibid., 58.

distinction in the field of science, for which China provided neither career paths nor supporting institutions. Most took work wherever they could find it.[6]

The T'ung Wen Kuan offered a more rounded, but less intensive course in science and other Western subjects. Although the original purpose of this school was to train interpreters, the curriculum was expanded under the leadership of the American missionary W. A. P. Martin, who served as president from 1870 to 1895, to include mathematics, science, economics, and international law. The faculty included several Chinese to teach the classics, one Chinese professor of math – Li Shan-lan, from 1869 until his death in 1882, and after 1886, Li's student, Hsi Kan – and foreign professors in each of the other subjects. Anatole Billequin, a student of the celebrated French chemist Jean Boussingault, taught chemistry from 1866 to 1894. The man Martin considered most responsible for introducing modern chemistry to China, Billequin also translated textbooks and conducted research on Chinese ceramics and other native arts. After his death in 1894, he was succeeded by Carl Stuhlmann, the first holder of a doctoral degree (from the University of Hamburg) to teach chemistry in China. A chemistry lab and science museum were erected in 1876, the earliest buildings of their type in China. Throughout the 1880s and 1890s, enrollments in chemistry were higher than for any other science.[7]

The course offerings at the T'ung Wen Kuan were adequate and the foreign faculty well trained, but there is little indication that their charges learned much about science. Standards were reportedly lax, and the teaching failed to inspire dull and lazy students. "With regard to astronomy, geography, mathematics and chemistry," one source observed, "they merely scratched the surface." The situation improved somewhat after the humiliating loss to Japan in 1895, an event that prompted widespread recognition of the need to learn from abroad, but even then, according to the memoir of one former student, Ch'i Ju-shan, few studied in earnest, their command of foreign languages was minimal, and the coverage of all subjects poor. After five years at the T'ung Wen Kuan, Ch'i came away with a low regard for what went on there:

After laying a foundation in one of the foreign languages, students were free to choose a science: chemistry, mathematics, astronomy or a few others, all of them

---

[6] Foochow graduates: Pao Tsun-p'eng, *Chung-kuo hai-chün shih* [History of the Chinese Navy] (Taipei: Hai-chün ch'u-pan-she, 1951), 225–6. Employment of Foochow graduates: Knight Biggerstaff, *Earliest Modern Government Schools*, 245, 248–50.

[7] T'ung Wen Kuan: Knight Biggerstaff, *Earliest Modern Government Schools*, 108, 126–9. Faculty: ibid., 144–5, 151; and W. A. P. Martin, *A Cycle of Cathay* (New York: Fleming H. Revell, 1900), 311. Billequin: *T'oung Pao*, 5 (1894):441–2; Knight Biggerstaff, *Earliest Modern Government Schools*, 120; W. A. P. Martin, *Cycle of Cathay*, 303. Martin's opinion: ibid., 314.

extremely haphazard. I chose chemistry. During the first month, the chemistry teacher just lectured on the funnel and its use, the melting vessel, alcohol lamp, glass tubing, thermometer, etc. and their uses. He lectured this way for two or three months. Naturally, this type of course did not require much thinking. Among [the science courses], the most serious was Chinese mathematics. The teacher was Hsi Han-po [Hsi Kan], an outstanding student of Li Shan-lan. His teaching method was very good.[8]

Government schools gave advanced training to a small number of men, most of whom made a lateral transfer from the examination ladder to become (in theory, if not fact) technical experts; a more extensive program of modern education was offered by foreign missionaries to any Chinese who would accept it. The missionary schools began at the primary level, offering traditional Chinese learning (the bait) and supplementing it with Bible studies, English, and other modern secular subjects. With the passage of time, some of these primaries produced teenage scholars who formed the core of expanded middle schools. By the turn of the century, four claimed the status of college. In 1900, the enrollment in 1,200 mission schools at all levels was around 25 thousand. In contrast to the arsenals and shipyards, which taught science and math as narrow technical skills to selected government employees, a number of missionaries shared the view that science should be part of a general liberal education to prepare ordinary Christians for roles in the modern world.[9]

Foremost among the missionary schools offering scientific studies was that founded at Tengchow in 1864 by an American Presbyterian, Calvin Mateer. During the 1880s and 1890s, the curriculum at Tengchow included algebra, geometry, trigonometry, calculus, surveying and navigation, one year each of chemistry, physiology, astronomy, and geology, and three years of physics. Zoology and botany were added in 1902. Science instruction was based on translations of standard Western texts. Mateer himself wrote and published Chinese-language textbooks on arithmetic, geometry, and algebra. Among his most remarkable achievements was the construction of a complete laboratory and

---

[8] One source: Ch'en Liang-i, ed., "Hsi-hsüeh" [Western studies], *Huang-ch'ao ching-shih wen san-pien*, chüan 2, cited in Liu Kuang-ting, "Chung-kuo hua-hsüeh chiao-yü fa-chan chien-shih" [Brief history of the development of Chinese chemical education], *Hua-hsüeh* [Chemistry] (The Chinese Chemical Society, Taiwan, China) 43, no. 4 (December 1985):A153. Ch'i quotation: Ch'i Ju-shan, "Ch'i Ju-shan tzu-chuan" [Autobiography of Ch'i Ju-shan], *Chung-kuo i-chou* [China Newsweek] 240 (November 1954): 18. Ch'i gives a devastating account of the T'ung Wen Kuan, before and during his tenure. For a summary, which helps put Ch'i's view in perspective, see: Knight Biggerstaff, *Earliest Modern Government Schools*, 141–7.

[9] Missionary schools: Jessie Lutz, *China and the Christian Colleges, 1850–1950*, (Ithaca: Cornell University Press, 1971), chap. 2. Enrollments: William Fenn, *Christian Higher Education in Changing China, 1880–1950*, (Grand Rapids, MI: W. Eerdmans, 1976), 25.

workshop that included "machines for turning, blacksmithing, plumbing, screw-cutting, burnishing, electroplating, casting, and so forth." According to an official history published in 1913, the College's storeroom held 360 instruments for the study of meteorology, astronomy, and physics, and chemistry equipment of even greater number and variety. The Tengchow program stands as one of the first successful attempts to bridge the gap between theoretical and practical study that has plagued Chinese education to the present day.[10]

Under Mateer's direction, most instruction in Western subjects, including the sciences, was by Chinese who had themselves graduated from Tengchow. Although no description of these classes survives, we do know something of the approach adopted by the founder. In 1874, Mateer taught a special course for three Chinese clerics who had completed their classical education and converted late in life. He described this course in his journal as follows:

I heard them a lesson every day – one day in philosophy (physics) and the next in chemistry. I went thus over optics and mechanics, and reviewed electricity, and went through the volume on chemistry. I practically gave all my time to the business of teaching and experimenting, and getting apparatus. I had carpenters and tinners at work a good part of the time.... In chemistry I made all the gases and more than are described in the book, and experimented on them fully. They gave me no small amount of trouble, but I succeeded with them all very well. I made both light and heavy carbureted hydrogen, and experimented with them. Then I made coal gas enough to light up the room through the whole evening. Altogether I have made for the students a fuller course of experiments in philosophy or chemistry than I saw myself.[11]

Three other mission colleges established in the 1880s followed the Tengchow model of combining classical, religious, and secular Western subjects. These were St. John's College (founded in Shanghai by the Episcopalians in 1882), Peking University (Peking, Methodist, 1888), and North China College (Tungchow, American Board of Foreign Missions, 1889). All offered courses in physics, chemistry, and math. Owing to its location, St. John's grew the fastest, erecting a separate science building in 1899, the first of its kind at a private college in China. By 1906, the school had four full-time professors in the fields of chemistry and physics, anatomy and physiology, biology, and math.[12]

---

[10] Tengchow: Irwin T. Hyatt, Jr., *Our Ordered Lives Confess*, 159–90. Curriculum: Wang Yüan-te and Liu Yü-feng, eds., *Wen-hui kuan-chih* [History of Tengchow College] (Weihsien, 1913), 28–31; and W. M. Hayes, "Course of Study – Tungchow College" (17 January 1891), *Correspondence of the Board of Foreign Missions of the Presbyterian Church, USA*, vol. 26, "Shantung Mission, 1891–92." NB, Hayes's report is actually on Tengchow College, although he uses a different transliteration. Mateer's view: Daniel W. Fisher, *Calvin Wilson Mateer*, 236, 244. Workshop: ibid., 241. Instruments: *Wen-hui kuan-chih*, 41–7.

[11] Quotation: Daniel Fisher, *Calvin Wilson Mateer*, 213.

[12] St. John's: Mary Lamberton, *St. John's University, Shanghai, 1876–1951* (New York: United Board for Christian Colleges in China, 1955), 11–18, 38, 66–8. Peking

Although an extraordinary man like Calvin Mateer might educate himself and his students in the sciences, most teachers in the mission colleges knew little about these subjects. Science courses were taught, if at all, by Chinese or foreigners who had received only the barest textbook learning and little or no laboratory training, and who were expected to offer several different courses, while performing a wide range of administrative and other duties. Owing to the shortage of foreigners at Tengchow, Chinese instructors taught the physical sciences, but as the college's annual report for 1890 pointed out, "they cannot with their limited opportunities for self-instruction keep up more than temporarily the present standard."[13]

Unlike those enrolled in the government training programs, most students in missionary schools were too poor to prepare for the civil service exams, and their prospects for employment in fields related to science and technology were similarly dim. I have found no data on the careers of mission school graduates during the nineteenth century, but after 1900, when the number of modern schools and demand for teachers grew exponentially, many entered teaching. A survey conducted in 1910 identified 180 surviving alumni of Tengchow College and their current occupations. Nearly 60 percent were teaching, 20 percent preaching, and most of the rest were in translating, commerce, or government service. Among the teachers, at least ten taught physics or chemistry and an equal number math. Several practiced medicine, but none was engaged in scientific or technical work related to industry, agriculture, or any other applied field.[14]

A few missionary schools focused on training doctors, nurses, and medical assistants. The practice of Western medicine amidst so much need and with such limited resources – fewer that 200 medical

Methodist University: Dwight Edwards, *Yenching University*, 19–20, 44. North China College: ibid., 30, 54; and Roberto Paterno, "Devello Z. Sheffield and the Founding of the North China College," *American Missionaries in China: Papers from Harvard Seminars*, ed. Kwang-ching Liu (Cambridge: Harvard University Press, 1966), 69, 73.

[13] Mission college faculty: Mary Lamberton, *St. John's University*, 38; Charles Corbett, *Shantung Christian University (Cheeloo)* (New York: United Board for Christian Colleges in China, 1955), 81–2. Training of missionaries; Mateer: Irwin Hyatt, *Our Ordered Lives Confess*, 143–4; Daniel Fisher, *Calvin Wilson Mateer*, 28–36, 238. Martin: Peter Duus, "Science and Salvation in China: The Life and Work of W. A. P. Martin (1827–1916)," *American Missionaries in China: Papers from Harvard Seminars*, ed. Kwang-ching Liu (Cambridge: Harvard University Press, 1966), 12–14; Ralph Covell, *W. A. P. Martin: Pioneer of Progress in China*, (Washington, DC: Christian University Press, 1978), 13–22. Young J. Allen: Adrian A. Bennett, *Missionary Journalist in China: Young J. Allen and His Magazines, 1860–1883* (Athens: University of Georgia Press, 1983), 4–10. Quotation: W. M. Hayes, "Report of Tungchow College, Jan 21st, 1890–January 17th, 1891," *Correspondence of the Board of Foreign Missions*.

[14] 1910 survey: *Wen-hui kuan-chih*, 55 6. List of alumni: ibid., 73–108. Tengchow graduates teaching science: Roberto Paterno, "Devello Z. Sheffield," 69; Dwight Edwards, *Yenching University*, 20; Irwin Hyatt, *Our Ordered Lives Confess*, 204, 229.

missionaries served in China from 1834 to 1890 – would not have been possible without native assistants, so most practitioners gave instruction to young men and women on the apprentice model. The center of this activity was Canton, home first to the American Peter Parker, followed by Englishman Benjamin Hobson, and American John Glasgow Kerr. By 1896, there were 462 Chinese trained or in training in missionary medical programs, 103 of whom studied in Canton. Most programs were small, enrolling fewer than ten students supervised by one or two missionaries. In the largest and best, Kerr's Canton school, students learned anatomy, physiology, pharmacology, and chemistry, all with at least some laboratory experiments. The great variety of textbooks translated by Kerr and other medical missionaries testifies to their determination to combine basic science with clinical applications.[15]

Despite these efforts, little chemistry or other scientific knowledge reached China through the medium of medicine. At this early date, Western medicine still relied on clinical experience more than laboratory studies, and it was natural that physicians in faraway China should fall back on homely solutions and common sense. Many medical missionaries specializing in surgery needed assistants and nurses who could perform skilled hands-on service, with or without the adornment of scientific theory. Students came to these training courses poorly prepared. Teachers were distracted by other obligations, most notably the treatment of patients. Laboratory facilities were limited or nonexistent. Finally, since Western medicine had not gained widespread acceptance in China, few talented Chinese were willing to enter this field or remain in school long enough to develop any expertise. By the end of the century, there were only 300 Chinese doctors of Western medicine in practice and 300 more in training. Most of them probably had little understanding of basic science and little need for it in their work.

Science and the marketplace

The best explanation for the meager development of scientific and technical education in nineteenth-century China lies in the marketplace. There was little demand for people with technical skills and little

---

[15] According to Henry Boone, president of the China Medical Missionary Association, between 1834 and 1890, 196 medical missionaries lived and worked in China. See: *CMJ* 4 (1890):110. Peter Parker: Edward V. Gulick, *Peter Parker and the Opening of China* (Cambridge: Harvard University Press, 1973), passim. Biographies of Hobson and Kerr: K. Chimin Wong, *Lancet and Cross*, 14–6, 23–8; K. Chimin Wong and Wu Lien-teh, *History of Chinese Medicine*, 364–6, 372–5, 391–3. Statistics on medical schools, students, and practitioners: *CMJ* 11, no. 2 (1897):89–91. Survey of medical schools: *CMJ* 23, no. 5 (1909):289–335.

evidence that prospects would improve any time soon. Students in modern government schools were distracted by the lure of classical studies, which promised the reward of an official degree, government job, or at least the honor of being considered an "educated man." Those in missionary schools were sometimes tempted by English, the shortcut to success in the treaty ports. But everyone could see that technical training was a bad bet.

The best evidence that classical studies remained a better career investment than technical training lies in the pattern of recruitment of officials to direct the institutions associated with self-strengthening. Here, if anywhere, the technically trained should have been preferred, but they were not. Li Hung-chang, who maintained the largest staff of experts and managers to administer his empire of arsenals, shipyards, shipping and telegraph companies, factories, academies, and hospitals, reserved the best jobs for men with general administrative and scholarly backgrounds, and when technical skill was required, he turned to foreigners. The only exception among Li's top advisers was Lo Feng-lu, who had graduated at the head of the first class of Foochow naval cadets in 1871 and studied chemistry in London, after which he served for many years as naval secretary and general factotum to Li in Tientsin. Elsewhere, on Li Hung-chang's staff and in arsenals and shipyards outside his control, the careers of men who relied on technical skills stagnated.[16]

Recent scholarship by Mary Rankin describes another sector, also outside the central bureaucracy, where new opportunities in functional specialties increased in the years following the Taiping Rebellion, but the advancement of men with technical training did not. Beginning in the 1860s, local elites in the Lower Yangtze established, financed, and managed "public," nongovernmental institutions designed to repair the damage wrought by the Taiping Rebellion and restore normal social and economic life. These included schools, welfare organizations, and offices to build and repair dikes, canals, roads, bridges, public buildings, and the like. In Chekiang, the focus of Rankin's study, "elite managers" were drawn from upper degree-holders, scholars, merchants, and members of prominent local families. Despite the importance of technical skill in the operation of many of these public works, none of the men identified by Rankin brought this type of training or experience to the job. It may be true that the organization and leadership of local society were changing

---

[16] Two studies detail the backgrounds of men recruited for service under Li Hung-chang and at arsenals and shipyards: Kenneth Folsom, *Friends, Guests, and Colleagues*, 133–57; and John L. Rawlinson, *China's Struggle for Naval Development, 1839–1895* (Cambridge: Harvard University Press, 1967), 84–94, 104–8, 157, 161, 164. Lo Feng-lu: Kenneth Folsom, *Friends, Guests, and Colleagues*, 140; and Knight Biggerstaff, *Earliest Modern Government Schools*, 252.

in the last decades of the Ch'ing, but these changes do not seem to have expanded the market for technical expertise.[17]

Finally, the limited opportunities open to men with technical training was a function of the slow pace of Chinese industrial development. Despite marginal changes in the military sector, the economy of late nineteenth-century China remained overwhelmingly traditional and agrarian. As late as the 1880s, manufacturing accounted for less than 4 percent of China's GNP, and almost all of this was by handicraft methods. According to the most widely cited authority on this subject, 103 foreign-owned and 108 Chinese-owned modern industrial enterprises (not counting the 19 government arsenals) were established in China prior to 1895. The foreign firms were devoted to the processing of silk, tea, and other products, light manufacturing, and ship construction and repair. The Chinese firms, both private and public-private combines, included 33 coal and metal mines and 75 factories engaged in silk reeling, cotton ginning and spinning, flour milling, and the manufacture of paper, matches, and other products. Most factories, both Chinese and foreign, were small, short-lived, clustered in and around the treaty ports, linked in some fashion to foreign trade, and distinguished from handicraft workshops only by their use of some steam or electric power. Although the data on this subject is incomplete, even if one takes the largest numbers and assumes the highest level of development, it is clear that the market for scientific and technical skills in the industrial sector of nineteenth-century China was small.[18]

What was true for the modern sector generally was particularly true for industries that relied on a knowledge of chemistry. Two products made or used by the arsenals – steel and gunpowder – created a potential demand for chemical inputs and knowledge of chemistry, but that demand did not materialize until after the turn of the century. Most of the materials used in the arsenals, including iron and steel, were imported. The refining of steel, which requires sulfuric acid, began on a small scale at three arsenals – Kiangnan, Tientsin, and Hanyang – in the 1890s, but there is no record that any of the acid was made in China and every reason to suspect both that it was imported and that the operations were supervised by foreign technicians. Beginning in the 1870s, black

[17] Mary Backus Rankin, Elite Activism and Political Transformation in China: Zhejiang Province, 1865–1911 (Stanford: Stanford University Press, 1986), 93–120.
[18] Modern industry in China, pre-1895: Albert Feuerwerker, "Economic trends in the late Ch'ing empire, 1870–1911," Cambridge History of China, 11:28–9, 32–3. GNP in 1880s: Chang Chung-li, The Income of the Chinese Gentry (1962), 296, cited in ibid., 2. Foreign and Chinese factories: Sun Yü-t'ang, comp., Chung-kuo chin-tai kung-yeh shih tsu-liao, ti-i chi, 1840–1895 nien [Source materials on the history of modern industry in China, first collection, 1840–1895], 2 vols. (Peking: K'o-hsüeh ch'u-pan-she, 1957). Foreign plants, 1894: ibid., 1:242–7. Chinese plants, 1872–1894: ibid., 2:1166–9, 1170–3.

and later brown or prismatic gunpowder were made at the Kiangnan, Tientsin, and Nanking arsenals. Black and brown powder, the most widely used explosives of the nineteenth century, are physical mixtures of saltpeter (potassium nitrate), charcoal, and sulfur, the manufacture of which involves no chemical transformations nor the consumption of prepared compounds. It was not until after 1900, when China began to make smokeless gunpowder, which requires the use of sulfuric and nitric acids in controlled chemical reactions, that one can speak of munitions as a "chemical" industry.[19]

Prospects for applying chemistry to other sectors of the economy were similarly bleak. A few products previously unknown in China, such as matches, were made in handicraft workshops, but these involved rather simple operations that offered little scope for scientific knowledge or technical skill. Some modern factories were set up to make textiles, paper, glass, and other products that consume chemical inputs. But these were few in number, small in scope, and we have no evidence that any employed men with technical expertise.[20]

Scientific publications and the literati

The preceding sections have focused on technical experts, the schools designed to produce them, and the market that might have employed them. The view from all sides shows why there was so little development in between. At the same time, there was another audience for the scientific translations and publications described: namely, members of the literati, scholars whose principal purpose was to master the classics and pass the exams, but who were also interested in broad cultural enrichment and useful device. These men were not looking for formal

[19] Manufacture of iron and steel: Thomas L. Kennedy, *The Arms of Kiangnan: Modernization in the Chinese Ordnance Industry, 1860–1895* (Boulder: Westview Press, 1978), 111–12, 134–5, 142–3; Thomas L. Kennedy, "Chang Chih-tung and the Struggle for Strategic Industrialization: The Establishment of the Hanyang Arsenal, 1884–95," *Harvard Journal of Asian Studies* 33 (1973):171–5; Wang Erh-min, *Ch'ing-chi ping-kung-yeh ti hsing-ch'i,* 77–103. Manufacture of gunpowder: Kennedy, *Arms of Kiangnan,* 52–3, 62–75, 114–18, 140–5. Kennedy, 115–18, notes that a nitric acid soaking plant for the production of guncotton was built at Tientsin in 1881, and a new plant for the production of acids was added in 1882. Sun Yü-t'ang, 1:335–43, describes Nanking Arsenal, 1881–97, where black gunpowder was made with foreign equipment and foreign technicians. *KCHP* 1, no. 2 (1877):8b–9a, reports that sulfuric and perhaps nitric acids were made at the Kiangnan and Tientsin arsenals during the 1870s and 1880s, although these were apparently small experimental efforts and played no part in actual production.

[20] Modern match and paper factories: Sun Yü-t'ang, cited in Albert Feuerwerker, "Economic trends in the late Ch'ing empire," 33.

credentials or means of employment. They were drawn to science as a hobby, a source of insight, a window on the world.

There is considerable evidence that men who fit this description read works produced by the Kiangnan Arsenal. The most reliable figures on the arsenal's publications are for the first decade, 1871–9, when more than 31,000 copies of the 98 titles then in print were sold, with sales of the more popular items reaching 800 to 1,000 copies each. The leading reformers, K'ang Yu-wei, Liang Ch'i-ch'ao, and T'an Ssu-t'ung, all read works translated at Kiangnan. Included in Liang's recommended list of Western books were twelve on chemistry, among them several transla-tions by Fryer and Hsü.[21] As one missionary-educator reported, the Kiangnan translations

exercised a very powerful influence upon large numbers who have not come in contact with our schools as students. The books prepared and sold by Dr. Fryer, especially, as well as those of others, are very widely read and circulated amongst the literati. We sometimes have men call upon us who are very well versed in science and mathematics, who have obtained their knowledge through the medium of books alone.[22]

Circulation of the journal Ko-chih hui-pien reached three to four thousand copies, most of which were sold in the eighteen treaty ports and twenty-one other cities that were home to the magazine's seventy sales agents. One measure of the magazine's popularity was the section "Questions and Answers," which appeared in each issue. This column provided readers with an opportunity to contribute local experience to the common pool of knowledge, obtain solutions to problems they faced in the laboratory and workshop, and simply satisfy their curiosity about the world around them. In the course of seven years, the magazine published 322 letters, the largest number of which emanated from Shanghai, followed by Canton, Hangchow, Tientsin, Amoy, and other treaty ports. Most inquired about scientific phenomena or their practical applications; a few about strange and novel things. One in ten dealt with some aspect of chemistry.[23]

---

[21] Kiangnan sales, 1871–9: John Fryer, "Science in China," 56–7. Individual titles: Adrian Bennett, John Fryer, 42, citing North China Daily New, 3 February 1880. K'ang, Liang, T'an: ibid., 42–5. Liang Ch'i-ch'ao, Hsi-hsüeh shu-mu piao [Bibliography of Western learning], rev. ed. (Chih-hsüeh chai, 1896), cited in ibid., 110.

[22] Rev. James J. Jackson, "Objects, Methods, and Results of Higher Education in our Mission Schools," CRMJ 24, no. 1 (January 1893):8.

[23] Circulation: Li San-pao, "Letters to the Editor," 742; and Roswell S. Britton, The Chinese Periodical Press, 1800–1912 (Shanghai: Kelly and Walsh, 1933), 61. For map showing location of the magazine's agents and a description of letters with many translated extracts, see: ibid., 743–67. Li gives the number of letters sent from each of the treaty ports, but no details on letters from other locations or for which no location was given: Shanghai, 52; Canton, 19; Hangchow, 16; Tientsin, 13; Amoy, 13; Ningpo, 12; Soochow, 12; Foochow, 11; Hankow, 11; Swatow, 5; Chefoo, 3; Nanking, 3; Newchwang, 3; Kiukiang, 1; Chinkiang, 1; Tamsui, 1; total, 176.

Among the correspondents interested in chemistry, several wanted to know about modern methods for manufacturing soap, ink, matches, vinegar, and explosives, and for bleaching, dyeing, and preparing photographic chemicals. A few asked for instruction in performing laboratory experiments, such as the preparation of hydrogen and oxygen and the extraction of carbon from sugar or wax. Others were just plain curious about why white wine turns urine yellow or carbon fails to make the air black (sic!). A minority of crasser types sought advice on purifying mixtures containing silver and gold. One dutiful son inquired about the best means of electroplating his deceased forbears. Several writers indicated that they had read books on chemistry, generally Kiangnan translations, and conducted experiments or tried to manufacture some product following the directions provided. Each question was answered, often in considerable detail. Topics that generated the greatest interest were dealt with in separate articles.[24]

There is no doubt that many late nineteenth-century scholars, particularly those in and around the treaty ports, read works on modern Western studies. But what did they learn? How did they understand these writings? How did they arrange the new pieces in refurnishing the Chinese mind? Two sources – the early writings of the reformer K'ang Yu-wei, and a selection of winning entries in an essay contest sponsored by the Shanghai Polytechnic Institute – help to answer these questions. Both suggest that late Ch'ing culture remained as vital and resistant to change as did the imperial state and society. For these scholars bent science in strange shapes to fit the language and structure of their traditional thoughts.

K'ang Yu-wei (1858–1927) is probably the most important and certainly the most complicated figure in the intellectual transition from traditional to modern China. Confucian scholar, political reformer, wide-ranging eclectic, and wide-eyed utopian, K'ang drew threads from many spindles to weave a rich tapestry of provocative ideas. Scion of a prominent Kwangtung family, he began his education with the usual diet of Confucian classics, but balked at the drill required to prepare for the exams, had what might be described as a nervous breakdown, and emerged with a vision of himself as the great sage whose mission was to save mankind. Thereafter, the scope of his scholarly interests widened to include Buddhism, Taoism, and Western studies, the last of which was

---

[24] This paragraph summarizes the content of thirty-seven letters dealing with chemistry, which appeared in *KCHP*, volumes 1–7. Manufacturing: 1, no. 2:8b–9a; 1, no. 3:10b; 1, no. 9:10b; 2, no. 1:15a; 2, no. 3:15a; 2, no. 6:15a–16a; 3, no. 4:16a–16b; 3, no. 9:16a; 3, no. 11:16a; 5, no. 2:50b–51b; 5, no. 3:48a–48b; 6, no. 1:42b; 6, no. 2:48a; 6, no. 3:49a; 7, no. 3:49b–50a. Laboratory experiments: 1, no. 11:11a; 2, no. 3:14b; 2, no. 4:12a; 2, no. 9:12b; 2, no. 10:14a–15b; 6, no. 2:48a. Electroplating corpses: 7, no. 1:49a.

enriched by a reading of the missionary works and translations already described. During the early 1880s, K'ang delved into acoustics, optics, chemistry, electricity, mechanics, and mathematics, topics he began to fit into his increasingly complex world view.[25]

Entries in K'ang's diary and other writings from this period show that he picked up several ideas from Western science – the principle of relativity, the notion of universal laws, and the theory of evolution.[26] Yet the point to make about K'ang Yu-wei is that he was not remotely a scientist, or even a serious student of science, but in this regard, as in others, an intellectual scavenger. K'ang selected from the grab bag of science those themes that served his political and intellectual agenda, while discarding others – natural selection, for example, which conflicted with the harmony of his coming "Great Unity" – that did not. As for the fundamental values and methods underlying modern science – the skepticism, rationalism, empiricism, naturalism, and interest in learning for its own sake – these are nowhere evident in the religious certainty, the teleology, the preoccupation with politics and morality that form the pillars of K'ang's essentially neo-Confucian thought.

To understand how these scientific concepts were adapted to K'ang's world view, the place to begin is by examining the term that was used during the latter half of the nineteenth century for science, namely *ko-wu* or *ko-chih*, which we have already met as Hsü Shou's "investigation of things" and in the title of Fryer's journal, *Ko-chih hui-pien*. This term comes from the *Ta-hsüeh* or *Great Learning*, one of the *Four Books* that made up the first canon of classical scholarship. The *Ta-hsüeh* is a very ancient text, elevated to a position of prime importance by the neo-Confucian synthesizer, Chu Hsi (1130–1220), that posits an eight-step program for self-cultivation leading, in theory, to good government and world peace. The final and most important step of this regimen is for the student to "extend [his] knowledge" [*chih-chih*] through the "investigation of things" [*ko-wu*].

Explaining the meaning of *ko-wu* was one of the central problems of neo-Confucian thought, debated by Chinese philosophers from Chu Hsi on. The legacy of these debates perforce colored the understanding of scholars who came upon this term, first as young men reading the *Ta-hsüeh* and again, later, when they took up the study of science. Part of the problem lay in the meaning of "things" [*wu*], which in Chu Hsi's own

25 Richard C. Howard, "K'ang Yu-wei (1858–1927): His Intellectual Background and Early Thought," *Confucian Personalities*, ed. Arthur F. Wright and Denis Twitchett (Stanford: Stanford University Press, 1962), 294–316.
26 Diary entries: Lo Jung-pang, *K'ang Yu-wei: A Biography and a Symposium* (Tucson: University of Arizona Press, 1967), 40–1. Other writings: K'ang Yu-wei, *Shih-li kung-fa* [Principles of truth and universal laws], 2a, cited in Richard Howard, "K'ang Yu-wei," 311. Other details: Frederic Wakeman, Jr., *History and Will: Philosophical Perspectives of Mao Tse-tung's Thought* (Berkeley: University of California Press, 1973), 127–30.

writings stood for the various categories of human affairs, a definition accepted by later neo-Confucians and left unchallenged in recent times. When in the nineteenth century the term *ko-wu* was borrowed as a translation for "science," there was naturally some confusion about the proper object of this study: Was it nature? or man? or documents? or all these "things"?

A further complication, at least as far as modern science is concerned, was introduced during the Ming dynasty by the "idealist" thinkers, led by Wang Yang-ming (1472–1529). Wang agreed with Chu Hsi on the definition of "things" [*wu*], but differed on the way these things or affairs were to be treated – that is, the meaning of *ko*. Chu Hsi took *ko* to mean "investigation" and held that man's understanding of universal principle, or *li*, would come through the accumulation of knowledge derived from study of the external (albeit social) environment. In Chu's view, knowledge came from the outside in: The way to extend one's knowledge to an understanding of universal principles lay in exploring objects and affairs of the real world. Wang Yang-ming believed, in contrast, that the most reliable source of knowledge and of moral or ethical insight was intuition. In this view, the proper meaning of *ko* was not to "investigate" or bring knowledge in from the outside, but to "rectify" or extend the correct and virtuous knowledge that came from within oneself to those external things or affairs that needed straightening out.[27]

In his interpretation of *ko-wu*, K'ang Yu-wei accepted Chu Hsi's broad definition of things, while following the idealist line of Wang Yang-ming by locating the ability to influence external objects within man. As a proponent of the New Text school, K'ang believed in the power of the sage (that is, K'ang himself) to transform the universe by extending human kindness, love, or *jen* outward to all things. This was the source of K'ang's utopian philosophy and his practical reformism.[28] It also made his interest in science, or more precisely his "scientism," subordinate to traditional concepts of the forces that govern the universe. K'ang was not interested in an empirical study of the material surroundings, but in the power of the force within himself, the "spiritual energy" that was already known to the ancients, to remake the world. "To whoever possesses consciousness," K'ang observed, this spirit "gives the power of attraction, like that of the lodestone, but how much more so in the case of man!"[29]

K'ang Yu-wei's vision had a powerful appeal for modern Chinese

---

[27] Fung Yu-lan, *A History of Chinese Philosophy*, 2 vols. trans. Derk Bodde (Princeton: Princeton University Press, 1953), 2:602–7.

[28] K'ang and Wang: Hao Chang, *Liang Ch'i-ch'ao and Intellectual Transition in China, 1890–1907* (Cambridge: Harvard University Press, 1971), 43–5. K'ang and New Text revival: Wakeman, *History and Will*, 108–11.

[29] K'ang Yu-wei, *Ta T'ung Shu* [Book of the great unity], part I, 6, cited in Fung Yu-lan, *History of Chinese Philosophy*, 2:685.

intellectuals, not in the 1880s and early 1890s when K'ang was little known and the faith of most men in the traditional order remained intact, but after 1895 when the whole system came unglued. His cosmological breadth, embracing man, nature, and the infinite in one grand, interconnected whole, struck most Chinese as the type of all-inclusive *Weltanschauung* that was needed to hold China's linked polity and culture together. His evolutionary dynamic, moving along a recognizable path toward a utopian goal, reduced Confucianism and the imperial institutions to mere stages in a train of progress that would eventually leave the grand tradition behind. And his eclecticism opened the door to sources of inspiration, including Western science, that were needed to build a better China in the future. This broad evolutionary cosmology proved to be the most vital and lasting framework for progressive Chinese thinkers of the next century. Ironically, it was the very engine of K'ang's vision – the spiritual link between man and the universe, the power of *jen*, and the introspective quest for self-cultivation and sagehood that lay at the heart of the neo-Confucian enterprise – that left him out of step with the new age. In the next chapter, we will see how he was replaced at the center of China's intellectual debate by men who shared his evolutionary cosmology, but described it in the more naturalistic language of modern science.[30]

K'ang Yu-wei's early works, which were virtually unknown at the time, bear striking similarities to the views of other less famous members of the literati who took an interest in Western science and technology during the same period. One selection of writings from the late 1880s shows how these men, like K'ang, tried to fit Western science and technology into traditional Chinese conceptions of knowledge. Part of the reason for this approach must be the conservatism of the literati themselves, part the persisting confusion over the meaning of the term *ko-chih*.

The writings in question are the prizewinning essays from a contest sponsored by the Shanghai Polytechnic Institute. This institute was founded in 1874 by John Fryer, Hsü Shou, and other Chinese and foreigners to promote the spread of scientific and technological information through its public library, reading room, exhibition hall, lectures, and courses. Although the institute attracted less attention than hoped for by the founders, one activity, a public essay contest, was a great success. The purpose of this exercise, Fryer explained, was "to try and induce the Chinese literati to investigate the various departments of

---

[30] Evolutionary cosmology: Charlotte Furth, "Intellectual Change: From the Reform Movement to the May Fourth Movement, 1895–1920," *The Cambridge History of China*, ed. by John K. Fairbank (Cambridge, Eng.: Cambridge University Press, 1983), 12:323–46.

Western knowledge with the view to their application in the Middle Kingdom." Every three months, beginning in 1886, a Chinese official was invited to set the questions, grade the entries, and contribute to the prize money. Since the monetary reward was small, the contestants must have been drawn primarily by an interest in the subject and the satisfaction of seeing their winning essays in print. Most of the questions addressed aspects of self-strengthening: the construction and operation of modern ships and guns; problems of foreign trade and foreign policy; reforms in banking, finance, law, transportation, and communication; and technical skills, such as surveying, the use of weights and measures, water conservancy, and land reclamation. A few dealt with basic science – the identification and naming of chemical substances, the cause of tides, behavior of heavenly bodies, and solution of mathematical problems – or with comparisons between Chinese and Western medicine and explanations of nature.[31]

Over the years, the number of contestants increased, from twenty-six to eighty-one per quarter. Data on seventy-two first-class prizewinners show that more than three-quarters came from the provinces of the Lower Yangtze, and almost all of these from the narrow confines of the Shanghai-Nanking-Hangchow triangle. Others lived in centers of reform activity in Kwangtung, Fukien, and Hunan. Only two were from northern China, none from Peking. Fifty-seven held the lowest official degree, *sheng-yüan*, or above, five were higher, *chü-jen*, and most of the rest were minor officials or students preparing for the exams. In sum, they represented that stratum of successful scholars and literati that the Kiangnan region produced in such abundance.[32]

Confusion over the meaning of science and its relation to traditional Chinese ideas about nature is evident in two sets of prizewinning essays, from 1887 and 1889, that compared *ko-chih* in China and the West. Each of the six winning essayists began in good examination style by summarizing the texts, beginning with the *Ta-hsüeh*, that set forth and explained the meaning of *ko-chih*, then compared the Chinese use of this term with what he understood to be the usage common in the West, based on a similar reading of translated foreign texts. In one way or

[31] SPI: Knight Biggerstaff, "Shanghai Polytechnic Institution and Reading Room," 141–2. Essay contest: ibid, 141–3; and Adrian Bennett, *John Fryer*, 56. Fryer quotation: John Fryer, "Report of the Chinese Prize Essay Scheme in Connection with the Chinese Polytechnic Institute and Reading Rooms," Shanghai, 1886–1887, cited in ibid. Winning essays appeared in: Wang T'ao, ed., *Ko-chih shu-yüan k'e-i* [Themes of the Shanghai Polytechnic Institute] (Shanghai: Ko-chih shu-yüan, 1886–93), which is available on two reels of microfilm, University of California East Asiatic Library, Berkeley, California. A review of the third volume of the *k'e-i* appears in *North China Herald*, 1 November 1889, 536–7. See: Bennett, *John Fryer*, 146 n41. See also: *North China Herald*, 20 July 1889, 85–6.

[32] Data gathered from *Ko-chih shu-yüan k'e-i*.

another, all agreed on two points. First, in China *ko-chih* has meant investigation into a wide range of issues, including both physical objects and human affairs, whereas in the West it has concentrated on the study of nature and the application of this knowledge to the solution of practical, material problems. Second, Chinese scholars have sought the underlying "principles" of things, the deep "roots" of learning, whereas Western scholars have been satisfied to obtain useful "techniques," the superficial "branches" of knowledge. P'eng Jui-hsi, a licentiate attached to a county school near Changsha in the southern province of Hunan, expressed these views, which were shared by most other prizewinners, as follows:

In China, the study of *ko-chih* includes everything. In its depth, it seeks the sources of nature and investigates the principles of life. In its breadth, it discerns the regulations and examines the way of peace and order, the maintenance of the law and the use of tools. Each name and each thing, every plant and every tree has its principle, which must be known. It [*ko-chih*] extends to the inside and outside, the fine and the coarse of all things and illuminates every corner of the mind. Therefore, we say that it includes both principles and techniques.

Westerners [on the other hand] devote their entire effort, their research and experimentation, to the study of instruments and numbers. Because of this, they can [understand] exhaustively the sun, the moon and the five planets, change the direction of the vapors, make use of water and fire, transform metals and stones, govern the winds and throw down lightning, measure the seas and traverse the skies. Their steamships cover a thousand *li* in the twinkling of an eye. They delve into every corner of the wind and rain, the cold and heat. And they are unbelievably skilled in the use of every type of tool. Therefore we say that Western studies are useful. . . .

In sum, China's tradition is to emphasize principles and make light of techniques, while the Western tradition is to emphasize techniques and make light of principles. From ancient times, order and disorder, security and danger have always been linked to the flourishing or decay of principles and not to the facility or clumsiness of techniques. Today, [however], the world is open to communication and exchange. Powerful neighbors are watching and waiting, holding fast to their strengths and sneering at our weaknesses. China must learn their superior skills in order to control them.[33]

Similarly, Chiang T'ung-yin, a low-level degree-holder from outside Shanghai, compared the Chinese quest for the "root" of knowledge with the Western focus on the "branch":

Important things have both root and branch. It is especially valuable to know their beginning and end. The ancient Confucians took *ko-chih* to be the means for opening things. They described the principle and sought the root. Westerners take *ko-chih* to be the source of clarifying things. They design their machines and only seek the branch. If you want to have both the root and the branch, then you

---

[33] *Ko-chih shu-yüan k'e-i*, Spring 1887, P'eng Jui-hsi, 1, 4.

must take Chinese studies to be the warp and Western studies to be the woof. Both are valuable, and to use both cannot be ridiculed as a trivial pursuit.[34]

Although recognizing these differences, most essayists would have agreed with the Hunanese Ko Tao-yin, holder of a degree by purchase, that in both China and the West "the principles of *ko-chih* are the same, while it is the objects of *ko-chih* that differ."[35]

Some essayists described the treatment Chinese and Westerners gave to each of several fields of nature studies. In these cases, the discussions were generally uninformed and confused. Chiang T'ung-yin reported that most of Darwin's work was in the field of "*hua-hsüeh*," apparently figuring that evolution must be a branch of China's own study of change [*hua*], and ignoring the fact that this term was now used as a translation for chemistry.[36] Other writers compared Chinese and Western astronomy in terms that were at best crude and at worst misleading.[37]

These essays, like the writings of K'ang Yu-wei, show how difficult it was for scholars reared in the classics to break loose from the traditional mental framework and accept new categories of thought. The notion of a special method of inquiry that posits testable hypotheses, subjects them to experimentation and observation, and expresses the findings in measurable mathematical terms, that maintains that knowledge susceptible to such methods is impermanent and certain to be replaced by something better, and that distinguishes this type of knowledge from others, not susceptible to the methods described – this notion is nowhere evident in the writings of late Ch'ing literati. Rather, we find here the view that the basic principles of *ko-chih* are the same in China and the West (indeed, several authors say that Western *ko-chih* originated in China!),[38] that they have been applied in different ways to different objects according to the interests and traditions of the two civilizations, and that these differences are evident in a number of otherwise logically distinct fields. The integrity and superiority of China was maintained by the insistence that each of these approaches had its strength – the Western to fashion useful "techniques"and the Chinese to understand the more fundamental "principles" that underlie all reality. What these writings reveal, in short, is a failure to recognize a modern conception of "science" that might compete with or even displace *ko-chih*, and an insistence on dealing with the challenge of the West by folding its discoveries into the fabric of traditional Chinese thought.

[34] Ibid., Spring 1889, Chiang T'ung-yin, 2.
[35] Ibid., Spring 1887, Ko Tao-yin, 1.
[36] Ibid., Spring 1889, Chiang T'ung-yin, 1.
[37] Ibid., Spring 1887, Ko Tao-yin, 1.
[38] Ibid., P'eng Jui-hsi, 7; Chiang T'ung-yin, 2.

Conclusion

Nathan Sivin, who has studied the transmission of Western science and mathematics to China in the seventeenth century, offers a sociological explanation for the extent and limits of science's earlier influence. Chinese astronomers of the seventeenth century applied knowledge from abroad to transform their view of the heavens in what Sivin calls "a conceptual revolution in astronomy," but dramatic changes in this discipline did *not* lead to a "scientific revolution" in Chinese thinking about nature as a whole.[39] Instead of spreading outward in new directions, this minirevolution served primarily to revive interest in traditional Chinese astronomy and mathematics. One reason for this inversion was that Chinese astronomers were members of an old elite, "who were bound to evaluate innovations in the light of established ideals that they felt an individual responsibility to strengthen and perpetuate."[40] China's breakthrough to modern science, Sivin concludes, had to await the profound social and political changes of the late nineteenth century:

By that time foreigners exempt from Chinese law and backed by gunboats had constructed new institutions and new career lines. These were most attractive to those they had educated, who had no other prospects. We can no longer talk about the encounter of the old and new astronomy. Social and political change had left nothing for the old to do.[41]

This study sheds new light on the relationship among science, state, and society that both confirms and refines Sivin's thesis. The careers of Hsü Shou and other scientific pioneers support Sivin's view: Converts to modern science came from among scholars whose progress through the examination system had stalled, they found refuge in an institutional and social niche created by foreign patrons, and as they shifted to this new base, they lost interest in China's ancient tradition of nature study, which offered nothing to their modern, Western commitment. What the present study adds to this picture is an awareness of how narrow and limited was this change: Much of the work was, in fact, done by foreigners. Many members of the Chinese elite remained hostile and many more indifferent to the new ideas. Those few, like Hsü Shou, who made the leap, failed to move beyond the stage of reading and writing about science. And scholars and educated laymen who followed the new literature were

[39] Nathan Sivin, "Why the Scientific Revolution did not take place in China – or didn't it?," *Transformation and Tradition in the Sciences: Essays in Honor of I. Bernard Cohen*, ed. Everett Mendelsohn (Cambridge: Cambridge University Press, 1984), 546–7.

[40] Nathan Sivin, "Wang Hsi-shan" *Dictionary of Scientific Biography*, 15 vols. (New York: Scribner's Sons, 1970–78), 14:161.

[41] Sivin, "Why the Scientific Revolution did not take place," 548–9.

confused by what they read or insisted on fitting it into traditional categories of thought

Science made little progress in nineteenth-century China primarily because the old order was able to contain the forces of change. By 1870, the Ch'ing dynasty had survived the twin challenges of rebellion and war. The "Restoration" was working, or appeared to be. Conservatives at court limited the size and scope of technical training at home and of scholarly delegations abroad. They confined the introduction of new technology to the military sector, while forestalling broad economic development. Most important, they kept control over the examinations, which remained unchanged in form and content. Without some undeniable shock to the system, there was no need for the incumbent elite to yield broadly to the challenge of new forces, and they did not.

Manchu officials and their conservative Chinese allies were not the only obstacle. Resistance also came from groups generally thought to favor change. After the death of Tseng Kuo-fan, Li Hung-chang and other self-strengtheners did little to advance the fortunes of technical experts. Similarly, the nongovernmental "public" enterprises created to restore the smooth functioning of south China after the rebellions were turned over to men with traditional academic and administrative backgrounds. There were few openings for people with scientific and technical skills, and no place at all for the "scientist" – no research institute, government agency, university department, industrial laboratory, or other enterprise to support people whose primary purpose was to investigate the workings of nature. As a result, few talented Chinese chose science, fewer still were rewarded for the effort, and no one who came of age prior to 1900 went on to the full-time study of science or won prominence in this field at home or abroad.

Finally, in the absence of political and social change, the commitment to traditional culture remained intact. In most cases, this meant simply studying the classics and preparing for the exams. Even among those who took an interest in science, however, traditional ideas and attitudes often overwhelmed and perverted the effort. This was true of the literati who read the Kiangnan translations and tried to square the notion of science with the "investigation of things." It was also true of the translators, like Hsü Shou, whose commitment remained primarily a literary affair. Few Chinese of this period, whatever their standing, spent much time in observation, experimentation, or scientific research. One finds a willingness to read and write about the subject. There were a great many translations, publications, letters to the editor, and minor degree-holders willing to submit essays in the familiar format of the written exam. But there is little sign of people digging into the stuff of nature, and it is therefore hard to call this "science" at all. The spread of science to China depended upon the creation of a constituency that was not adequately provided by inserting the thin wedge of imperialism into the still vital timber of traditional China. First, the great tree had to fall, a process that began in 1895.

PART II

The interregnum
1895 – 1927

During the period from 1895 to 1927 – which I have chosen to call the "interregnum," because it was characterized by weak political and social structures and bracketed before and behind by comparatively stable, unified regimes – science sank stronger roots in China, but failed to grow a healthy, harmonious vine. From the decline of the empire at the end of the nineteenth century until the emergence of a viable republic in 1928, China's political, social, and cultural order crumbled. Institutions that had lasted a millennium and more were abandoned, class relationships were reshuffled, the established faith was discredited, and revolutions of every description rushed in to fill the void. The brake on science in the previous century had been too much authority of the wrong kind. The great opportunity for but also the chief impediment to the development of science in the early twentieth century was the absence of authority of any kind at all. To create in the midst of chaos the new social roles, cultural ideals, and institutional forms appropriate to science was the difficult, half-met challenge of the age.

This period began with one of the most important, traumatic events in modern Chinese history: the defeat in the Sino-Japanese War of 1894–5. Unlike earlier losses to the European powers, this defeat was a clear signal that China's response to the West had failed and that refusal to accept more fundamental change had brought submission even to the lowly "dwarf-pirate" Japanese. After 1895, an ever-larger segment of China's elite proposed the adoption of ever-deeper changes, beginning with the remodeling of government institutions in the Hundred Days Reforms of 1898. For the last time, the conservatives succeeded in blocking these efforts and made a final, desperate attempt to preserve the status quo by throwing their weight behind the nativist Boxer Rebellion (1900). When this failed, the Empress Dowager agreed to abolish the classical examinations (1905), undertake a series of reforms, and try to save the dynasty by modernizing it. It was too late. The Revolution of 1911 replaced the empire with what pretended to be a republican form of government. After a brief search for order under military strongman Yuan Shih-kai (1912–16), however, the experiment with self-government failed. During the next decade (1916–27), China fell victim to the clash of warlord armies, no combination of which could hold the country together or remain satisfied with only one part of it.

Despite their manifold troubles, successive governments of the interregnum – Manchu, Chinese, republican, and warlord – did their best to build a new order that was modern, effective, and therefore scientific. Each administration added to the nationwide system of modern schools, opened research facilities to map and make better use of China's resources, and supported public and private industries that employed up-to-date techniques. Yet in each instance the will and ability of the state fell short of the task, while the spread of chaos left the public arena, the place where one might learn about, engage in, and make use of science, up for grabs.

Many forces were sucked into the resulting vacuum. Three played leading roles in the development of science. First were members of the new urban intelligentsia. These were men who had abandoned classical learning, went abroad to study, and returned with varying degrees of knowledge about the modern world. A few with the best scientific training made their mark as teachers, scholars, and engineers. More prominent were the pundits, organizers, and entrepreneurs, many of whom had only a passing exposure to science, but applied what they knew in reshaping Chinese society and culture along modern, "scientific" lines. Second, missionaries and other foreigners, who had been crucial to the development of science in the nineteenth century, played an even larger role during the first quarter of the twentieth. Peking Union Medical College, a creation of the Rockefeller Foundation, towered above the landscape of East Asia in scientific research. With the injection of Rockefeller money and recruitment of better faculty from the United States, mission colleges and universities emerged as China's leading centers of scientific education. Remission of the American share of the Boxer indemnity built Tsinghua University and provided scholarships for hundreds of Chinese students in the United States. Through a variety of channels, public and private, the future leaders of Chinese science were trained in the leading universities of the West. Finally, private Chinese entrepreneurs led in the building of modern industries, the introduction of new technologies, and the employment of skilled personnel. With the breakdown of central authority, native capitalism flourished, permitting innovations in industrial research and engineering to enter China through the private sector.

The result of all this was the creation of several substantial, but in the end disconnected initiatives. Through their propaganda and organization, the scientifically minded intelligentsia helped foster a culture and society attuned to the values of scientism, but did little to promote the study or practice of science per se. Chinese who studied science and went on to do research often reached the top of the scholarly ladder, only to discover that they had lost touch with the people and problems at the bottom. Chinese industry grew too slowly to provide jobs for the many young people who had acquired new skills, and Chinese agriculture, which could have used more help, was abandoned by those who moved to the cities, went abroad to study, and sought to fulfill personal goals in the upscale modern world. Absent was a government with the strength, vision, and will to pull the fragments together, enforce choices, and forge a working system. Science, like much else during the interregnum, lay about in bright, broken pieces.

4

First-generation scientists: makers of China's New Culture

The collapse of the old order made way for new forces that would promote the study and practice of science. Foremost among these was the modern urban intelligentsia, which included China's first generation of "scientists." The chief contribution of these men (there were as yet no women) lay in bringing science to bear on Chinese culture and creating new forms of organization to accommodate the growth of scientific expertise. In some ways, their efforts helped establish a stronger, more vibrant China. In other ways, they sowed the seeds of conflict within Chinese society and between society and the state.

The Chinese intelligentsia of the early twentieth century – centered in coastal cities, Western oriented and often foreign educated, iconoclastic, innovative, restless, looking for ways to save China – numbered many young men who had been trained in science at various levels of sophistication. They constituted, in quantity and quality, a critical mass of experts that had been lacking in the previous century. It is appropriate to call them China's first "scientists." Still, this label belongs in quotation marks, for the most prominent figures of this generation did no research or other work that qualifies them for membership in the scientific profession. Their contribution was rather as a radical advocacy group, who popularized the ideas and built the organizations that made a place for science in the Chinese scheme of things. This was the age of scientific propagandists, organizers, and entrepreneurs.

Under their guidance, science played a leading role in the making of modern China. Spokesmen for this generation found in science a common ground of ideas and values that would bind together Chinese society, at a time when China's elite was abandoning Confucian ethics and seeking new spiritual and ethical ground. During the 1910s and 1920s, science, or more properly "scientism," emerged as a pillar of China's "New Culture." At the same time, members of the first generation built organizations, such as the Science Society of China, whose mission was to support the work of China's scientists and spread knowledge of science among young, attentive, urban Chinese. They hoped their efforts would engender a more unified, effective, modern China. In fact, they produced both new bases of unity and sources of conflict.

Scientists of the first generation

Biographical data on over one hundred chemists active in China during the early twentieth century provides a profile of China's first generation of scientists.[1] These men were drawn from the same social and geographic origins that supplied the elites of late traditional China. What distinguished them from their predecessors was the rapidity with which they abandoned the classics and turned to foreign learning. Most members of this generation continued the established preference for humanistic and literary subjects. But some took up the study of science, and it is on this smaller group that our attention fastens.

Born in the period 1880–95, most scientists of the first generation came from the southern coastal provinces of Kiangsu, Chekiang, Fukien, and Kwangtung. These provinces, particularly Kiangsu and Chekiang, had long served as the cradle of classical scholarship, and the proximity to urban centers, also an advantage in earlier times, became especially important after the opening of the treaty ports, China's windows on the modern world. Data on family backgrounds, although less plentiful, also suggest the importance of privilege. There is little evidence to pinpoint the source of wealth that supported China's first scientists. A few young men from decidedly humble beginnings won fellowships to foreign universities. In most cases, however, families supported their sons and later daughters through the long years of study in China and abroad. The importance of private wealth seems to have increased with the passage of time.

Privilege, too, characterized the education of these men. China's first chemists began with classical studies, shifted, when events dictated, to modern Chinese schools in the cities nearest home, and completed their

---

[1] This data on Chinese chemists comes from: *Chiao-yü-pu hua-hsüeh t'ao-lun-hui chuan-k'an* [Report of the Chemistry Forum of the Ministry of Education] (Nanking: Kuo-li pien-i-kuan [National Institute for Compilation and Translation], 1932), 283–7; "Chung-kuo k'o-hsüeh-yüan hua-hsüeh-pu hsüeh-pu wei-yüan chien-chieh" [Brief introduction to the members of the Department of Chemistry of the Chinese Academy of Sciences], *HHTP* 9 (1981): 552–64; and biographies of chemists from *HHTP*, *CKKCSL*, and other sources listed in the bibliography. Peter Buck, *American Science and Modern China, 1876–1936*, (Cambridge, Eng.: Cambridge University Press, 1980), 99–116, found similar data on seventy-nine early members of the Science Society of China.

undergraduate and in a few cases graduate training abroad. Some, including the chemists Jen Hung-chün, Chou Tsan-quo, Ting Hsü-hsien, and Cheng Chen-wen, passed the lowest-level examinations, before these were abolished in 1905. None of these pioneers attended a mission school, most of which were then located in remote, rural areas and whose clientele were primarily children of poor peasants, lacking in other options. Experience in modern schools undoubtedly fostered greater awareness of events at home and alternatives available abroad, but neither the quantity nor quality of instruction in these schools was high, and in the sciences it was limited or nonexistent. The real education of this generation had to await their journey to the West.[2]

The place of overseas study in the preparation of China's elite changed dramatically after the turn of the century. The number of students going abroad increased from a handful in the 1870s and 1880s to over fifteen thousand in 1906. Whereas the former were government-sponsored technicians sent to learn the secrets of self-strengthening, the latter included scholarship students along with many others who were wealthy or lucky enough to go. The country of preference by a wide margin was Japan, chosen on the basis of proximity and cost. And the objects of study were as varied as the whims of the students themselves. The small minority that chose to study science was among the best prepared and best financed, and went to the United States or Europe, where the opportunities for studying science were greatest.[3]

The careers of these men after they returned to China reflected the limited demand for scientific and technical expertise. A few achieved the highest academic awards from the world's leading universities and went on to perform original, in some cases even significant, research. A much larger number, whose formal education fell short of the doctorate, helped establish the departments of science in China's colleges and universities. Some found work in the handful of industries that had begun to introduce modern technology. Still, many prominent figures of this generation made their mark, not by scientific achievements narrowly

---

[2] Jen Hung-chün: Howard L. Boorman and Richard C. Howard, eds., *Biographical Dictionary of Republican China*, 4 vols. (New York: Columbia University Press, 1967–71), 2: 219–22. Chou Tsan-quo: Kao I-sheng, Chu Jen-hung, and Hsieh Yü-yüan, "Wo-kuo Chung-ts'ao-yao hua-hsüeh yen-chiu ti hsien-ch'ü-che – Chao Ch'eng-ku chiao-shou" [Pioneer in the study of Chinese medicinal herbs – Professor Chou Tsan-quo], *HHTP* 3 (1980):178–81. Cheng Chen-wen: Li Ch'iao-p'ing, *Chung-kuo hua-hsüeh shih*, 2:784–90. Ting Hsü-hsien: Fan Chi-hsing and K'o Kuei-hua, "Hua-hsüeh-chia hua-hsüeh-shih-chia Ting Hsü-hsien chiao-shou ti i-sheng" [The life of Professor Ting Hsü-hsien, chemist and chemical historian], *HHTP* 6 (1979):547–52. Cheng Chen-Wen: Li Ch'iao-p'ing, *Chung-kuo hua-hsüeh shih*, 1:784–90.

[3] The most complete study of Chinese students abroad is Y. C. Wang, *Chinese Intellectuals and the West, 1872–1949* (Chapel Hill: University of North Carolina Press, 1966). For the period 1872–1911, see pages 41–98.

defined, but because the times offered scope to those with even limited technical expertise.

It was the scientific organizers, ideologues, and entrepreneurs who emerged as the leaders of and left the greatest imprint on Chinese science of the 1910s and 1920s. Most notable was Ting Wen-chiang, holder of a bachelor's degree in geology (Glasgow, 1911), founder of the National Geological Survey, sometime scholar but more often pundit and political operative, who reigned as China's leading statesman of science until his death in 1936. Second to Ting was Jen Hung-chün, a native of Szechwan province, who studied chemistry in Japan and at Cornell (B.S., 1916) and Columbia universities (M.S., 1917), and served as the first president of the Science Society (1914–23), executive director of the China Foundation (1929–34, 1942–8), and secretary general of the Academia Sinica (1939–42). Wu Chih-hui and Li Shih-tseng, leaders of the anarchist movement, studied and wrote on biology, paleontology, and evolution, and had a profound influence on the intellectual and philosophical debates of the 1920s. There were many other members of this generation with similar competence and commitment.[4]

It would be interesting to know how and why these men came to choose science as a career. Several factors instrumental in other times and places were present in early twentieth-century China, particularly in the areas where most scientists lived and worked. Among these, attention might focus on the demonstration of foreign examples, commercialization of the economy, introduction of new technologies and rise of the merchant class, radical political activism, and the search for new ways to "save the nation." It is possible to cite biographies that show how the life choices of Chinese scientists were shaped by one or more of these factors, but data of this type are too sparse to support a general theory of causation. We can say only that the environment encouraged many young people to pursue new outlets, and some of them opted for science.

Introducing the language of "science"

The first challenge to those who would spread this faith in China was linguistic, for an understanding of science depended on the invention of language, free from traditional connotations, to describe the new ideas.

---

[4] Ting: Charlotte Furth, *Ting Wen-chiang: Science and China's New Culture* (Cambridge: Harvard University Press, 1970). Jen: Boorman and Howard, *Biographical Dictionary*, 1:219–22. Wu and Li: James Pusey, *China and Charles Darwin* (Cambridge: Harvard University Press, 1983), 373–4, notes 189 and 193.

Table 4.1. *Chinese scientific publications, 1851–1936*

Field	1851–1911		1911–36	
	Number	%	Number	%
General	44	9.4	62	12.5
Astronomy and meteorology	12	2.6	33	6.7
Mathematics	164	35.0	141	28.5
Physics			52	10.5
Chemistry			51	10.3
Physics and chemistry	98	20.9		
Natural history	92	19.7		
Biology			139	28.1
Earth sciences	58	12.4	17	3.4
Total	468	100.0	495	100.0

Source: Chou Ch'ang-shou, "I-k'an k'o-hsüeh shu-chi k'ao-lüeh," 431–70.

At the heart of the matter was the question of what expression the Chinese would use to stand for "science" itself? Use of the term *ko-wu* or *ko-chih* had caused enormous confusion over the nature of science and its supposed connection to traditional Chinese thought. The solution to this problem came, after the turn of the century, from a source that had previously made little contribution to Chinese culture – namely, Japan.

After 1900, several changes occurred in the translation of foreign texts into Chinese. First was the tremendous growth in volume: The average annual number of translations produced during the first half of the twentieth century was ten to fifteen times greater than in the latter half of the nineteenth. Second was a change in subject matter: As politics came to dominate other concerns, attention shifted from science and technology to history, geography and social studies. The share of translations devoted to pure and applied sciences declined from 70 percent for the period 1850–99, to 32 percent for 1902–4, and 25 percent for 1912–40. Still, the overall growth in translation was such that the annual output of scientific works increased from two or three volumes in the late nineteenth century to between twenty and thirty in the early twentieth. Third was the greater independence of the Chinese: In contrast to previous experience, science translations of the twentieth century were executed almost exclusively by Chinese and the work widely dispersed, along with the study of science generally, rather than restricted to a few locations. Fourth was the rise in level of sophistication: Whereas most translations of the nineteenth century had been secondary school texts, those produced after 1900 included more of the latest scholarly monographs (see Table 4.1).[5]

Finally, after 1895, Chinese interested in all aspects of modern Western culture turned to Japan, and Chinese translators looked increasingly to Japanese texts, which were conveniently written in the common ideographic script. During the second half of the nineteenth century, half of all Chinese translations had been based on British sources; between 1902 and 1904, 60 percent were taken from the Japanese. In the pure and applied sciences, the Japanese share jumped from 15 percent in the earlier period to 58 percent in the latter. The first science periodical published by the Chinese themselves, the *Ya-ch'üan tsa-chih* [Ya-ch'üan Magazine], which appeared from 1900 to 1901, was composed almost exclusively of articles translated from Japanese. This flirtation was a mixed blessing for Chinese science, since the Japanese themselves were just beginning to understand the new world, and most of their books were translations of European texts, which were retranslated at the cost of some precision into Chinese. Later Chinese translators worked directly from American and British texts, which accounted for more than half of all translations and two-thirds of those in science and technology in the years 1912 to 1940. During the first decade of the century, however, Japan was China's chief source of inspiration, a short-lived but intense experience that affected every field of learning.[6]

The translation of Japanese texts brought with it the introduction of Japanese terms. A recent survey estimates that over three-fourths of the new Chinese vocabulary adopted during this period was taken from Chinese-character compounds that had become standard Japanese.[7] In chemistry, where the terminology worked out by Fryer, Hsü, and others proved workable and gained widespread acceptance, the Chinese preserved their established system. The names of elements chosen by Japan's pioneer chemical translator, Udagawa Yōan, remain in use in that country, whereas the Chinese have kept or revised the names devised at Kiangnan. In other fields, such as geology and physics, where Chinese translators failed to lay a firm foundation, much new vocabulary was

---

[5] Annual averages: Tsuen-hsuin Tsien, "Western Impact on China through Translation," 315, 319–20. Chou Ch'ang-shou, "I-k'an k'o-hsüeh shu-chi k'ao-lüeh," 431–4, reports annual average scientific publications of 1.5 for 1851–97, 9.0 for 1898–1901, and 33.8 for 1902–11.

[6] Sources of translations: Tsuen-hsuin Tsien, "Western Impact on China through Translation," 315, 319–20. *Ya-ch'üan*: Chang Tzu-kao and Yang Ken, "Chieh-shao yu-kuan Chung-kuo chin-tai hua-hsüeh-shih ti i-hsiang ts'an-k'ao tsu-liao – 'Ya-ch'üan tsa-chih,'" 55–9.

[7] Three-quarters: Sanetō Keishū, *Chūgokujin Nihon ryūgaku shi*, 378, cited in Marius Jansen, "Japan and the Chinese Revolution of 1911," *Cambridge History of Modern China*, 11:362.

Table 4.2. *Chinese terms for physical structures*

	Element	Compound	Atom	Molecule
Fryer-Hsü	*yüan-chih*	*tsa-chih*	*chih-tien*	*tsa-tien*
	原質	雜質	質點	雜點
From Japanese after 1900	*yüan-su*	*hua-ho-wu*	*yüan-tzu*	*fen-tzu*
	分子	元素	化合物	原子

Source: Chang Ch'ing-lien, "Hsü Shou yü 'Hua-hsüeh chien-yüan'" [Hsü Shou and the "Hua-hsüeh chien-yüan"], *CKKCSL* 6, no. 4 (1985):55.

introduced from Japan.[8] In the case of basic physical structures, for example, the Chinese abandoned expressions invented during the nineteenth century in favour of Japanese terms (Table 4.2).

Among the most important Chinese borrowings from Japan was the term for "science." For the most part, scientific terminology is technical and arbitrary. Its usefulness depends only on its simplicity, consistency, and adherence to the accepted paradigm, and it is of little interest to those outside the discipline. This is true, for example, of the names of elements, compounds, and other terms of art described in Chapter 2. There are, however, a small number of expressions that go to the heart of the paradigm, whose meanings are profound and subtle, and whose influence extends beyond the boundaries of the craft, to educated people as a whole. Such is the case with science, a concept that touches most branches of modern thought and had to be grasped by all Chinese who sought entry to the new world.

As we have seen, the term in general use in China before 1900, *ko-chih* or *ko-wu*, was burdened by its traditional meaning – investigation of things – and liable, therefore, to distract scholars who had been reared in the classics from a proper understanding of modern science. What was needed was a fresh neologism, free from established connotations, that could stand for the new idea. The Japanese had come to grips with this problem several decades before it was appreciated in China. Some Japanese, influenced by the same neo-Confucian texts that guided scholars on the mainland, chose the terms *kakubutsu* [*ko-wu*] and *kakuchi* [*ko-chih*] to describe "science," and in later years "physics" (Table 4.3). But more important was the choice made by Rangaku [Dutch Studies] scholars of the late eighteenth century to represent "science" with another neo-Confucian expression, *kyūri* [in Chinese, *ch'iung-li* or *chiu-li*], that means to study "exhaustively the principles" of

[8] Japanese influence on Chinese geological terms: Yang Tsui-hua Lee, "Geological Sciences in Republican China, 1912–37," 39–47.

Table 4.3. *Japanese and Chinese scientific terms*

Japanese	Chinese	Characters
kakubutsu	ko-wu	格物
kakuchi	ko-chih	格致
kyūri	chiu-li	究理
kyūri	ch'iung-li	窮理
kagaku	k'o-hsüeh	科學

things. This linguistic decision testifies to the sophistication of the Japanese, who recognized that science involves an understanding of *ri* [*li* in Chinese] or "principles," in contrast to the Chinese, who clung stubbornly to the notion that their own *ko-wu* was rooted in "principles" and Western science was a mere grab-bag of "techniques." By the middle of the nineteenth century, *kyūri* was in general use as the Japanese equivalent for "science" in the paradigmatic sense of "natural philosophy."[9]

Kyūri, although superior to the Chinese *ko-chih* or *ko-wu*, also suffered from the association with neo-Confucianism. All these terms, borrowed from the lexicon of the classics, introduced an inevitable confusion between traditional philosophical and humanistic concerns and the scientific study of nature. To overcome this problem, Japanese scientists fastened on a new expression, *kagaku*, a character combination previously unknown in China or Japan that first appeared in 1871 in an essay by Inoue Kowashi. The first character of this compound, *ka* [k'o in Chinese], has the original meaning of "class, kind, or variety." Thus *kagaku* might be rendered "classified learning," a designation that is not limited to the natural sciences, but may include any single discipline or group of disciplines. As the historian Nakayama Shigeru points out, this term successfully captured the specialized and institutionally differentiated character of science or *Wissenschaft* as it was understood in the late nineteenth century. During the Meiji period (until 1912), *kagaku* was applied to the collection of nonphilosophical, technical disciplines, while *kyūri* continued to be used for the cognitive "natural philosophy."[10]

[9] *Kyūri*: Albert Craig, "Science and Confucianism in Tokugawa Japan," *Changing Japanese Attitudes Toward Modernization*, ed. Marius Jansen (Princeton: Princeton University Press, 1965), 139–41; Masayoshi Sugimoto and David L. Swain, eds., *Science and Culture in Traditional Japan, A. D. 600–1854*, (Cambridge: M.I.T. Press, 1978), 303–6; Shigeru Nakayama, *Academic and Scientific Traditions in China, Japan, and the West*, 203–5.

[10] Ibid., 208–9. James Curtis Hepburn, *A Japanese–English and English–Japanese Dictionary*, 3d ed., (Tokyo: Z. P. Maruya & Co., 1886), calls *kyūri*, "natural philosophy," and *kagaku*, "scientific studies, science in general."

This etymology is telling, for after the turn of the century, Chinese writers adopted the Japanese *kagaku* [*k'o-hsüeh*] in the sense of "classified learning," "disciplines," or "sciences," while maintaining *ko-wu* or *ko-chih* in the paradigmatic sense. Perhaps the first appearance of the new term was in Yen Fu's translation of John Stuart Mill's *System of Logic*, which was executed during the period 1900–2 and published in 1905. A comparison of the two texts shows that Yen rendered "science" or "sciences" as *k'o-hsüeh*, explaining *k'o* as a "class" of studies, as in the passage, "all experimental [studies] are *k'o*." Where Mill speaks of a broader, conceptual "natural philosophy" or "investigation of nature," however, Yen reverted to *ko-wu*. In similar fashion, the regulations for the establishment of modern schools issued in 1903 provided for the teaching of *ko-chih*, a collective term that included math, astronomy, physics, chemistry, biology, and geology, while the various "sciences" or "disciplines" were labeled *k'o-hsüeh*. School curricula from this period continued to use the phrase *ko-wu* to describe their science offerings. The first Chinese to use *k'o-hsüeh* in the dual sense of paradigm and discipline may have been Wu Chih-hui. The inaugural issue of Wu's journal, *New Century* [*Hsin shih-chi*], published in Paris in 1907, celebrated the discovery of "scientific laws" [*k'o-hsüeh kung-li*] and the merging of human society with the "laws of nature" [*tzu-jan chih kung-li*], expressions that pointed toward the use of *k'o-hsüeh* to embrace a systemic "principle" or "law." Around the same time, Wu wrote of the discrete "physical, chemical and mechanical sciences [*k'o-hsüeh*]," indicating that the same term could stand for the several classes of learning. This expression was also used to describe the application of new techniques to Chinese industry, for one 1911 publication refers to their "scientific" [*k'o-hsüeh ti*] methods.[11]

A survey of dictionaries published in China between 1900 and 1916 shows that Wu Chih-hui was somewhat ahead of his time, however,

[11] Yen Fu, *Ming-hsüeh* [Logic], translation of John Stuart Mill, *System of Logic*, in Yen Fu, *Yen i ming-chu ts'ung-k'an* [A collection of Yen Fu's translated works] (Shanghai: Shang-wu yin-shu-kuan, 1931), 2:64–6. Compare with: John Stuart Mill, *System of Logic*, vol. 1 (London: Longmans, Green, Reader, and Dyer, 1872), bk. 2, chap. 4, par. 5, 251–3. School regulations, 1903: Jen Hung-chün, "Wu-shih nien lai ti k'o-hsueh" [Science during the past fifty years], *Wu-shih nien lai ti Chung-kuo* [China during the past fifty years], ed. P'an Kung-chan (Chungking: Sheng-li ch'u-pan-she, 1945), 187. School curricula: Ruth Hayhoe, "Towards the Forging of a Chinese University Ethos: Zhendan and Fudan, 1903–1919," *China Quarterly* 94 (June 1983):330–1. *New Century*: *Hsin shih-chi* [New century] 1: (22 June 1907):1. "Physical, chemical and mechanical sciences": Wu Chih-hui, "Ta jen shu" [Answer to a letter] (28 September 1907), *Wu Chih-hui hsüeh-shu lun-chu* [Selections of Wu Chih-hui's academic writings] (Shanghai: Ch'u-pan ho-tso-she, 1926), 237. 1911 publication: Wang Ching-yü, comp., *Chung-kuo chin-tai kung-yeh shih tsu-liao, ti-erh-chi, 1895–1914 nien* [Source materials on the history of modern industry in China, 2d collection, 1895–1914], 2 vols. (Peking: K'o-hsüeh ch'u-pan-she, 1957), 1:320.

because common usage followed Yen Fu's two-track terminology. In the most authoritative lexicons, *ko-chih* and *ko-wu* were rendered as "science," "natural philosophy," or as Herbert Giles (1912) preferred, "researches, especially of natural science." Less often, *chiu-li* (from the Japanese, *kyūri*) was also equated with "natural philosophy." *K'o-hsüeh*, if mentioned at all, was "science" in the sense of "discipline," including the mental, moral, and political as well as physical sciences, or as the *Hsüeh-sheng tzu-tien* [Student dictionary] published in Shanghai in 1915 put it, "Any learning which is systematic and can be independent is called *k'o-hsüeh*." In no instance was *k'o-hsüeh* accorded a broad paradigmatic meaning.[12]

In sum, during the early years of the twentieth century, there were significant changes in the language with which Chinese spoke of science, changes that pointed toward more modern catagories of thought. Still, as late as the mid-1910s, China lacked a single term to stand for "science" in the several senses of an intellectual paradigm, a method of acquiring knowledge, and a group of disciplines whose purpose was to explain physical reality. But this was the turning point, for after 1915 the discussion of science intensified, and one expression, *k'o-hsüeh*, emerged as the common label for a new, profound and broadly conceived way of looking at the world.

The rise of scientism

Linguistic development was part of a broader transformation of Chinese culture. It is telling, indeed necessary, that the two should change at the same time and in the same direction. *K'o-hsüeh* brought together the disciplines through which people examine their material environment, the methods by which that examination is made, and the models that explain how nature works. It provided a tight verbal package whose meaning was precise, powerful, and modern; it held the one key that opened doors all around. These were just the virtues required by Chinese who wanted to raise up "science" as a model for China's culture. They

[12] See, for example: *Technical Terms: English and Chinese*, ed. the Educational Association of China (Shanghai: Presbyterian Mission Press, 1904); Yen Wei-ching, *An English and Chinese Standard Dictionary*, 2d ed. (Shanghai: Commercial Press, 1908); John Harrington Gubbins, *A Dictionary of Chinese–Japanese Words in the Japanese Language*, 2d ed. (Tokyo: Maruya & Co., 1908); Herbert Giles, *A Chinese–English Dictionary* (Shanghai, 1912), reprint (Taipei: Ch'eng-Wen Publishing Company, 1967); Wu Chin et al., eds., *Hsüeh-sheng tzu-tien* [Student dictionary], (Shanghai: Shang-wu yin-shu-kuan, 1915); and Karl Hemeling, *English–Chinese Dictionary of the Standard Chinese Spoken Language* (1916), reprint (Freeport, NY: Books for Libraries Press, 1973).

hoped that "science," or more precisely "scientism," would replace the discredited Confucian morality, provide the basis for a new public ethic, and attract recruits to the study of nature per se. With language appropriate to the task, China's first generation scientists could get on with building the New Culture.

The earliest and most important purveyor of scientism was the aforementioned translator, Yen Fu, who around the turn of the century introduced to China the Darwinian view of natural and social evolution. Unlike his predecessor, K'ang Yu-wei, who had been in the first instance a Confucian scholar and received foreign influences second hand, Yen Fu's classical studies were interrupted at an early stage. He turned for his education to the Foochow Naval Academy, and later journeyed to England for advanced training in naval science. While abroad, Yen discovered the intellectual underpinnings of Western wealth and power, a lesson he proceeded to describe in translations and essays published after 1895.

Because he was less constrained by traditional language and concepts and understood better the nature of science and its place in Western thought, Yen Fu made two important advances over K'ang Yu-wei. First, Yen gave to K'ang's evolutionary cosmology a wholly secular, naturalistic explanation that owed its inspiration to Darwin rather than to traditional Chinese categories of thought. K'ang's model was teleological, the final "Great Unity" known and preordained, and the engine driving man and nature toward this goal was a mystic "spirit" that could be understood and guided by the mind of the neo-Confucian sage. Yen Fu, on the other hand, admitted to no spirits, sages, or minds and saw no object toward which history was headed. Yen's great achievement was to drive the neo-Confucian spirit-mind from the cosmos, which was left to run by the forces of nature itself.[13]

The second contribution of Yen Fu, a main theme in his translation of Mill's *Logic*, was the notion that scientific thinking is essentially inductive and that its powerful empirical methodology can be applied to all things. This bias ignored the prominent role deduction, mathematics, and plain speculation had played in the story of modern science, but Yen's approach suited himself and his readers. Induction was important to the Chinese of the post-Confucian era, because it distinguished their new scientistic faith from what they saw as the blind alley of introspection that characterized traditional metaphysics – the Sung and Ming idealism that supposed that "one could know the world without stepping outside one's door." "Heaven," Yen Fu explained to the Shanghai Young Men's Association, in 1906,

---

[13] Charlotte Furth, "Intellectual Change," *Cambridge History of China*, 12:336–40.

in giving birth to man, endows him with consciousness. It does not equip him at birth with any a priori (*yü chü*) intuitions. If one wishes to acquire knowledge, one must derive it by induction from that which is on the surface and close at hand.... In induction one must rely on facts.[14]

By the 1910s, belief in a naturalistic evolutionary cosmology, whose particular manifestations could be known and larger structure confirmed by empirical methods, was a prominent feature of the Chinese intellectual landscape. It was the mission of this generation to spread the faith more widely and make it the basis of what some called the New Culture. Jen Hung-chün, looking back on his school days around the time of the 1911 Revolution, recalled that young people of that era were "confused, vacillating and unable to decide which way to turn." The collapse of the Ch'ing dynasty, rather than freeing Chinese to scale new heights, led to political chaos and a deepening cultural malaise. "It seemed," mused Jen, "that the most important thing was to create a new faith. Only when we had a new faith would we know which path to follow."[15]

The faith offered by Jen Hung-chün and his generation was "scientism," the application of scientific models and methods to human affairs. Two journals, *Science* [*K'o-hsüeh*] and *New Youth* [*Hsin ch'ing-nien*], both of which began publication on 1915, led the way in presenting science (now referred to exclusively as *k'o-hsüeh*) as a panacea for China's ills. Ch'en Tu-hsiu, the founder of *New Youth* and prolific writer on this theme, told his eager readers that "the future true faith and course of action for mankind must follow the unbending path of science."[116] Wu Chih-hui, another spokesman for the scientific worldview, offered the seminal statement of a mechanistic chain of being that began long ago with an explosion of the original "thing":

You could say that this explosion was the fulfillment of its own will. In an instant, it gave rise to a billion universes, a trillion, trillion, trillion, trillion selves. The

---

[14] Sung and Ming idealism: Benjamin Schwartz, *In Search of Wealth and Power: Yen Fu and the West* (New York: Harper & Row, 1964), 190. Yen Fu quotation: ibid., 187.

[15] Quotation: Jen Hung-chün, "Jen Shu-yung hsien-sheng chih chiang-yen" [Speech of Mr. Jen Shu-yung], in "Fu Ch'uan k'ao-ch'a-t'uan tsai Ch'eng-tu ta-hsüeh yen-shuo lu" [Record of speeches made at Chengtu University by the Investigating Committee to Szechwan], *KH* 15, no. 7 (1931):1168. For a discussion of scientism in the May Fourth period, see: D. W. Y. Kwok, *Scientism in Chinese Thought, 1900–1950* (New York: Biblo and Tannen, 1971), passim; Charollete Furth, *Ting Wen-chiang*, 94–135; and Peter Buck, *American Science and Modern China*, 171–85.

[16] Ch'en Tu-hsiu, "Tsai lun K'ung-chiao wen-t'i" [Again discussing the problem of Confucianism], *HCN* 2, no. 5 (1 January 1917):1st article, 1. For similar treatment of this subject, see Ch'en's other articles: "Ching-kao ch'ing-nien" [Warning youth], *HCN* 1, no. 1 (15 September 1915); "Chin-jih chih chiao-yü fang-chen" [Education policy for today], *HCN* 1, no. 2 (15 October 1915); and "Tang-tai erh ta k'o-hsüeh-chia chih ssu-hsiang" [The thought of two great scientists of the modern era], *HCN* 2, no. 1 (1 September 1916).

change was quite simple. It was none other than the transformation of an "unimaginable" elemental force into units of matter. These units formed electrons, the electrons formed atoms, and the atoms formed stars, planets, suns, moons, mountains, rivers, grasses, trees, birds, beasts, insects and fish. You may call this evolution, or you may call it illusion. In either case, the change goes on; there is nothing which has reached its fulfillment and is not in permanent change.[17]

To this impersonal, evolutionary system, spokesmen for scientism added the scientific method – rational, skeptical, empirical, and, because reality was one, applicable to all things. Nothing, not even the deepest faith, value, or emotion, escaped the cold eye of science. Having rejected Confucian bonds and the other ties that held together traditional society, many idealistic youth championed love as the key to individual happiness and social order. The Harvard-trained psychologist T'ang Yüeh reminded them, however, that this impulse too was subject to scientific scrutiny:

We view love as a mystery, because it is more complicated than other psychological phenomena and for the moment cannot be fully analyzed, because it is the happiest of feelings and we are unwilling to subject it to rational analysis and thereby destroy it, and because men of letters have idealized it and made us believe it is "mysterious, sacred and inviolable." In fact, love is like fire. Its essence is neutral and its effects may be either good or bad in the same measure as it is governed by the power of reason.[18]

Deeply disillusioned with China's Confucian past and its warlord present, many young urban intellectuals saw in this scientistic vision a formula they could apply in building a better future. In addition to the utilitarian role assigned to it in the nineteenth century, science was now to serve as the basis of a whole new culture.

Still, there were dissenters. Critics on the political and cultural right charged that this mechanistic view of man and nature reduced life to meaningless matter in motion. Among the naysayers were traditionalists, who thought that too much of China was being sacrificed to the West and too much spiritualism and humanism to the idol of the machine. They were joined by some notably modern figures, men who saw that the West itself was in crisis and attributed this crisis to an excess of materialism. Liang Ch'i-ch'ao, a pioneer reformer and early champion of modern Western thought, returned from a tour of Europe in 1919 to report on the physical and spiritual destruction of that continent. "The

[17] Wu Chih-hui. "I-ke hsin hsin-yang ti yü-chou-kuan chi jen-sheng-kuan" [The cosmology and philosophy of life of a new belief], *K'o-hsüeh yü jen-sheng-kuan* [Science and the philosophy of life], ed. Ch'en Tu-hsiu and Hu Shih (Shanghai: Ya-tung, 1923), reprinted as *K'o-hsueh yü jen-sheng-huan chih lun-chan* [Debate on science and the philosophy of life], ed. by Wang Meng-tsou (Hong Kong: Chinese University of Hong Kong, 1973), 523. This source is cited hereafter as *KHYJSK*.

[18] T'ang Yüeh, "I-ke ch'ih-jen ti shuo-meng" [Gibberish of an idiot], *KHYJSK*, 391.

Europeans had a grand dream of the omnipotence of science," Liang told his countrymen, "but now they are proclaiming it bankrupt. This is a major turning point in contemporary thought."[19]

The debate between these groups was joined, in 1923, on the question, Can science govern a philosophy of life? Was science alone sufficient to answer all the questions one might ask, or at least all the questions for which one could expect to find answers? Or was there another source of wisdom? Scholars and pundits in Peking squared off for a fight over China's soul. The first salvo was fired by the idealists, who challenged the monopoly of science by positing a dualistic universe, with a physical and therefore scientific realm on one side, and a human, spirtual realm on the other. Tsinghua University philosopher Chang Chün-mai told his students that science could not govern a philosophy of life, because the former was objective, logical, analytical, and relied on cause and effect and the uniformity of objects, whereas the latter was subjective, intuitive, synthetic, open to free will and to the uniqueness of personality. Because of these differences, Chang concluded, "no matter how science develops, it cannot solve the problems of the philosophy of life, which depend entirely on man himself and nothing else."[20] On the opposite side, defenders of science held fast to the faith in a unitary order. Their purpose was precisely to eliminate subjectivity from all fields of thought and substitute a kind of truth that could be recognized and accepted by all. The key was to establish induction as the sole gateway to knowledge. "The omnipotence of science," explained Ting Wen-chiang, "lies not in its subject matter, but in its methods."[21]

This debate ground on through the spring and summer of 1923, generating a thick volume of essays before exhausting the imagination of the speakers and the patience of their audience. Most historians agree that the proponents of science won the argument, in the sense that their views enjoyed the broadest and deepest support among the small but significant segment of Chinese who were listening. This victory was possible because the overwhelming majority of urban, educated Chinese wanted to move the nation forward, and although they differed on many things, they shared an interest in crushing the revival of traditionalism in all its forms.

On the morning after their victory, however, the inevitable split among the "scientists" surfaced. The reason for this split lay in a contradiction,

[19] Liang Ch'i-ch'ao, "Ou-yu chung chih i-pan kuan-ch'a chi i-pan kan-hsiang" [General observations and impressions on a European journey], *Liang Jen-kung wen-chi* [Collected works of Liang Jen-kung (Liang Ch'i-ch'ao)] (Hong Kong: San-ta ch'u-pan-she, n.d.), 23.

[20] Chang Chün-mai, "Jen-sheng-kuan" [The philosophy of life], *KHYJSK*, 9.

[21] Ting Wen-chiang, "Hsüan-hsüeh yü k'o-hsüeh – ta Chang Chün-mai" [Metaphysics and science – a reply to Chang Chün-mai], *KHYJSK*, 260.

inherent in the ideology of scientism since the time of Yen Fu, between the poles of evolutionary cosmology and empirical method. The cosmology was an object of faith, not fact. It was beyond the reach of simple proofs – or disproofs – and those who embraced it had no reason to let go. The methodology, on the other hand, was not a vision of the whole, but a means of investigating the parts. True empiricists denied sweeping, a priori conclusions and agreed to accept whatever individual bits of data told them. Scientism offered a choice between embracing the total system, an all-encompassing natural evolutionary cosmology, or insisting on a test of its elements, one by one. On close examination, it could not do both.

In the second round of the debate, which continued at least until the mid-1930s, the cosmologists won out over the empiricists. After 1923, Chinese Marxists took over the mantle of evolutionary cosmology and buttressed it with a dynamic materialist rhetoric that in the end carried the day. Two explanations have been offered for this outcome. Charlotte Furth argues from a cultural and historical perspective that Chinese have always preferred some vision that locates all forms of truth in a single system of understanding. In this view, Chinese Marxists fell heir to the all-encompassing Confucian-Taoist metaphysic, which had a special hold on the Chinese mind. Danny Kwok, on the other hand, offers a political interpretation. Cosmology bested empiricism, Kwok explains, because the battle was waged in a public forum, where patience was limited, emotions strong, and simple, absolute, resolute arguments had the greatest appeal.[22] Less persuasive, or perhaps with too little time to demonstrate their case, Chinese empiricists, disciples of the American pragmatist John Dewey, rejected all-embracing "isms" in favor of the notion that society's problems should be understood and overcome, one at a time, through the application of patience and reason. As the world crumbled around them, Hu Shih and other champions of this liberal creed saw their audience dwindle and their influence recede. The fast, unsure pace of this age cried out for some new source of certainty; most Chinese preferred a version of science that offered not questions, but answers.

Scholarly organization: the historical legacy

The second arena where China's early scientists made their mark was in erecting organizations to house, give identity to, and promote the interests of science. This was, for China, a novel and therefore difficult

---

[22] Charlotte Furth, *Ting Wen-chiang*, 133. D. W. Y. Kwok, *Scientism and Chinese Thought*, 184–5.

task. Organizations that provided a focus or enclave for men with similar backgrounds, functions, or purposes were few and weak in traditional China. The imperial order welcomed participation by families and individuals through authorized channels, the examinations and the bureaucracy, or through informal arrangements that performed the functions of local government. But voluntary association along political or professional lines was discouraged, and in extreme cases prohibited. Scholarly societies had become so powerful by the end of the Ming dynasty that the new Manchu rulers outlawed academies and scholarly associations of all types. During most of the Ch'ing, the organization of scholars in any form not authorized by the state played little part in the Chinese scheme of things.[23]

By the late nineteenth century, as the power of the dynasty declined, autonomous forces began to separate from the linked state-and-society, and new forms of organization began to emerge. The debacle of 1895 furthered the disintegration of the established order and the rise of new social groups. After a hiatus of more than two centuries, the crisis atmosphere of the late 1890s gave scholars the chance to come together for common association and concerted action. The result was the creation of "study societies" [hsüeh-hui], formed by like-minded men to investigate and spread ideas on how to save China. The first of these, the Ch'iang hsüeh-hui [Society for the Study of (Self)-strengthening], was founded in Peking by K'ang Yu-wei in 1895, after which the concept spread rapidly. During the next three years, scores of hsüeh-hui were established throughout China.[24]

In some ways, the hsüeh-hui were a novel form of organization. They had charters, membership lists, officers, and other formal bureaucratic paraphernalia. Members included people with and without official degrees, in and out of office, traditional scholars and those finding their way in modern learning. Rules were approved and officers chosen, ostensibly at least, by democratic means. Activities included translation

[23] Traditional scholarly organization: William Atwell, "From Education to Politics: The Fu She," The Unfolding of Neo-Confucianism, ed. William T. deBary (New York: Columbia University Press, 1975); John Meskill, "Academies and Politics in the Ming Dynasty," Chinese Government in Ming Times: Seven Studies, ed. Charles O. Hucker (New York: Columbia University Press, 1969); and Hsieh Kuo-chen, Ming-Ch'ing chih chi tang-she yün-tung k'ao [Examination of the movement of associations during the Ming-Ch'ing period] (Shanghai, 1934, reprinted in Taipei, 1967), 251–4. Manchu ban on scholarly association: Benjamin A. Elman, From Philosophy to Philology: Intellectual and Social Aspects of Change in Late Imperial China (Cambridge: Harvard University Press, 1984), 114.

[24] Two recent studies that treat the changes in social organization during the late Ch'ing are: Joseph Fewsmith, Party, State, and Local Elites in Republican China (Honolulu: University of Hawaii Press, 1985), 16–25; and Mary Rankin, Elite Activism and Political Transformation in China, 1–20. Ch'iang-hsüeh-hui: Richard Howard, "K'ang Yu-wei," 415ff.

of books and periodicals, publication of journals, establishment of museums, libraries, and study centers, and regular meetings for the discussion and dissemination of new ideas. Several societies were devoted to science and technology, and a few issued specialized journals: The *Nung-hsüeh pao* [Journal of Agriculture], published in Shanghai from 1897 to 1906, introduced information on agriculture, forestry, and animal husbandry; the *Hsin-hsüeh pao* [New Study Journal], a monthly first appearing in 1897, covered math, medicine, and natural history; and the *Ko-chih hsin-pao* [Journal of Science], published in Shanghai in 1898, focused on math, physics, and chemistry. "There are limitations to the view of one man," K'ang Yu-wei explained,

but there are no bounds to the knowledge of a large body of people. Thus for study, we should seek out friends with whom to investigate matters by discussion. Every ten days hold a meeting to discuss and argue back and forth.... The advantages to be gained are inexhaustible.[25]

Yet the *hsüeh-hui* were not true "scholarly" societies, not at all devoted to science, and in the end not able to hold their own against governmental power. Their purpose was political, to reform Chinese state and society by influencing the government, either directly through dialogue between scholars and officials or indirectly by shaping the views of the larger polity. In this regard, they were more akin to lobby organizations or protopolitical parties than disinterested professional groupings. Membership was not limited to people with special knowledge, training, or skills, but open to everyone interested in or committed to a particular cause. Those societies that paid special attention to science disseminated information and supported the adoption of new techniques, but none maintained a program of research. In any case, the *hsüeh-hui* were shortlived: Identified with the reform group, they were forced, after the return of the conservatives in 1898, to disband or carry on their activities under some other guise. In the short run, at least, the government, which had authorized the formation of these new groups, was able to undo them.[26]

[25] *Hsüeh-hui* and publications: P'an Chün-hsiang, "Wu-hsü shih-ch'i ti wo-kuo tzu-jan k'o-hsüeh hsüeh-hui" [Scientific societies in our country at the time of the 1898 reforms], *CKKCSL* 4, no. 1 (1983):28–30; T'ang Chih-chün, *Wu-hsü pien-fa shih lun-ts'ung* [Collected essays on the history of the 1898 reforms] (Wuhan: Hu-pei jen-min ch'u-pan-she, 1957), 243–55. Quotation: K'ang Yu-wei, "Kuei-hsüeh ta-wen," 412–13, in Richard Howard, "K'ang Yu-wei," 426. The *hsüeh-hui* were succeeded, following the Boxer Rebellion, by a second generation of societies, such as the Ya-ch'üan hsüeh-kuan, publisher of the *Ya-ch'üan tsa-chih*, the first science journal published by the Chinese alone. See: Chang Tzu-kao and Yang Ken, "Chieh-shao yu-kuan Chung-kuo chin-tai hua-hsüeh-shih ti i-hsiang ts'an-k'ao tsu-liao – 'Ya-ch'üan tsa-chih,'" 55–9.
[26] Wang Yü-sheng, ed., *Chung-kuo k'o-hsüeh chi-shu shih-kao* [Draft history of Chinese science and technology], (Peking: K'o-hsüeh ch'u-pan-she, 1982), 299–300.

Ambivalence on the part of both government and society toward new forms of social organization continued. Particularly instructive in this regard is the formation of professional associations, during the late Ch'ing and early Republican periods. By this time, the process of modernization had thrown up a new strata of professional educators, lawyers, engineers, journalists, businessmen, and bankers. Anxious to coopt these groups, get them to accept responsibility for self-regulation, and assign them quasi-governmental functions, late-Ch'ing officials promulgated laws that required the formation of *fa-t'uan* – literally, "associations established by law" – in each profession. The creation of chambers of commerce in 1903 began a process that led to the establishment of similar organs for education (1906), agriculture (1907), law (1912), and banking (1915). Just as the imperial government recruited and thereby controlled the traditional gentry, so Chinese governments of the early twentieth century would recruit and control the educated classes of their time.[27]

By the 1910s, the question of the structure of China's moden polity had been raised but not answered. During the preceding decades, state and society had been pulled further and further apart. 'Chinese society had grown more complex and resistant to government control. Some progressive spokesmen argued that the traditional unity of state-and-society should give way to diversity, the creation of spheres of autonomy where like-minded citizens could assert their own interests, that the new China should be assembled from the building blocks of local authority and voluntary association. At the same time, the state had its means for keeping hold: the persistence of the Confucian ideal that the educated man should serve society through government, and where that failed, raw power. Whatever their reasons, teachers, bankers, lawyers, and other professionals had joined authorized *fa-t'uan* and agreed to work through approved channels. The questions of how scholars should assemble, for what purpose, and in what way this might affect the balance between state and society were all unsettled when the first generation of scientists came of age.

China's first scientific organizations

The chief contribution of the first-generation scientists in the field of organization was the Science Society of China (SSC), which helped

---

27 *Fa-t'uan*: Andrew J. Nathan, *Peking Politics, 1918–1923: Factionalism and the Failure of Constitutionalism* (Berkeley, University of California Press, 1985), 13–15; R. Keith Schoppa, *Chinese Elites and Political Change: Zhejiang Province in the Early Twentieth Century* (Cambridge: Harvard University Press, 1982), 34–39.

Leading "first-generation" scientists, Board of Directors, Science Society of China. Front row, left to right: Chao Yüan-jen, Chou Jen. Back row: Ping Chih, Jen Hung-chün, Hu Ming-fu. (K'o-hsüeh, 1915.)

promote the study of science and reshaped Chinese society in dramatic and unsettling ways. The SSC was founded in 1914, in Ithaca, New York, by Chinese students at Cornell and other American universities, several of whom – Hu Shih, Chao Yüan-jen, Jen Hung-chün, Yang Ch'üan, and Hu Ming-fu – played important roles in China's subsequent history. The initial purpose of the society was to propagate among Chinese at home and abroad knowledge of science and its potential benefits to industry, agriculture, and other fields. At first this was to be done through publication of the journal *K'o-hsüeh* [Science]. Soon, however, the founders recognized that writing alone would not achieve their purposes. Under a new charter adopted in 1915, two changes were introduced to make the SSC a proper "scholarly society": Voting membership was limited to those who "do scientific research or engage in scientific professions," and the program was expanded to cover a wider range of activities, including the conduct of scientific research.[28]

It would be interesting to know what model the founders had in mind for their new organization, but the sources offer little guidance. Jen Hung-chün, the society's first president, ruled out the influence of Western examples, which first came to Jen's attention in 1916, after they had already "built the cart behind closed doors" (that is, without blueprints). Although he fails to mention any positive models he and others may have drawn on, Jen is explicit about the negative example they set out to repudiate, and this tells us as much as anything about their intentions. In describing the society's early history, Jen recalled the determination to reject practices characteristic of China's old "study societies" [hsüeh-hui] and to create in their place a "scholarly society" [hsüeh-she] of the modern type. In terms of subject matter, the old "study societies discussed only moral and ethical principles from China's ancient books and classics," whereas "the modern scholarly society discusses experimental science and its applications." In structure, the study societies were built around "a single master, a man of virtue and talent, who, because of his great learning and lofty reputation, attracted swarms of students," whereas the "modern scholarly society is formed by the mutual assent of specialists, similar in learning and knowledge, who want to improve themselves through discussion." Finally, Jen explained, these changes in organizational form were made necessary by the greater complexity of modern science – "as the boundaries of science grow wider and deeper and its disciplines become more and more finely divided, a single person, no matter how brilliant, can no longer embrace all knowledge" – and by the high cost of scientific research, so that "if we want knowledge to advance and not be limited by material [shortages], we must have an organization that can supply equipment and materials to the research scientist."[29]

It was generally agreed that the purpose of the society was to spread knowledge of science among the Chinese public, but the founders differed on whether or not its activities should include the conduct of

---

28 Knowledge of the early history of the SSC comes primarily from the writings of the society's first president, Jen Hung-chün: "K'o-hsüeh-she kai-tsu shih-mo" [Complete account of the reorganization of the Science Society], KH 2, no. 1 (1916):127–32; "Wai-kuo k'o-hsüeh-she chi pen-she chih li-shih" [The history of foreign science societies and our society], KH 3, no. 1 (1917):2–18; "Chung-kuo k'o-hsüeh-she chih kuo-ch'ü chi chiang-lai" [The past and future of the Science Society of China], KH 8, no. 1 (1923):1–9; "Tao Hu Ming-fu" [Eulogy to Hu Ming-fu], KH 13, no. 6 (1928):822–6; and "Chung-kuo k'o-hsüeh-she she-shih chien-shu" [Brief account of the history of the Science Society of China], CKKCSL 4, no. 1 (1983):2–13. For an account in English, see: Science Society of China, ed., The Science Society of China: Its History, Organization, and Activities, (Shanghai: The Science Press, 1931).
29 Jen Hung-chün, "Wai-kuo k'o-hsüeh-she chi pen-she chih li-shih," 1–3, 17.

scientific research. Some argued that research was peripheral to education and should, in any case, be performed by the universities. But Jen replied that China's universities were not yet equal to the charge and that the enlightenment of China could not be achieved by reading and writing alone. Jen's advice recalls the limits of China's earlier experience with translation and the failure of nineteenth-century Chinese scientists to engage nature directly. "The advancement of science cannot be achieved by writing a few essays or speaking a few empty words," Jen explained. "It is something that can be achieved only by hard work in the field or laboratory."[30] It is to the credit of Jen Hung-chün and those who shared his vision that the society established, in 1922, the Biological Research Institute, which was the first private center of scientific research created by the Chinese themselves and played a leading role in the study of China's flora and fauna from that time forward.

In 1918, as many of the founding members completed their studies and returned home, the SSC moved its operations to China. Membership grew from 35 in 1914, to 363 in 1918, to more than 1,000 in 1930, to over 1,500 in 1937, the last year before the war, making it the largest and most comprehensive scientific organization in China. The society's most important activity was publication in Chinese of the journal *K'o-hsüeh*, a monthly that appeared with few interruptions from 1915 to 1949 and occupied a position in China similar to *Nature* in Britain or *Science* in the United States. During the early years, from 1915 to 1920, this journal carried articles addressed to a lay audience, explaining science and its practical applications. Later, as this function was taken over by other popular publications, *K'o-hsüeh* shifted to technical subjects, including reports on the research of Chinese scholars. Through this and other publications devoted exclusively to research, annual meetings, lectures, exhibitions, prizes, and branch societies, the SSC helped to propagate and legitimize the study of science among Chinese at large and to bind together and raise the spirits of the members themselves.[31]

China, unlike Britain, France, or Russia, but like America at a similar stage of development, was a large sprawling country that had no single cultural, economic, and political capital. Returned students from different countries lacked cohesion, a common language, and purpose. Several attempts to expand their parochial groupings established abroad into comprehensive, national organizations foundered. Whereas most men and women trained in the sciences congregated in a few major cities, a substantial minority – 40 percent of all SSC members in 1923 – ended up in provinces with fewer than ten members. Wherever they were located, almost all worked in small, primitive institutions, where

---

[30] Ibid., 18.
[31] Membership: *The Science Society of China*, 5–7.

teaching and administrative responsibilities were heavy and the opportunity to do research limited or nonexistent. Under these circumstances, there was a danger that the sense of participation in a larger profession and cultivation of the scientific ethos would be perverted or lost. The most important function of the society was to maintain contact among China's scientists and keep their faith alive.[32]

There was during this period another type of scientific society, whose origin and purpose contrast with those of the SSC and which constitute therefore a second line of organizational development. These were the professional associations in the various disciplines that were created when a critical mass of scholars was assembled, invariably around or at the initiative of some government agency. The first of these, the Geological Society of China, was established in 1922 by Dr. H. T. Chang, chief geologist at the National Geological Survey, and his colleagues in and around Peking. Two years later, the Chinese Meteorological Society and the Société Astronomique de Chine were formed in similar fashion by staff of the government observatories and associated faculty from government universities.[33]

Organization of the chemical profession also benefited from a kindred government agency, although private industry played a more important role in this field, and the pure scholars established their identity somewhat later. The Bureau of Industrial Research, set up in Peking in 1915 under the Ministry of Agriculture and Commerce, helped promote industry by analyzing and developing native products, including wines, hides, dyes, and cosmetics. North China, already blessed with traditional chemical industries, saw the rise of a modern salt refinery and associated soda factory during and after World War I. Encouraged by these developments, in the spring of 1922 two Peking University professors convened a meeting of chemists and chemical technicians in the capital to form the China Society of Chemical Industry. The purpose of this society was to promote development of this field by establishing closer relations

[32] America: John Greene, "Science, Learning and Utility: Patterns of Organization in the Early American Republic," *The Pursuit of Knowledge in the Early American Republic*, ed. Alexandra Oleson and Sanborn C. Brown (Baltimore: Johns Hopkins University Press, 1976). SSC members in China: Peter Buck, *American Science and Modern China*, 186–7. Forty percent: Jen Hung-chün, "Chung-kuo k'o-hsüeh-she chih kuo-ch'ü chi chiang-lai," 5.

[33] Geological Society: Yang Tsun-yi, "Development of Geology in China since 1911," *China Institute Bulletin* 3, no. 5 (February 1939):131–7. Meteorological society: Chiang Ping-jan, "Erh-shih-nien lai Chung-kuo ch'i-hsiang shih-yeh kai-k'uang" [Chinese meteorology during the past twenty years], *KH* 20, no. 8 (1936):623–42. Société Astronomique: Peter Buck, *American Science and Modern China*, 210. The earliest Chinese professional society in a field related to science was the National Medical Association [*Chung-kuo i-hsüeh-hui*], publisher of the *National Medical Journal of China* [*Chung-kuo i-hsüeh tsa-chih*], both of which were founded in 1915. See: K. Chimin Wong and Wu Lien-teh, *History of Chinese Medicine*, 604–5.

between academic and industrial chemists. Members were drawn from throughout the country, and the society was expanded in 1924 by the addition of branches in Shanghai and Tientsin, the latter headed by Fan Hsü-tung, founder of the Chiu-ta Refined Salt and Yungli Soda companies. The society published a journal, *Chemical Industry* [*Chung-hua hua-hsüeh kung-yeh-hui hui-chih*, 1923–6; *Hua-hsüeh kung-yeh*, 1929–41], in Chinese for those interested in chemical problems of an applied nature and, beginning in 1946, *Chemical World* [*Hua-hsüeh shih-chieh*], a popular magazine for students and laymen. In chemistry, as in most other disciplines, formation of an association of pure scholars had to await the next generation, which provided a larger number of practicing scientists.[34]

Conclusion

Members of the first generation of scientists were devoted to science and to China, and they had reason to hope that their work would contribute to both. By raising up the ideals of "science," they hoped to contribute to a more effective culture that would replace bankrupt Confucianism as the standard for the new China. By designing and building new forms of organization, they hoped to give a shape to Chinese society that would enable men (and eventually women) to serve the nation by performing modern technical functions and professional roles. In both respects, they achieved some notable successes, and at the same time unwittingly triggered conflicts that have marked the development of science in China ever since.

The short-term result of introducing science into the debate over culture was a victory for the purveyors of "scientism." But members of the first generation who urged the acceptance of scientific ideals did so because they believed science would provide an objective, universal basis for the reintegration of Chinese society and culture – and in this respect, they failed. Rather than knit China together, science and culture came to represent the poles of an intramural conflict that weakened the nation's unity and resolve. During the 1930s and 1940s, first the conservative wing of the Kuomintang, then the radical Communists struck against the assumptions that had been so easily accepted during the May Fourth era.

The creation of new scientific organizations, the Science Society of China and the various associations representing individual disciplines,

---

[34] BIR: Li Ch'iao-p'ing, *Chung-kuo hua-hsüeh shih*, 266, 293. CSCI: Ch'en Hsin-wen, "Chung-kuo hua-kung hsüeh-hui" [China Society of Chemical Industry], *CKKCSL* 3, no. 3 (1982):57–62.

introduced a second feature into the Chinese social order – the professional ideal. The initiative came in all cases from outside the government, and in some from Chinese outside of China. The intention was to address the Chinese people directly, bypassing government. And the purpose was to foster an independent enterprise that would help remake China, irrespective of who ruled the country. Underlying these efforts was a conception, unstated and perhaps unconscious, of the scientific role: The scientist's purpose was to pursue knowledge, apply it in useful ways, and communicate it freely to others. His commitment was to an autonomous activity, separate from politics, yet serving in a disinterested way the public good. This ideal struck at the roots of Chinese tradition, which favored government service as the only, or at least the most appropriate, means of fulfilling the responsibility of the educated man. Instead, the new society would be built up of free agents, each making a special contribution, while the system was held together by the unseen power of its own logic.

Not everyone accepted this vision, however, and when their time came the rulers of China rejected it. In part, the problem lay with the sheer chaos of China itself: There was neither time nor opportunity for isolated decisions or technical solutions to take hold. The architecture had to be established before the bricks could be laid. More to the point, powerful Chinese on both the left and the right mistrusted the very notion of a society built up of autonomous centers of specialization. To separate state and society and to divide society into islands of expertise would, in their view, hasten the disintegration of China and undermine the ability of central authorities to restore order. The professional ideal and scholars who supported it were attacked by the right wing of the Kuomintang in the 1930s, and the radical Maoists in the 1940s. In the end, neither party could accept the idea of professional autonomy, scientific or otherwise.

Still, the model of science – scientism – influenced a wide range of activities in early twentieth-century China. It reinforced trends in literature toward a more simplified, useful language and in history, philology, and the humanities toward greater skepticism with regard to established truths and more critical and empirical approaches to scholarship. It hastened the introduction of statistical methods and planning in government and colored politics with the notion that scientific approaches were more modern and therefore better than the alternatives. In all areas, it helped undermine traditional values and the authority of the ancients and hastened the development of a more rational, secular culture. There seems, in short, much in the intellectual and cultural environment of early twentieth-century China to explain the growing popularity of scientific ideas.

But what about science – the hard-nosed investigation of the material world – itself? Many members of the first generation would have agreed with Jen Hung-chün that the success of their enterprise depended on the degree to which this enterprise moved from the pulpit to the laboratory. Did the popularity of scientism in the New Culture of the 1920s translate to a greater practice of science in education, research, and industrial application? The following chapters attempt to answer this question by taking up each of these areas in turn.

5

Learning about science

"The study of science [*ko-chih*] and manufacturing," wrote Li Tuan-fen, president of the Board of Ceremonies, in an 1896 memorial that led to the establishment of China's first national university,

cannot be precise without the use of experimentation and measurement, nor correct without foreign travel and investigation. At present, [however], our various offices have not provided charts and instruments, nor have they sent [scholars] abroad for study, but continue to sift through old stacks of papers and end up with only empty talk and no useful results.[1]

The inertia of the old order, which so frustrated Li Tuan-fen and others like him, came under harsh attack following the defeat of China in 1895. Governments of the interregnum, beginning with the Ch'ing, did their best to revive the spirit of learning, while introducing modern subjects, including science and math. On the whole, however, their efforts fell short of the mark. At the very time Peking tried to broaden the range of its activities, the Chinese state was losing its ability to govern. In the absence of native leadership, foreigners – missionaries, scholars, and philanthropists – built the best colleges and universities in China and did the best job of teaching science to the Chinese. Yet the results in both Chinese- and foreign-run schools were limited by traditional values and habits of mind, which proved remarkably resistant to the needs of science. As late as the 1920s, it still proved difficult to translate the broad spirit of "scientism" into hard learning about "science."

Modern primary and secondary education

Nothing has distinguished modern China more clearly from its traditional predecessor than the nature, scope, and purpose of their approaches to learning. To begin, education has become a public affair. In imperial China, the state administered the examinations, but left education in

---

[1] Statement by Li Tuan-fen, quoted in Tai An-pang, "Chin-tai Chung-kuo hua-hsüeh chiao-yü chih chin-chan" [Development of modern Chinese chemical education], *Hsü Shou ho Chung-kuo chin-tai hua-hsüeh shih*, 241. Biography of Li: Arthur Hummel, *Eminent Chinese of the Ch'ing Period*, 63, 674.

private hands. In modern China, the government, as much as possible, has run the schools. Second, the scope of education has greatly expanded. Whereas the empire recruited only a few "men of talent," after 1900 the swelling tide of nationalism prompted demands to educate the entire citizenry. Enrollments in modern government schools expanded from 13 hundred in 1903, to 9 million in 1929 and 20 million in 1936 (Appendix, Table A.2). By 1930, 21 percent of Chinese children between the ages of six and nine were in modern schools of some type. Third, the curriculum of the schools changed, with the classics giving way to history, geography, foreign languages, math, and science.[2]

Despite efforts to maintain uniform standards, education in early twentieth-century China varied enormously with time, place, and level of instruction. The best training was at the higher levels, in the cities, particularly those on the coast, and came after the establishment of a unified government in 1928. Most primary schooling was left to the villages, underfinanced, and dominated by old literati who knew little of the new world. Colleges and universities, concentrated in Peking, Nanking, Shanghai, and a few other cities, were favored in every respect. The children of comfortable, educated, urban families enjoyed the best opportunities; boys and girls in remote villages had virtually none.

Mission schools remained a factor throughout this period, although their importance shrank as public education expanded. In 1922, the Protestants operated 7 thousand primary schools with 180 thousand students and 300 middle schools with 15 thousand students, or 3 and 13 percent, respectively, of the national totals. At the same time, there were 3 thousand Catholic schools with 144 thousand students. Perhaps as many as half of all missionary schools closed during the period 1925–30, owing to the rise of violence and antiforeign sentiment. Thereafter, the situation settled and the schools reopened, albeit on a reduced scale, until the Communist takeover in 1949. Mission schools were generally better financed and more advantageously located than those run by the government, but by all accounts the nature and quality of instruction, particularly in the sciences, were the same. With few exceptions, the following comments pertain to public and private schools alike.[3]

---

[2] 1903 enrollment: Kuo Ping Wen, *The Chinese System of Public Education* (New York: Teachers College, Columbia University, 1915), 108. 1936 enrollment: *China Year Book* (1937–43), 395. 21 percent: League of Nations' Mission of Educational Experts, *The Reorganization of Education in China* (Paris: League of Nations' Institute of Intellectual Co-operation, 1932), 78.

[3] Mission school enrollments: *Christian Education in China: A Study Made by an Educational Commission Representing the Mission Boards and Societies Conducting Work in China* (New York: Committee of Reference and Counsel of the Foreign Missions Conference of North America, 1922), 415–16; and Paul Monroe, "A Report on Education in China," *The Institute of International Education Bulletin* 4 (20 October 1922):27–8. Wang Tsu-t'ao, "Chung-kuo chin-tai ti hua-hsüeh chiao-yü" [Modern

Primary schools, which were organized and paid for at the county level or below, reflected the uneven distribution of resources and low cultural climate of which they were a part. Throughout the early twentieth century, primary schools accounted for more than 95 percent of all students in China. In 1922, however, the ratio of expenditures per student for lower primary to universities and other high schools was 1:100, and by 1931 the gap had widened to 1:200, in contrast to the ratios of 1:8 to 1:10 common in European countries at this time. The standard curricula issued by the Ministry of Education for use in primary schools focused on Chinese language and literature, history, and geography. Arithmetic and nature study were assigned 10 percent of class time each. All instruction was in Chinese, and after 1912 textbooks were written in the vernacular "national language" [*kuo-yü*]. Foreign languages were not offered at the elementary level.[4]

By far the most serious problem with primary education, in science and other fields, was the poor quality of instruction. At the beginning of the century, there were few qualified teachers to fill the thousands of posts created by the explosion of modern education. With the passage of time, the new normal schools (of secondary rank) graduated men and women for this task. The standard curriculum of the three-year normal program included mathematics in all years and one year each of biology, chemistry, and physics. The normal schools did not teach foreign languages, which had no place in the primary curriculum.[5]

Still, few Chinese, even those trained in pedagogy, had any enthusiasm for teaching – a situation made worse by the poor treatment accorded members of this profession. A 1922 survey of primary school teachers, most from the wealthy provinces of Kiangsu and Chekiang, showed that

Chinese chemical education], *CKKCSL* 5 (1984):93–4, reports similar percentages for mission enrollments in 1917. *Christian Education in China* provides detailed descriptions of: teaching methods, 79; curricula, 89–91; teachers, 138–41; and language of instruction, 342. For further details on science instruction in mission primary and secondary schools, see: Edgar Knight, "Christian Education," *Laymen's Foreign Missions Inquiry Fact-Finders' Reports: China*, vol. 5, pt. 2, ed. Orville A. Petty (New York: Harper & Bros., 1933), 409–18; Ida Belle Lewis, "A Study of Primary Schools," ibid., 618–52; William Adolph, "Science in the Middle School," *Educational Review* (Shanghai: China Christian Educational Assoc.) 10, no. 2 (April 1918):120–1; and Andrew Allison, "Middle School Science," *Educational Review* 12, no. 2 (April 1920), 135–8.

4 Ratio of expenditures per student, 1922: *China Year Book* (1921–2), 556; *China Year Book* (1924–5), 251. Charles Edmunds, "Modern Education in China," Department of the Interior, Bureau of Education, *Bulletin* 44 (Washington: GPO, 1919), 66–70, reports a ratio of 1:100, for 1918. Ratio, 1931: League of Nations' Mission, *Reorganization of Education in China*, 50–1. European ratios: ibid., 51. Standard primary curriculum for 1912: Kuo Ping Wen, *Chinese System of Public Education*, 173. 1922: *China Year Book* (1929–30), 522. 1929: *China Year Book* (1933), 530.

5 Teachers: Kuo Ping Wen, *Chinese System of Public Education*, 151. Normal school curricula: ibid., 175; Wang Shih-chieh, "Education," *The Chinese Yearbook, 1937* (Shanghai: Commercial Press, 1937), 1047; *China Year Book* (1938), 314.

the average salary, even after several years experience, was only one-half the amount required to support a family of four. Teachers had no union, civil service, or other protection. Most people who made it through secondary school found teaching unattractive, and many looked elsewhere for jobs. By 1928, China had less than one-fourth the number of primary school teachers needed to meet the goal of providing four years of compulsory education – and the gap between supply and demand was widening.[6]

The lack of commitment from above, however, was nothing compared to the hostility with which modern schools were greeted by many parents of children slated to attend them. Chinese of the early twentieth century, particularly those living in the countryside, considered modern education strange, inappropriate, and expensive. "The people hate these schools and those who manage them," explained the *North China Herald* in 1910, "because of the heavy taxation and in some instances illicit taxation." In some rural areas, peasants continued to support schools of the old type, even though these schools were outlawed and the examinations that had provided the reward for classical learning were long defunct. Nor was this practice limited to the countryside: One survey conducted in 1923 showed that in Nanking and Canton there were more students in traditional schools than in schools of the modern type.[7]

Secondary education was better than primary in all respects, including the sciences, although many of the same problems remained as one moved up the educational ladder. Middle schools, which were established at the provincial level and usually located in cities, took students out of their homes and taught them new languages and new ideas. At the center of the secondary curriculum were Chinese language, history, and geography. Prior to 1922, foreign languages, primarily English, and mathematics were required in each year of middle school, along with three years of "nature study" and one year each of physics and chemistry. English, math, and general science were offered in the junior middle schools set up after 1922, and under the "progressive" policies of this

[6] 1922 survey: Cheng Tsung-hai, "Elementary Education in China," Chinese National Association for the Advancement of Education, *Bulletin* 2, no. 14 (1923):27. Teacher status: League of Nations' Mission, *Reorganization of Education in China*, 56. 1928 figures: *China Year Book* (1929–30), 526.

[7] *Herald*, 1910: *North China Herald*, 15 July 1910, p. 138, cited in James Cole, *Shaohsing: Competition and Cooperation in Nineteenth-Century China* (Tucson: University of Arizona Press, 1986), 144. Education in rural areas: Liao T'ai-ch'u "Rural Education in Transition: A Study of the Old-fashioned Chinese Schools (Szu-shu) in Shantung and Szechuan," *Yenching Journal of Social Studies* 4, no. 2 (February 1949):19–69. 1923 study: "Statistical Summaries of Chinese Education," Chinese National Association for the Advancement of Education, *Bulletin* 2, no. 16 (1923):ii. For other reports on the continued strength of traditional education, see: Philip Kuhn, "The Development of Local Government," *The Cambridge History of China*, 13:339.

period students could elect an even more intensive math and science track at the senior level. They could also opt out of science, and many did. One survey of sixteen middle schools in Kiangsu, educationally the leading province of China, found that there was in fact no uniform science curriculum. Normal schools of secondary rank offered less math and science and no foreign language. Vocational schools taught hands-on skills, usually unadorned by academic study.[8]

Most middle and normal schools had few facilities for the study of science. According to one report, the typical Chinese middle school science room of the 1920s "differs not at all from any of the other classrooms. Seldom does one see a convenience of any kind such as a Western science teacher would consider almost indispensable." The science rooms had no supply of gas, water, or electricity, no stock of specimens, reagents, or other materials. Visitors to these schools found only a few inferior Chinese or Japanese copies of Western scientific instruments, which were used improperly if at all, poorly maintained, and often worthless.[9]

Science teachers in middle and normal schools received the best training China had to offer, but most reportedly lacked a commitment to teaching and left the field at the first opportunity. A 1923 survey of 183 middle school science teachers found that almost all were graduates of higher institutions, who had trained for careers in science or engineering and entered teaching only because they could not find work in their chosen fields. Since they were paid by the hour, many left tasks for which they were not paid – such as demonstrating the use of scientific equipment! – to the janitors. Most teachers held multiple jobs. Many spent their time looking for employment elsewhere. The average length of service was about six years, and the turnover correspondingly high.[10]

Coming from primary schools and villages where scientific principles and habits of mind were unknown, exposed to few artifacts that bespoke an interest in nature, and taught by men and women with limited interest in their jobs, most Chinese middle school students of the 1910s and 1920s learned little about science. The large majority stayed within the bounds of literature and philosophy, a safe haven for those reared in a culture that cherished the humanities as the scholarly ideal. Students who

---

8 Curriculum, pre-1922: Kuo Ping Wen, The Chinese System of Public Education, 174. Post-1922: China Year Book (1929), 524–5. Survey: Wang Chin, "Ch'u-chi chung-hsüeh chih hun-ho tzu-jan k'o-hsüeh chiao-hsüeh wen-t'i" [The problem of teaching general science in lower middle schools], KH 13, no. 8 (1929):1094.
9 Typical science room: George Twiss, Science and Education in China (Shanghai: Commercial Press, 1925), 293. Visitors: ibid., 293–322; L. G. Morgan, The Teaching of Science to the Chinese (Hong Kong: Kelly and Walsh, 1933), 85–91. Laboratory facilities: Wu Ch'eng-lo, "Ch'üan-kuo k'o-hsüeh chiao-yü she-pei kai-yao" [Survey of science teaching facilities in the entire country], KH 9, no. 8 (1924):950–2, 976–7.
10 George Twiss, Science and Education in China, 149–64, 326–31.

made it to college were, as one critic found, ill-prepared for the rigors of science:

These [college] students are not only deficient in knowledge of the elementary scientific facts and theories, but they are also lacking in the habits of accuracy, the ideals of thoroughness and exactitude, and the appreciation of careful and methodical observation, experimenting, and thinking that are the very essence and foundation of scientific study. Furthermore, they lack this knowledge and these qualities to such an extent that *they are not even aware of their deficiencies*.[11]

Chinese higher education

Colleges and universities established in China during the first quarter of the twentieth century were favored in every respect over primary and secondary schools. In science, as in other fields, the system was top-heavy. Even so, erecting higher institutions of learning on so flimsy a base was a difficult task. By 1927, few observers, Chinese or foreign, were satisfied with the results.

The first institutions under Chinese management that aspired to the status of "university" on the model common in Europe and America were established around the turn of the century as part of the same reform movement that spawned the larger system of public education. The Imperial University of Peking, founded in 1898, escaped, along with other educational initiatives, the conservative reaction against the Hundred Days Reforms and opened its doors on the last day of the year. This school, which occupied the pinnacle of a planned nationwide system, was conceived as an expanded version of the T'ung Wen Kuan for the similar purpose of training degree-holders in modern subjects to prepare them for government service. W. A. P. Martin was the first dean of the Western faculty. Science courses (referred to as *ko-chih*) were taught by Dr. Robert Coltman, the only foreigner on the faculty with formal training in science, and several graduates of Mateer's school at Tengchow. According to one report, the new university boasted a modern plant, containing "room after room with the latest maps hanging from the walls, and where all kinds of instruments for experimental purposes in physics, geometry and chemistry were stored in glass cases and on shelves." This glory was short-lived, however, for the glass cases and everything in them were smashed by Boxer rebels who rampaged through the capital in the summer of 1900.[12]

[11] Ibid., 341.
[12] Imperial University: Renville C. Lund, "The Imperial University of Peking," (Ph.D. diss., University of Washington, 1956), 97, 107, 118−22, 133. Tengchow graduates: W. A. P. Martin, *The Awakening of China* (New York: Doubleday, 1907), 285. Ko-chih: Liu Kuang-ting, "Chung-kuo hua-hsüeh chiao-yü fa-chan chien-shih," 155.

When the university reopened in 1903, Martin and the other Western faculty were dismissed, as the Chinese turned for guidance in education, as in other fields, to Japan. During the next decade, Japanese professors taught most courses on science and technology, although apparently with little success. They were joined near the end of the decade by two Chinese. Wang Feng-tsao, a native of Kiangsu, *chin-shih*, graduate of the T'ung Wen Kuan, and assistant to Anatole Billequin and other foreigners in the translation of scientific textbooks, served as head of the science department from 1909 to 1911. He was followed by Sung Fa-hsiang from Fukien, who had studied at the Anglo-Chinese school in that province and in the United States (M.S., Ohio Wesleyan, 1906) and served as professor of chemistry from 1908 to 1912. Sung's commitment to this field was limited, however, for he later shifted to the Ministry of Finance and abandoned all interest in science.[13]

During its brief history, the Imperial University failed to venture beyond the narrow mandate to train minor officials for teaching and government service. The level of instruction was low, reflecting a lack of preparation on the part of students and teachers alike. From 1903 to 1911, about 1,000 men attended the university; 330 completed the program of study. Only one graduate of this period, Ping Chih (Ph.D., Cornell, 1918), the foremost Chinese zoologist of his generation, later gained prominence in the field of science.[14]

Despite its lofty status, the university at Peking was no match for Shansi Imperial University at Taiyuan. The latter was founded in 1901 with funds from the indemnity levied against Shansi for the massacre of Christians by the Boxers. Timothy Richard, a Baptist missionary, served as chancellor of the university and head of the Western College (there was also a Chinese College under Chinese direction) during its first decade. The faculty numbered six to eight foreigners, including E. T. Nystrom (B.S., Stockholm and Upsala), who taught applied chemistry and engineering. A three-year preparatory course covered Chinese, English, math, chemistry, and physics in each year, along with other Western subjects. The four-year university course was divided into four sections: law, mining, engineering, and science, the last of which was devoted primarily to chemistry. Instruction was in Chinese. Textbooks, including one on chemistry by Nystrom, were translated by members of the faculty and their Chinese assistants. The governor of Shansi selected

---

[13] University: Renville Lund, "Imperial University," 159, 189, 230, 313, 323. Japanese influence: Tai An-pang, "Chin-tai Chung-kuo hua-hsüeh chiao-yü chih chin-chan," 246. Wang Feng-tsao: Knight Biggerstaff, *Earliest Modern Government Schools*, 152. Sung Fa-hsiang: Ernest Dewitt Burton, "Journal and Record of Interviews and Observations, University of Chicago, Oriental Education Investigation, 1909" (Unpub. ms., Missionary Research Library, Union Theological Seminary, New York), 564.

[14] Renville Lund, "Imperial University," 236–46, 329.

the students, all of whom were minor degree-holders from the same province. In 1909, there were seventy-nine students in the college, sixteen of whom studied science and thirteen engineering. The Oriental Education Commission of the Univeristy of Chicago, which visited all major colleges in China in 1909, ranked Taiyuan at or near the top:

There are excellent chemical laboratories, both elementary and advanced, good tables for physical laboratory work, a fair beginning in engineering appliances, an electric plant, etc. . . . The work that is being done by the Western College seemed to us good; probably none better is being done in China.[15]

By most accounts, the center of Chinese academic life at the turn of the century was Tientsin, the site of several higher schools, including Peiyang University, successor to the *Pei-yang hsi-hsüeh*, one of the first modern government schools established after the debacle of 1895. The *hsi-hsüeh* offered two consecutive four-year programs in English, math, and in the upper grades chemistry, astronomy, mining, and physics (the last of which was referred to as *ko-wu*). According to the Chicago commission of 1909, aside from Taiyuan, which was in fact more a missionary than a Chinese institution, Peiyang University was the best of the lot. By 1917, Peiyang had a faculty of 13 Americans, 1 Englishman, and 10 Chinese, teaching 150 students in the preparatory course and another 150 in the departments of law, mining, and civil engineering. The laboratories were said to be well equipped, but not in good repair. In later years, the Peiyang engineering program enjoyed a favorable reputation, although one inspection of the facilities in 1933 revealed that the physics and chemistry labs were very poor and the teaching of basic sciences "could not be rated much higher than entirely useless."[16]

Chinese higher education improved after the establishment of the Republic in 1912, but on the whole the problems facing Chinese educators remained greater than the solutions available to them. In those areas where the government was most active, its influence was sometimes

---

[15] Western College: Timothy Richard, *Forty-five Years in China*, (New York: Frederick A. Stokes, 1916), 299–302; Ernest Dewitt Burton and Thomas Chrowder Chamberlin, "Report of the Oriental Educational Commission of the University of Chicago, Part VI, China, (December 1909)," (Unpub. ms., Missionary Research Library, Union Theological Seminary, New York), 594–600. Nystrom translation: Liu Kuang-ting, "Chung-kuo hua-hsüeh chiao-yu," 155. Quotation: Burton and Chamberlin, "Report," 599–600.

[16] Tientsin: Ernest Dewitt Burton, "Journal and Record," 492; Ernest Burton and Thomas Chamberlin, "Report," 623. Chicago Commission: ibid. *Hsi-hsüeh*: Liu Kuang-ting, "Chung-kuo hua-hsüeh chiao-yü," 155; and E-tu Zen Sun, "The Growth of the Academic Community, 1912–1949," *The Cambridge History of China*, 12:375. 1917 report: Ralph D. Whitmore, "Engineering Education in China," *The Tsing Hua Journal* 2, no. 5 (March 1917): 13. 1933 inspection: W. E. Tisdale, "Report of Visit to Scientific Institutions in China (September–December 1933)," RG 1, ser. 601, box 40, Rockefeller Archive Center Tarrytown, NY (hereafter abbreviated as RAC), pp. 18–19.

harmful. Beginning in 1912, public universities and middle schools were directed to teach theory, scientific and otherwise, arts, and humanities in line with traditional scholarly concerns. Lacking a mandate to study physical or natural phenomena, these institutions erected few laboratories and received little scientific apparatus. Technical schools, on the other hand, were given a great many machines and other devices, most of which had direct industrial applications, but they made no attempt to teach scientific theory or methods. As late as the 1920s, technical schools continued to enjoy the best equipment, while academically oriented middle schools, colleges, and universities held a monopoly on the teaching of basic science. This separation of theory and practice hindered education on both sides of the divide.[17]

More commonly, Chinese governments of the warlord era lacked the wherewithal to support and direct the development of higher education, which grew like Topsy – rapidly and without discernible logic or direction. The progressive spirit of the 1920s encouraged independence and initiative by students, teachers, and school administrators to create the kind of education each saw fit, leading to a proliferation of schools that called themselves colleges or universities, often without valid claim to the title. The number of "universities" in China jumped from seven in 1916, to forty-seven in 1925. Beginning from a handful of schools with no more than 1 thousand students at the beginning of the century, by 1933 there were 109 institutions of higher learning with an enrollment of nearly 47 thousand. Given the choice, most educators demonstrated little interest in the study of science, and in the case of those few who did, all the prerequisites – laboratories, trained teachers, well-prepared students, regular funds, and an atmosphere of political and psychological stability – were conspicuously lacking.[18]

China's first institution of higher learning established without direct foreign assistance and with a notable program of scientific education and research, was National Southeastern University (NSU). Southeastern was founded in 1915 as Nanking Higher Normal College, became a university in 1921, and took the name National Central University in 1928. It prospered while other schools foundered, because of the vision of its first president, Kuo Ping-wen, a graduate of Columbia Teachers College, and the favorable situation of Kiangsu Province, which provided protection and support during these difficult years. By 1922, NSU had over 700 students, chosen on the basis of highly selective entrance exams in Chinese, English, math, and science, and 86 full-time teachers, most

---

[17] Lab facilities: Wu Ch'eng-lo, "Ch'üan-kuo k'o-hsüeh chiao-yü she-pei kai-yao," 950–1, 976.
[18] Number of universities, enrollments: Wang Shih-chieh, "Education," 1030–3; *China Year Book* (1935), 282.

returned students from America and almost all native Chinese. (With few exceptions, no foreign professors taught in Chinese universities during the Republican period.) Following his tour in 1922, the American educator Paul Monroe judged this the best public university in China.[19]

The chemistry department at Southeastern was founded by one of the first notable Chinese teachers of science, Wang Chin. Wang was born in Fukien Province in 1888 and studied in the translation department of Peking Imperial University before going to the United States for further study. After graduating from Lehigh University in 1914 with a bachelor of science degree in chemistry, he returned to China and was appointed professor of chemistry and chairman of the department at the Nanking Higher Normal School. When Wang began teaching, no school under Chinese administration (save Tsinghua, which was in all ways a special case) had decent laboratory facilities or trained students in experimental techniques. A chemical historian who investigated the native industries, using both textual and experimental methods, Wang introduced his students to the theory and practice of chemistry. By 1922, Southeastern enrolled twenty-five students in fourteen chemistry courses, covering a range of academic specialties, plus introductions to industrial and agricultural applications. Texts and reference materials were in English, instruction in some combination of English and Chinese. Wang's was the first program to combine classroom lectures with direct hands-on experience. Many of his students and contemporaries regard him as the founder of the Chinese chemical profession.[20]

The most serious problems Southeastern faced in the early years were the inexperience of its faculty and the poverty of its physical plant. Laboratories for biology and physics were overcrowded, materials in short supply, and the general situation, in the view of one qualified observer, "impossible." The chemistry department was in better shape, with five labs and adequate equipment for students to perform most experiments described in the introductory texts. But the labs themselves were badly constructed, and there was no supply of electricity, gas, or running water. Stanley Wilson, an American chemist brought to China

[19] NSU. N. Gist Gee, "Southeastern University, Nanking," December 1922, CMB, ser. 2, box 83, RAC. Paul Monroe: Yoshi S. Kuno, *Educational Institutions in the Orient with Special Reference to Colleges and Universities in the United States*, part 2, *Chinese Educational Institutions* (Berkeley: University of California Press, 1928), Appendix A, i–ii.

[20] Wang Chin: Yang Kuo-liang and Cheng T'ang, "Wo-kuo hua-hsüeh-shih ho fen-hsi hua-hsüeh yen-chiu ti k'ai-t'a-che – Wang Chin chiao-shou" [Pioneer in our country's chemical history and analytical chemical research – Professor Wang Chin], *HHTP* 9 (1982):553–8. Enrollments and texts: N. Gist Gee, "Southeastern." Regarded by many: Tseng Chao-lun, "Erh-shih nien-lai Chung-kuo hua-hsüeh chih chin-chan" [Development of chemistry in China during the past twenty years], *KH* 19, no. 10 (October 1935):1517.

by the Rockefeller Foundation, took a dim view of NSU's chemistry program as he found it in 1921, reporting that "the students were getting the worst kind of habits in this laboratory.... I should prefer to take [them] before they had had these courses rather than afterwards." One year later, a kinder critic praised the program faintly: "While conditions are not ideal by any means, we feel that decided advance has been made recently in the department and that the men in charge are doing their best to make good with materials which they have." The library was weak in all sciences. Some periodicals began arriving in the early 1920s, but there were few reference books and no back files of the major journals.[21]

Among the public colleges and universities of early Republican China, the one that received the most preferential treatment was Tsinghua. In 1908, the U.S. government remitted to China $10.8 million of the unpaid balance (then totalling $24.4 million) of the indemnity owed by China for damages stemming from the Boxer Rebellion, on the condition that the money be used for education. Fellowships were created to train Chinese students in America, and Tsinghua College was set up in 1911 to prepare these students for their future studies. Throughout its history, Tsinghua bore the character and privilege of these origins. In the early years, over half the teachers were Americans. Later, most were Tsinghua graduates who had completed advanced study in the United States. Except for courses in Chinese language and culture, all instruction was in English and followed the American model. Generously endowed, Tsinghua's annual per student expenditure was equal to that of the three other leading Chinese universities combined. These advantages sheltered the college from the disruptions of the early Republican and warlord periods and freed students and faculty to concentrate on academic concerns.[22]

In 1920, Stanley Wilson gave the Tsinghua science program mixed reviews. Biology and physics were poorly equipped, and the faculty of these departments in disarray. Chemistry was doing better, in part because of the influence of the University of Wisconsin, which served as the model for this department. Laboratory facilities were in plentiful

[21] NSU facilities: N. Gist Gee, "Southeastern," 14–21. Stanley Wilson, "Science Work in Schools and Colleges of the Nanking-Shanghai Region, December 12, 1921," RG 4.2.B9, box 49, PUMC DR 48, RAC, pp. 10–11. Lab texts in use in 1922 were: Alexander Smith, *Experimental Inorganic Chemistry*; A. A. Noyes, *Qualitative Analysis*; and Talbot, *Quantitative Analysis*.

[22] Tsinghua history: Ch'ing-hua ta-hsüeh hsiao-shih pien-hsieh-tsu [Committee for Editing and Writing the History of Tsinghua University], *Ch'ing-hua ta-hsüeh hsiao-shih kao* [Draft history of Tsinghua University] (Peking: Chung-hua shu-chü, 1981), passim; Y. C. Wang, *Chinese Intellectuals and the West*, 111–14; *China Year Book* (1916), 422; *China Year Book* (1921–2), 558; *China Year Book* (1924–5), 230; *China Year Book* (1926–7), 419–20. Per student expenditures: Paul Monroe, "A Report on Education in China," 34; *Ch'ing-hua ta-hsüeh hsiao-shih kao*, 56.

supply, and the faculty boasted several excellent men, including Chang Tzu-kao (M.S., M.I.T., 1915), chairman of the chemistry department, C. A. Pierle (Ph.D., Wisconsin, 1919), and Pierle's assistant, Yang Kuang-pi (M.S., Wisconsin, 1917), whom Wilson described as "one of the best in China." The chemistry courses, in Wilson's view, were "the equivalent of first-class university courses in America."[23]

National Peking University, successor to the Imperial University, was the epicenter of China's cultural quake, the place where science as a model for the New Culture was most celebrated. It speaks volumes about modern Chinese education that Pei-ta, as it was known to contemporaries, was so weak in the study of science proper. At the outset, the university had a college of engineering, but this school was removed around 1918 to other jurisdiction, leaving colleges of law, literature, and science. For the next decade, only geology was well represented among the sciences, owing to the special attention paid this field by the Peking government and the presence of the American paleontologist A. W. Grabau. Physics was the best of the rest, but there was only one laboratory, and some students reportedly spent several years at the university without doing a single experiment. Chemistry was worse. There was no biology at all until 1925 – this, despite the fact that several Pei-ta professors led the campaign to popularize social Darwinism. The graduating class of 1922, which might be taken as representative, included only three to five students in chemistry, compared to fifty-seven in law. By the end of the warlord period, the campus was in constant turmoil, a site of regular clashes between student demonstrators and police, while classical scholars, trained for the imperial examinations, controlled much of the academic program, such as it was.[24]

Throughout the late Ch'ing and early Republican periods, science advanced slowly and haltingly in colleges and universities under Chinese control. In part this was because these institutions were themselves in chaos. The exceptions, Tsinghua and Southeastern, were insulated from external events, well financed, and staffed by American-trained scholars. More typical was Peking University, the eye of the political and intellectual storm, which did little to promote the study of science. One observer explained this difference by pointing out that the Peking faculty

[23] Stanley Wilson, "Science Work," 27–30. Chang Tzu-kao: Chou Hsin, Yang Ken, and Pai Kuang-mei, "Yu-hsiu ti chiao-yü-chia hua-hsüeh-shih-chia – Chang Tzu-kao chiao-shou" [Outstanding educator and historian of chemistry – Professor Chang Tzu-kao], HHTP 10 (1980):622–6.

[24] Removal of the engineering college: Ralph Whitmore, "Engineering Education in China," 13. Physics: Pei-ching ta-hsüeh hsiao-shih [History of Peking University], ed. Su Ch'ao-jan, et al., (Shanghai: Shang-hai chiao-yü ch'u-pan-she, 1981), 138–9, 199–200. Biology: ibid., 137. Class of 1922: ibid., 155. End of warlord period: Allen Bernard Linden, "Politics and Higher Education in China: The Kuomintang and the University Community, 1927–37," (Ph.D. diss., Columbia University, 1969), 86.

was dominated by Chinese trained in Japan and influenced by the European emphasis on philosophy and literature, whereas most of the Southeastern faculty were from American schools and oriented toward professional education. Whatever the differences and reasons for them, Chinese scientific education of the 1920s remained a "dreary" affair, as Roger Greene, one of China's most sympathetic critics, described it.[25]

Finally, a surprising feature of this education was the bias against practical and applied studies. Prior to 1930, no Chinese institution of higher learning had a decent program in medicine or agriculture, and the demand for courses in chemistry and biology was correspondingly low. The Chinese did take the lead in engineering, which spawned work in mathematics and physics, but the first department of chemical engineering was not established until 1928, by which time the Nationalists intervened to change the whole direction of Chinese education. During the warlord era, most schooling was left, by choice or default, to the intellectuals, whose preference for literary, philosophical, and political subjects prevailed. Meanwhile, the mission schools did a much better job of teaching science and, ironically, gave it a more useful, practical bent.

Science in missionary higher education

The decade 1915–25 saw the emergence of mission colleges and universities as the leading centers of Chinese higher education, particularly in the sciences. There were several reasons for this development. At the beginning of the period, the mission schools had already established the best foundation of students, teachers, facilities, and finances, and throughout the decade they remained insulated from most of the chaos that disrupted Chinese public education. At the same time, professional teachers and scholars with advanced training in the sciences and other fields began to arrive from the United States to bolster the programs established by older, less qualified missionaries. Finally, there was an unprecedented injection of funds, both from the established mission boards and from a new source, the Rockefeller Foundation, which introduced a program of modern medical education that brought with it financial and other support for science teaching, most of which went to biology, chemistry, and physics departments of the leading Protestant institutions.

The Rockefeller Foundation, which was established in New York in 1913, had an enormous impact on the development of scientific

---

[25] One observer: Paul Monroe, "A Report on Education," 15. Roger Greene: Roger Greene, "Aspects of Science Education," *There is Another China* (New York: King's Crown Press, 1948), 101.

education and research in China. The men who guided the foundation through its early years believed that mankind needed better health more than other things, and that this goal could be achieved by bringing modern science more fully to bear on the teaching and practice of medicine. The scope of their activity was worldwide, and in 1914 and 1915 they sent two commissions to China to assess the state of medical education and practice in that country and recommend ways of improving them. The members of both commissions included leading scholars and educators, such as William Welch, the principal architect of the elite Johns Hopkins University medical school and recognized "dean of American medicine." These men judged the situation in China in the light of their own experience at the pinnacle of American science and recommended a course of action that would put China on the high road of international medical practice.

The first commission reviewed the work of more than twenty medical schools, including eight under Chinese management, twelve union (i.e., managed by two or more mission bodies) colleges, and others under Japanese, German, and British control. Several of these schools were trying to take a scientific approach to medicine. They accepted only middle school graduates with previous science training, and they taught chemistry, physics, and biology in specially designed preparatory courses and in the first year of the medical school itself. In the opinion of the foreign observers, however, they did the job poorly. The commission judged the Chinese schools "absolutely worthless," and, even taking into account the programs under foreign control, concluded: "It is evident that there is no medical school now in China which is adequately equipped and no school which is adequately manned."[26]

In view of these findings, the foundation decided to build its own medical college in China, so the second commission focused on the ability of schools in that country to provide premedical training, which the Rockefeller men took to mean a two-year university-level course in English, Chinese, and the basic laboratory sciences. With a few notable exceptions – all mission colleges – the schools in China were again found

[26] First commission: "Report of the China Medical Commission to the Rockefeller Foundation," New York, 21 October 1914, RG 1, ser. 601, box 27, fld. 243, RAC, pp. 22, 40. When this report was published, under the title *Medicine in China* (New York: China Medical Commission of the Rockefeller Foundation, 1914), the phrase, "absolutely worthless," was deleted. For background on Western medicine in China at this time, see: Harold Balme, *China and Modern Medicine: A Study in Medical Missionary Development* (London: United Council for Missionary Education, 1921), 108–16; John Z. Bowers, *Western Medicine in a Chinese Palace: Peking Union Medical College, 1917–1951* (New York: Macy Foundation, 1971), 9–28; Peter Buck, *American Science and Modern China*, 42–3; Ralph Croizier, *Traditional Medicine in Modern China: Science, Nationalism, and Tensions of Cultural Change* (Cambridge: Harvard University Press, 1968), 38–40.

*Peking Union Medical College, students in "old school," circa 1910.
(Courtesy of the Rockefeller Archive Center.)*

lacking. College laboratories, where they existed, were small, poorly equipped, and badly managed. The faculties included only a handful of individuals with advanced training abroad. Instruction was largely by lecture with few demonstrations, and there was little laboratory work by the students themselves. The commissioners concluded that in order to establish a satisfactory premed course, they must first upgrade the laboratory sciences "as regards both equipment and teachers."[27]

Based on these findings, the Rockefeller Foundation launched its program of modern medical education and research in China. The centerpiece of this effort was the Peking Union Medical College (PUMC) built on the foundations of an older missionary school of the same name.

[27] Second commission: "A Supplementary Report of the Commission that Visited China, August to December 1915," RG 1, ser. 601, box 27, fld. 246, RAC. Conclusion: "Report of the Second China Medical Commission to the China Medical Board of the Rockefeller Foundation. Itinerary and Findings. Special Reports, 1915," RG 1, ser. 601, box 27, fld. 244, RAC, p. 13.

*Peking Union Medical College, students in "old school," circa
1910. (Courtesy of the Rockefeller Archive Center.)*

From the outset, PUMC was meant to be a "Johns Hopkins for China."
Equipment and facilities were the best money could buy. The faculty,
largely foreign in the early years but Chinese later on, was composed of
internationally recognized scholars who were offered high pay and
excellent working conditions and expected to produce notable research
results. Students were recruited by examination, required to complete a
rigorous three-year premed program, and given two more years of
laboratory sciences in the four-year medical course. Since the goal was to
meet international standards, English was the sole medium of instruction. Graduates of PUMC were expected to fill the leading ranks of
China's medical profession, and many did.[28]

In the beginning, China had no students qualified to enter a medical
college of this caliber, so in 1917 PUMC opened its own premed
program, from which most early recruits to the medical school were
chosen. The premed faculty was selected from graduates of leading
American universities, such as Stanley D. Wilson (Ph.D., Chicago, 1916),
who served as professor of chemistry. The curriculum covered Chinese,
English and one other foreign language, physics, chemistry, biology, and

[28] PUMC: Mary Brown Bullock, *An American Transplant: The Rockefeller Foundation
and Peking Union Medical College* (Berkeley: University of California Press, 1980),
chap. 2–5; John Bowers, *Western Medicine*, chap. 5–7.

Peking Union Medical College, first-year premed students, 1920.
(Courtesy of the Rockefeller Archive Center.)

math. After 1925, when PUMC began to draw students from other premed programs, the admission standard remained the same: three years of college; Chinese, English, and one other language; one year of biology, one year of physics, and two years of chemistry, all including lab work. The first two years of the medical college proper were spent in lab courses on anatomy, physiology, biochemistry, pathology, pharmacology, and bacteriology. During the period 1924–41, PUMC graduated 313 medical doctors, many of whom went on to teach and do research in other medical schools in China.[29]

As important as PUMC was, the Rockefeller Foundation had an even greater impact on the study of science through grants to Chinese colleges and universities. From 1915 to 1933, the foundation gave over U.S.$38 million, nearly $18 million of which went to PUMC, while most of the rest, around $1.1 million per year, was distributed by the China Medical Board (CMB), a subsidiary of the foundation set up to disburse funds in China. CMB grants went to erect buildings, purchase equipment, pay the salaries of biology, chemistry, and physics teachers at ten Christian and three Chinese universities, and support sixty to ninety fellowships per year for medical students and teachers to do advanced study at PUMC or abroad.[30]

[29] Premed faculty: ibid., 64. PUMC graduates: Mary Bullock, *American Transplant*, 108–33. PUMC scholars: John Bowers, *Western Medicine*, 93–120.
[30] Rockefeller, 1915–33: S. M. Gunn, "China and the Rockefeller Fund," January 1934, RG 1, ser. 601, box 12, fld. 130, RAC, p. 59.

One way to assess the importance of the Rockefeller contributions is to compare them with the budgets of the recipient institutions. In 1914, the total budget of Shantung Christian University (SCU) was Mex$20,000, and CMB made annual grants, between 1916 and 1921, of Mex$40,000. By 1922, when SCU's expenditures increased to Mex$143,000, the CMB grant was Mex$33,000. In 1922, the budget of Yenching University was Mex$90,000; the CMB made a one-time gift of Mex$150,000 for building and equipment plus annual grants of Mex$7,500 for on-going operations. The entire PUMC premedical program – students, teachers, buildings, and equipment – were handed over to Yenching in 1925, raising the total Rockefeller contribution to nearly U.S.$800,000 and assuring this institution a position of leadership in science education in China. Nankai, the most prominent private Chinese university, spent Mex$117,000 in 1922, the year it received Mex$125,000 to erect new facilities and the first annual subsidy of Mex$7,500. The largest Chinese university, Southeastern, with expenditures of over Mex$600,000, received grants equal to those made to Nankai. In sum, financial assistance from the China Medical Board made a major contribution to higher scientific education in China at a time when facilities in these schools were minimal and little support was available from other sources.[31]

The mission colleges improved, too, because they were able to recruit more and better teachers. After the turn of the century, the balance, which had previously favored the evangelical missionaries, tilted toward proponents of the Social Gospel. By the 1910s, the new generation of mission workers included dozens of men and women with advanced degrees in the sciences, many of whom taught and did research in China and used their home leaves for continuing study, rather than church work. In the field of chemistry, they were led by William Adolph (Ph.D. Pennsylvania, 1915), Stanley Wilson (Ph.D., Chicago, 1916), and Earl Wilson (M.S., M.I.T., 1928) of Yenching University, James C. Thomson (Columbia, M.S., 1917, Ph.D. 1932) of Nanking, and Henry Frank (M.S., Pittsburgh, 1922) of Lingnan. These men and in a few cases women built the science departments of the 1920s, gradually turning them over to their own students, who returned with advanced degrees from American and European universities in the 1930s.[32]

With more money and better teachers, the mission schools improved rapidly, surpassing in most respects their Chinese counterparts. A study entitled "Christian Higher Education in China," compiled by Earl H.

[31] SCU: *Rockefeller Foundation, China Medical Board, Sixth Annual Report* (1920), 43. Yenching, Nankai, NSU: Ibid., *Eighth Annual Report* (1922), 6–10.
[32] Faculty recruiting: Jessie Lutz, *China and the Christian Colleges*, 190–200; and Alice H. Gregg, *China and Educational Autonomy* (Syracuse: Syracuse University Press, 1946), 215–19.

Table 5.1.   *Semester hours taught in Chinese mission colleges, 1926*

Subject	Semester hours	% total semester hours
Sciences		
Chemistry	1,024	
Biology	816	
Physics	458	
Math	422	
Other	60	
Total	2,780	35
Languages	2,156	27
Social sciences	1,083	14
Vocational	982	12
Arts	940	12
Total	7,941	100

Source: E. H. Cressy, "Higher Education," 40.

Cressy on behalf of the China Christian Educational Association in 1926, provides a detailed picture of the mission colleges and universities at their peak. By this date, the Protestants operated sixteen colleges and universities in China, the largest of which had between 200 and 500 students. Enrollments in these schools had expanded from fewer than 200 in 1903 to more than 4,000, or about one-fifth of all university students in China, in 1926. The faculties of the mission schools were well appointed: one teacher for each eight or nine students. Forty percent of the teachers were Chinese, although most were confined to the lower ranks and/or departments of language and culture. Few gained higher appointments in the science departments until after 1930. Over half the mission college teachers had advanced degrees, compared to one-third of the faculty of twenty-four leading Chinese universities.[33]

All sixteen mission colleges included in Cressy's study had departments of chemistry and physics; all but two or three, biology and math. Most offered a major in science, although students were generally required to spread courses across several disciplines, since there were too few offerings in any one. Measured in semester hours taught, chemistry was

[33] Enrollments: Earl H. Cressy, "Christian Higher Education in China, A Study for the Year, 1925–26," China Christian Educational Association, *Bulletin* 28 (1928):7, 29. One-fifth: Wang Shih-chieh, "Education," 1031. Chinese faculty: Earl Cressy, "Christian Higher Education," 99; Jessie Lutz, *China and the Christian Colleges*, 196. Higher degrees: Earl Cressy, "Christian Higher Education," 105, 296. Cressy notes that in a representative sample of American colleges of comparable size, 64 percent of the faculty had advanced degrees.

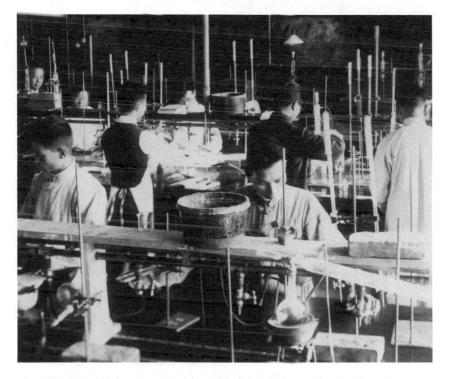

Nanking University, organic chemistry laboratory, 1925. (Courtesy of the Rockefeller Archive Center.)

surpassed only by English, and as a group the sciences ranked first (Table 5.1).[34]

Facilities in the better mission colleges were adequate, at least for work at the undergraduate level. In prescribing minimum equipment standards, American accrediting agencies of this period set values of $10 thousand for chemistry and biology and $12 thousand for physics. Ten mission colleges met this standard in chemistry, nine in biology, and six in physics. The libraries of the better schools had between 10 thousand and 20 thousand English-language volumes, including in most cases the basic textbooks and journals in the core sciences. In contrast to the Protestants, the leading Catholic colleges of early twentieth-century China, Futan and Chen-tan (or L'Aurore), offered little instruction in the sciences.[35]

In view of the (generally accurate) charge that Chinese education of

[34] Ibid., 40, 53, 58–9, 86.
[35] Equipment: ibid., 229–31. Libraries: ibid., 225–7. Catholic colleges: Ruth Hayhoe, "Towards the Forging of a Chinese University Ethos," 330–7.

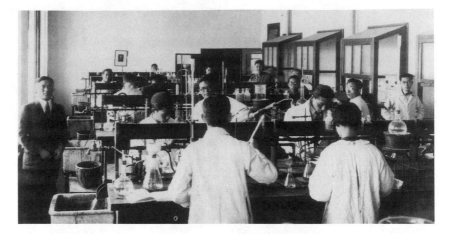

Yenching University, organic chemistry laboratory, 1929. (Courtesy of the Rockefeller Archive Center.)

this era failed to equip students with practical skills, it is worth noting how much of the thrust behind missionary education, particularly in the sciences, rested on the application of knowledge to useful ends. Since the middle of the nineteenth century, Western missionaries had promoted the study of science as preparation for the practice of medicine. This was also the motive behind the Rockefeller initiative. For other missionaries and foreigners, the purpose of science was to improve Chinese agriculture. The modern study of agriculture had begun during the last decade of the Ch'ing at the Imperial university and other Chinese government schools. But this work, which was directed by Japanese instructors using their own language, texts, and specimens, bore little relation to actual conditions in China and left no significant legacy. More lasting was the formation in 1914 of the Department (later College) of Agriculture and Forestry at Nanking University, one of the largest and most successful missionary institutions. During the course of the next quarter century, Nanking became the country's foremost center for agricultural education and research. The Cooperative Crop Improvement Program between Nanking and Cornell universities brought top American professors to China to direct research on and development of improved strains of cotton, rice, wheat, and other crops. Meanwhile, in the southern port of Canton, Chinese businessmen joined missionary educators to form the Lingnan College of Agriculture. Nanking and Lingnan both had large faculties and active research programs, primarily in the fields of plant breeding, agricultural economics, horticulture, and animal husbandry. Chemistry, it should be noted, played only a small part in these programs, and there was little or no research in agricultural chemistry per se.[36]

Most faculty of the leading mission colleges – Yenching, Nanking, and Lingnan – specialized on problems of an applied sort. At Yenching, Stanley Wilson and William Adolph worked in the field of nutrition and food chemistry and did research on glutamic acid (the basis for the Chinese seasoning monosodium glutamate) and the medicinal drug *Ephedra vulgaris*. Earl Wilson, later joined by two returned students, Ts'ai Liu-sheng (Ph.D., Chicago, 1932) and Feng Yün-hao (Ph.D., Ohio State, 1931), a woman, taught industrial chemistry and conducted research on tanning. The senior Chinese chemist at Lingnan, Ch'iu Yan-tsz (Ph.D., Cornell, 1927), studied soybean milk, wine, tea, and other food products, while his colleague Huang Wen-wei (Ph.D., Ohio State, 1930), analyzed Kwangtung waters. The founder of the Nanking chemistry department was James Thomson, whose dissertation dealt with the composition of an important Chinese native product, tung oil. Other members of the Nanking faculty – Ch'en Yu-gwan (Ph.D., Columbia, 1922), Tai An-pang (Ph.D., Columbia, 1931), Lee Fang-hsuin (Ph.D., Northwestern, 1931), and Jeu Kia-khwe (Ph.D., Princeton, 1932) – also worked in organic chemistry, which bore particular relevance to the local scene. Y.C. Tao (M.S., Cornell) taught industrial chemistry and specialized in the study of dyeing and tanning.[37]

Research projects conducted by students at mission colleges followed this practical bent. Of the thirteen master's theses submitted to the Yenching department of chemistry during the years 1927–32, eight dealt with aspects of the Chinese diet (iodine, starches, tea, etc.), four with industry (tanning, ceramics, dyes, and soda), and one with medicinal drugs. A list of fifty-nine Yenching bachelor's theses demonstrates a similar preference: nineteen dealt with tanning, thirteen diet and

36 Nanking University: Jessie Lutz, *China and the Christian Colleges*, 180; W. E. Tisdale, "Report of Visit," 11, 49–50; Earl Cressy, "Christian Higher Education," 137, 142; T. H. Shen, "First Attempts to Transform Chinese Agriculture," *The Strenuous Decade: China's Nation-Building Efforts, 1927–37*, ed. Paul K. T. Sih (New York: St. John's University Press, 1970), 211–21. Lingnan: W. F. Tisdale, "Report of Visit," 12; Charles Hodge Corbett, *Lingnan University* (New York: Trustees of Lingnan University, 1963), 73–5.

37 Yenching faculty: Letter, N. Gist Gee to R. M. Pearce, 5 February 1929, RG 1, ser. 601, box 40, fld. 333, RAC. Ts'ai and Feng: *Yenching Natural Science News* 1 (1933):14, RG 1, ser. 601, box 41, fld. 342, RAC. Yenching Research: Chang I-tsun, "Chung-kuo ti hua-hsüeh" [Chinese chemistry], *Chung-hua min-kuo k'o-hsüeh chih* [Record of science in the Republic of China], ed. Li Hsi-mou (Taipei: Chung-hua wen-hua ch'u-pan shih-yeh wei-yüan-hui, 1955), 6. Lingnan faculty: Letter, N. Gist Gee to L. W. Jones, 9 February 1932, RG 1, ser. 601, box 38, fld. 309, RAC, pp. 18–21; letter, N. Gist Gee to H. A. Spoehr, 25 September 1930, RG 1, ser. 601, box 38, fld. 307, RAC, pp. 5–8; W. E. Tisdale, "Report of Visit," 42–3. Nanking faculty: letters, Gee to Spoehr and Gee to Jones, cited above; W. E. Tisdale, "Report of Visit," 48–9; Stanley Wilson, "Science Work in Schools and Colleges of the Nanking-Shanghai Region," 5–7.

nutrition, nine soil and water, five mining, and five other chemical industries. A smaller sample of bachelor's theses, submitted by the class of 1932 in the chemistry department at Nanking University, shows that these students also favored such topics as the polymerization of glue, manufacture of ink, and synthesis of glutamic acid.[38]

An evaluation of Chinese scientific education

In his autobiography, Kuo Mo-jo, one of the leading Chinese literary figures of the 1920s, recalled his early education in a modern middle school in Szechwan province during the last years of the empire. One day the botany teacher, a graduate of the local normal school, misread the characters for "natural conditions" [t'ien-jan ching-hsiang] as "heavenly-dragon conditions" [t'ien-lung ching-hsiang] and proceeded to lecture the class on the difference between the flying dragons of the heavens and the manifest dragons of the fields. Kuo's experience was not unique. After 1900, the enthusiasm of many Chinese for modern education burned brightly, while their ability to deliver on this promise smoldered. And by all accounts, the most serious problem lay at the heart of the enterprise – the teaching.[39]

Chinese and foreign observers found the same faults with China's science education, which were evident at all levels, times, and places.[40]

---

[38] Yenching theses: Yenching Natural Science News 1 (1933):18-23. Nanking theses: letter, Gee to Jones, 5.

[39] Kuo Mo-jo, Wo-ti yu-nien [My youth] (Shanghai: Ch'üan-ch'iu shu-tien, 1947), 159.

[40] Long before Western pedagogical experts visited China in the 1920s, Chinese educators identified the main problem of science education in their country as poor teaching, particularly the lack of observation and experimentation by the students themselves. The principal journals of the Chinese teaching profession – Chiao-yü tsa-chih [Chinese educational review], monthly, 1909-48; Chung-hua chiao-yü chieh [Chung-hua educational review], monthly, 1913-50; and Hsin chiao-yü [New education], monthly, 1919-25 – contain many articles describing the shortcomings of Chinese science education and recommending solutions, all of which stress the need for more direct contact between students and the physical world. See, for example: Mo Ssu, "Lun kai-liang li-hua chiao-shou-fa" [On improving physics and chemistry teaching methods], CHCYC 1, no. 2 (1913):23-4; Ku Hsing, "Li-k'o chiao-shou ko-hsin chih yen-chiu" [Research on the reform of science teaching], CYTC 10, no. 1 (1918), bk 24:13,076-84; Ch'en Chien-shan, "Chung-hsüeh-hsiao chih po-wu-hsüeh chiao-shou" [Natural history teaching in middle schools], CYTC 14, no. 6 (1922), bk 35:19,927-42; Hu Heng-ch'en, "Ch'u-chi chung-hsüeh ti li-hua chiao-hsüeh-fa" [Physics and chemistry teaching methods in lower middle schools], CYTC 17, no. 6 (1925), bk 44:26,171-82; Pai T'ao, "She-shih erh-t'ung k'o-hsüeh chiao-yü ti hsin lu-hsien" [Constructing a new road of science education for children], CHCYC 20, no. 12 (1933):23-32; Kuo Ping Wen, Chinese System of Education, 166. Western educators who visited China after 1920 made the same criticisms. Those whose views were widely published in Chinese and English include: Paul Monroe, George Twiss, L. G. Morgan, Edgar Knight, Ida Belle Lewis, and the League of Nation's Mission of Educational Experts. See bibliography for citations.

To begin, instruction was almost exclusively by lecture. There was little or no discussion, questioning by teachers or students, or independent work by individuals or small groups. In the view of some Chinese who had spent time on both sides of the podium, the dominance of the teacher and passivity of the students demonstrated the persistence of old habits. The teacher talked, because that was the role of the teacher and the posture least threatening to anyone unsure about the new subjects. As for the students, one ex-teacher explained, they maintained their silence for many reasons:

Some are shy, some are lazy, and some are cowed by the unfriendly attitude of the teacher. These problems stem from the fact that there is no habit of free discussion and no close feelings between student and teacher. To rectify this situation, we must replace the haughtiness of the teacher with a more egalitarian spirit and foster an atmosphere of warmth and friendship between the two parties, so that the present separation will disappear.[41]

Content of the lectures was derived from books – often word for word. The purpose of the whole exercise was more or less to memorize the written and spoken texts. Middle school students attended class for thirty to thirty-five hours per week, so there was little time to read or think about the material on one's own, activities that were not considered important in any case. The premise, in the sciences as in other fields, was that knowledge came from books, and the content of these books must be committed to memory.

Given this faith in the written word, observation and experimentation served mainly as illustration, devoid of the spirit of discovery. Primary classrooms contained no specimens, tools, or special apparatuses, and students took no field trips to look at things outside. Although some middle schools provided for experiments to be done by the students themselves, more often the teacher performed demonstrations in front of the class, while students looked on, uninvited to handle or even observe the specimens and instruments at close range. Worse still, these demonstrations were invariably done in a perfunctory, even sloppy manner, leaving students with the impression that science had little to do with accurate measurement or precise reasoning and technique. As one former student recalled, the teaching of science was foreign to the whole spirit of the Chinese classroom:

When they began teaching science in Chinese schools, it was as though they were savoring a "foreign eight-legged essay." The teacher lectured from the text, and the students followed along in their books. Although they did a few experiments, these were copied directly from the text, while the teacher repeated a few passages and the students just looked on. When the author was in middle school, every time we went to the physics or chemistry class, I would hear the students

[41] Hu Heng-ch'en, "Ch'u-chi chung-hsüeh," 44:26,174.

scream: "Let's go see Professor So-and-so do his tricks!" Then, in the classroom, we would watch the professor do a few tricks, after which he read a passage from the textbook, and the class was over. The next day, it was the same. A semester, a year went by, we finished the book, the course was over. Most of the instruments and specimens were purchased from abroad. The teacher used them to impress upon the students the ready-made knowledge of others. He never thought of getting the students to do things for themselves. The school wasted a great deal of money on foreign instruments and specimens, which were put on display so that the school authorities could flaunt the wonders of modern education.[42]

During the 1920s, Chinese educators looked to America for leadership, and what American educators told them was that their approach to the sciences was all wrong. Paul Monroe, a prominent figure at Columbia Teachers College, reviewed the work of China's schools in 1922, and found the most serious failing to be the teaching of science in the middle schools, pointing specifically to the heavy reliance on textbooks and lectures and the absence of direct laboratory experience by the students themselves. Taking Monroe's criticism to heart, the Chinese National Association for the Advancement of Education hired an American advisor, George Ransom Twiss, who had taught science at the high school level and studied science teaching in American public schools. Twiss's survey of Chinese primary and middle schools, carried out between 1922 and 1924, was published in the association's journal and widely discussed in Chinese educational circles. His findings confirmed many of the criticisms already noted by the Chinese themselves and traced their origin to the training of teachers in Chinese colleges and universities:

All of them [middle and normal school science teachers] have been taught mostly, and most of them wholly, by lecture and textbook methods similar to those they are using. They have had very little or no training in the art of careful experimenting, and in careful, close reasoning from observed facts to logically valid conclusions. They have had no correct and faultless examples of scientific experimenting and teaching to use as standards against which they might measure their own teaching performances.[43]

---

[42] Quotation: Pai T'ao, "Erh-t'ung k'o-hsüeh chiao-yü," 25. Recent articles in the PRC publication, Hua-hsüeh t'ung-pao, demonstrate a continuing concern that Chinese science students do too little experimental work. See: "Shih-yen kung-tso yao chia-ch'iang," [We must strengthen experimental work], HHTP 1 (1982):13–15; Wang Shu-p'ing and Ch'iao Shih-te, "T'an-t'an chia-ch'iang hua-hsüeh shih-yen wen-t'i" [Discussing the problem of strengthening chemical experimentation], HHTP 2 (1982): 81–2; and Li Ch'ung-hsi, "Chia-ch'iang shih-yen chiao-hsüeh shih tang-wu chih chi" [Strengthening experimental teaching is an urgent task], HHTP 2 (1982):83.

[43] Monroe views: Paul Monroe, "Report on Education," 22–3. Monroe's views widely reported: Wang Hsiu-lu, "Chung-hsüeh chih k'o-hsüeh chiao-yü" [Middle school science education], KH 7, no. 11 (1922):1121. Impact of Monroe and Twiss: Wu Ch'eng-lo, "Ch'üan-kuo k'o-hsüeh chiao-yü she-pei," 952. Quotation: George Twiss, Science and Education, 245–6.

Conclusion

In assessing the record of Chinese education during the interregnum, it is wise to proceed with caution. Neither source material nor secondary literature on this subject abounds in quantity or quality. Official materials – standard curricula, statistics, plans, and proclamations – reflect more wishful thinking than real results. On the other hand, accounts by eyewitness observers, Chinese and foreign, denigrate China's schools in a way that suggests unfair comparisons with models seen (or imagined?) elsewhere. Taken alongside Göttingen, Cambridge, or M.I.T., Chinese universities of the 1920s were surely a paltry lot. But they were immensely better than anything that Chinese of the previous generation had known.

The first thing to note about China's attempt to establish a nationwide system of modern schools is the immense, even ludicrous gap between the size of the problem and the means available for solving it. China occupied most of a continent. Its hundreds of millions of people were tucked away in remote villages and its best minds wrapped in the wisdom of the ancients. To move a mountain of rice, the Chinese, weakened by rebellion, revolution, and war, came armed with a pair of broken chopsticks. Progress was measured by the grain.

Collapse of the old order left the field open to new forces, and leadership in scientific education came from abroad. The foreigners were few in number, but strong in organization, knowledge, money, and purpose. By concentrating these resources in the cities, along the coast, and among the upper classes, foreign missionaries, scholars, and philanthropists were able to offer the best scientific training available in China, at least until the Nationalists took power in 1928. The case of education demonstrates the general point that the first rapid strides in the development of science occurred when the heavy hand of authority, the state, and the society it helped maintain, were removed.

How good was this education? Almost everyone who visited China's laboratories and classrooms reported that it was not very good at all. Independent observers, Chinese and foreign, criticized the excessive reliance on lectures, the rote memorization of written texts, and the lack of questioning, debate, and direct hands-on experience in the laboratory, workbench, or field. It is hard to deny the truth of these reports. Still, there are the nagging questions: How much better was the situation in the United States or Europe? or in other countries at China's stage of development, which few of the judges had ever seen? And how is it that so many Chinese scientists, who earned doctorates from the world's leading universities and went on to do outstanding research, began their education in these schools?

A more favorable judgment is given by one study from this period that set out to assess the quality of China's higher institutions. In 1928, Yoshi Kuno, a professor at the University of California, surveyed American university officials who had experience with Chinese students, in order to guide others in the placement of new arrivals. Educators questioned by Kuno agreed that a small number of the best universities – Peking, Nanking, Southeastern, and Yenching – along with an equal number of engineering schools produced graduates who were no more than one year behind their American counterparts. (Tsinghua was expected to join this group, when it became a full-fledged university in 1929.) A larger number of schools, including most of the mission colleges, were ranked somewhat lower: Their graduates might enter American colleges at the junior or senior levels. What this means is that the first and second echelons of Chinese higher education were no more than two years behind the best American universities – and some of that gap must be attributed to the need for additional language study. Although Kuno says nothing about the sciences per se, his conclusions make Chinese education look much better than observers in China gave it credit for.[44]

Finally, what about quantity: Was there too much or too little training in science? There are no statistics from this period on enrollments by department in schools under Chinese administration. The study by E. H. Cressy shows that during the 1925–6 school year, 35 percent of semester hours taught in mission colleges and universities were devoted to science, which was a significant share of their resources. But students did not flock to these courses – at least not as much as the current of "scientism" suggested or the educators hoped. Writing in 1926, Timothy Ting-fang Lew, dean of the Yenching School of Religion, noted the failure of students to choose scientific and technical subjects and explained it by their preference for traditional literary and philosophical fare:

The fact still remains that the students who specialize in science have not increased in considerable numbers under the present strenuous effort to advocate science.... With the exception of certain technical institutions, students in colleges of arts and sciences have not chosen subjects in the sciences as they have literature, law, politics, history, sociology, economics, and the like. There is still lurking in their minds the notion that physical sciences are somewhat philistine compared with literary work.[45]

From the students' point of view, there may have been too much science. Given the meager development of China's industrial and modern sectors, the uninspiring pace of foreign investment and trade, and the inability of the government to design and execute a program of economic development, there were few jobs for men and women with scientific

---

[44] Yoshi Kuno, *Educational Institutions in the Orient*, 55–68 and Appendix A.
[45] Lew quotation: Jessie Lutz, *China and the Christian Colleges*, 188.

training and little prospect that the situation would improve any time soon. The market for lower technical skills was somewhat better, but no Chinese with the wherewithal to advance up the educational ladder wanted to be a craftsman or skilled laborer. Unemployment among university graduates was a chronic problem, which grew worse with the passage of time. Yet if one was to be unemployed, it was better to have a general, liberal education that might help obtain a position in government, business, or public administration, or at least provide a respectable excuse while waiting for a job to come along.

6

The beginning of chemical research

In the case of scientific research, as in education, only modest advances were made during the interregnum, and the greatest accomplishments came more as a result of foreign assistance than from developments within China itself. The Chinese laid some foundation stones amidst the rubble of their shattered empire. The government in Peking supported research in a few fields, among which geology produced the best results. The Science Society of China established a biological laboratory and museum, where scholars began to collect and analyze samples of Chinese flora and fauna. Chinese industrialists set up their own lab to work on problems of applied chemistry. University-based scientists introduced a more rigorous approach to the study of local problems. In these ways, Chinese officials, philanthropists, scholars, and entrepreneurs took the first steps toward building a national research program.

Their efforts paled, however, alongside the contribution of the foreigners, whose money, organization, knowledge, and skill were matched by a will to introduce scientific research to China. The Rockefeller Foundation and Peking Union Medical College elevated work in the medical sciences far above anything previously known in East Asia. Mission colleges and foreign scholars extended a high standard of scientific theory and technical skill to a broader audience. Most important was the training in European and American universities of Chinese students who would lead their country's scholarly enterprise in the 1930s. Research, the highest stage of scientific development, was the last to be freed from foreign control.

The beginning of modern scientific research in China

The first Chinese to take an interest in science in the nineteenth century engaged in little or no research, in the proper sense of the term. In chemistry, as in most other fields, scholars were preoccupied with understanding the meaning of foreign texts and translating these texts

132

into Chinese. On occasion they performed experiments described in the books or compared nature as they found it with the illustrations and diagrams, but these were the actions of students trying to learn the truth, not scholars seeking to discover it. Research performed by Chinese in China, using the models and methods of modern science and having as its object the creation of new knowledge, began in earnest in the 1920s.

The Chinese government deserves some credit for initiating this effort. Leaders of the young Republic recognized, better than their predecessors, that the "wealth and power" needed to save China depended on an understanding of and command over nature. The first institutions for the scientific study of the earth, the minerals below, and the skies above were established in Peking during the 1910s. Officials responsible for these agencies had little money to spend and a narrow vision of the possibilities, but they set China on the right track.

Many Chinese intellectuals of the early twentieth century embraced ideals that favored a scientific approach to knowledge and learning. A heady nationalism drove some to regain control of their country from foreign interlopers, including those who would capture and carry off knowledge of China's natural wonders. A harsh iconoclasm prepared them to reject delivered truths, including traditional models for explaining and studying nature. And a new spirit of empiricism and rationalism led them to seek something better to replace the old cosmology. Not only in laboratories and libraries, but in factories, offices, and homes, Chinese of a young, urban, intellectual stripe were demanding new answers and willing to accept new ways of getting them. Although scholarly products of the 1920s were of uneven quality and unevenly distributed, the new approach to learning was everywhere in evidence.

Despite these positive signs, however, Chinese state and society fell short of the challenge. Prior to 1928, China lacked an authority that could provide the physical security, economic support, and psychological confidence needed to sustain a program of scientific research. Plans were drawn up and budgets approved, but in the end the money went to the warlords, whose armies wreaked havoc on everything they touched. Most Chinese spent these years running from pillar to post, while the best young minds found no repose.

Nor were Chinese intellectuals of the 1920s likely to choose careers in science. In the crisis atmosphere of the early twentieth century, recalled the chemist Tseng Chao-lun, "the minds of most youth remained filled with the thought, 'politics first,' while science continued to be neglected."[1] There was a preference for social and political studies in

[1] Tseng Chao-lun, "Chung-kuo hsüeh-shu ti chin-chan," 57.

China's past and an urgency for social and political change in its present. This was not the time for gazing leisurely through a microscope or waiting for the next generation of hybrids to reveal their traits.

Given the shortcomings and disabilities of the Chinese themselves, it is hardly surprising that foreigners should lead in bringing scientific research to China. Beginning in the nineteenth century, explorers, naturalists, and other pioneers came to study the earth, plants, animals, and people of East Asia and to report their findings to a willing audience back home. In the twentieth century, foreign scholars took up residence in mission colleges and universities, taught courses, and did research in their spare time. Grants from foreign governments and foundations supported the growth of scientific institutions, research projects, and expeditions. Those Chinese who joined this enterprise did so as students and assistants. This was the age of their apprenticeship.

In the early part of this century, a few Chinese began to do reserach using the models and methods of modern science in those disciplines that rely primarily on observation and description. These were the fields where Chinese could take control of their own treasures and place them at the service of the nation. The study of rocks and rivers, plants and animals could be carried out by novices using simple devices, described in everyday language, and fit into conceptual frameworks accessible to laymen and new recruits. Finally, data on a relatively unknown corner of the world were eagerly received by the international scientific community, providing recognition for fledgling scientists and a source of funds to carry on their work. For all these reasons, the first achievements of Chinese scientists lay in reporting on the natural wonders in their own backyard.

The earliest and most widely recognized results were in geology. Interest in the contours of the earth's surface was deeply rooted in traditional geomancy, but this had little to do with the scientific study of this subject, which was driven, in China as elsewhere, by the pressing needs and persuasive theories of more recent times. Collection and interpretation of Chinese data within the framework of modern geology were begun by Europeans, most notably the German Ferdinand von Richthofen, who traveled throughout China during the 1870s to produce a massive study and atlas that remained the standard, until Chinese scholars revised his work in the early twentieth century. In 1916, Peking established the National Geological Survey to assess and retake control of the nation's mineral resources, exploitation of which had been handed over to foreign powers in treaties signed under duress. Several returned students, including Ting Wen-chiang, H. T. Chang, and Wong Wen-hao, and a handful of foreign scholars, led by J. G. Andersson and A. W. Grabau, surveyed and mapped each of the provinces, studied China's iron, coal, and petroleum deposits, analyzed its soils, and explored the geological structures, fossils, and seismology of a hitherto neglected portion of the earth's crust. The National Survey, the Geological Society (established in 1922), and the Institute of Geology of the Academia Sinica (1929) published memoirs, bulletins, and other periodicals that reported the findings of Chinese geologists to audiences at home and abroad. Although mostly descriptive in nature (theoretical musings were limited to a few foreigners, led by Grabau), much of this work won high marks at the time and remains an important source for the study of this region even today.[2]

The development of meteorology followed a similar path. By the 1870s, concern for the security of trade and travel led the foreign-controlled Imperial Maritime Customs Service to set up weather stations along the rivers and coasts. These stations sent daily telegrams to the Zikawei [Hsü-chia-hui] Observatory, erected near Shanghai by French Jesuits in 1865, where foreign scholars compiled periodic weather reports and maps. After 1912, the Chinese government entered this field by establishing in Peking the Central Observatory, which constructed its own maps, based on data from the outlying stations, and published twice-daily reports. By the early 1920s, the training of basic-level technicians and the creation of more advanced programs at two national universities enabled the Chinese to expand their network of stations into the interior and increase the quantity and quality of meteorological research.[3]

Unlike geology and meteorology, whose use in mining and navigation attracted government support, the life sciences, which served the less valued fields of agriculture and medicine, were neglected. The study of biology in China began as a school subject, taught through foreign textbooks and translations that had little to do with the local setting. As in the case of geology, foreign explorers of the late nineteenth and early

---

[2] Chinese earth studies in the premodern era: Joseph Needham, *Science and Civilisation in China*, 2:359–63, and 3:497–680. Richthofen: Ferdinand von Richthofen, *China: Ergebnisse eigener Reisen und darauf gegründeter Studien* [China, the results of my travels and the studies based thereon], 5 vols, (Berlin: D. Reimer, 1877–1912). The most complete account of earth studies in Republican China is Tsui-hua Yang Lee, "Geological Sciences in Republican China, 1912–1937." Ting Wen-chiang: Boorman and Howard, *Biographical Dictionary*, 3:278–82; and Charlotte Furth, *Ting Wen-chiang*. Wong Wen-hao: Boorman and Howard, *Biographical Dictionary*, 3:411–2. For other details, see: A. W. Grabau, "Paleontology," *Symposium on Chinese Culture*, ed. Sophia H. Chen Zen (New York: Paragon, 1969); W. H. Wong, "Chinese Geology," ibid.; H. T. Chang, "On the History of the Geological Science in China," *Bulletin of the Geological Society of China* 1, no. 1 (1922):4–7; *The National Geological Survey of China, 1916–1931* (Peiping: n.p., 1931); and Yang Tsun-yi, "Development of Geology in China since 1911," 131–7.

[3] Premodern history: Joseph Needham, *Science and Civilisation in China*, 3:462–96. Modern history: Chiang Ping-jan, "Erh-shih-nien-lai Chung-kuo ch'i-hsiang shih-yeh kai-k'uang," 623–42. Zikawei: Yen Lin-shan and Ma Tzung-liang, "Hsü-chia-hui t'ien-wen-t'ai ti chien-li ho fa-chan" [Establishment and development of the Zikawei Observatory], *CKKCSL* 5, no. 2 (1984):65–72.

twentieth centuries helped popularize the study of China's wildlife, after which Chinese trained in Western universities gradually joined in collecting and classifying the flora and fauna of their homeland. China's first generation of modern naturalists, led by the zoologist Ping Chih (Ph.D., Cornell, 1918), and the botanist Hu Hsien-su (Sc.D., Harvard, 1925), guided expeditions throughout the country to collect specimens that were displayed in the institutes, museums, and university departments set up during the 1920s to support biological research. Results of their findings were reported in both foreign and Chinese journals. In many cases, specimens found in China had to be sent abroad for identification by better-trained scholars who enjoyed access to larger collections and more complete reference libraries. The overwhelming majority of Chinese biologists of the Republican period concentrated on taxonomy; others took up descriptive work in morphology, anatomy, cytology, or embryology. It was not until after 1930 that a few undertook experimental work in genetics and other fields. Important biomedical research was begun in the 1930s at the Peking Union Medical College and the Henry Lester Institute in Shanghai. Agricultural studies, neglected in the early years, also began in earnest in the 1930s.[4]

Experimental research, the approach most characteristic of the physical sciences, took root somewhat later and for reasons opposite those that placed observation first. The life and earth sciences are "local," their objects of study particular in place and time, whereas the behavior of physical phenomena is the same everywhere. Chinese who studied physics and chemistry had no special data to offer and had to compete for laurels with better trained and equipped scholars in the West. The theoretical and mathematical models of these disciplines were abstruse, less accessible to novices, and the required facilities were prohibitively expensive and difficult to obtain. Talented young Chinese who wanted to serve their country and make a mark in science were unwise to risk a career in the laboratory. Those who did found the road a hard one.

Physics was the last of the major scientific disciplines in which Chinese performed independent, publishable research. Despite the attention paid in the nineteenth century to mechanics and engineering, the formal study of physics trailed behind most other subjects in the schools set up after

---

[4] Biological research: Lu Yü-tao, "Erh-shih-nien-lai chih Chung-kuo tung-wu-hsüeh" [Twenty years of Chinese zoology], *KH* 20, no. 1 (1936):41–8; Hu Hsien-su, "Erh-shih-nien-lai Chung-kuo chih-wu-hsüeh chih chin-pu" [Progress in Chinese botany during the past twenty years], *KH* 19, no. 10 (1935):1555–9; Ping Chih, "Kuo-nei sheng-wu k'o-hsüeh chin-nien-lai chih chin-chan" [Recent development of biological sciences in China], *TFTC* 28, no. 13 (1931):99–110. Genetics: Laurence Schneider, "Genetics in Republican China" (Paper presented to a conference sponsored by the Rockefeller Foundation, Pocantico Hills, NY, May 1984).

1900. Peking University established the first department of physics in 1918. During the next decade, instruction was offered at several other sites, but most teaching in this field was undistinguished. The first Chinese to publish independent research in physics was Woo Yui-hsun (Ph.D., Chicago, 1925), professor at Tsinghua, whose work on the scattering of x-rays appeared in the British journal Nature in 1930. By the end of 1935, fewer than ninety articles had been published by Chinese physicists in journals at home or abroad. Although some of this work was respectable, none commanded attention outside of China.[5]

Progress in the natural sciences was accompanied by a more rigorous approach to the study of human affairs, which underwent profound changes in China after 1919, a period marked by skepticism, even hostility, toward the tradition and the application of models and methods borrowed from modern science. In the field of history, a younger generation of scholars led by Hu Shih and Ku Chich-kang used inductive and empirical methods to reexamine traditional historiography. The most important papers of this school, published in 1926 under the title Ku Shih Pien [Critical reviews of ancient history], addressed in various ways the shared admonition to "Show your proof." Even though criticized for shirking the responsibility to be "relevant" and reducing scholarship to textual nitpicking, Ku and company helped put an end to the unquestioned acceptance of delivered truth and pointed the way to a more critical study of China's past.[6]

In addition to the assault on ancient texts, the study of history and prehistory was broadened by the use of new types of evidence and new methods of obtaining it. For centuries, Chinese had shown an interest in ancient artifacts – stones, coins, bronzes, and the like. But as Chang Kwang-chih points out, "this tradition had never gone much beyond the scope and spirit of antiquarianism, before [China's] contacts with the west." When, in the early 1920s, J. G. Andersson established the existence of a neolithic culture in northern China, Chinese began to take up the study of archaeology. The digs at Anyang in northern Honan, directed by the leading Chinese archaeologist of this generation, Li Chi (Ph.D., Harvard, 1923), and carried out by scholars from the National Research Institute of History and Philology, transformed our understanding of early Chinese history and established an independent

---

[5] Physics in Republican China: Chi-ting Kwei, "The Status of Physics in China," *American Journal of Physics* 12, no. 1 (February 1944):13–18; Yen Chi-tz'u, "Erh-shih-nien-lai Chung-kuo wu-li-hsüeh chih chin-chan" [Development of Chinese physics during the past twenty years], *KH* 19, no. 11 (1935):1705–16; Yen Chi-tz'u, "Chin shu-nien-lai kuo-nei chih wu-li-hsüeh yen-chiu" [Physics research in China in recent years], *TFTC* 32, no. 1 (1935):15–20.

[6] Laurence Schneider, *Ku Chieh-kang and China's New History* (Berkeley: University of California Press, 1971).

tradition for this science in China, which continues to surprise the world with the buried splendors of its past.[7]

The social sciences, particularly sociology and economics, made some headway during the 1920s, although progress in these fields was not as rapid as in history or some branches of natural science. Chinese scholars had always taken a keen interest in man and society, but their approach was normative rather than empirical. Field work, introduced by foreign scholars and carried out with the support of foreign foundations, opened new vistas to their Chinese students and assistants. In 1930, graduates returning from abroad created the Chinese Sociological Society and the *Journal of Sociology*. During the 1930s, sociology courses were introduced into several Chinese universities, while Chinese teachers and students, research institutes, and government agencies applied statistical survey techniques to the study of local society. By this period, according to one authority, outside of western Europe and North America, "China was the seat of the most flourishing sociology in the world."[8]

Empirical and statistical approaches also transformed the study of economics. Chinese students under the direction of professors Franklin Ho and H. D. Fong at the Nankai University Institute of Economics surveyed factories and workshops in the Tientsin area and subjected their findings to modes of analysis employed by economists in the West. The institute also compiled and published statistical series of wholesale and retail, import and export prices. After 1928, these data and studies based on them were published in the institute's journal, *The Nankai Economic and Social Quarterly*.[9]

Earliest chemical research in China

A closer examination of one discipline brings into sharper focus several features of the Chinese research enterprise. As one of the universal physical sciences, chemistry was as far removed from China as physics and attracted equally scant attention among scholars in that country, at least at first. By the early twentieth century, however, the applications of

[7] History of Chinese archaeology: Li Chi, "Archaeology," *Symposium on Chinese Culture*; Chang Kwang-chih, *Shang Civilization* (New Haven: Yale University Press, 1980), 42–3; Chang Kwang-chih, *The Archaeology of Ancient China* (New Haven: Yale University Press, 1968), 3–6. Quotation: ibid., 3.

[8] Social science: Wang Yü-ch'uan, "The Development of Modern Social Science in China," *Pacific Affairs* 11, no. 3 (September 1938), 345–62. Sociology: Siu-lun Wong, *Sociology and Socialism in Contemporary China* (London: Routledge & Kegan Paul, 1979), 1–36. Quotation: Maurice Freedman, "Sociology in and of China," *British Journal of Sociology* 13 (1962):113.

[9] J. B. Condliffe, "The Nankai Institute of Economics," *There is Another China* (New York: King's Crown Press, 1984), 68–70.

chemistry to industry, agriculture, and medicine were widely recognized, so it is surprising that the Republic did so little to promote research in this field. The explanation seems to be that Peking could see as far as getting the minerals out of the ground, and therefore supported geology, but was not prepared to take the next step to support research on manufacturing, which was left to others. In the absence of government leadership and native scholarly interest, the major breakthroughs in chemistry were made by foreigners – first by teachers at the missionary colleges, and later by philanthropists who built the Peking Union Medical College and helped make it a center of world-class biomedical research.

The first scholars to do research in China using the methods and models of modern chemistry were foreign teachers with graduate degrees from American and European universities who began to arrive in the 1910s. The most notable of these were the Americans: at Yenching, William Adolph, Stanley Wilson, and Earl Wilson; at Lingnan, Henry Frank; and at Nanking, James Thomson. An Englishman, Bernard Read, worked at the Henry Lester Institute of Medical Research in Shanghai and a Frenchman, M. Tarle, at the North China Chemical Research Institute in Tientsin. With few exceptions, these men worked on problems of a local, applied sort. The most productive of them, Adolph, did field studies and laboratory experiments on the nutritive value of the Chinese diet. Beginning in 1920, he published more than two dozen articles that appeared in both Chinese and foreign journals. Read worked on the chemistry of medicinal herbs and the vitamin content of Chinese plants, Stanley Wilson on herbal drugs, Earl Wilson on industrial chemistry, Frank on x-rays, and Thomson on tung oil. With the exception of Adolph, none of these men gained recognition beyond the borders of China. Their principal achievement outside of teaching was to present the model of laboratory research to their Chinese students and colleagues.[10]

Compared with the mission schools, Chinese colleges and universities of this period did little to promote scientific research. Wang Chin, chairman of the department at National Southeastern University in Nanking, stands virtually alone among the faculty of Chinese institutions who performed what might be called original chemical research. During the years 1919 to 1923, Wang published articles on the Fukien camphor industry and on traditional techniques used in the manufacture of ceramics, spirits, and fragrances. His most extensive study, on Chinese coins, included laboratory analysis of alloys containing copper, lead, tin,

[10] Research by Chinese and foreign chemists in China: Chang I-tsun, "Chung-kuo ti hua-hsüeh," 1–17; Tseng Chao-lun, "Erh-shih-nien-lai Chung-kuo hua-hsüeh chih chin-chan," 1519–21.

and zinc and the development of a typology based on the proportions of these metals to identify the dynasty in which a particular coin was minted. This work, although crude in comparison to what was being done elsewhere at this time, gave countless students their first hands-on laboratory experience and stands as the earliest chemical research performed and published by a Chinese scholar in China.[11]

Giant of Chinese chemistry: Wu Hsien

These meager efforts by Chinese and foreign scholars were dwarfed by the accomplishments of one man. Wu Hsien was without doubt the greatest chemist and perhaps the greatest scientist in China during the first half of the twentieth century. When he published his first work in 1919, there was no chemical research of any description in China. When he cut the ties to his homeland in 1949, the study of chemistry had spread throughout the country, and a research tradition was firmly in place. During the intervening years, no one contributed more to this enterprise than Wu Hsien. His success, particularly during the 1920s, was made possible by the Boxer Indemnity Fund, the Rockefeller Foundation, and the Peking Union Medical College, all of which share a place on the stage of Wu's story.[12]

Wu Hsien was born in 1893 in Foochow, capital of Fukien Province, the second of three children in a family that boasted some official and scholarly prominence. Like many young men of his time, Wu studied the classics under a family tutor, took and failed the lowest civil service exam, and began to look elsewhere for his future. In this instance, geography was on his side. Owing to its excellent harbor, propitious location on the coast opposite Taiwan, and selection as the site of China's first modern shipyard, Wu's native Foochow emerged during the latter half of the nineteenth century as an important center for the study of science, technology, and other Western subjects. Many graduates of the Foochow Naval Academy found positions teaching in modern schools, such as the Fukien Provincial High School, where Wu received his introduction to foreign learning. Following graduation in 1910, he passed the government examination and won a Boxer indemnity scholarship to study in the United States. In the fall of 1911, after one

---

[11] Camphor: *KH* 6, no. 2 (1921): 239–40. Fermentation: *KH* 6, (1921):270–82. Ceramics: *KH* 6 (1921):869–82. Fragrances: *KH* 4, no. 10 (1919):935–42. Coins: Wang Chin, "Wu-chu-ch'ien hua-hsüeh ch'eng-fen chi ku-tai ying-yung ch'ien, hsi, hsin, la k'ao" [Chemical composition of Chinese coins and investigation of the use of lead, tin and zinc in ancient (coins)], *KH* 8, no. 8 (1923):839–54.

[12] Unless otherwise noted, the following is based on Daisy Yen Wu, *Hsien Wu, 1893–1959: In Loving Memory* (Boston: Author, 1959) and on interviews with Mrs. Wu in New York, February 1984. I would like to thank Mrs. Wu for her kind assistance.

semester of language and other preparation, the first class of sixty-two Boxer Fellows sailed for San Francisco.

Wu enrolled in the Massachusetts Institute of Technology with the intention to study naval architecture, an interest kindled by his childhood in Foochow and the influence of his teachers from the Naval Academy. During his first summer vacation on a New Hampshire farm, however, he was inspired by reading Huxley's essay, "The Physical Basis of Life," and from this moment fastened his sights on biochemistry. Following his graduation from M.I.T. in 1917, he entered the laboratory of Professor Otto Folin at the Harvard Medical School and just two years later completed the Ph.D.

Wu's dissertation won worldwide acclaim, and his career proceeded on a triumphal march. He returned to China in 1920, joined the faculty of Peking Union Medical College, and received early promotion to department chairman (1925) and full professor (1928). At home, he helped organize the Chinese Physiological Society (1926), was chosen the first president of the Society, and later served as editor of its *Chinese Journal of Physiology*, one of the pioneer scientific journals in China and a model for many that followed. He found time to write a popular textbook, *Principles of Physical Biochemistry*, which was published in English in 1934. And he helped create the journal, *Tu-li p'ing-lun* [Independent critic], a platform from which the leading figures of the Peking academic establishment pronounced on the great issues of the day. Wu's global reputation, established early and sustained by later work, earned him invitations to attend international conferences, membership in Western scientific associations, and listings in the various biographies of prominent scientists. Constant travel and close work with foreign scholars helped cement his identity as a bridge between two cultures.[13]

Unlike most leading Chinese scientists, who joined the retreat to southwestern China after the Japanese invasion of 1937, Wu remained at PUMC, which enjoyed the special status accorded to foreign institutions, until 1942, when the occupying army forced it to close. In 1944, he moved to Chungking to help set up an institute for the study of nutrition. This began the last phase of Wu's career in China, as a scholar-bureaucrat in the field of public health. After the war, he was appointed to a special commission to study reconstruction, named director of the projected National Institute of Health, and represented China at various

---

[13] For a sample of Wu Hsien's writings on current affairs, see: Wu Hsien (T'ao-ming), "Chung-kuo ti ping ying-kai tsen-ma chih" [How can (we) cure China's illness], *TLPL* 51 (1933):32–6; Wu Hsien (T'ao-ming), "Ting Hsien chien-wen tsa-lu" [Notes on a visit to Ting Hsien], *TLPL* 4 (1932):13–18; and Wu Hsien (T'ao-ming), "Ts'ung hsiao-hai shuo tao ta-jen" [Discussing adults from the point of view of children], *TLPL* 57 (1933):19–21. For background on *Tu-li p'ing-lun*, see Charlotte Furth, *Ting Wen-chiang*, 196ff.

Peking Union Medical College, chemical research laboratory, 1924.
(Courtesy of the Rockefeller Archive Center.)

international conferences.[14] He was in the United States in 1949 when the Communists seized power, and his family left China to join him. In 1959, after serving for a time on the faculty of the University of Alabama, Wu suffered his second heart attack and died.[15]

Wu Hsien's ascent to the fast-track of the scientific profession derived from the immediate and widespread adoption of the Folin-Wu method of blood analysis, which was based on Wu's Ph.D. dissertation published in 1919. Until this time, the measurement of a single constituent of blood required a sample of 100 cubic centimeters (cc) or more, making analysis difficult in all cases and impossible in many clinical situations. Under Folin's direction, Wu developed techniques for preparing a protein-free blood filtrate, suitable for the quantitative determination of all the important constituents of blood, from a sample of only 10 cc. The key innovation was the use of tungstic acid to precipitate proteins from the blood, while leaving other substances intact. The authors also developed

[14] For an example of work from this period, see: Hsien Wu, "Nutritional Deficiencies in China and Southeast Asia," *Fourth International Congress on Tropical Medicine and Malaria*, (Washington, DC: Department of State, 1948), 1217–23.

[15] Wu's final work, which contains musings on science and its relationship to Chinese culture, was published posthumously. See: Wu Hsien, *A Guide to Scientific Living* (Taipei: Academia Sinica, 1963).

Peking Union Medical College, biochemistry research laboratory.
(Courtesy of the Rockefeller Archive Center.)

improved methods for measuring each important blood constituent. The method for determining sugar in particular was more convenient and accurate than previous means and could be performed with a single drop of blood. Taken together, these techniques offered a compact system of blood analysis that was accurate and saved the time and trouble of everyone involved. They were a boon to clinicians, who could obtain quick and reliable information even from patients who could spare little blood. The procedure for determining sugar also aided the study of diabetes and connectedly insulin, a recently discovered hormone that attracted considerable attention in the early 1920s. For more than seventy years, the Folin-Wu method has occupied an important place in the study and practice of medicine.[16]

After leaving Folin, Wu linked up with another senior scholar on the brink of a major discovery. Early in 1922, he began work with Donald Van Slyke, a biochemist at the Rockefeller Institute in New York, who was on the trail of the chemical mechanisms that govern the exchange of

[16] Wu's dissertation: Otto Folin and Hsien Wu, *JBC* 38 (1919):81–110. Blood sugar: Otto Folin and Hsien Wu, *JBC* 41 (1920):367–74. For evaluations of the Folin-Wu method, see: Alexander Marble, "Otto Folin: Benefactor of Diabetics through Biochemistry," *Diabetes* 2 (1953):503–5; and Philip Shaffer, "Otto Folin (1867–1934)," *Journal of Nutrition* 52 (1954):3–11.

Peking Union Medical College, Department of Physiological Chemistry and Pharmacology, 1923. Wu Hsien, front center. (Courtesy of the Rockefeller Archive Center.)

oxygen, carbon dioxide, and other substances in the blood during the respiratory cycle. That autumn they moved to PUMC, where Wu trained Chinese technicians to carry out the research designed by Van Slyke, who like other visiting scholars discovered that the technical work at PUMC was unsurpassed. "In those three months at Peking," Van Slyke later testified, "I completed experiments that would have required a year in any other laboratory, including the Rockefeller Institute." The results of this research, published in 1923 by Van Slyke, Wu, and Franklin McLean, set forth the mathematical formula that describes the exchange between red cells and plasma of electrolytes and water under varying pressures of oxygen and carbon dioxide. This paper stands as a landmark in our understanding of the chemical basis for respiration and demonstrates the first application of the Gibbs-Donnan law of heterogeneous equilibrium, a deduction from thermodynamics, to the study of physiology. At the age of thirty, Wu Hsien had published two papers, either of which might crown the career of a successful research chemist.[17]

Whereas primary credit for these works went to the senior partners, Wu's greatest personal achievement came a decade later when he articulated a theory to explain the denaturation of proteins. Denaturation is a change in the natural protein molecule by which it loses its solubility in water and other solvents. The significance of this process was widely recognized but little understood when Wu Hsien began his investigation in the mid-1920s. He was the first to offer a valid explanation, although many ignored it at the time, and the implications of his theory helped to confirm the emerging picture of the structure and nature of proteins.

Wu was first attracted to the problem of denaturation after his return to China, as he sought to extend the study of blood proteins begun at Harvard. His first paper on this subject, coauthored with Daisy Yen, his assistant at PUMC and later his wife, was published in a Japanese journal in 1924. In applying Folin's methods, which involved treating plasma proteins with dilute acids and alkalis, Wu and Yen were unable to explain the results, "and it occurred to us," they wrote, "that the phenomenon under observation might be related to the process of denaturation." Chemists of this era knew that the coagulation of proteins occurs in two steps: first, denaturation, or the loss of solubility that results from treatment by acids, alkalis, heat or other agents; followed by flocculation, or the clumping together of suspended particles, causing them to precipitate. The importance of coagulation of proteins had been

---

[17] Research paper: Donald D. Van Slyke, Hsien Wu, and Franklin C. McLean, *JBC* 56, no. 3 (1923):765–849. Assessment of this work: Lawrence J. Henderson, *Blood: A Study in General Physiology* (New Haven: Yale University Press, 1928), 91–114. Van Slyke's views of PUMC: John Z. Bowers, *Western Medicine in a Chinese Palace*, 99.

recognized for more than three decades, and because the second step in this process was better understood than the first, many chemists were drawn to the challenge of explaining denaturation. In 1931, Wu provided the answer.[18]

Like many loftier treatises that mark great turning points in the history of science, the power of Wu's theory was that it offered a simple explanation for a wide range of observed, but hitherto disconnected and obscure phenomena. During the late 1920s, Wu and other chemists had established that denaturation could be brought on by treating proteins with acids, alkalis, metallic salts, alcohols, ethers, or urea by applying heat, ultraviolet light, pressure, or sonic waves, or simply by shaking them up. The denatured protein retained the molecular weight and functional groups of its parent, but showed a change in viscosity, antigenic property, tryptic digestibility, and acid- and base-binding power. Previously, Wu had maintained that denaturation must result from hydrolysis, or separation of the amino acids that comprise the protein chain. In 1931, he proposed a new explanation – namely, disorganization:

The compact and crystalline structure of the natural protein molecule, being formed by virtue of secondary valences, is easily destroyed by physical as well as chemical forces. Denaturation is disorganization of the natural protein molecule, the change from the regular arrangement of a rigid structure to the irregular, diffuse arrangement of the flexible open chain.[19]

Most of what was known about the denaturation of proteins fit this theory.

Wu's research was recognized by other chemists working on this problem, his articles cited, and his contribution noted in the *Annual Review of Biochemistry* for 1932, which recorded the major achievements in this field during the previous year. Yet it was not until several years later that the significance of his theory was fully appreciated. The missing piece was an explanation of the intramolecular forces that hold the native protein in its characteristic shape. This explanation was first offered by Alfred Mirsky and Linus Pauling in 1936, but for some reason, Mirsky, who was the leading figure in the study of denaturation, did not discover Wu's paper until two years later. Perhaps the fact that the work was done in China and reported in a relatively obscure journal helped to shield it from view. In any case, the clarity and correctness of Wu's insight has since placed his work in high esteem. With recognition

[18] Hsien Wu and Daisy Yen, *Journal of Biochemistry* (Japan) 4 (1924):345–84. State of field in 1920's: Joseph Furton, *Molecules and Life: Historical Essays on the Interplay of Chemistry and Biology* (New York: John Wiley & Sons, 1972), 144–5. Wu's theory: Hsien Wu, *CJP* 5 (1931):321–44. A preliminary report of this theory appeared in: Hsien Wu, *American Journal of Physiology* 90 (1929):562–3.
[19] Hsien Wu, *CJP* 5 (1931):335.

that the character of proteins derives not only from the number and type of amino acids they contain, but also from the distinctive "folding" of their long chains, Wu's description of the process of denaturation has emerged as one of the earliest and clearest statements of the important role played by large-scale architecture in the function of protein molecules.[20]

Concurrent with his research on denaturation, Wu also helped to remove one of the most serious obstacles to the understanding of immunology in a study that won more immediate and widespread esteem. By the mid-1920s, it was known that the immunity of an organism against infection proceeds by a reaction between an invading *antigen* and a defensive *antibody* to produce a *precipitate*, which is removed from the bloodstream. There was little precise information on the nature of this reaction, however, in large part because there was no way to measure the composition of the precipitate. How much antigen did it contain? how much antibody? how much, if any, of something else? And what were these components made of? Many scholars assumed that they were proteins, but this had not been firmly established.

Wu's contribution to this line of research, published in 1928, again derived from his study of blood. Several years earlier he had developed a method for determining very small amounts of hemoglobin. Now, using a measured quantity of hemoglobin as the antigen, the amount of antibody could be calculated as the difference between the total protein precipitate and the hemoglobin. Since the use of hemoglobin was found to have certain limitations, the following year Wu published further results demonstrating the use of a colored tracer, iodo-albumin, for the same purpose. The significance of this research was immediately recognized, and further quantitative study of the immunological reaction followed.[21]

Most of Wu's research was at the forefront of knowledge about the basic mechanisms underlying the chemistry of life. Surrounded by the comfort and security of PUMC, he acted as member of the international scientific profession who by coincidence worked in China. But he was

---

[20] *Annual Review of Biochemistry*, 1 (1932):161–2. Alfred Mirsky and Linus Pauling, *Proceedings of the National Academy of Sciences* 22 (1936):439–47. Mirsky's discovery of Wu: A. E. Mirsky, *Cold Spring Harbor Symposium on Quantitative Biology* 6 (1938):152. Although Mirsky did not recognize his debt to Wu, others did. See: *Annual Review of Biochemistry* 6 (1937):184. Assessments of Wu's theory of denaturation: Joseph Furton, *Molecules and Life*, 146–7; Felix Haurowitz, *Chemistry and Biology of Proteins* (New York: Academic Press, 1950), 126.

[21] Studies of immunological reaction: Hsien Wu et al., *PSEBM* 25 (1928):853–5; and Hsien Wu et al., *PSEBM* 26 (1929):737–8. Contemporary recognition of Wu's contribution: Heidelberger and Kendall, *Journal of Experimental Medicine* 50 (1929):809–23; Boyd and Hooker, *Journal of General Physiology* 17 (1934):341–8. Evaluation by a historian of science: Felix Haurowitz, *Chemistry and Biology of Proteins*, 291.

also a Chinese, committed to his homeland, and this concern was evident inside and outside his laboratory. Guided by his wife, Daisy, whose principal interest was in food chemistry, Wu devoted a growing portion of his time to the problems of nutrition and the Chinese diet.

For more than a decade, Wu and his colleagues at PUMC conducted experiments on laboratory rats to determine the effects of a vegetarian diet similar to that common in northern China – primarily cereals, legumes, and vegetables with little fruit, meat, or dairy products. They found that rats on the vegetarian diet were as healthy and active and lived as long as the omnivorous variety, but they grew more slowly, particularly when young. Biological assays demonstrated that the vegetarian diet was deficient in vitamins A, D, and G and calcium. The authors concluded that the small stature of Chinese, like their laboratory rats, was due to insufficient nutrition during the period of lactation.[22]

Wu's laboratory research made no substantial contribution to the study of food or food chemistry, but his popular writings on the subject were a huge success. The most important of these was the book *Ying-yang kai-lun* [Principles of nutrition], first published in 1929. The poverty of life in general and diet in particular were universally acknowledged features of early twentieth-century China, but no one had thought to write about this problem until Wu Hsien returned home in 1920. Despite his heavy commitment to teaching and research, he produced a book addressed to students and educated laymen that explained the scientific basis for nutrition and presented a wealth of information about the Chinese diet. Wu advocated no particular program of action, but as he noted in the preface, "If, on account of this book the nutrition of 400 million Chinese is improved a little, then the author's wish shall be fulfilled." For more than fifty years, this modest essay remained one of the most widely read books on the subject.[23]

Other chemical research

Besides Wu Hsien, there were a few other important chemical researchers in China during this period, all of whom worked at PUMC. Two of the

---

[22] See articles by Wu Hsien et al.: *CJP* 2 (1928):173–94; 3 (1929):157–70; 6 (1932):251–6; 9 (1935):119–24; 16(1941):229–40, 309–10.

[23] Wu Hsien, *Ying-yang kai-lun* [Principles of nutrition] (Shanghai: Commercial Press, 1929). The quotation is from the preface to the first edition. For details on its publication history, see the 1973 edition published in Taipei by the T'ai-wan shang-wu yin-shu-kuan. For other popular works by Wu on this subject, see: Wu Hsien "Chinese Diet in the Light of Modern Knowledge of Nutrition," *Chinese Social and Political Science Review* 2 (1927):56–81; Wu Hsien, "Wu kuo-jen chih ch'ih-fan wen-t'i" [The food problem of our people], *TLPL* 2 (1932):15–19; and Wu Hsien, "Tsai lun ch'ih-fan wen-t'i" [More on the food problem], *TLPL* 205 (1936):14–16.

first four prizes for scientific research awarded by the China Foundation for the Promotion of Education and Culture went to PUMC chemists Chen Ko-kuei and Chou Tsan-quo for their analyses of the active ingredients in Chinese herbal medicines. In each case, these studies followed the same steps: A medicinal plant was ground to a fine powder and dissolved in dilute acid, percolating alcohol, or other solvent. This solution was then subjected to filtration, distillation, extraction, recrystallization, and other procedures in order to isolate and purify the constituent *alkaloids*, compounds that were believed responsible for the drugs' curative powers. Each alkaloid was analyzed to establish the characteristic physical constants – size and shape of crystals, melting point, specific rotation, and molecular formula – of the compound or its derivatives. (These measurements were sufficient to establish the identity of a given alkaloid, but not to determine its precise chemical structure. The latter objective, which at that time could be accomplished only by synthesis of the compound from known starting materials, was not achieved in China until after 1949.) Finally, the most promising alkaloids were subjected to physiological studies, in which a solution of the compound was injected into laboratory animals and the effect on blood pressure, respiration, heartbeat, and other vital signs observed. As research progressed, comparisons were made of the types and effects of alkaloids drawn from different parts of a single plant or from the same plant picked in different locations or seasons. The goal was to isolate a compound that was safe and convenient to use and demonstrated the desired medicinal effect.[24]

The most important study of this type was the analysis of ephedrine, the active principle of the plant *Ephedra vulgaris*, known to the Chinese as *ma-huang* – literally "astringent-yellow" for its taste and color. According to the *Pen Ts'ao Kang Mu* [Great pharmacopoeia], the most authoritative of Chinese *materia medica*, compiled in 1596, *ma-huang* improves circulation, causes sweating, stops coughs, reduces fever, and appears as an ingredient in many well-known prescriptions. In 1887, the Japanese chemist Nagai Nagayoshi extracted from this plant an alkaloid that he named *ephedrine*. Other studies of this compound followed in Japan and Europe, where varieties of the plant are found, and the physiological effects observed. A synthesis of ephedrine was performed in 1920, establishing its exact chemical structure. Still lacking were a complete statement of the medicinal value of ephedrine and a demonstration of its efficacy as a clinical drug, tasks that were taken up by scholars in China.[25]

[24] China Foundation prizes: Jen Hung-chün, "Chung-chi-hui yü Chung-kuo k'o-hsüeh" [The China Foundation and Chinese science], *KH* 17, no. 9 (1933):1524.
[25] Nagai Nagayoshi, *Pharmazeutische zeitung-nachrichten* 32 (1887):700. Synthesis of

The man who moved *ma-huang* off the shelves of the Chinese apothecary and into the arsenal of Western medicine was the chemist Chen Ko-kuei. Chen was a native of Shanghai, Boxer indemnity scholar, and student of physiology at the University of Wisconsin, where he earned his doctorate in 1923 for a study of muscle autolysis. Called home by his mother's illness, he accepted a post as senior assistant at PUMC. During trips to Shanghai to visit his mother, Chen developed an interest in Chinese medicine and asked the family herbalist for a list of the ten most potent native drugs. At the top of this list was *ma-huang*. Back in Peking, Chen and a colleague, Carl Schmidt, purchased samples of the plant from local drugstores, extracted the ephedrine, and undertook studies of the physiological effects on laboratory animals and later human patients. These studies, published in 1924-5, demonstrated to the satisfaction of scholars and the delight of the medical world that

ephedrine is of considerable value as a circulatory stimulant in surgical shock, as a bronchodilator in asthma, as a mydriatic [pupil dilator], and as an apparently specific remedy in Addison's disease. Since ephedrine is effectively absorbed from the gastrointestinal tract, and since solutions of the alkaloid are very stable, it appears to be a drug of definite clinical value.

Because it could be taken orally and had prolonged effect, ephedrine was preferred to the established drug, adrenaline, in many applications. It gained immediate acceptance and remains in use today for relief of asthma, hay fever, and low blood pressure. After completing this work, Chen left China to study medicine and later served as director of pharmacological research for the Eli Lilly Company in Indianapolis, Indiana. He is one of the few prominent Chinese scientists of this generation who left China prior to 1949 to seek his fortune abroad.[26]

Another chemist, Chou Tsan-quo, had conducted similar studies of Chinese herbal medicines. Chou's 1914 doctoral dissertation at the University of Geneva was based on an investigation of *Cordyalis* (known to the Chinese as *yen-hu-so*). During the 1920s, he worked at PUMC and later at the Institute of Materia Medica of the National Academy of Peiping, where he carried forward Chen's study of ephedrine and extracted and identified the alkaloids of other medicinal plants, including the lily *Fritillaria* [*pei-mu*], and the yellow jasmine, *Gelsemium* [*kou-wen*].[27]

ephedrine: Spath and Gohring, *Monatshefte für chemie*...61, (1920):319. For other details on ephedrine research, see K. K. Chen et al. idem., *PSEBM* 21 (1924):351–4; idem., *Journal of Pharmacology and Experimental Therapeutics* 24 (1924):339–57; idem., *China Medical Journal* 39 (1925):982–9; idem., *Journal of the American Pharmaceutical Association* 14(1925):189–94, and 15 (1926):625–39.

[26] Quotation: *J. Am. Pharm. Assn.* 14 (1925):189. Other details: John Bowers, *Western Medicine in a Chinese Palace*, 103–5.

[27] Review of Chou's work: Hsü Chih-fang, "Chung-kuo hua-hsüeh-chia tui-yü yao-wu

Research on the industrial applications of chemistry also began around the same time, albeit under different circumstances. The earliest applied research was carried out in the Geological Survey and the Bureau of Industrial Research, both government-sponsored research institutes established in 1916 under the Ministry of Agriculture and Commerce, the former to investigate China's mineral resources and the latter to foster development of traditional industries. Another important step was taken in 1922, when Fan Hsü-tung, founder of China's first modern salt and soda companies, set up alongside his factories in Tangku the Golden Sea Research Institute of Chemical Industry. Fan transferred the research facilities and staff of his Chiu-ta Salt Company, an initial grant of 100,000 *yüan*, and the income due him as founder of Chiu-ta to launch this new enterprise, which became the country's most important private laboratory for industrial research. During the 1920s, Golden Sea researchers concentrated on problems related to the manufacture of salt and soda, although their activities gradually expanded to other industries as well. The institute's director was Sun Hsüeh-wu, a graduate of St. John's University in Shanghai, Harvard Ph.D. in chemical engineering, and former chief chemist of the Kaiping Coal Mines, who accepted a substantial cut in salary for the privilege of helping to build China's first private industrial lab. The lack of biographical data on Golden Sea researchers suggests that they were probably trained locally and engaged in relatively pedestrian work. As we shall see, this separation in the experience and training of China's academic and industrial researchers persisted down to 1949.[28]

Conclusion

In scientific research, even more than education, the Chinese found it difficult to make a start, and the greatest achievements of the

hua-hsüeh ti kung-hsien" [Contributions of Chinese chemists to pharmacological chemistry], *HHSC* 10, no. 7 (July 1955):296–306. Studies of alkaloids: Chou Tsan-quo et al., *CJP* 2 (1928):203–18; 3 (1929):67–74, 301–5; 5 (1931):349–52; 6 (1932):265–70; 7(1933):35–44; 8 (1934):155–60; 9 (1935):267–74; 10 (1936):79–84, 507–12; and 13 (1938):167–72.

28 The first major project of the Bureau of Industrial Research was on the manufacture of Shaohsing wine. See: Li Ch'iao-p'ing, *Chung-kuo hua-hsüeh shih*, 2:266, 293. Golden Sea Institute: Lu Pin, "Huang-hai hua-hsüeh kung-yeh yen-chiu-she" [Golden Sea Research Institute for Chemical Industry], *CKKCSL* 2, no. 1 (1981):56–60; Ou-yang I, "P'ing-tung hua-hsüeh kung-yeh k'ao-ch'a-chi" [Record of an investigation of the chemical industries east of Peiping], *HHKY* 5, no. 1 (February 1930):91; Ch'ai Ching-hsü, "Chin Ku T'ang-shan nan-Man kung-yeh ts'an-kuan pao-kao" [Report on an inspection of the industry of Tientsin, Tangku, Tangshan, and southern Manchuria], *HHKY* 5, no. 2 (May 1930):111; Tseng Chao-lun, "Erh-shih-nien-lai Chung kuo hua-hsüeh chih chin-chan," 1523; Hsüeh-wu Sun, "The Hwang-hai (Golden Sea) Research Institute of Chemical Industry," *Science and Technology in China* 1, no. 4

interregnum were made under foreign direction, support, and advice. Research poses the greatest challenges in science: It requires knowledge, skill, vision, and the facilities to put these attributes to use. The Chinese were ill-equipped in all areas, and the disruptions of this era made it difficult to fill their many needs. The foreigners supplied everything – successful examples of how research was done, training of Chinese students, first at home then abroad, jobs and facilities to return to, and a connection with the scientific community in the West. In chemistry, all the notable results of this period flowed through these channels.

This foreign influence contributed, in part, to a second feature of China's early efforts in modern scientific research, which was their separation from Chinese surroundings. Science, in its pure form, is a quest for knowledge that knows no national boundaries. The foreign mentors of China's first-generation scientists prepared them to take part in this global enterprise, while Peking Union Medical College also supported basic teaching and research. Other factors, indigenous to China, help explain the penchant for pure research. First was a scholarly tradition that tolerated the disconnected pedant, the man with long fingernails and longer sleeves who disdained physical labor of any sort. Second was the trend among China's modern intelligentsia to move into cities and up professional ladders, while breaking ties with the society they left behind. Third was the absence of an effective government that might have pulled Chinese scholars back to more practical concerns. In the end, these indigenous elements were probably more important than foreign influences in tilting China's agenda toward basic research.

Finally, the most important feature of this story is the success – in a few cases, the immense success – of men who moved in one lifetime between two cultures. Foreigners could build institutions in China that made the conduct of science possible, but Chinese scholars occupied this space and gave it life. Wu Hsien, one who did so, was a remarkable scholar who stood out in a time and place where few others prospered. But Wu was not alone. Other members of his generation – the chemist Chen Ko-kuei, the zoologist Ping Chih, the botanist Hu Hsien-su, the geologist Wong Wen-hao, to name a few – also distinguished themselves in this way. What the careers of these men show is that within China's traditional culture were a respect for learning and habits of mind that, given the chance, could be applied with success to the study of science. These scientists were the hearty plants raised in a foreign hothouse when all around was snow and ice. More like them would bloom after 1928, when the climate for Chinese science improved.

(August 1948):69–70; and Fang Hsin-fang, Wei Wen-te, and Chao Po-ch'üan, "Huang-hai hua-hsüeh kung-yeh yen-chiu-she kung-tso kai-yao" [Summary of the work of the Golden Sea Research Institute for Chemical Industry], HHTP 9 (1982):559–64.

7

Chinese entrepreneurs and the rise of the chemical industry

In the case of industry, as with education and research, governments of the interregnum tried to introduce new methods that would strengthen and enrich China, but the scope of the problem mounted as their ability to deal with it declined. Central and provincial authorities created more arsenals, equipped them with better machines, and brought in foreign advisers who knew how to get the best results. At the same time, Peking offered loans, tax breaks, and other incentives to spur private investment in areas deemed vital to China's economy. If these effects had come earlier when the Chinese state was still capable of effective action, they might have done more good. By the early 1900s, they were too little, too late.

More important than government in the development of industry during this period was the role of private investors and entrepreneurs. The interregnum was the great age of Chinese capitalism. As the power of the state declined, the commercial talents of the Chinese people came to the fore. In the present example, this meant introducing new technologies and skills that built China's first chemical industries. The beginnings were modest, but the implications for the future profound.

The state sector: chemicals for military use

The first factories in China to use power-driven machinery and other modern devices were the state-run arsenals, which were established beginning in the latter half of the nineteenth century. As for chemicals, the introduction of rapid-fire guns at the turn of the century called for a new type of smokeless gunpowder. Smokeless, as distinct from the traditional black, gunpowder is made by a chemical process that involves the application of sulfuric and nitric acid to cotton fibers. These acids were the first chemicals made in China by modern means.

Virtually the only explosive used during the nineteenth century, as well as the millennium that preceded it, was a mixture of saltpeter (potassium nitrate), charcoal, and sulfur, commonly called black gunpowder, or simply gunpowder. Invented in China in the ninth century, gunpowder made its way westward through the Middle East to Europe. There, production techniques and applications to warfare and mining were refined. In the seventeenth century, the improved product and the guns to fire it were reintroduced to China, where these military skills had lapsed. Chinese manufacture of gunpowder by machine methods began in the 1870s at the Lung-hua Powder Plant, attached to the Kiangnan Arsenal, and at other arsenals in Tientsin and Nanking. By the late 1870s, Tientsin, the largest of these plants, produced 275 tons annually.[1]

Whereas the rapid reaction among substances contained in it produces a bang, black gunpowder is a physical mixture whose manufacture involves neither chemical transformations nor the consumption of prepared compounds. The application of industrial chemicals and chemical processes came later, with the introduction of smokeless gunpowder. The basic component of smokeless powder is nitrocellulose, or guncotton, which is made by treating cotton fibers with sulfuric and nitric acids. The techniques for manufacturing guncotton were developed in Germany, France, and Britain during the latter half of the nineteenth century. The product of this research, smokeless powder, came into general use by European armies after 1900 and was the chief propellant used in World War I. The advantage of smokeless over black gunpowder is that the former is almost entirely converted to gas, so that a smaller charge produces greater force. The absence of smoke also made it easier to operate quick-firing guns and magazine rifles introduced around the same time.

The man most responsible for developing the manufacture of smokeless gunpowder in China was Hsü Chien-yin, the eldest son of Hsü Shou, who had followed his father to the Anking and Kiangnan arsenals. John Fryer, a friend of the young Hsü during their years at Kiangnan, called him the "cleverest Chinaman I ever met," and considered himself "a child compared to him [Hsü] in many respects." The two Hsüs may have begun to experiment with guncotton as early as 1867 at Kiangnan. Later, Chien-yin moved to Tientsin, and in 1876 sulfuric acid was

---

[1] Invention of gunpowder: Joseph Needham, *The Grand Titration: Science and Society in East and West* (London: Allen & Unwin, 1972), 65–71. Chinese production: Thomas L. Kennedy, *The Arms of Kiangnan*, 62–75, 114–18, 130–4. Throughout this book, the term *ton* means *metric ton* and is equal to 1,000 kilograms or 2,205 pounds avoirdupois. Chinese weights have been converted as follows: 1 *catty* equals 1.33 pounds; 1 *picul* equals 133 pounds.

[2] Manufacture of gunpowder: Ormond Lissak, *Ordnance and Gunnery* (New York: Wiley, 1915), 1–15.

reportedly being made at both sites by father and son respectively. Although there is no evidence that this acid was used to make explosives, the subject of guncotton was discussed in the journal *Ko-chih hui-pien*, and it is unlikely that acid would have been made by powder factories for any other purpose. These early efforts produced no concrete results, in part because the Chinese lacked adequate technical knowledge, something they could acquire only through direct experience in the West. It was, therefore, a matter of good fortune that Hsü Chien-yin should be selected as China's first "technical attaché" in Europe. In 1879, he was assigned to the Chinese embassy in Germany and spent the next five years traveling throughout the continent. His observations, recorded in the book *Random Notes on Travels in Europe* [*Ou-yu tsa-lu*], include a detailed account of the visit to a German powder factory and the process used there to make nitrocellulose.[3]

On his return to China in 1884, Hsü Chien-yin suffered a falling out with the powerful viceroy and self-strengthener Li Hung-chang, after which Hsü was excluded from work at Kiangnan, where he might otherwise have tried out his newly acquired knowledge. Interest in smokeless powder picked up in 1890, when Kiangnan introduced a modified British Lee rifle that required the new type of propellant. In 1895, the arsenal purchased a complete set of German equipment for the manufacture of nitric acid, guncotton, and smokeless powder. A German technician engaged to put this machinery into operation failed, claiming that the equipment would not work in the climate of China. Chinese staff at Lung-hua took over, managing to make nitrocellulose on an experimental scale, although they too failed to achieve large-scale production. The effort to make acid and smokeless powder at Kiangnan stalled.

Meanwhile, Li Hung-chang's principal rival, Viceroy Chang Chih-tung, purchased similar German equipment for the Hanyang Arsenal in Hupeh, where he intended to make rapid-fire rifles and smokeless powder. Chang's interest in powder making increased after the Boxer Rebellion of 1900, when the foreign powers imposed a military and economic blockade of China, halting the import of guns and ammunition. As a result, he called Hsü Chien-yin to Hanyang to help install and operate the new machinery. According to the official report of the

[3] Fryer on Hsü Chien-yin: John Fryer, letter of July 1868, cited in Adrian Bennett, *John Fryer*, 28. Manufacture of sulfuric acid at Kiangnan and Tientsin: *KCHP* 1, no. 2 (1876):8b–9a. Discussion of guncotton: *KCHP* 2, no. 6 (1877):15a; and 4, no. 1 (1881):11b. Biography of Hsü Chien-yin: Chi Hung-k'un, "Hsü Chien-yin yü Chung-kuo wu-yen huo-yao ti yen-chih" [Hsü Chien-yin and research and development on Chinese smokeless gunpowder], *TJKHSYC* 4, no. 1 (1985):90–8; Chang Tzu-kao and Yang Ken, "Hsü Shou fu-tzu nien-p'u" [Annual record of Hsu Shou and son], *CKKCSL* 2, no. 4 (1981):55–62; Chao Erh-hsün et al., ed., *Ch'ing-shih kao*, 13, 930–1.

Kiangnan Arsenal, which became the chief center for the manufacture of smokeless powder in the early twentieth century, credit for establishing this industry belongs to Hsü Chien-yin. Hsü was killed in an explosion in 1901 while doing research, but the following year, his colleagues at Hanyang achieved results. By 1905, production of smokeless powder at Kiangnan reached twenty-seven tons per year. In the process employed there, nitrates were reacted with sulfuric acid to produce nitric acid, and these acids were used to treat cotton. Temperature, pressure, and density of the liquids were monitored at each step to ensure successful results. At first, the sulfuric acid used in this process was imported. Later, probably before the outbreak of World War I, both the Kiangnan and Hanyang arsenals began to make their own acid. During the 1920s, the manufacture of modern explosives spread to government arsenals throughout China.[4]

Until 1927, when more modern techniques were introduced, the Chinese made sulfuric acid by the lead chamber method. This method was invented in England in the middle of the eighteenth century and came into widespread use along with the growth in demand for sulfuric acid as an intermediate in the manufacture of soda, bleaching powder, and other products. The only description I have been able to find of a lead chamber in China shows that they were crude copies of similar devices used in the West. In Europe, the chambers were made of riveted lead sheets enclosing a space of several hundred cubic feet or more and built atop sand or stone foundations. Saltpeter and sulfur-bearing substances, generally iron pyrites or brimstone, were placed in the chamber and ignited. Gases from the combustion condensed on the walls of the chamber and were absorbed by a layer of water that covered the floor. After repeated burnings over a period of weeks, the solution was withdrawn and boiled down to condense the acid. The chambers were easy to construct, the raw materials widely available. The product, as high as 66 percent pure, was strong enough for most requirements and could be concentrated by other techniques when necessary. Sulfuric acid was combined with potassium or sodium nitrate and heated to produce nitric acid.[5]

---

[4] Chang purchases equipment: Wang Ching-yü *Chung-kuo chin-tai kung-yeh shih tsu-liao*, bk 2, pt 2, 425–6, 430. Boxers/Chang: Chi Hung-k'un and Wang Chih-hao, "Wo-kuo Ch'ing-mo ai-kuo k'o-hsüeh-chia Hsü Chien-yin" [Patriotic scientist in our country during the late Ch'ing, Hsü Chien-yin], *TJKHSYC* 4, no. 3 (1985):192. Official record: Wei Yün-kung, *Chiang-nan chih-tsao-chü chi*, 2:37, 3:76, 9:7–10. Role of Hsü Chien-yin: Chi Hung-k'un, "Hsü Chien-yin yü Chung-kuo wu-yen huo-yao," 96. Other details: *North China Herald*, 26 April 1895, 612; 16 July 1897, 130–3; and Minami Manshū tetsudō kabushinki gaisha, Tenshin jimusho chōsaka [South Manchurian Railway Company, Tientsin Work Investigation Section], *Shina ni okeru san sōda oyobi chisso kōgyō* [China's acid, soda, and nitrogen industries] (Tientsin, 1937), 123–4. This source is cited hereinafter as *Shina kōgyō*.

[5] Chamber method: L. F. Haber, *The Chemical Industry During the Nineteenth Century:*

Despite the sensitive nature of their work, Chinese arsenals relied heavily on foreign machinery, supplies, and technicians. Prior to World War II, all the major Chinese powder factories were built with British or German equipment and operated under the guidance of foreign managers and technicians. Attempts to replace the foreigners with Chinese were never wholly successful. During the 1920s, China's largest arsenal, at Mukden, and others at Taiyuan, Hanyang, and elsewhere depended heavily on foreign equipment and supervision.[6]

On the whole, the arsenals were not a great success. The manufacture of ordnance was expensive, the products of uneven quality, and the output limited. Those responsible for China's defense generally preferred to buy arms rather than make them. The Hanyang Arsenal, whose peak production reached 120 tons of sulfuric acid and 20 tons of gunpowder per month, declined along with the Chinese iron industry in the early 1920s, after which Hanyang began importing all its metals and explosives. By the mid-1930s, the Kung-hsien (Honan) and Canton arsenals were still in operation, but many of the other powder factories in China had shut down. As was the case with modern Chinese industry generally, the arsenals made a qualitative contribution through the introduction of new ideas and techniques, but their quantitative impact on China's economy was small.[7]

The private sector: chemicals for industry and consumption

Modern machine manufacture in the civilian sector trailed the establishment of arsenals and shipyards by several decades. Although a few factories were created in the 1870s and 1880s, broad-gauged industrial development, under both Chinese and foreign ownership and covering a wide range of products for general use, began after the defeat by Japan in

A Study of the Economic Aspect of Applied Chemistry in Europe and North America (Oxford: Clarendon Press, 1958), 1–8. Manufacture of acids in China: Ch'en T'ao-sheng, "Hu Han hua-hsüeh kung-yeh k'ao-ch'a-chi" [Report on an investigation of the chemical industries of Shanghai and Hankow], HHKY 2, no. 1 (January 1924):234–40; Wei T'ing-ying, "Chi-nan, Ch'ing-tao, Te-chou, T'ang-shan, T'ang-ku hua-hsüeh kung-yeh k'ao-ch'a-chi" [Report on an investigation of the chemical industries of Tsinan, Tsingtao, Tehchow, Tangshan, and Tangku], HHKY 2, no. 2 (July 1924):212.

6 Tseng Kuo-fan and Li Hung-chang: Thomas Kennedy, Arms of Kiangnan, 35–6, 40–1, 52–3, 62–4, 72–5, 114–15, 134, 142. Mukden, Taiyuan, and Hanyang: Anthony Chan, Arming the Chinese: The Western Armaments Trade in Warlord China, 1920–28 (Vancouver: University of British Columbia Press, 1982), 110–12. Other arsenals: China Year Book (1933), 544–6; and (1936), 428–30.

7 Purchase vs. manufacture of arms: Anthony Chan, Arming the Chinese, 112–14; Hsi-sheng Ch'i, Warlord Politics in China, 1916–1928 (Stanford: Stanford University Press, 1976), 116–20. Hanyang: Wu Ch'eng-lo, "Chung-kuo hua-hsüeh kung-yeh she-chi chi yüan-liao wen-t'i" [Question of plans and materials for the Chinese chemical industry], HHKY 4, no. 2 (October 1929):5.

Table 7.1. *Chinese-owned factories that use chemical inputs,*
inaugurated 1895–1913

Industry	Number of enterprises	Percentage of total	Initial capital, Ch$1,000	Percentage of total	Avg. capital, Ch$1,000
Cotton weaving and dyeing	27	4.9	1,261	1.1	47
Glass	10	1.8	3,429	2.9	343
Matches	26	4.7	3,444	2.9	132
Candles, soap	18	3.3	805	0.7	45
Paper	14	2.6	5,929	4.9	424
Printing	6	1.1	1,160	1.0	193
Tanning	11	2.0	4,608	3.8	419
Subtotal	112	20.4	20,636	17.2	184
Total	549	100.0	120,288	100.0	219

Source: Wang Ching-yü, *Chung-kuo chin-tai kung-yeh shih tsu-liao,*
2:869–920, calculated by Albert Feuerwerker, "Economic Trends in the Late
Ch'ing Empire, 1870–1911," 35.

1895, when foreigners secured the right to set up factories in the treaty
ports, and Chinese themselves recognized the need for wider, deeper
change. The manufacture in China of those consumer products that
depend on chemical inputs – such as textiles, soap, paper, glass, leather,
and cosmetics – dates from the turn of the century. At first, the basic
chemicals, acids and sodas, were imported. After World War I, the
Chinese began to make these intermediate products on their own.

Between 1895 and 1913, the pace of industrialization quickened.
Spurred by the legalization of foreign industry in the treaty ports, at least
136 new foreign and Sino-foreign manufacturing and mining enterprises
were established with total initial capital of over Ch$100 million. Most
engaged in the processing of silk, tea, flour, sugar, eggs, and other agricultural products. A few made soap, candles, paper, matches, and leather.
Investment by Chinese private and joint public-private firms also picked
up. According to one estimate, at least 549 Chinese-owned manufacturing and mining enterprises with initial capitalization of Ch$120
million were inaugurated during this period. Most of these firms were in
coal and metal mining, public utilities, textiles, food processing, and
other light manufactures. Statistics on those light industrials that
depended on chemical inputs are presented in Table 7.1.[8]

[8] Albert Feuerwerker, "Economic Trends in the Late Ch'ing Empire, 1870–1911," *The*
Cambridge History of China, 11:29–36. For details on foreign manufacturers, see:

World War I had important, albeit mixed effects on the development of Chinese industry. On the one hand, the reduction of European imports was a boon to Chinese firms that made consumer goods. On the other hand, the growth of these enterprises was limited, so long as machinery and other capital goods could not be obtained from abroad. The greatest spurt came just after the war, when Chinese manufacturers, poised to meet the foreign competition on their own turf, were able to get the necessary inputs. Data for the early Republican period is even sparser than for the late Ch'ing, but two studies covering 1913 and 1920 show that the numbers of factories (including arsenals and utilities) and industrial workers in China more than doubled during these years.[9]

Chinese industry grew rapidly during the early Republic, but because this growth issued from such a low base, there was little change in the structure of the Chinese economy as a whole. Excluding the recession of 1921–2, during which time industrial production dropped by nearly 14 percent, the average annual increase in the gross value of industrial output for the years 1912 to 1928 was a hefty 12.5 percent. Still, by 1933, the combined output of handicrafts, factories, mining, and utilities occupied only 10.5 percent of China's net domestic product, while handicrafts accounted for 67.8 percent, factories 20.9 percent, mining 7.0 percent, and utilities 4.3 percent of the industrial sector. In sum, modern power-driven machine manufacture represented just over 2 percent of China's total product.[10]

The manufacture of basic industrial chemicals by the Chinese was a product of economic and political changes brought on by World War I. By 1914, there were dozens of factories in China, both Chinese- and foreign-owned, that depended on imported acids and sodas to make textiles, glass, soap, paper, and the like. The war disrupted the supply of these consumer products from abroad, stimulating domestic output while making scarce the intermediate materials and capital goods needed

Wang Ching-yü, *Chung-kuo chin-tai kung-yeh shih tsu-liao*, 2:279–319. And on Chinese-owned consumer "chemical" industries established during the high-tide of investment, 1905–8: ibid., 811–21.

[9] Ch'en Chen and Yao Lo, eds., *Chung-kuo chin-tai kung-yeh-shih tsu-liao* [Source materials on China's modern industrial history], 4 vols. (Peking: San-lien shu-tien, 1957–61), 1:55–6, cited in Albert Feuerwerker, *Economic Trends in the Republic of China, 1912–1949* (Ann Arbor: Center for Chinese Studies, University of Michigan, 1977), 17.

[10] Annual industrial growth rates for 1912–28 are based on an index of fifteen mining and manufacturing products, covering 50 percent of industrial output and employing 1933 price weights, compiled by John Chang, *Industrial Development in Pre-Communist China: A Quantitative Analysis* (Chicago: Aldine, 1969), 60–1. Data for 1933 are from Ta-chung Liu and Kung-chia Yeh, *The Economy of the Chinese Mainland: National Income and Economic Development, 1933–1959* (Princeton: Princeton University Press, 1965), 66. Both studies are cited in Albert Feuerwerker, *Economic Trends in the Republic of China*, 12, 26–7.

for expansion. The creation of new markets through foreign intervention, followed by successful import substitution on the part of local manufacturers, emerged during this period as the dominant pattern of Chinese industrial development. Whereas Chinese industrialists were driven primarily by economic incentives, a new spirit of nationalism spurred them to seize control of China's fate. On both economic and moral grounds, the "made-in-China" movement attracted many enthusiasts.

Chinese who joined the battle for economic and political independence soon discovered that the problems of setting up and operating a factory in China were enormous, particularly in fields that required unfamiliar and sophisticated techniques. In many cases, industrial pioneers, who had only the most rudimentary knowledge of their trade, tried to build plants without blueprints and run them without the aid of foreign advisors. One report on a sulfuric acid factory in central Shantung, where raw materials were plentiful and cheap, demonstrates the difficulty of this task:

On arrival, our surprise was exceeded only by a certain amount of pathetic admiration. A couple of thousand dollars, all the money the investors possessed, had been expended in the erection of a small plant for the manufacture of sulfuric acid by the chamber process. There were four little lead chambers in a row, a pyrite burner was located at one end and at the other end was an outlet pipe for the excess gases. The manager, who was a high-school graduate and had studied a half year of chemistry in high school, stepped proudly forward, and explained that he had designed the plant unaided and that the completed plant was an exact copy of the diagrammatic sketch which was to be found in his high school textbook. The plant had failed to produce satisfactory acid. The "company" was profoundly disappointed when shown that there were a number of important details which had been omitted in the crude sketch. This is not an isolated case of patriotic enthusiasm.[11]

Fan Hsü-tung and the chemical industry in northern China

The difficulties experienced by these hapless tinkerers make even more stunning the success of the men who founded China's first privately owned chemical industries: Fan Hsü-tung in northern China, and Wu Yün-ch'u in the south. Fan was the greatest of these chemical pioneers, creating the first modern factory in China to make an industrial-grade material, refined salt, the first to make an industrial chemical, sodium carbonate, by synthetic means, and the first to erect a large, integrated chemical plant for the manufacture of ammonium sulfate, on the eve

[11] William H. Adolph, "Chemical Industry in China," *Journal of Industrial and Engineering Chemistry* 13, no. 12 (December 1921):1099.

of World War II. After the outbreak of the Second World War, he duplicated several of these efforts in the remote southwest, where Chinese carried on their resistance against Japan. But it was during the First World War, when Chinese knew nothing of this business, that Fan made his earliest and most remarkable breakthroughs.

Fan Hsü-tung (also known as Fan Jui) was born in 1882 in the market town of Hsiang-yin near Changsha, capital of Hunan Province. His elder brother, Fan Yüan-lien (Fan Ching-sheng), was a follower of the reformer Liang Ch'i-ch'ao. The two brothers attended the Academy of Current Affairs in Changsha where Liang served as dean of students. After the aborted Reforms of 1898, they followed Liang to Japan, where Hsü-tung completed middle school and studied chemistry at one of the imperial universities. They returned to China in 1912. Fan Yüan-lien, an influential figure in the new Republic, used his official connections to promote the fortunes of his younger brother – first, in arranging for him to join a delegation to Europe to study industrial development and government administration of the production and marketing of salt.[12]

Fan Hsü-tung's European tour, particularly his visit to Germany, had a catalytic effect on China's industrial development, similar to the travels of Hsü Chien-yin a generation earlier. In 1913, Yuan Shih-kai, the military strongman who dominated the Republic in its early years, put up the nation's salt tax as collateral on a foreign loan, the terms of which required reform of the salt administration. This problem was hardly new to the Chinese, whose debates on the subject date at least to the famous *Discourses on Salt and Iron* of the first century B.C. In the present instance, however, their solution had to satisfy the European bankers; the delegation on which Fan served was charged with finding a solution acceptable to all sides. Fan's travels introduced him to the technology and management of salt refining, as well as the larger problems of industrialization. He returned to China determined to apply what he had learned. By the 1910s, foreign importers had introduced refined salt into China and demonstrated the profitability of its sale for both personal consumption and industrial use. After a visit to the Changlu salt fields in northern China, Fan resolved to establish his own company to compete for a share of this market.

[12] Biography of Fan Hsü-tung: Ch'en Chen and Yao Lo, *Chung-kuo chin-tai kung-yeh-shih tsu-liao*, 1:513–20; Chih Feng, "Fan Hsü-tung: wo-kuo hua-hsüeh kung-yeh ti t'o-huang-che" [Fan Hsü-tung: trailblazer of our country's chemical industry], *CKKCSL* 1, no. 3 (December 1980):2–9. These sources and *Shina kōgyō*, 5, agree that Fan studied chemistry, but each places him in a different Japanese imperial university. Fan Yüan-lien (Fan Ching-sheng) was actually Hsü-tung's cousin, raised in Hsü-tung's home after the death of Yüan-lien's father. In biographies of Hsü-tung, Yüan-lien is referred to as the elder brother, and the two boys probably viewed their relationship in this way. For details on Fan Yüan-lien, see: Boorman and Howard, *Biographical Dictionary*, 2:14.

Chiu-ta Refined Salt Company, Tangku. (Sung-ho Lin, Factory Workers in Tangku.*)*

The Chiu-ta Refined Salt Company was established in 1914 as a joint-stock company with an initial capital of 50 thousand *yüan*. According to one source, the money was raised through the influence of Fan Yüan-lien and other government officials – a plausible assertion, since the manufacture and sale of salt in China had been a government monopoly, and creation of a private corporation for this purpose would require official sanction. Located in Tangku, on the coast of northern China near Tientsin, the Chiu-ta factory drew on the techniques of an established industry that produced salt by the evaporation of seawater. The crude brine was refined using soda ash and lime to precipitate the impurities, a method that differed little from traditional practices, except that it used high-grade imported soda rather than raw ash. The favorable location enabled Chiu-ta to transport its products by water or rail to major cities throughout China. This business was an instant success; rapid expansion followed. By 1930, the company's assets exceeded two million *yüan*.[13]

Some refined salt (sodium chloride) was sold for personal consumption, but the much larger market for this substance were the manufacturers of industrial chemicals, especially soda. Soda, or soda ash (sodium carbonate), and the other chemicals related to it – sodium hydroxide (caustic soda), sodium bicarbonate (baking soda), sodium silicate, and sodium sulfide – are used in the manufacture of a wide range of consumer products, including soap, paper, glass, textiles, dyes, drugs, and foodstuffs. The explosive growth of these industries in nineteenth-century Europe was made possible by the invention of methods for making soda

---

[13] Role of Yüan-lien: Chih Feng, "Fan Hsü-tung," 4. Details on Chiu-ta: Ou-yang I, "P'ing-tung hua-hsüeh kung-yeh k'ao-ch'a-chi," 91–5. Traditional salt process: Li Ch'iao-p'ing, *The Chemical Arts of Old China* (Easton, PA: Journal of Chemical Education, 1948), 54–65.

on a grand scale. By 1914, annual world production of sodium carbonate was around three million tons.[14]

Although the manufacture of synthetic soda is a quite recent development, natural soda has been known and used since prehistoric times. The name *soda ash* derives from the fact that sodium carbonate was first obtained by burning seaweed or other plants and leaching the ashes in hot water to produce a brown lye, used in domestic laundering. Later, large quantities of natural soda were found in crystalline or powdered form on the earth's surface and dissolved in lakes in arid regions. By the nineteenth century, local inhabitants were cutting blocks of ice from the frozen surfaces of lakes in Manchuria and Inner Mongolia and transporting them to Changchiakou, along the Great Wall west of Peking, for refining and marketing throughout north China. At the turn of the century, annual Chinese production of natural soda reached 200 thousand tons. Because of its low quality (only 45 to 50 percent sodium carbonate) and the high cost of transportation and refining, however, the natural product could not compete with synthetic soda for use in modern industry.[15]

Until the early 1920s, all of China's industrial-grade soda came from foreign suppliers, primarily Brunner-Mond and Company of England, the world's largest producer. After 1914, however, Britain cut back on exports of this strategic material, reducing the supply to China, whose imports had been running around 20 thousand tons annually. During the war, the price of soda in China rose to seven or eight times its prewar level, forcing many manufacturers of glass, soap, paper, and other secondary products out of business. The loss of foreign soda, in the short run a devastating blow to those industries that depended on this chemical, gave the Chinese a chance to enter this field on their own.[16]

No one was in a better position to seize this opportunity than China's foremost maker of salt, Fan Hsü-tung. At the outset, Fan's problems were legal and financial. The early Republican government maintained the traditional Chinese view of salt as a consumer product, to be tapped as a source of revenue, and taxed refined salt at ten to twenty times its market price, making it too expensive to be used as an industrial raw material. Finally, in 1917, through the influence of officials connected to the salt administration, the government announced an exemption of tax on salt for industrial purposes. With this concession in hand, Fan assembled a group of investors (including some of those officials who had

---

[14] Soda production figures: Hou Te-pang, *Manufacture of Soda with Special Reference to the Ammonia Process* (New York: Hafner, 1969), 41.

[15] Ibid., 14–30; and Kuo Pen-lan, "Chang-chia-k'ou chih liang ta hua-hsüeh kung-yeh" [Two great chemical industries of Changchiakou], *HHKY* 1, no. 2 (July 1923):137–9.

[16] Chinese soda imports: see Table 11.2. Impact of the war: Ou-yang I, "P'ing-tung hua-hsüeh kung-yeh k'ao-ch'a-chi," 97–8.

Yungli Soda Company, Tangku. (Sung-ho Lin, Factory Workers in Tangku.*)*

been responsible for the change in the tax law) to establish the Yungli Soda Manufacturing Company. The company's initial capital of 300 thousand *yüan* came primarily from profits on the sale of Chiu-ta salt and investments from two banks: the Kincheng Bank of Tientsin and the Shanghai Commercial and Savings Bank. The founders of these banks, Chou Tso-min (Kincheng) and Ch'en Kuang-fu (Shanghai), sat on the Yungli board of directors and played a major role in the subsequent development of China's chemical industries.[17]

[17] Yungli Soda Company: Ch'en Chen and Yao Lo, *Chung-kuo chin-tai kung-yeh-shih tsu-liao,* 1:514–16; Chih Feng, "Fan Hsü-tung," 6; Ou-yang I, "P'ing-tung hua-hsüeh kung-yeh k'ao-ch'a-chi," 97–106; Liu Ta, "Wo-kuo chih-chien kung-yeh yü Ying-kuo Pu-nei-men kung-szu ti tou-cheng" [Struggle between our country's soda industry and the Brunner-Mond Company of England], *CKKCSL* 1, no. 2 (October 1980):101–3; Fan Jui (Fan Hsü-tung), "Yung-li chih-chien kung-szu ta-shih-chi" [Description of the Yungli Soda Manufacturing Company], *HHKY* 2, no. 1 (January 1924):253–60.

Fan Hsü-tung planned to locate the Yungli factory alongside the Chiu-ta plant in Tangku, close to the raw materials and transportation facilities, but there remained the problem of finding the right equipment and technical expertise. From the outset, all agreed that Yungli should adopt the Solvay or ammonia process. Developed by the Belgian Ernest Solvay in the 1860s, this process had displaced the older Le Blanc method, and by the end of World War I accounted for virtually all manufactured soda. The Solvay cartel was controlled by a small number of very large producers, led by Brunner-Mond, whose dominant position assured a high degree of secrecy in all phases of the business. Since Yungli was not a member of the cartel, it could not obtain license, equipment, or technical assistance to build and operate a plant of this type. Fan reportedly paid $20 thousand for stolen blueprints of an American Solvay factory, but these proved too crude to be of any use. In the end, the Chinese had to figure things out for themselves.[18]

Determined to avoid the expense and dependency of establishing a joint venture or bringing in foreign advisers, the Yungli group recruited its own technical personnel. In addition to Fan himself, one-third of the company's founders were men who had received some technical training in China or abroad. During the early years, they sent recruiters to middle schools and colleges in China and the United States in search of people to staff the new enterprise. Among those hired in America were several future leaders of Chinese chemical engineering, including Wu Ch'eng-lo, Leo Shoo-tze, and Hou Te-pang. The last of these, Hou Te-pang, succeeded in ferreting out the secrets of the Solvay industry and applying them successfully in China.[19]

There is a remarkable similarity in the background, education, and careers of Hou Te-pang and Wu Hsien: Both were born in the early 1890s in Fukien Province, were Boxer fellows and classmates at M.I.T. (B.S., 1917), and eventually became the dominant figures of their generation in chemical engineering and chemical research, respectively. Hou was born in 1890 in a village south of Foochow, to a farming family of modest means. He began his education in the village before moving to Shanghai, where he graduated in 1908 from the Fukien-Anhwei Railroad School. After two years as a railroad construction supervisor, Hou passed the national examination to qualify for a Boxer indemnity

[18] History of soda industry: Hou Te-pang, *Manufacture of Soda*, 1–13. Unless otherwise noted, dollars ($) means U.S. dollars. For conversion of American dollars to *yüan*, or Chinese dollars (Ch$), see Cheng Yu-kwei, *Foreign Trade and Industrial Development of China* (Washington, DC: University Press of Washington, DC, 1956), xi, 262–3.

[19] Yungli technical personnel: Ch'en Hsin-wen, "Chi Yung-li hua-hsüeh kung-yeh kung-szu p'ei-yang jen-ts'ai ti tso-fa" [Record of the methods of developing human resources of the Yungli Chemical Industry Company], *CKKCSI*, 2, no. 3 (September 1981):28–34.

fellowship. This led to two years of preparatory study in Peking, followed by a bachelor's degree from M.I.T. and a doctorate from Columbia University (1921), both in chemical engineering. After graduation, Hou joined the Yungli corporation as its chief engineer. During the next decade, he traveled between the United States and China, studying the soda industry, designing and purchasing equipment, and overseeing the construction and operation of the plant in Tangku. The knowledge acquired in this work formed the basis for his book, *Manufacture of Soda with Special Reference to the Ammonia Process*, which was published in English by the American Chemical Society in 1934 and remains a classic in this field.[20]

Construction of the Yungli plant began in the spring of 1921, using equipment imported from the United States. The Peking government continued to promote this venture, granting the company a monopoly on the manufacture of soda in the Tangku area and an exemption from tax on soda products to go with the earlier concession on salt. Encouraged by these moves, the stockholders approved additional levies to expand operations to the very large scale required for profitable soda manufacture. In March 1922, the new factory held its first experimental production run. What the experiments showed, however, was that Yungli's problems had just begun.

In its essentials, the Solvay process treats salt with ammonia and carbon dioxide, forming ammonium chloride and sodium bicarbonate. When the latter substance is heated, carbon dioxide and water are driven off, leaving a residue of sodium carbonate, or soda ash. The ammonium chloride can be combined with lime to yield calcium chloride and ammonia, and the latter substance recycled. The elegance of the process derives from the fact that the most expensive material, ammonia, is reused and the waste product, calcium chloride, is innocuous. The technological requirements of the system are high. Large machines made of special alloys are needed to withstand the effect of highly corrosive materials under high temperature and pressure. The process tolerates no interruption, and the flow rates of liquids and gases must be precisely controlled. Enormous quantities of raw materials – salt, limestone, coal, and coke – as well as the product itself, must be transported, regularly and rapidly, into and out of the yard. Yet owing to the continuity and automatic control of the operation, one well-trained operator and a number of manual laborers can attend to an entire plant.[21]

[20] Biography of Hou Te-pang: Sung Tzu-ch'eng, "Hou Te-pang ch'eng-kung chih lu" [Hou Te-pang's road of success], *CKKCSL* 1, no. 1 (May 1980):26–39; Kao Su, "Chung-kuo chih-chien kung-yeh ti hsien-ch'ü – Hou Te-pang po-shih" [Pioneer of China's soda industry – Dr. Hou Te-pang], *HHTP* 5 (1979):461; and Boorman and Howard, *Biographical Dictionary*, 2:84.

[21] Operation of Solvay plant: Hou Te-pang, *Manufacture of Soda*, 403, 431, 516.

When introduced to China, this procedure met with two problems. First, the Yungli factory used sea salt, which, although cheaper, contained more impurities than the rock salt or subterranean brine used in Solvay factories in the West. More precisely, sea brine contains magnesium and calcium salts, which form a sticky "mud" that clogs the system. Hou Te-pang's first achievement was to develop a method of treating the sea brine with lime and soda, converting these impurities to the corresponding sodium salts that could be used to produce still more soda. Hou showed that even though some soda was consumed in the refining process, this method was more economical than the established technique of evaporating the seawater and separating the salts by physical means. The second problem arose from the fact that, lacking a supply of crude ammonia, Yungli engineers had substituted ammonium sulfate, which caused the iron on the inside walls of the vats to pollute the soda, turning it red. Hou's solution was to add sodium sulfate, which reacted to form a protective crust of iron sulfate on the surface of the vats, resulting in the production of snow-white crystals.[22]

With the solution of these technical problems, regular production began in 1924. Yungli was the first factory in the Far East to manufacture soda by the Solvay process. In 1926, its "Red Triangle" brand won a gold medal at the International Products Exhibition in Philadelphia. Production rose to 14 thousand tons in 1927, and over 50 thousand tons in 1937. In 1931, exports from China, primarily to Japan, exceeded 7 thousand tons. Yungli was a technical and financial success, paying handsome dividends to its stockholders and favorably affecting China's balance of trade.[23]

Wu Yün-ch'u and the chemical industry of southern China

While Fan Hsü-tung was building chemical factories in northern China, another captain of industry, Wu Yün-ch'u, was laying his base in the south. These undertakings illustrate two poles of Chinese entrepreneurship: Fan milked political connections in the northern capital, binding together public and private interests in a manner reminiscent of the *kuan-tu shang-pan* [official supervision—merchant management] system of the late Ch'ing. His products – salt and soda – were basic chemicals that fed the industrial machine from the top. Wu worked in the freewheeling atmosphere of the treaty ports, where a sharp eye and

[22] Technical problems: ibid., 46–69; Ou-yang I, "P'ing-tung hua-hsüeh kung-yeh k'ao-ch'i," 103; Sung Tzu-ch'eng, "Hou Te-pang ch'eng-kung chih lu," 30–1.
[23] Gold medal: ibid., 31. Production and export figures: *Shina kōgyō*, 14–16, 81, 100. Profits: ibid., 24.

willingness to take risks were often more useful than friends in high places. Eventually, Wu's empire extended to the manufacture of acids and sodas, but he began at the bottom of the market created by the introduction of new consumer products from abroad.

Wu Yün-ch'u was born in 1890 to a poor household in a village outside Shanghai. He attended local schools for a time, then shifted to the industrial training school attached to the Kiangnan Arsenal, where he was introduced to foreign languages and science, primarily chemistry. Among Wu's teachers was Hsü Hua-feng (1858–c. 1930), the third son of Hsü Shou, who taught chemistry at Kiangnan and later served as director of a soap factory. A regular visitor to the Hsü home, Wu and the youngest son became close friends – another thread linking Chinese chemistry to the larger fabric of nineteenth- and twentieth-century China.[24]

After completing the four-year course, Wu worked as a technician at Kiangnan and later at the Hanyang Arsenal, where he rose to head the chemical and powder plant. During World War I while still at Hanyang, he noted the shortage of chemicals required to make matches, a product introduced from Europe and later made scarce by the interruption of trade. In 1918, with 5 thousand *yüan* of his own money, he set up China's first factory for the manufacture of match-making materials (potassium chlorate). When the enterprise prospered, Wu resigned from the arsenal and went to Shanghai to seek his fortune. In close touch with the match business, he next discovered the demand for glue used to fasten the combustible materials to the stick. In 1921, using capital raised from Shanghai match merchants, Wu opened his second business. With a keen eye he scanned the horizon for the next opportunity.[25]

By the early 1920s, Shanghai, a center of international commerce, was awash with new fashions and products. Among these was a condiment, monosodium glutamate or MSG, whose popularity had penetrated even the bastion of Chinese cuisine. The meatlike taste of monosodium glutamate was first noted by the Japanese chemist Kikunae Ikeda in the course of his study of *Laminaria japonica*, the Japanese seaweed used as a flavoring. In 1908, Ikeda and a colleague patented the manufacture of MSG, which Japanese factories began to make in bulk. The popularity of this seasoning spread throughout the East, particularly to China, where by the mid-1920s consumption exceeded $1 million per year.

Wu Yün-ch'u recognized the potential of this industry in China and undertook to master the method of manufacture. In 1922, he raised 50

[24] Biography of Wu Yün-ch'u: Ch'en Chen and Yao Lo, *Chung-kuo chin-tai kung-yeh-shih tsu-liao*, 1:521–9. Hsü Hua-feng: Chi Hung-k'un, "Hsü Hua-feng shih-lüeh" [Brief biography of Hsü Hua-feng], *Hsü Shou ho Chung-kuo chin-tai hua-hsüeh shih*, 347.

[25] Match and glue factories: Ch'en T'ao-sheng, "Hu Han hua-hsüeh kung-yeh k'ao-ch'a-chi," 242.

thousand *yüan*, primarily from the Kincheng Bank, to establish the T'ien-ch'u Monosodium Glutamate Factory. Japanese manufacturers protested that this enterprise violated their legal rights, but Wu was able to secure patents in the United States, Britain, and France, and production proceeded. By 1928, the value of T'ien-ch'u MSG exceeded that of the imports. Because it required special knowledge and skill, the manufacture of this product did not spread beyond Shanghai, but it served during the 1930s as an important catalyst for the chemical and food industries of that city.[26]

Development of the MSG industry in China depended in large part on the supply of two key raw materials. The first of these was "crude gluten," which contained the essential protein and was obtained as a byproduct in the manufacture of starch from wheat. In the 1920s, the only source of gluten was spinning factories, which made starch to stiffen yarn and cloth. Their limited production, however, led some experts to conclude that an adequate supply of gluten would have to await the development of other starch-consuming industries, such as the manufacture of glucose and alcohol. They may have been right: Production of both MSG and ethyl alcohol grew rapidly during the early 1930s, although I have found no evidence for a causal connection between these developments. The second key substance, hydrochloric acid, which was reacted with the gluten to form MSG, was first imported from Japan. In order to displace the Japanese as supplier of the T'ien-ch'u plant, Wu established the next element of his chemical complex, a factory to make hydrochloric acid, the description of which belongs to the next part of this story.[27]

Conclusion

The case of the chemical industry demonstrates both the extent and limits of change in the economy, science, and technology of interregnum China. In industry, as in other fields, 1895 marked a turning point, after which central authority declined and new forces took charge. This meant the bubbling up, here and there, of new products, methods, forms of management, and expertise that in the long run would shake China to the

---

[26] History and manufacture of MSG: Wang Shih-mo, "Tan-pai-chih chung chih ku-suan yü ku-suan-na t'iao-wei-fen" [Glutamic acid and monosodium glutamate seasoning in proteins], *KH* 17, no. 7 (July 1933):1018–48; and Tseng Chao-lun and Hu Mei, "Gluten Hydrolysis and Preparation of d-Glutamic Acid Hydrochloride," *JCCS* 3, no. 2 (June 1935):154–72. Patent: "T'ien-ch'u wei-ching chih-tsao kai-k'uang" [Summary of T'ien-ch'u monosodium glutamate manufacturing] *HHSC* 3, no. 6 (1948):9.

[27] Motive for making HCl: *China Industrial Handbooks: Kiangsu* (Shanghai: Ministry of Industry, 1933), 655–6.

core. In the short run, however, the scope of change was limited. A few dozen factories made "chemical" products, mostly simple consumer goods, by modern means. Only one firm, Yungli, manufactured industrial chemicals, salt and soda, for the civilian market. Some government arsenals produced sulfuric and nitric acid, but this was done largely under foreign management and direction. These changes were more of quality than quantity – a sign of things to come.

Young Chinese of this period who might have a taste for scientific and technical studies could see both the hope of the future and the bleak reality of the present. By the early 1920s, they could look to several examples of successful careers that could not have existed a generation before. Fan Hsü-tung's work in salt and soda demonstrated how the partnership between public and private sectors could make a man rich and famous, while serving the nation as a whole. Hou Te-pang's story showed that fame and fortune could come from a knowledge of science, engineering, and experience in the West. Wu Yün-ch'u taught his countrymen that in the unformed clay of warlord China, anyone with skill, imagination, and courage could create the future. The point of all these examples was the same: The world was changing, opportunities abounded, and knowledge of science was one key for entering the new age. But the record of the chemical industry also proved the opposite: namely, that change in the structure of the Chinese economy was slow, that there were few jobs for skilled workers at any level, and that the immediate prospects for students of chemistry, or any other branch of science, were dim.

This misfit between industry and education also characterized relations between industry and research. On the one hand, the example of the Yungli conglomerate and the Golden Sea Institute for Chemical Research showed that some Chinese recognized the connections among research, development, and industrial growth, particularly in the chemical sector, and were willing to invest in the future. Golden Sea had excellent facilities, a large number of trained scholars, and performed research on a wide range of problems that went beyond the immediate needs of the parent salt and soda factories. Elsewhere, however, the reality was quite different. Most "chemical" factories in interregnum China were small, crude operations that processed agricultural goods or made simple consumer products. Descriptions of these factories suggest that their "research and development," if any, consisted of improvements made on a trial-and-error basis by the resident technician, who was at the same time owner, manager, and jack-of-all-trades. There was a bridge between scholarship and production, but it was narrow, weak, and already overcrowded.

The first decades of China's industrial development produced limited results, and these in turn had a limited impact on the study and practice of science. On the other hand, qualitative changes in this sector established models for the future and inspired enterprising young scholars and industrialists to help make them real. Here, as elsewhere, individual achievements gave cause for pride and optimism. But in the absence of an established order, the whole was less than the sum of its parts. The progress made during the interregnum in separate sectors of China's society and culture could reach fulfillment only under a unified system of government, which was achieved, at least in part, during the Nanking Decade.

PART III

The Nanking Decade
1927–1937

During the Nanking Decade (1927–37), so named because in these years the Nationalist Government under the Kuomintang Party made its capital in this city on the Lower Yangtze, the Chinese took charge of science in their country and carried it to new heights. Two factors explain this success. First, establishment of a viable central authority, the first in China since the fall of the Ch'ing, offered a framework for effective, purposeful action that had been lacking under the warlords. Second, the return from abroad of China's second generation of scientists provided the leadership needed in teaching, research, and the application of new techniques. Owing to these political and social developments, Chinese science advanced rapidly during the 1930s and did so under direction of the Chinese themselves.

Creation of a viable central government was one key to this success. The collapse of the old order, following the debacle of 1895, had removed those institutional and cultural impediments that slowed the development of science in the nineteenth century and opened avenues for new initiatives during the first quarter of the twentieth. By the late 1920s, however, the age of creative disorder had passed, the further advances depended on the establishment of secure, stable patterns of behavior. It is important to note that the Nationalists never succeeded in their goal of unifying China. During the Nanking Decade, Manchuria was lost to the Japanese, the Chinese civil war ground on without resolution, most parts of rural China paid allegiance to Nanking in name only, and even in the areas under their control the Nationalists remained fragmented, factionalized, and weak. Still, the Nanking government provided to the coastal, urban, and privileged sectors of China a degree of stability, security, and direction that had been unknown in that country for more than a generation.

The Nationalists had a hand in all the scientific developments of this period. Science education was expanded and improved under the new Ministry of Education. More and better research was performed in China's national universities, academies, and institutes. Nanking encouraged the introduction of new technologies in public and private industry. In all areas, the emphasis was on the practical application of knowledge to fulfill national goals. Science became a responsibility of the state and an instrument with which the Nationalists sought to change China.

Just as important as politics were social developments – highlighted by the return from abroad of China's second generation of scientists. The scientific cohort of the 1930s was more numerous and better trained than its predecessors. This group included scores of young men, and now also women, in every discipline, with doctorates from the world's leading universities and published writings to their credit, poised to take over positions in teaching, research, and engineering and to perform work on the standard they had learned in the West. Previous Chinese elites, both traditional literati and modern intelligentsia, had pursued literary and humanistic studies to prepare for leadership in the social and political realms. Scientists of the second generation conceived of themselves as members of a profession, whose careers and contributions lay within the confines of their special expertise. Unlike the pundits, organizers, and entrepreneurs who preceded them, the scientists of the 1930s were ready to serve China in classrooms, laboratories and factories.

The combination of a strong central authority and abundant human resources enabled the Chinese to take charge of their own house. During the Nanking Decade, public colleges and universities replaced the mission schools at the forefront of higher education, while China's universities, academies, and institutes led in every branch of research. Returned students dominated all these activities; foreigners were pressed to the sidelines or off the field entirely. New industrial technologies were introduced by Chinese enterprises, under native management and direction. Foreign universities continued to train Chinese at the doctoral level, and foreign foundations and governments provided support. But after 1930, in contrast to before, the teaching, scholarship, and other professional tasks, as well as the management and most of the financing, were in Chinese hands.

As the quality of Chinese science improved, it came to resemble what one would expect to find at any point on the periphery of this global enterprise. Students in China, trained through the bachelor's level, advanced to doctoral programs in the leading European and American universities, where many performed with distinction and some were invited to stay on as assistants or junior faculty. Back in China, young scientists continued the work they had begun as graduate students and gradually moved to design and execute their own research, the best of which was published in the world's leading scientific journals and cited by colleagues abroad. Factories in China purchased machinery and copied technology from industrial leaders in the West and in some instances developed or improved techniques in their own workshops and labs.

The twin developments that explain China's success – the growth of state power and professional expertise – also contained within them the seed of conflict. The state, or elements of it, wanted to use science for its own ends. Conservative members of the Kuomintang favored the application of knowledge to increase production and strengthen defense, and they expected Chinese scientists, educators, and scholars to support these goals. Proponents of central authority viewed science as a tool and the scientist as a technician who would apply his skills to the machinery of the nation-state.

Chinese scientists reacted with ambivalence to this approach. They were glad to see the government increase support for science, which had been neglected during the chaos of the 1920s. Many were fervent nationalists, who cared enough about their country to return home when they could have found attractive, well-paying jobs in the West. And as professionals, they were hesitant to venture beyond the limits of their expertise into a political debate over the uses of science outside the lab. On the other hand, China's scientists were part of a larger enterprise, whose purpose was not to make more and better things, but to discover or create new knowledge. Their standing and sense of worth as members of the scientific profession were threatened by the pressure to divert attention and resources to mundane, practical ends. Few were willing to step out of the laboratory and defend science against the threat of expanding state power. But many welcomed the efforts of others, particularly the older liberal intellectuals who championed science as the model for a culture that should be independent from political control and built up, piece by piece, from the separate choices of free individuals. This conflict – science as an instrument of power and scientists as servants of the state versus science as a quest for knowledge and scientists as members of an autonomous profession – sharpened, even as its elements, authority and expertise, combined to raise up science in Chinese hands.

8

Science and the state during the Nanking Decade

The advance of Chinese science during the Nanking Decade was made possible by the return to China of large numbers of well-trained scientists and the efforts of a government determined to place science and technology at the service of the nation. The second generation of scientists were more numerous than their predecessors, better prepared, and more committed to careers in the laboratory, classroom, and factory than in the public arena. The Nationalists, better equipped for the task than Chinese rulers before them, pursued policies designed to apply scientific knowledge to economic production and national defense. In some cases, scientists and officials cooperated to produce noteworthy results. In others, they clashed over the nature of science and its relation to human needs. The problem of balancing the interests of state and society proved crucial and controversial in the development of science during the Nanking Decade.

Scientists of the second generation

Scientists of the second generation were distinguished from their predecessors in three ways. First, they were younger, born around the turn of the century and still green at the time of the Nationalist takeover in 1928. Second, they were better educated, most having completed doctorates in European and American universities and many having their first scholarly publications behind them. Third, they were more committed to research and teaching as their professional responsibility, and less interested in the broad political and cultural concerns that had preoccupied intellectuals of the May Fourth era.

A collective biography of this generation points up the changes that had occurred since the beginning of the century. To begin, these men and women enjoyed the privileges of their predecessors, but more so. Data from various sources show that representation from Kiangsu and Chekiang, culturally the most advantaged provinces, increased over time.

Among the first members of the Academia Sinica, chosen in 1948 as the cream of China's scientific community, 41 percent of those born in 1895 or before and 53 percent born thereafter were from Kiangsu or Chekiang. In a sample of 106 chemists active in China during the first half of the century, 35 percent of those born before 1910 and 53 percent born in 1910 or later were from these provinces. Similarly, Chinese born in China and listed in the 1955 edition of *American Men of Science* include 111 individuals born before 1920, of whom 33 percent were from Kiangsu and Chekiang, and 47 born in 1920 or later, 53 percent from these provinces. In each case the increase came at the expense of all other provinces, including coastal Fukien and Kwangtung. Since many students spent time in one or more of the modern schools of the Lower Yangtze before going abroad, the role of this area was in fact greater than suggested by province of origin alone. This pattern helped to tighten the geographic focus of scholarly achievement and social mobility, as the traditional order, which had allowed for recruitment of intellectuals from and their return to remote areas, gave way to the process of modernization and the concentration of new elites in the cities and on the coast. Making high marks in the universities of Europe and America, a prerequisite for success in the sciences more than in any other field, depended upon the economic and cultural background of the students, both of which were favored by residence on the Lower Yangtze.[1]

It is also probable, although less demonstrable, that the later generation drew a larger share of its members from privileged families. Y. C. Wang points out that with the passage of time, Chinese government funds were shifted from fellowships for study abroad to schools in China, causing a decline in overseas students on fellowship with the result that during the 1920s and 1930s, a growing percentage of these students came from relatively wealthy business, professional, and official households. Poor students could still afford to study in Japan, where scientific and technical programs were less developed, but substantial backing was required to go to Europe or America, which retained a hegemony in these fields. Within China, the situation was much the same: Surveys of students in Chinese institutions of higher learning during the 1930s show that over 70 percent were from commercial, educational, and professional backgrounds, while those from farm families declined from 26 percent

---

[1] Academia Sinica, 1948: *Kuo-li chung-yang yen-chiu-yüan yüan-shih-lu* [Roster of members of the Academia Sincia], vol. 1, ed. Chu Chia-hua (Academia Sinica, 1948). *American Men of Science*, 9th ed., ed. Jaques Cattell (Lancaster, PA: Science Press, 1955). Data on chemists presented here has been assembled from: *Chiao-yü-pu hua-hsüeh t'ao-lun-hui chuan-k'an*, 283–7; "Chung-kuo k'o-hsüeh-yüan hua-hsüeh-pu hsüeh-pu wei-yüan chien-chieh," 552–64; and biographies of chemists from *HHTP, CKKCSL*, and other sources listed in the bibliography. This is the same sample that was cited in Chapter 4, note 1.

in 1931 to less than 20 percent in 1935. Although data on individual scientists are sparse, two chemists of the Nanking era, Sah Pen-t'ieh and Tseng Chao-lun, were reportedly from prominent families; none that I have encountered was said to be of peasant or working-class stock, and the documented increase of Kiangsu and Chekiang natives supports the notion that the economic origins of these students were also on the rise. Scholarships, particularly those from the American and British Boxer indemnity funds, opened the door for some future scientists with limited means, but most Chinese who made it in this field apparently came from the right families as well as the right regions.[2]

What most distinguished members of the second generation, however, was their schooling, beginning with preparation in China. As late as the 1910s, there was no education in China, particularly in the sciences, approaching the standard common in the West. Lacking a firm base at home, few Chinese from the earlier period reached the top of the academic ladder abroad. By the 1920s, on the other hand, graduates of many Chinese colleges and universities were able, after a year or two of language and other preparation, to enter leading European and American doctoral programs. Most did their undergraduate degrees in institutions under Chinese administration, although a sizeable minority attended mission schools, which were somewhat better in the sciences.[3]

After preliminary studies in China, usually through the bachelor's degree, future scientists invariably went to the United States or western Europe for advanced training. Despite Nanking's efforts to channel students into scientific and technical fields, members of this generation were *no* more likely than their predecessors to choose careers in science. Evidence compiled in China shows that by 1933 a majority of students leaving for study abroad announced their intention to specialize in some technical subject (Appendix, Table A.3). According to Y. C. Wang, who gathered his data from Chinese students in the United States, however, during the 1920s and 1930s the percentage who chose to major in pure or applied science in fact *decreased* over time (Appendix, Table A.4). This apparent contradiction can be explained by the persistent gap between government policy and student interest: In granting visas for

[2] Y. C. Wang, *Chinese Intellectuals and the West*, 150–5. Survey of students, 1931: *Chung-kuo chiao-yü nien-chien, ti-i-tz'u* [The first China education year book], ed. Chiao-yü-pu [Ministry of Education] (Shanghai: K'ai-ming shu-tien, 1934), 4:58. Survey, 1935: Kuo Tze-hsiung, "Higher Education in China," *Information Bulletin* 3, no. 2 (21 January 1937):42.

[3] Entry of Chinese college graduates into American universities: Yoshi Kuno, *Educational Institutions in the Orient*. Four of thirty-seven members of the 1948 Academia Sinica, eleven of forty-four chemists born between 1896 and 1920, and 35 percent of native Chinese listed in the 1955 edition of *American Men of Science* first attended missionary colleges or universities in China. For reasons that I cannot explain, the mission schools produced a greater proportion of biological than physical scientists.

foreign study, Nanking favored applicants who indicated the intention to enter scientific or technical fields. But after reaching the West, many of these students reverted to the established perference for humanities, social studies, or the arts. The migration of Chinese students in Europe and the United States into departments of science and engineering began only in the 1940s and continued into the early 1950s, suggesting that it may have been the decision to leave China that persuaded these young men and women to acquire skills that would help them to get jobs abroad.[4]

Although the percentage of Chinese overseas students who opted for science declined during the prewar era, the rise in absolute numbers and improvement in the training these students received before going abroad resulted in a substantial increase in the number who earned advanced degrees. There were in all the world before 1920 just thirty Chinese with doctorates in the pure or applied sciences. The numbers of Ph.D.'s in all fields increased sharply during the years 1920–4, and steadily for each of the five-year periods that followed, before declining somewhat after the outbreak of World War II (Appendix, Table A.5). During the 1930s alone, nearly 700 Chinese were awarded doctorates in science, engineering, agriculture, and medicine. It should be noted that these figures represent all graduates with Chinese surnames, and thus include an indeterminate number of overseas Chinese with little or no connection to China. The extraordinary number of doctorates in medicine, primarily from German universities, also tilts the balance that actually prevailed among degree-holders who returned to live and work in China. Even after discounting for these and other distortions, however, it remains that after 1920 an abundance of young Chinese trained in the world's best universities were available to take over the scientific enterprise begun by the previous generation.

It is striking, moreover, how many Chinese students of this generation chose to return home. Today, the "brain drain" from developing countries is taken to be the norm, but this was not the case with China in the years before World War II. In chemistry, Wu Hsien, Hou Te-pang, Sah Pen-t'ieh, Chuang Chang-kong, and others at the forefront of their profession could have secured positions in Europe or the United States, but only one prominent chemist, Chen Ko-kuei, left China prior to 1937 for a job in the West. Idealism, a desire to serve the nation, undoubtedly helped persuade many young people to return home. The improvement of conditions in China also played a role. The establishment of a viable government committed to the advancement of science, greater security,

---

[4] Specialties of Chinese students approved for study abroad in 1930–2: *Chung-kuo chiao-yü nien-chien* (1934), 3:3. In 1933: Kuo Tze-hsiung, "Higher Education in China," 46–7.

and a higher standard of living; the creation of societies, journals, institutes, and other prerequisites of professional life; and developments in the modern sector of the economy – all these factors invited the return of Chinese youth hoping to build productive and rewarding careers for themselves and their country.[5]

These returnees were favored by an expansion in the number of academic and research positions in China and the mobility of teachers and scholars among the available posts. After 1928, the Nanking government created the Academia Sinica, Peiping Academy, and several research institutes and expanded the science faculties of the major universities, opening up new seats for aspiring scholars. Meanwhile, there was a lively competition among these institutions for the best people. In contrast to the almost total absence of professional mobility in post-1949 China, scientists of the Nanking Decade moved once, twice, or even three times in the scramble for talent and jobs.

On the whole, the best and brightest of this generation were well rewarded. Data on 122 chemists active in China during the period 1920–49 show that almost all had some overseas experience and most held advanced degrees (82 doctorates and 22 master's). Three-quarters of the degrees were from American universities, most of the rest from Europe, and only a handful, all bachelor's, from Japan. A snapshot of China's leading chemists, who gathered in Nanking in the summer of 1932 for the government-sponsored Chemistry Forum, presents a similar picture. The list includes forty-five individuals, all male, ranging in age from twenty-seven to forty-four, with an average age of thirty-three. Three-fifths were natives of Kiangsu, Chekiang, or Kwangtung. All had studied abroad: three-fourths in the United States, the rest divided equally between Europe and Japan. Most had advanced degrees (twenty-three doctorates, nine master's) and an average of six years postgraduate experience. Thirty were employed in universities, ten in government agencies, four in industry, and one in a research institute.[6]

Since teaching was the principal means of employment for most Chinese scientists, a review of the nation's university faculty provides another look at this generation's leaders. W. E. Tisdale, who surveyed the top Chinese public and private universities in the fall of 1933, found 109 science teachers with higher degrees from foreign universities, most concentrated in the departments of biology, chemistry, and physics of the seven or eight leading institutions, and a somewhat larger number of

---

[5] Return of students: Chang I-tsun, "Chung-kuo ti hua-hsüeh," 3. Daisy Wu, wife of the late Wu Hsien, told the author in an interview in 1984 that it never occurred to her or any of the Chinese students she knew in the United States during the 1920s to do anything but return to China.

[6] One hundred twenty-two chemists: See sources in note 1. Snapshot: *Chiao-yü-pu hua-hsüeh t'ao-lun-hui chuan-k'an*, 283–8.

junior faculty and assistants with bachelor's degrees from schools in China or abroad. Faculty of the Chinese institutions, which by this date had eclipsed their missionary rivals, accounted for 60 percent of the higher degrees, over 90 percent of which were from American universities. The professors ranged in age from twenty-five to forty-eight, with an average age of thirty-four and five years postgraduate experience.[7]

Unfortunately, there are few figures on the number and disciplines of technically trained people in pre-1949 China. The best guide for the 1930s is the membership of the Science Society of China (SSC), the largest and most representative organization of its kind. The roster of the SSC grew from around 1,000 in 1930 to more than 1,500 in 1937, while the distribution of members by field remained approximately the same (Table 8.1).

If the brightest and luckiest students flourished after their return to China, the less talented or fortunate found the situation difficult. The problem for most was the shortage of jobs that could make use of their skills. All the available data show that the great majority of Chinese trained in technical subjects wanted to work in research, development, or engineering, but because such positions were scarce, most ended up teaching. Y. C. Wang found that during the years 1917 to 1937, over half of all returned students entered education or government service. A 1925 survey shows that engineers, the largest student group, were among the hardest hit – less than one in three was able to find employment in a technical field – while an even larger percentage of those trained in the sciences taught school, in colleges and universities when possible, and, as these jobs filled up, at the secondary level. Graduates of universities in China followed the same path. Surveys conducted in the 1920s and 1930s confirm that most middle school science teachers considered their jobs temporary necessities and planned to leave as soon as they found positions in government or industry.[8]

The gap between hope and reality widened with the passage of time. In 1932, the geologist Wong Wen-hao responded to the many students and recent graduates who had written to him expressing their pessimism about the country's future and their own prospects, with the advice that they move inland or to the countryside where their skills were in short

[7] W. E. Tisdale, "Report of Visit."

[8] Y. C. Wang and 1925 survey: Y. C. Wang, Chinese Intellectuals and the West, 168–74, 514–16. Graduates of Chinese universities: Wang Chung-t'ien, "Kuo-nei ta-hsüeh chi chuan-men hsüeh-hsiao pi-yeh-sheng chiu-yeh ch'ing-k'uang ti i-ke tiao-ch'a" [Investigation into the employment situation for graduates of Chinese universities and higher technical schools], CHCYC 22, no. 6 (1934):49–60. Middle school teachers: George Twiss, Science and Education in China, 156–7; and Tai An-pang, "Chung-kuo hua-hsüeh chiao-yü ti hsien-chuang" [Present situation of Chinese chemical education], KH 24, no. 2 (February 1940):89–96.

Table 8.1 *Membership of the Science Society of China by fields, 1930*

Field	Number	Percentage
Physical sciences	232	22.9
Chemistry	99	
Physics	47	
Mathematics	40	
Geology	35	
Astronomy	6	
Meteorology	5	
Biological sciences	201	19.8
Agriculture	74	
Biology	69	
Medicine	58	
Engineering sciences	343	33.9
Civil	106	
Electrical	70	
Mechanical	69	
Mining	48	
Chemical	42	
Textile	8	
Social Sciences	169	16.7
Economics	61	
Education	35	
History and philosophy	31	
Political science and sociology	28	
Psychology	14	
Unknown	68	6.7
Total	1,013	100.0

Source: Science Society of China, *The Science Society of China: Its History, Organization, and Activities* (Shanghai: The Science Press, 1931), 6–7.

supply. But few educated Chinese wanted to leave the bright lights and big cities, and the government did little to make other options attractive to them. Thus, the perennial programs to find suitable employment for recent university graduates foundered on the simple fact that there were too many people seeking too few jobs. In the spring of 1936, the National Academic Work Advisory Board (NAWAB) estimated that 2,000 more university graduates had been added to the unemployment roles. After the outbreak of the war, the problem grew worse. All parties shared the blame for this situation: The government failed to attempt meaningful rural reforms; short-sighted entrepreneurs sought quick profits rather than invest in research and development; colleges and universities ignored the need for vocational training; and students shunned practical studies while in school and refused to take low-status manual jobs after graduation. Whatever the reasons, the result was the same: Most Chinese with technical training could not find positions that suited them or contributed to the development of China.[9]

These problems produced tension, disillusionment, and the dissipation of intellectual energy among scientists of the second generation. The first generation had included many men with wit and skill, but little capacity to do independent research. At first this was no handicap, for there was narrow scope for research in China anyway and great need for teachers, organizers, propagandists, and other types who could help build a place for science in the Chinese scheme of things. Whatever their contributions, these men moved into a virtual vacuum, occupying positions in colleges, universities, and other institutions that flourished amidst the economic and cultural growth of the 1920s. In these early years, foreign scholars with advanced training in the sciences joined the faculty of the mission colleges to strengthen programs that the older missionaries recognized as beyond their ken. When the second generation of Chinese with doctorates from the world's leading universities returned home, many of the best jobs were taken. And the inequity was exacerbated, at least one disgruntled observer noted, by the incompetence of many senior scholars and the habit of some to protect their positions by forming cliques based on personal connections, rather than rewarding excellence.[10]

Many young scholars who had made a successful start as graduate students and junior researchers lost their enthusiasm for science after returning home. While abroad, they enjoyed access to the most modern facilities, the encouragement of mentors, and the incentive of pursuing

[9] Wong Wen-hao: Wong Wen-hao, "I-ke ta-p'o fan-men ti fang-fa" [A method for overcoming depression], *TLPL* 10 (24 July 1932):2–5. NAWAB: Ch'eng Ping-hua, "Chiu-chi shih-yeh ta-hsüeh-sheng chung ying chu-i ti chi-tien" [A few noteworthy points on the relief of unemployment among university students], *TLPL* 219 (1936): 17–19. Problem grows worse: Tseng Chao-lun, "Chih-te chu-i ti ta-hsüeh-sheng ch'u-lu wen-t'i" [The problem of the future of university students deserves attention], *Hsin min-tsu* [The new nation] 2, no. 1 (17 July 1938):5–7. Explanations: ibid.; Ch'en Tai-sun, "Kuan-yü ta-hsüeh pi-yeh-sheng chih-yeh wen-t'i i-ke chien-i" [A proposal on the problem of employment of university graduates], *TLPL* 211 (1936):8–9; Ku Yü-hsiu, "Chuan-men jen-ts'ai ti p'ei-yang" [Cultivating technical personnel], *TLPL* 71 (1933):10; Ts'ao Hui-ch'ün, "Ta-hsüeh-sheng yü hua-hsüeh kung-yeh" [University students and chemical industry], *HHKY* 8, no. 2 (July 1933):1–7; and Wang chung-t'ien, "Kuo-nei ta-hsüeh chi chuan-men hsüeh-hsiao ti i-ke tiao-ch'a," 49–60.
[10] Criticism of older generation: Wang Ching-hsi, "T'i-ch'ang k'o-hsüeh yen-chiu tsui ying chu-i ti i-chien-shih – jen-ts'ai ti p'ei-yang" [In support of the most important aspect of scientific research – cultivation of human talent], *TLPL* 26 (1932):11–12.

higher degrees, all of which helped build confidence and hope. Back in China, everything changed. Yesterday's high-spirited youth found themselves alone in a society that neither understood nor appreciated their work, with little equipment to perform research, limited opportunity to apply their knowledge, poorly prepared students, and few colleagues with whom to share their dreams. In these circumstances, many decided that rather than struggle on amidst hardship, they should begin to reap the benefits of their long apprenticeship. "In this scientifically immature China," Ping Chih, the country's leading zoologist, explained,

where our studies are rudimentary in every respect, many believe there is no harm in taking it easy or pursuing material comfort for themselves. Leaning on the weak reed of their foreign degrees, they seek comfortable positions in education and government [and let their commitment to science lapse]. This mentality is characteristic of the so-called scientists in China today [1934]. It has held back the development of science in this country and is a disgrace to us all.[11]

If the situation was difficult for the returned students, it was even worse for those who remained at home. In European and American universities professional advancement in the sciences follows a well-worn path from undergraduate to graduate student, to junior and finally senior professor. The graduate student is the linchpin of this system, serving simultaneously as trainee in a research occupation above and laboratory instructor to the undergraduates below. In China and in other scientifically dependent areas, however, graduate students are often removed to foreign institutions, denying them the experience of training on their home turf and leaving local educators to create something in their stead. To deal with this problem, Chinese colleges and universities of the Republican period established "teaching assistants" [chu-chiao], laboratory instructors who had the closest contact to, best understanding of, and sometimes greatest influence on the students assigned to them, but no prospect for promotion to a regular teaching post. Professorships were reserved for returned students, who may have taken their turn as graduate assistants in foreign universities, but had little experience with students and institutions in China. Separation of these roles was one of the more unfortunate results of the continued dependence of Chinese on higher education in the West.[12]

[11] Quotation: Ping Chih, "K'o-hsüeh tsai Chung-kuo chih chiang-lai" [The future of science in China], KH 18, no. 3 (1934):302.
[12] Teaching assistants: Sun Hsüeh-wu, "T'i-ch'ang tzu-jan k'o-hsüeh chiao-yü ti chi-chien chi-wu" [Urgent tasks for promoting natural science education], TLPL 34 (1933):9–12. Problem of promoting from within: Meng Chen [Fu Ssu-nien], "Kai-ke kao-teng chiao-yü chung chi-ko wen-t'i" [Some questions on the reform of higher education], TLPL 14 (21 August 1932):5.

Chemists of the second generation

There is unfortunately little information on individual scientists of the second generation. Unlike many of their predecessors, who left a trail of paper recording their involvement in public affairs, members of the second generation were distinguished by a professional commitment to laboratory and classroom that left little time for outside activities. The product of these true scientists lies in their published research. But the more we know about their scholarship, the less we know about their lives.

We have, for example, only the sketchiest information on the two most productive chemists of this generation, Chuang Chang-kong and Sah Pen-t'ieh. Chuang was born in Fukien Province in 1894. He earned both bachelor's (1921) and doctoral (1924) degrees from the University of Chicago, after which he returned to China and taught at Tungpei University in Manchuria. When the Japanese annexed this territory in 1931, he accepted an invitation to spend a year in the Gottingen laboratory of Adolph Windaus, who had won the 1928 Nobel Prize for work on sterols and sex hormones, subjects that Chuang began to investigate on his own. After returning to China in 1933, he was named head of the National Institute of Chemistry, a post he held until 1937, when the institute retreated inland to escape the Japanese invasion. During the first part of the war, Chuang stayed in the International Settlement of Shanghai and continued his research. When the enemy occupied this part of the city in 1941, he rejoined his colleagues in Kunming. During the late 1930s, Chuang published several articles on the synthesis of sterols and sex hormones in the leading German chemical journal, confirming his reputation as an important figure in this branch of chemistry. He also worked on Chinese herbals, establishing the structure of two new alkaloids. After a brief stint as president of National Taiwan University in 1948, Chuang returned to the mainland, where he served as director of the Institute of Organic Chemistry of the new Chinese Academy of Sciences and became a prominent figure in the People's Republic of China (PRC) scientific establishment. He died in 1962.[13]

Sah Pen-t'ieh, also known as Peter Sah, was born in 1900 to a family of comfortable means and scholarly prowess, which included Peter's younger brother, Adam Sah or Sah Pen-t'ung, one of China's leading

---

[13] Life and work of Chuang Chang-kong: Liu Kuang-ting, "Chu-ming yu-chi hua-hsüeh-chia Chuang Ch'ang-kung chien-chuan" [Brief biography of the famous organic chemist, Chuang Chang-kong], *Chuan-chi wen-hsüeh* [Biographical essays], vol. 39, no. 4, 31–3. T'ien Yü-lin, Kao I-sheng, and Huang Yao-tseng, "Wo-kuo yu-chi hua-hsüeh ti

physicists. Sah Pen-t'ieh earned his doctorate at Wisconsin in 1926, after which he returned to China to take a position in the department of chemistry at Tsinghua University. During the next decade, he produced nearly one hundred research articles, by far the largest number of any Chinese chemist of this period, and trained dozens of students in modern laboratory techniques. In the mid-1930s, he worked in the laboratory of Adolf Windaus, previously visited by Chuang Chang-kong, where he was introduced to the study of sterols and sex hormones. During the war, Sah remained in Peking, continued his research at Fu-jen University, and published the results in European journals. After the war, he left China and completed his career on the faculty of the University of California at Davis. Neither Chuang nor Sah left writings on nontechnical subjects, and none of their contemporaries thought to record the biographies of two of China's leading scientists.[14]

Only one chemist of this generation ventured beyond the confines of his profession to play a larger role in society and politics, and thereby left behind a more colorful life story. Tseng Chao-lun (1899–1967) was stuff for a novel. A bohemian scholar in faded blue gown and cloth shoes, Tseng fought with his wife over his bathing habits and other things and found a home among students, who regarded him as a man who was "at ease with poverty and rejoiced in the Way." He was born in Hunan Province in 1899, a descendant of the great viceroy Tseng Kuo-fan, and later married, somewhat unhappily, into the family of the prominent Nationalist general Yü Ta-wei. After graduating from Tsinghua in 1920, he won a Boxer fellowship to study at M.I.T., where he earned his doctorate in chemical engineering in 1926. Back in China, Tseng created two of the country's most important centers of chemical education and research – first at National Central University in Nanking, and after 1931 at Peking University – helped found the Chinese Chemical Society, served as the first editor of its *Journal*, and wrote on scientific matters for both professional and lay audiences. A nonconformist and endless kibitzer, Tseng was a popular figure on campus and in the teahouses that surrounded it. He was just the man to attract capable and idealistic young Chinese to the study of science.[15]

Most of Tseng's research was on the fundamental problems of chemistry. Although substantial enough to win the respect of his peers, it was published only in China and gained no recognition abroad – the ultimate mark of professional success, earned by such figures as Wu

hsien-ch'ü – Chuang Ch'ang-kung chiao-shou" [Pioneer of our country's organic chemistry – Professor Chuang Chang-kong], *HHTP* 4 (1979):365–72.

[14] Career of Sah Pen-t'ieh: Chang I-tsun, "Chung-kuo ti hua-hsüeh," 1–6. Windaus: Sah Pen-t'ieh, *SRNTU* 3 (1935):95–108.

[15] Tseng's personal manner and relations with students: *Hsin Hua Jih Pao* [New China

Hsien, Chuang Chang-kong, and Sah Pen-t'ieh. On the other hand, Tseng was more deeply involved than many of his contemporaries in problems of a local variety. He did pioneering studies on the manufacture of the flavoring MSG, and was particularly adept at the manufacture of laboratory devices, a crucial skill for scholars in this still remote area. Given his interest in politics and close identification with students in Peking, who were in constant turmoil over the encroachment of Japanese armies and the failure of the Chinese government to combat them, it is no surprise that Tseng should be among the first scholars to turn to problems of national defense. In his laboratory at Peking University, he developed techniques to manufacture TNT and detonating devices using local materials. And in the classroom he lectured on the "chemistry of national defense," which included instruction on how to deal with poisonous gases, the manufacture of explosives, purification of drinking water, and construction of latrines.[16]

After the outbreak of the war, Tseng joined the retreat to the hinterland. As a professor at the wartime university in Kunming, he taught basic chemistry and the application of this science to local needs, while doing research on topics ranging from the rather esoteric measurement of dipole moments to the manufacture of acid from vinegar. In his spare time, he explored the rugged mountains of southwestern China, recording his experience in a journal that was later published in two volumes. Finally, at the end of the war, he emerged as an important figure in the Democratic League, which offered a moderate alternative between the extremes of the Kuomintang and Communist parties. Tseng's informal style fit well with the hard life of the interior and contributed to the creative, rough-and-tumble atmosphere of education in wartime China.[17]

Tseng Chao-lun never joined a political party. He was a fervent Chinese nationalist who served both Kuomintang and Communist regimes, but his strong sense of duty led him to take a critical stance toward authority of all types. Tseng remained in China after 1949 and

daily] (Chungking), 8 August 1944, 2; Lin Ying, "Lien-ta pa-nien" [Eight years at Associated University], *Ta Kung Pao* (Shanghai), 26 November 1946, 7; *Lien-ta pa-nien* [Eight years at Associated University] (Kunming: Hsi-nan Lien-ta hsüeh-sheng ch'u-pan-she, 1946), 173–4; Chang Ch'in, "Pu-shu Lien-ta ti yü-le-i-shih-chu-hsing" [More on education, recreation, clothing, food, housing, and activities at Associated University], *Ch'ing-hua hsiao-yu t'ung-hsün* [Tsinghua alumni bulletin] 77 (31 October 1981):12. Career at NCU and Pei-ta: Chang I-tsun, "Chung-kuo ti hua-hsüeh," 4–5. Role in CCS: Li Ch'iao-p'ing, *Chung-kuo hua-hsüeh shih*, 512–5, 524–6.

16 Reports on improved apparatuses: *JCCS* 1 (1933):37–43, 143–82; *SQNUP* 4 (1934):250, 283–332. Research on explosives: *HHKC* 2 (1935):128–32; 4 (1937):59–64, 65–7. Chemistry of national defense: Private communication of interview conducted by John Israel.

17 Research: Chang I-tsun, "Chung-kuo ti hua-hsüeh," ll; *JCCS* 11 (1944):65–7. Journal:

wrote to colleagues overseas, urging them to return home. His star rose in the early 1950s, when he served in a number of important political and administrative posts, including vice minister for higher education. But he got into trouble during the Hundred Flowers movement in 1957, when he published a declaration demanding better treatment and greater independence for scientists, and was branded a rightist during the bitter counterattack that followed. After recovering in the early 1960s, Tseng was again attacked, this time bodily, and died in 1967 of injuries inflicted by the Red Guards.[18]

Science and the state during the Nanking Decade

At the time the second generation scientists were returning home, a new, more powerful and purposeful government was taking charge – at least in those areas of China where science was serious business. In some respect, the interests of the scientists were compatible with those of the state. In the fields of language reform, organization, and finance, the state provided structure, leadership, and support, which left the scientists ample room to pursue their special professional goals. In other areas, most notably education and research, the approaches of the two sides were different and the conflict between them inevitable.

The Nationalists had a hand in every major scientific initiative of the Nanking Decade. Their boldest venture was the National Central Academy of Sciences, or Academia Sinica, which included institutes for research in each branch of natural and social science. A second, more modest effort, the National Academy of Peiping, performed a similar role in the north. For the applied sciences, the Ministry of Industry established the National Bureau of Industrial Research and the National Agricultural Research Bureau, both of which operated laboratories and extension services to promote research, development, and economic growth. The National Geological Survey, a creation of the old Peking government, housed a new laboratory for research on fuels. Government bureaus for

Tseng Chao-lun, *Ta-liang-shan I-ch'ü k'ao-ch'a-chi* [Notes on an investigation of the Yi region of Taliangshan] (Shanghai: Tu-shu ch'u-pan-she, 1945); and Tseng Chao-lun, *Tien-k'ang tao-shang* [On the Yunnan-Sikang Road] (Kweilin: Wen-yü shu-tien, 1943). Democratic League: "Tseng Chao-lun," *Chung-kung jen-ming-lu* [Biographies of Chinese Communist personalities] (Taipei: Institute of International Relations, 1978).

[18] Post-1949 career: Wang Chih-hao and Hsing Jun-ch'uan, "Chih-ming hsüeh-che, hua-hsüeh-chia Tseng Chao-lun chiao-shou" [Famous scholar and chemist, Professor Tseng Chao-lun], *HHTP* 9 (1980):559–67; Liu Kuang-ting, "Ts'ung Tseng Chao-lun ti tsao-yü t'an-ch'i" [On the vicissitudes of Tseng Chao-lun's life], *Chung-hua tsa-chih* [China magazine] 19, no. 218 (Semptember 1981):15–16; Stephen Salaff, "A Biography of Hua Lo-keng," *Science and Technology in East Asia*, ed. Nathan Sivin (New York: Science History Publications, 1977), 226–9.

meteorology, navigation, transportation, communication, standards, and other technical functions were created or existing agencies expanded and given a broad mandate to conduct research and promote development in their respective fields.

Nationalist educational policy also stressed the study of science and its application to practical ends. The Nanking Ministry of Education, more powerful and purposeful than any China had known in modern times, remolded Chinese schools to serve the goals of production and defense. New colleges and departments of science, engineering, agriculture, and medicine were created and old ones expanded, and students and funds were channeled into these fields. By the end of the decade, a majority of all Chinese university graduates had majored in some branch of pure or applied science. As a result of these efforts, public colleges, universities, and middle schools eclipsed their missionary counterparts as the country's leading centers of education and research.

Finally, the government encouraged innovation and the application of new techniques to agriculture, industry, and national defense. Nanking directly supported industrial research, while adopting indirect measures, such as tariffs, taxes, rebates, and loans, that favored the introduction of new technologies in the private sector. More broadly, the Nationalists created a sense of stability and confidence that persuaded individuals and corporations to invest in the future.

The mission of the Nationalists was to reunify the country, restore its sovereignty, and revive the pride and independence China had known before the misfortunes of the past century. This called for a strong, effective, purposeful state. It also required the development of science, technology, and the application of knowledge to economic production and national defense. The following chapters describe Nanking's efforts to achieve these goals in the areas of education, research, and industry. Here, we will take up some briefer examples that show how the Nationalists took the lead and the scientists followed in solving problems of scientific language, organization, and finance.

THE REFORM OF CHEMICAL TERMINOLOGY

The same factors that had delayed the reform of Chinese scientific language in the 1910s and 1920s, assured its success in the 1930s. First and most important was the formation of a government dedicated to promoting science, and thus to finding a medium that would make work in this area easier. In 1932, Nanking established the National Institute for Compilation and Translation (NICT) to set standards for Chinese language and writing, produce materials that adhered to these standards, and regulate publishing to ensure that others did the same. One of the

first challenges the NICT faced was to formulate authoritative vocabularies for each branch of science. The government's ability to work its
will was favored by a second factor, the destruction of the Commercial
Press in the January 1932 bombing of Shanghai by the Japanese.
Although this enterprise eventually recovered, the setback removed it for
a time as a competitor in the field of translation. Temporarily out of
work, Commercial Press science editor, Cheng Chen-wen, accepted a job
with the NICT, after which he placed his considerable knowledge and
prestige at the service of the official campaign to standardize scientific
Chinese. Finally, by the early 1930s, Chinese scientists were returning in
large numbers from European and American universities to take up
careers in teaching and research. With new opportunities before them,
the most talented young scientists were interested less in inventing
systems of nomenclature than in finding one they could use for other,
loftier purposes. They were ready, even anxious for someone to impose a
linguistic solution.

Nowhere was the challenge to language reformers greater or success
more striking than in chemistry, whose nomenclature remained only
partially established following the death of Hsü Shou and the departure
of John Fryer at the end of the previous century. Identifying and gaining
acceptance for a common linguistic standard was one objective of the
Chemistry Forum, a meeting of China's leading chemists, convened by
the Ministry of Education in August 1932 as part of a general program to
mobilize the scientific community and strengthen the nation against the
rising Japanese threat. Before the meeting, Cheng Chen-wen of the NICT
drafted a proposal that was presented, debated and approved. The final
product, entitled "Principles of Chemical Nomenclature" [*Hua-hsüeh
ming-ming yüan-tse*], was promulgated by the Ministry of Education in
November 1932 and published in June 1933. Less an invention than the
confirmation of reforms adopted in various quarters during the preceding
half-century, this document survives with appropriate amendments as
the constitution of Chinese chemical terminology.[19]

The "Principles" set forth general guidelines, explained how they
should be applied in naming elements and compounds, and provided a
long glossary giving English and Chinese equivalents for thousands of
substances. It included an updated list of the ninety-two natural elements
and authorized one Chinese character for each. Forty-four of the original
sixty-four names selected by Fryer and Hsü were approved and remain in

---

[19] Chemistry Forum: *Chiao-yü-pu hua-hsüeh t'ao-lun-hui chuan-k'an*. "Principles":
Hua-hsüeh ming-ming yüan-tse (tseng-ting-pen) [Principles of chemical nomenclature
(revised and enlarged edition)], ed. Kuo-li pien-i-kuan [National Institute for Compilation and Translation], promulgated by the Ministry of Education, November 1932
(Taipei: Cheng-chung shu-chü, 1965).

use today. Ten characters were assigned new phonetic particles to make their pronunciation conform to the Latin originals. Eight were changed so that their radicals would reflect the properties of the elements themselves. These included the gases – hydrogen, nitrogen, chlorine, fluorine, and oxygen – now represented by characters formed from the gas [*ch'i*] radical and a phonetic particle taken from the old label. Two elements, beryllium and niobium, had to be assigned new characters, because the Latin names themselves had been changed. In all of these cases and in the selection of characters to represent the twenty-eight new elements, the list adopted in 1932 followed the system first devised at Kiangnan (Appendix, Table A.1).

The authors of the "Principles" agreed with Fryer and Hsü that inorganic compounds should be named by their molecular formulas, expressed in characters rather than symbols. At the same time, they authorized parallel forms, similar to those used in the West, that lend themselves to normal speech, a feature lacking in the Kiangnan nomenclature. According to the new system, all binary compounds were to be named by linking the negative ion (x) to the positive ion (y) with an appropriate character or characters (Table 8.2). Compounds in which the elements combine in only one valence or oxidation state, such as sodium, barium, and aluminum chloride, were to follow the formula "*x-hua-y*." Compounds in which the elements can combine in more than one valence, such as ferric and ferrous chloride, nitric oxide, nitrous oxide, and nitrogen peroxide, were to follow formulas that correspond to their English names: Those with the suffix -ic would be named "*x-hua-y*;" those with the suffix -ous, "*x-hua-ya-y*;" and those with the prefix per-, "*kuo-x-hua-y*." Similarly, the salts, such as sodium phosphate and ferrous sulfate, were to be named by the formulas "*x-suan-y*" and "*x-suan-ya-y*." Expanding from this base and with provisions for more complex structures, the "Principles" laid down a blueprint for creating common names for all inorganic compounds.

The greatest contribution, however, came where it was needed most, in the naming of organic compounds. By the early 1900s, a workable system of organic nomenclature had been adopted throughout the West. The task for Chinese chemists was to establish its Chinese equivalent. Although none of the schemes proposed before 1930 gained general acceptance, one by Yü Ho-ch'in (1908) suggested a terminology based on organic structures rather than on the component elements or the sound of foreign names, which pointed to the eventual solution. This system was given final form at the Chemistry Forum.[20]

---

[20] Yü Ho-ch'in: Yüan Han-ch'ing, *Chung-kuo hua-hsüeh-shih lun-wen chi*, 299–300.

Table 8.2. *Selected terms from the "Principles of Chemical Nomenclature" (1932)*

English	Chinese	Characters
Sodium chloride	lü-hua-na	氯化鈉
Barium chloride	lü-hua-pei	氯化鋇
Aluminum chloride	lü-hua-lü	氯化鋁
Ferric chloride	lü-hua-t'ieh	氯化鐵
Ferrous chloride	lü-hua-ya-t'ieh	氯化亞鐵
Nitric oxide	yang-hua-tan	氧化氮
Nitrous oxide	yang-hua-ya-tan	氧化亞氮
Nitrogen peroxide	kuo-yang-hua-tan	過氧化氮
Sodium phosphate	lin-suan-na	磷酸鈉
Ferrous sulfate	liu-suan-ya-t'ieh	硫酸亞鐵
Alkane	wan	烷
Alkene	hsi	烯
Alkyne	ch'üeh	炔
Benzene	pen	苯
Naphthalene	nai	萘
Methane	chia-wan	甲烷
Ethene	i-hsi	乙烯
Propyne	ping-ch'üeh	丙炔
Undecane	shih-i-wan	十一烷
Cyclopentane	huan-wu-wan	環戊烷
Methyl propane	chia-(wan)-chi-(tai)-ping-wan	甲烷基代丙烷
Methyl cyclopropane	chia-(wan)-chi-(tai) huan ping wan	甲烷基代環丙烷
Ethyl benzene	i-(wan-chi-tai)-pen	乙烷基代苯

Most organic compounds are composed of relatively few elements, primarily carbon, hydrogen, and oxygen. The types and even the numbers of atoms of each element may be the same in different compounds. What distinguishes them is the structure or spacial arrangement of their component parts. The corresponding names, therefore, must reflect this molecular architecture. The authors of the "Principles" began by selecting characters to represent each of the simplest structures, such as the types of carbon–carbon bonds that distinguish the alkanes, alkenes and alkynes. In each case, they borrowed a traditional character containing the fire radical [*huo*], chosen because it appears in the symbol for carbon, and a second particle to suggest the appropriate structural feature. Thus the character for alkanes, whose carbon bonds are saturated with hydrogen, contains the particle *wan*, meaning "complete." The character for alkenes, which include double or partly unsaturated bonds, deficient in hydrogen, contains the particle *hsi*, meaning "few" or "scarce." And the character for alkynes, which include triple or fully unsaturated bonds with few or no hydrogens attached, contains the particle *ch'üeh*, which means "lacking." Characters of this type, chosen because of the meaning represented by each of the component parts, are called *hui-i*, and their pronunciation follows that of the secondary particle (Table 8.2).

A different logic entered into selection of characters to represent the nuclei of cyclical or ringed hydrocarbons. Since no obvious meaning such as "complete" or "lacking" could be attached to these structures, the choice proceeded from the opposite direction, that is, the sound. Each cyclical structure was assigned the grass [*ts'ao*] radical, chosen because it appears in the expression *fang-hsiang*, which means "aromatic," the term that applies to one common class of ringed hydrocarbons, and a second particle, the pronunciation of which most nearly matched the first sound of the English name. In some cases, such as benzene, an appropriate character, *pen*, could be found among those already in existence. In other cases, such as naphthalene, *nai*, a character having the desired features, had to be created. Both cases represent the *hsieh-sheng* type character, formed by a radical chosen for its meaning and a second particle representing the sound.

These types of bonds or rings appear in many different compounds whose names must be equally varied and precise. In the case of an open-chain hydrocarbon, for example, the first requirement is to describe its length, the number of carbons in the longest chain. The solution adopted in the "Principles" was to place a numerical indicator – one of the first ten heavenly stems for chains of one to ten carbons and the standard Chinese numerals for those greater than ten – in front of the character that indicates the nature of the bonds. Thus methane, the one-carbon alkane, is called, *chia-wan*; ethene, the two-carbon alkene, *i-hsi*; propyne, the three-carbon alkyne, *ping-ch'üeh*; undecane, the eleven-carbon alkane, *shih-i-wan*; and so forth. Closed chain or cyclic hydrocarbons were identified by the prefix *huan*, meaning "ring." Thus the term for cyclopentane is *huan-wu-wan*. Once the main chain has been identified, the various branches and subgroups attached to it, their type, and location must also be reflected in the name. The Chinese solved this problem by naming the appendages in the same fashion as the primary structures, adding the suffix -*chi*, or "radical," to indicate its dependency. Compounds containing a side chain (*s*) attached to the main chain (*m*) were named by the formula, "*s-chi-tai-m*." Superfluous characters (shown in parentheses) could be deleted. Thus, methyl propane was written as *chia-(wan)-chi-(tai)-ping-wan*; methyl cyclopropane, *chia-(wan)-chi-(tai)-huan-ping-wan*; ethyl benzene, *i-(wan-chi-tai)-pen*; and so forth (Table 8.2).

Given the limits on space and the reader's patience, I have described only the first few steps in a much larger system of nomenclature designed to deal with the compounds known to chemists in the 1930s. The Chinese modeled their chemical terminology on that already established in the West. Indeed, there was little choice in the matter, for the names one applies to chemical substances reflect the theories by which one understands them, and Chinese of this period were learning both the theory and the language. This system was adequate for the needs of the 1930s, and has proven capable of expansion and adaptation, something that all chemical nomenclatures have experienced over the last half-century. In fashioning a language that would enable them to participate in the international culture of science, the Chinese rejected Western symbols and methods of notation, preferring to use traditional ideographs and piece them together by means of their own invention.[21]

Language reform was made possible by the confluence of those political and social trends already described: the emergence of a political authority bent on applying science to China while bringing this process under Chinese control, and the return to China of a generation of scientists who were more interested in doing science than talking about it. Both features also played a role in the establishment during the 1930s of new professional associations.

SCIENTIFIC ORGANIZATIONS

The formation during the Nanking Decade of professional societies in chemistry and other branches of science echoes the themes of rising state power and declining interest among scientists in social and political affairs. Scientists of the first generation had established their own organizations, separate from the government and in some ways in opposition to it. Scientists of the second generation, preoccupied with proper scientific work, ignored the business of organization, which was taken up, when needed, by the state.

[21] Two recent studies that describe the subsequent development of Chinese chemical nomenclature include: Viviane and Jean-Claude Alleton, *Terminologie de la Chimie en Chinois Moderne* [Modern Chinese chemical terminology], Maison des Sciences de l'Homme, Materiaux pour l'Étude de l'Extreme-Orient Moderne et Contemporain, Études Linguistique, 1 (Paris: Mouton, 1966); and Stephane Rolland, *Le Langage Chimique Chinois* [The Chinese chemical language] (Taipei: European Languages Edition, 1985). Other details: Tseng Chao-lun, "K'o-hsüeh ming-ts'u chung ti tsao-tzu wen-t'i," 3–4.

The first professional scientific association created after 1927 was in chemistry. Before this date, China's chemists had demonstrated little organizational aptitude or skill. Except for the Chinese Society of Chemical Industry (CSCI), founded in 1922 and devoted to industrial applications, there was no gathering place for Chinese chemists. At the beginning of the century, Chinese students in France, Germany, and the United States established separate chemical societies and tried to expand these parochial groupings into national organizations after their return to China. None succeeded, however, because the sectarian attitudes of overseas students were too strong. With close personal relations based on long, shared experiences and operating in separate foreign languages, Chinese who studied in different countries found it difficult, once back in China, to bridge these gaps. The one overseas enterprise that did survive, the Chinese Society of Chemical Engineers (CSCE) [*Chung-kuo hua-hsüeh kung-ch'eng hsüeh-hui*], founded in the United States in 1930, did so because it was without competition from students in other countries. This society published a journal, *Chemical Engineering* [*Hua-hsüeh kung-ch'eng*], which first appeared in 1934, the year the society moved to China.[22]

Despite the apparent need for some organization to facilitate internal communication and represent their common interests, neither the chemists nor scientists in other disciplines were quick to act. The initiative had to come from Nanking. The Chemistry Forum of August 1932, which produced the language reform already described, was convened by the government to mobilize China's chemists behind national goals, at a time when China had no private organization that could perform this task. Because they had an interest in channels that could be used to direct the scientific community, those officials who called the forum urged the chemists to stay on after the end of the formal sessions to set up an association of their own. This extended conference produced the Chinese Chemical Society (CCS), the first professional scientific association created in China since the early 1920s and a model for successors in other fields.

Once established, the Chinese Chemical Society disassociated itself from political concerns and operated on a pattern similar to scholarly societies in the West. Governance of the society, the ratification and amendment of its charter and the election of officers, followed

[22] CSCI and CSCE: Ch'en Hsin-wen, "Chung-kuo hua-kung hsüeh-hui," 57–62. Early societies in Europe and America: Yüan Han-ch'ing, *Chung-kuo hua-hsüeh-shih lun-wen chi*, 296–7; and Tseng Chao-lun, "Erh-shih-nien-lai Chung-kuo hua-hsüeh chih chin-chan," 1531–2. Failure of overseas organizations to survive in China: *Chiao-yü-pu hua-hsüeh t'ao-lun-hui chuan-k'an*, 13.

Table 8.3. *Chinese scientific societies*

Name	Founded	Membership in 1936
Geological Society of China	1922	200+
Société Astronomique de Chine	1924	
Chinese Meterological Society	1924	
Chinese Chemical Society	1932	1,004
Chinese Physical Society	1932	200+
Chinese Botanical Society	1933	140
Chinese Geographical Society	1934	200+
Chinese Zoological Society	1934	185
Chinese Mathematical Society	1935	70

Source: Tseng Chao-lun, "Chung-kuo k'o-hsüeh hui-she kai-shu" [Summary of Chinese scientific societies], *KH* 20, no. 10 (October 1936): 798–830.

democratic procedures. Membership, which was open to men and women in all branches of chemistry, grew from fewer than 400 in 1933, to nearly 1,300 in 1937, and over 3,000 in 1949. The society met every year from 1933 to 1946, and despite the problems of starting up and keeping the group going during the war, an average of thirty-four papers was read at each meeting. Beginning in 1933, the *Journal of the Chinese Chemical Society* [*Chung-kuo hua-hsüeh-hui hui-chih*], a quarterly in Western languages, carried articles on chemical research in China for an international audience. After 1934, the CCS also published *Chemistry* [*Hua-hsüeh*], a quarterly in Chinese, aimed at popularizing this discipline among students and laymen. The latter journal included a section entitled "Chemical Abstracts" that contained approximately five hundred items in each issue, or more than fifty thousand during the years 1934 to 1952, and constitutes the most authoritative record of chemical research in pre-Communist China. For internal communications, members relied on the *Chemical Bulletin* [*Hua-hsüeh t'ung-hsün*], which reported the activities of the society and its members.[23]

Formation of the CCS, which represented the largest number of

[23] CCS membership and annual meetings: "Chung-kuo hua-hsüeh-hui ti nien-hui huo-tung" [Annual meetings of the Chinese Chemical Society], *HHTP* 9 (1982):552. Publications and other details: Liu Hui, "Chung-kuo hua-hsüeh-hui" [Chinese Chemical Society], *CKKCSL* 3, no. 3 (1982):53–61. Abstracts: Shen Kuo-ch'iang, "'Chung-kuo hua-hsüeh ts'o-yao' ti chu-yao ch'eng-chiu ho ying-hsiang" [Essential accomplishments and influence of "Chinese Chemical Abstracts"], *CKKCSL* 4, no. 2 (1983):89–92.

Chinese scientists in any one field, inspired similar organizations in other disciplines (Table 8.3). The enormous size, social diversity, and poor internal communications of China slowed development of that country, and nowhere was this lack of integration more troubling than in the sciences. Scientific societies were formed late and with some difficulty, yet after their formation they played an important role in knitting together the scholarly community. Each society held annual national and more regular regional meetings and published separate journals to report on research (in Western languages) and to introduce its subject to the general public (in Chinese). In part because of these efforts, the number of scientific journals increased from 8 in 1914 to 127 in 1930; the publication of books on science expanded at a similar pace.[24]

PAYING FOR SCIENCE

Finally, the compatibility between state interests and the scientists in Nanking China is demonstrated by the greater government expenditures on behalf of science, technology, and related activities. Not that the Nationalists were all that generous, for their budget was devoted primarily to the army, the bureaucracy, and servicing the national debt, while funds earmarked for education and culture were often diverted to other purposes. But the figures show that in the course of the Nanking Decade, the Chinese eclipsed the foreigners as the principle source of funds for scientific activity in China.

Foreign contributions to Chinese science and education, which had played a dominant role during the 1920s, rose even higher after 1928. The mission schools fell on hard times, owing to their dependence on American sources, which were hurt by the depression, and their failure to develop alternatives in China. Still, in 1932–3, the sixteen mission colleges had a combined income of Ch$4.1 million, more than half of which came from abroad. The annual Rockefeller Foundation contribution to Chinese premed programs increased from less than U.S.$100,000 during the 1920s to over U.S.$300,000 from 1929 to 1932. Remissions from the various Boxer indemnity funds accounted for even more. In 1924, Washington agreed to give back to China the remaining U.S.$12 million owed the United States. (The first installment of U.S.$10.8 million had been returned in 1908 to build Tsinghua College and provide fellowships for Chinese students in the United States.) Several parties in

---

[24] Scientific journals: Peter Buck, *American Science and Modern China*, 209. David C. Reynolds, "The Advancement of Knowledge and the Enrichment of Life: The Science Society of China and the Understanding of Science in the Republic, 1914–1930," (Ph.D. diss., University of Wisconsin, 1983), 107, 217–24, reports that the number of science books published in China increased from 102 in 1920, to 326 in 1930.

Table 8.4 *China Foundation expenditures, 1926–1937 (1,000 silver dollars, Ch$ or Mex$)*

Year	Science teaching	Fellowships/ prizes	Research professorships	Institutions (science & technology)	Other (nontechnical)	Total
1926–7	153	16		329	543	1,041
1927–8	192	51		264	495	1,002
1928–9						
1929–30	233	60		456	1,252	2,001
1930–1						
1931–2	210	130	28	959	803	2,130
1932–3	234	119	28	682	688	1,751
1933–4	101	106	33	769	657	1,665
1934–5	28	110	32	602	514	1,286
1935–6	62	100	32	507	499	1,200
1936–7	78	122	32	596	507	1,335

Note: Expenditures were made in both silver (Chinese or Mexican) and gold (U.S.) dollars. During the period 1920–37, the value of one U.S. $ fluctuated between two and five Ch$ or Mex$. To make comparisons, I have converted all amounts to silver dollars, following the table in Arthur Young, *China's Nation-Building Effort, 1927–1937* (Stanford: 1981), 469.
Source: Report of the China Foundation for the Promotion of Education and Culture (Peking and Shanghai, various years).

China sought to gain control over these funds, but Washington insisted that the money be assigned to the newly created China Foundation for the Promotion of Education and Culture, whose board was made up of ten Chinese and five Americans, and used solely for "the development of scientific knowledge."[25]

Beginning in 1926, the China Foundation made grants to support a variety of scientific activities (Table 8.4). To improve science teaching, professorships were established in physics, chemistry, botany, zoology, and educational psychology in each of seven Chinese normal universities. A smaller number of research professorships were set up for leading scientists in universities and research institutes. During the period 1928 to 1945, 735 fellowships were awarded for use in China or abroad. About half of these were in life sciences, 40 percent in the physical sciences, and 10 percent in the earth sciences, and almost all were to support basic rather than applied research. Among the most numerous recipients were the chemists, although a comparison of the numbers of applications made and grants given shows that the foundation favored development of the life and earth sciences, fields in which the Chinese could contribute to the common body of scientific knowledge. Prizes were awarded for outstanding research, including two in the field of chemistry: to Chen Ko-kuei for his work on the Chinese drug *Ephedra vulgaris* [*ma-huang*], and Chou Tsan-quo for isolating the active ingredient of the herbal Cordyalis [*yen-hu-so*]. The largest sums went to purchase equipment and provide general support for China's leading universities, academies, and other scientific institutions.[26]

Other countries followed the United States in returning unspent portions of the Boxer indemnity. A joint Anglo-Chinese board, similar to that which governed the China Foundation, was established in 1931 with a mandate to use the funds due London to make railroad construction loans in China and donate the interest on the loans to worthy educational and cultural purposes. Beginning in 1934, this board made annual grants for scientific work in Chinese colleges, universities, and research

[25] History of China Foundation: *Report of the China Foundation for the Promotion of Education and Culture*, first report (1926), 1–9; Jen Hung-chün, "Chung-chi-hui yü Chung-kuo k'o-hsüeh," 1521–4; and Jen Hung-chün, "Shih-nien-lai Chung-chi-hui shih-yeh ti hui-ku" [Look back at ten years of work of the China Foundation], TFTC 32, no. 7 (1935):19–25.

[26] Research fellowships and support for Chinese geology and biology: *Summary Report of the Activities of the China Foundation for the Promotion of Education and Culture, from 1925 to 1945*, (n.p., December 1946):6, 10–11; and Jen Hung-chün, "Shih-nien-lai Chung-chi-hui shih-yeh ti hui-ku," 21–22, 24. Comparison of grants and applications: Hoh Yam Tong, "The Boxer Indemnity Remissions and Education in China," (Ph.D. diss., Columbia Teachers College, 1933), 235. Failure of overseas organizations to survive in China: *Chiao-yü-pu hua-hsüeh t'ao-lun-hui chuan-k'an*, 13. Other details: *Report of the China Foundation*, various years.

Table 8.5 *British Boxer indemnity grants, 1934–1937 (Ch$1,000)*

Year	Institutions	Fellowships	Total
1934	420	180	1,200
1935	420	180	1,340
1936	980	420	2,800
1937	1,225	525	3,500

Note: Total includes grants for other than science and technology.
Source: Report of the Board of Trustees for Administration of Boxer Indemnity Funds Remitted by the British Government (Nanking, 1934–37).

Table 8.6 *Funds for education and research, 1934–1937 (Ch$1,000,000)*

Year	Central government funds		U.S. and UK Boxer funds		Boxer fund share, b/(a + b) (%)
	Total	Education and culture (a)	Science	Total (b)	
1934	896	13	1.6	2.9	18.2
1935	1,031	32	1.4	2.6	7.5
1936	1,182	37	2.1	4.0	9.8
1937	1,251	42	2.6	4.8	10.3
Total	4,360	124	7.7	14.3	10.3

Source: Government: Arthur Young, *China's Nation-Building Effort*, 436–9.
Boxer fund: Tables 8.4 and 8.5.

academies and awarded twenty to thirty fellowships each year for postgraduate study in Britain (Table 8.5).[27]

Remission of the indemnities was not the same as foreign aid. The money belonged to China in the first place, and the foreigners were simply returning it on terms of their own choosing. Still, the administered funds had to be used for scientific and scholarly activities, and there can be little doubt that more money reached these targets under this arrangement than would otherwise have been the case. Throughout these years, the foreign financial contribution to Chinese science remained large and its

[27] *Report of the Board of Trustees for Administration of Boxer Indemnity Funds Remitted by the British Goverment* (Nanking, 1931–6).

impact enhanced by the focus on a small number of activities, disciplines, and institutions.

In spite of all this, foreign giving was eclipsed by even faster growth in China's own expenditures. According to one estimate, Chinese government spending on education was ten times higher in the 1930s than in the 1920s. By 1935, state funds for research approached 4 million *yüan*. During the years for which we have reliable figures, 1934–7, Nanking's contribution to education and culture exceeded the combined U.S. and U.K. Boxer indemnity funds by nine to one (Table 8.6). Other sources, including the Rockefeller Foundation and the various mission boards, also continued to give, but the foreign share of overall spending on Chinese science almost certainly declined.[28]

The politics of science during the Nanking Decade

Strong governmental leadership in some areas of science – the standardization of language, formation of professional societies, and financial support – was welcomed all around. But in other areas, most notably education and research, state and society clashed. Conservative proponents of expanded state power favored policies that would place science and the scientists at the service of economic production and national defense. Liberal intellectuals, holdovers from the 1920s, defended the notion of an autonomous social and cultural order, within which scholars would be free to define and pursue their own goals. Caught between these two forces were the professional scientists, who wanted to avoid the distraction of politics and get on with teaching and research, but were unable to keep the conflict out of their classrooms and labs. As the decade progressed, this debate over science and science policy sharpened.

On one side of the debate were liberal scholars and educators, veterans of the New Culture era, who had grown weary of the warlords and saw in the Kuomintang a promise for restoring peace, unifying the country, and reviving the nation. This group included several of China's most prominent intellectuals – educators Ts'ai Yüan-p'ei and Chiang Monlin, historians Tsiang T'ing-fu and Fu Ssu-nien, philosopher Hu Shih, and

---

[28] 1920s versus 1930s: Arthur Young, *China's Nation-Building Effort, 1927–1937* (Stanford: Hoover Institution Press, 1971), 79. Four million *yüan*: Ting Wen-chiang, "Chung-yang yen-chiu-yüan ti shih-ming" [Historic mission of the Academia Sinica], *TFTC* 32, no. 2 (1935):5. Jen Hung-chün, "K'ang-chan hou ti-k'o-hsüeh" [Science after the War of Resistance], *TFTC* 37, no. 13 (1940): 23, estimated that in 1940, government research institutes spent more than Ch$2 million, private research institutes less than Ch$3 million, and the total budget was Ch$1.6 billion, making the share for research around 0.2 percent.

geologist Wong Wen-hao. Initially joined by their hopes for Nationalist success, the liberals eventually fell apart, as Nanking's behavior satisfied some and alienated others. Despite these differences, however, they remained unified by a common conception of knowledge and learning, which placed them increasingly at odds with more conservative elements of the Kuomintang.[29]

In the view of the liberals, the proper object of learning was to develop the capacities and fulfill the aspirations of the individual, a goal that could be achieved only within a culture that was independent of politics and defined by the creativity of the participants themselves. The production and dissemination of ideas, the work of research and education, should be free of external political or religious control. Scholars should devote their careers to discovering truths or creating symbols – art for art's sake, science for science's sake – and making these products available to the public at large, rather than seeking wealth and power or following the dictates of the state. Science, in the opinion of these men, was not so much a technical mode of inquiry limited to the material objects of nature, but a model for the culture as a whole. A culture defined by the free and unfettered search for truth on the basis of empirical methods and rational thought would enrich and strengthen all Chinese. Encroachment by political or other authority could only obstruct the pursuit of knowledge and in the end harm China and all mankind.

These liberal opinions were anathema to the men who came to dominate educational and cultural policy under the Kuomintang. The better-educated and more sophisticated elements of the Kuomintang right looked at the role of ideas and the process of learning through the opposite end of the microscope: The purpose was not the fulfillment of each individual cell, but service to the entire organism. Society, left to its own devices, would not generate a unified, purposeful culture, but become divided, misguided, and self-destructive, trends they saw in the China of their day. Rather than permit chaos through neglect, they believed the state should take responsibility for molding the thoughts of its citizens and mobilizing them behind a common cause. In this way the nation would be made richer, safer, and more glorious.

Viewed through this conservative lens, science had an important but limited role. Conservatives rejected entirely the notion that science should form the basis for a New Culture. Rather, the values of social responsibility and service to the group were adequately provided by traditional Chinese philosophy, updated in the form of Sun Yat-sen's

---

[29] Eugene Lubot, *Liberalism in an Illiberal Age: New Cultural Liberals in Republican China, 1919–1937* (Westport, CT: Greenwood Press, 1982), 61–6, 94–107.

"Three People's Principles," which should remain the touchstone for the moral and political training of the young. Science had nothing to offer in this respect. Worse, it was infested with Western liberalism, which had demonstrably confused Chinese youth and led them astray. Whereas it must be kept separate from social and political thinking, science would help create those instruments of wealth and power needed to defend and enrich the collective whole. As in the late Ch'ing dynasty, so again in the Nanking Decade, an influential group of government leaders asserted the possibility and necessity for grafting modern, foreign "techniques" onto an unchanged Chinese "essence."

By returning to the principles of self-strengthening, Chinese conservatives revived the duality between utility and value, which had come and gone during the intervening years. Nineteenth-century reformers had posited a separation between useful Western "functions" [yung] and an eternal Chinese "essence" [t'i]. This attempt to maintain separate yet compatible cultural elements had been set back by a competing Chinese habit of mind, which was the insistence that reality be explained in terms of a single all-embracing worldview. The Confucian-Taoist cosmology had served this purpose in traditional times. At the end of the nineteenth century, as the old order collapsed, K'ang Yu-wei tried to stretch this fabric to wrap the larger, more varied corpus of his Sino-foreign eclecticism. The problem for the keepers of the tradition, however, was that when Western science demonstrated to the satisfaction of most Chinese its superiority in explaining the workings of nature, the cosmological imperative could be saved only by making science, or more properly scientism, the standard for explaining all things. The 1923 debate on science and metaphysics marked the ascendance of a naturalistic, evolutionary cosmology as the unifying vision for intellectuals of the May Fourth era. To combat this new orthodoxy, defenders of traditional values were forced to take refuge in "dualism" – the idea that science might explain the material world, while a "spirit" or "metaphysic" remained the necessary and proper guide for human affairs. This theme was prominent among historians of the "national essence" [kuo-ts'ui] school, who sought the roots of a unique Chinese culture in the traditions of its people, race, and land; it informed the writings of the philosopher Chang Chün-mai, who attacked scientism in the 1923 debate; and it provided the framework for conservative Chinese thinkers of the interwar period. After 1927, the Kuomintang right continued along this dual track: Science was the proper tool for manipulating the material objects of nature, but it must remain firmly in hands guided by a spirit that was uniquely and proudly Chinese.[30]

[30] This theme has been developed by Charlotte Furth, "Intellectual Change," 12:351–61.

The leading spokesmen for this approach were the Ch'en brothers, Kuo-fu and Li fu, namesakes of the CC (Chen and Chen) clique, which played a dominant role in Kuomintang organizational and cultural affairs. Both men had a background and interest in science: Li-fu studied engineering in the United States (M.S., Pittsburgh, 1925) and began his career as a mining engineer; Kuo-fu promoted the study and practice of traditional Chinese medicine, which he defended on the grounds of its compatibility with scientific principles. Yet their appreciation for science was purely instrumental: It was a technique that could be applied to create wealth and power, but offered no lessons for broader Chinese culture, which should remain firmly rooted in the past.[31]

The Ch'ens' social philosophy was explicitly Confucian, stressing the responsibility of the individual to serve the group, and they argued that the fulfillment of this historic commitment lay in Sun's "Three People's Principles," which in their view meant obedience to the Kuomintang. This traditionalism led them to revive some elements of ancient Chinese cosmology, even though these were incompatible with science. In his collection of essays entitled *Vitalism* [*Wei-sheng lun*], an instructional booklet used in the conservative New Life Movement of the mid-1930s, Ch'en Li-fu described a universe that followed the forces of *yin* and *yang*, the five elements and other archaic principles, but omitted mention of modern science. Similarly, Kuo-fu's promotion of traditional medicine, although in some ways based on scientific stardards, also revived ancient theories and practices.[32]

For the most part, scientists of the second generation stayed out of the debate between May Fourth liberals and Kuomintang conservatives. Most scientists were young, inexperienced, and hesitant to challenge the authority of older men of established reputation. Just back in China after several years abroad, they were unfamiliar with the terms of debate that veterans of the New Culture era had worked and reworked in their tireless polemics. More to the point, the primary commitment of the second generation was to teaching and research. Their place was in the classroom and the laboratory, and they left public affairs to the scholarly entrepreneurs who preceded them.

Finally, many of the younger scientists were probably uncertain about which side to favor, for the choice was not a simple one. The case could be made that the long-range interests of Chinese science lay with the

---

[31] Biographies of Ch'en Kuo-fu and Ch'en Li-fu: Boorman and Howard, *Biographical Dictionary*, 201–11.

[32] Ch'en Li-fu's original work, *Wei-sheng lun* [Vitalism], published in 1934, was later revised and translated: Ch'en Li-fu, *Philosophy of Life*, trans. Jen Tai (New York: Philosophical Library, 1948). Ch'en Kuo-fu on Chinese medicine: Ralph Croizier, *Traditional Medicine in Modern China*, 92–102, 109, 147–8. Other details: Allen Linden, "Politics and Higher Education in China," 173–7.

liberal vision of knowledge and its relation to society. The freedom of the individual and the independence of culture, the right to pursue learning for its own sake, the upholding of values associated with the development of science in the West – all these claims must have made the liberal agenda attractive to young scientists fresh from the leading centers of learning in Europe and America. It made sense that Chinese scholars could look forward to successful careers, only if China came to embrace the ideals of the modern scientific order. The free and unfettered pursuit of research would create new knowledge, satisfy personal aspirations, and contribute to China and to the world.

On the other hand, the conservative, statist approach also had its appeal. Scientists of the second generation, including many who could have found comfortable positions in the West, chose to return to China, demonstrating that their commitment was to the nation no less than to science. Many of these young men and women saw in the Nationalist government a promise of order, stability, and direction that would enable them to satisfy personal goals, while helping to rebuild their homeland. Few could ignore the fact that the liberals, for all their lip service to science as a model for the New Culture, had done little to promote the study of nature per se. The Kuomintang, while taking a narrower view of culture, promised to commit more resources to education and research. If the liberals held out grandiose visions for the future, the conservatives offered laboratories in the schools, equipment on the shelves, and concrete plans for using science to save China. Viewed in this way, science and nationalism were compatible, even interdependent ideals.

Conclusion

After 1927, the story of Chinese science unfolded within a new political and social environment. The Nanking regime was bigger, stronger, and more determined than any of its predecessors to effect changes in this area. But government action was subject to political forces, and there was a sharp, unresolved difference between liberal advocates of an autonomous culture infused with the values of science, and conservative spokesmen for enhancing state power by the application of science to production and defense. Between and beneath these forces were the scientists themselves. By this date, a second generation, larger in number, better trained, more committed and competent than their predecessors, was taking over China's laboratories, museums, libraries, and classrooms. Their inclination was to avoid conflict on matters of public policy. But whatever the government and cultural leaders decided to do about science, these professionals would have to carry it out, and in this way their influence would be felt. All three groups played a part in the major arenas of scientific and technical development – education, research, and industry – that are the subjects of the following chapters.

9

Scientific education: the balance achieved

One political battlefield in Republican China, as throughout the twentieth century, was education. When the Nationalists took power in 1928, they found much to criticize on China's campuses: the lack of discipline, the impracticality of the subject matter, the absence of order, standards, and direction. Of particular concern was the weakness of scientific and technical training and the negative effect this had on economic production and national defense. Nanking's response was to train more students in the application of science and train them better. Leading the campaign for more useful studies was Ch'en Kuo-fu and his conservative Kuomintang allies.

Not everyone agreed with these objectives, however. Liberal educators, men who had made their mark during the New Culture era, rejected Ch'en's utilitarian, statist approach. Less vocal, but ultimately more important in deciding the character of Chinese education, were scientists and other educators who patterned their courses on the model of the European and American universities where they had been trained. Both groups favored a more academic curriculum that included the study of basic theory and knowledge, a broad foundation of learning that would equip students to deal with the unseen needs of the future.

This conflict, although at times quite bitter, ended in a compromise that was good for the development of Chinese science. The conservatives captured the Ministry of Education (MOE) and policy-making at the national level, giving them control over the allocation of resources, which they directed toward the study of science in general and the applied fields in particular. Nanking was wise, however, to reject the more draconian proposals of the Kuomintang right and leave the scientists alone to teach the courses as they saw fit, with the result that most middle and higher schools continued to offer a standard academic fare. The result was a workable balance that enlarged the scope and improved the quality of science in schools throughout China.

The debate over educational policy, 1927–1933

The first battle between the New Culture liberals and Kuomintang conservatives was over control of China's schools. Liberal educators, led by former Peking University president Ts'ai Yüan-p'ci, wanted a decentralized system, in which the professoriate would retain maximum influence, and the devotion to free thought and cultural autonomy would reign supreme. Conservatives favored centralization under the Ministry of Education, through which they could instill discipline, direction, and the practical skills needed to increase production and strengthen national defense. This struggle, already engaged in the late 1920s, sharpened following Japan's attack on Manchuria in 1931. Chinese university students protested against Nanking's weak response to the Japanese; party conservatives blamed the unruly youth for adding to the nation's woes. "Student tides" alternated with police crackdowns in a descending spiral of repression and dissent.

These events prompted an attack by the conservatives on the state of Chinese education in general and the teaching of science in particular. From the perspective of the right, the unruliness of the campuses was due in part to the fact that too many students were studying literature, politics, humanities, and the arts. Attention to these topics, particularly under the direction of liberal, Western-oriented scholars, fostered individualism, discontent, and an unhealthy interest in things that were no business of the young. Although preoccupied with politics, students learned nothing of practical use, with the result that after graduation they added to the swelling ranks of overeducated, underskilled, restless urban unemployed. The solution, some Party members believed, was to shift students from arts and humanities to science and technology, to give them training in useful, practical skills, diverting their attention from politics and preparing them for more productive work.

Ch'en Kuo-fu set forth the case for transferring resources to technical studies in his "Preliminary Plan for Educational Reform," which was presented in May 1932 to the powerful Kuomintang Central Political Council.[1] Ch'en's premise, hotly disputed by many educators, was that "the main goal of our country's education is to train human talent in order to meet the needs of society." In recent years, Ch'en argued, Chinese higher education had tilted in the wrong direction:

Owing to the surplus [of graduates in literature and law], the number of unemployed steadily rises, producing all sorts of disruptions in society, while

---

[1] Ch'en Kuo-fu, "Kai-ke chiao-yü ch'u-pu fang-an" [Preliminary plan for educational reform], *Ch'en Kuo-fu hsien-sheng ch'üan-chi* [Complete works of Mr. Ch'en Kuo-fu], vol. 1 (Taipei: Cheng-chung shu-chü, 1952), 169–70.

owing to the shortage [of graduates in agriculture, engineering and medicine], various reconstruction projects cannot find the technical personnel they need to advance.

Ch'en proposed measures to reverse this trend: For a period of ten years, universities and colleges should stop admitting students into the departments of literature, law, and the arts, while the funds previously designated for these programs should be shifted to agriculture, engineering, and medicine. During the same period, students sent abroad at government expense should be restricted to the study of applied sciences. Middle schools should emphasize courses on math, physics, and chemistry, to prepare students for higher technical studies. Finally, classroom instruction at all levels should be supplemented with practical experience in factories, hospitals, and other real-life settings.

Leading members of the Peking liberal establishment resisted these measures. Most accepted Ch'en Kuo-fu's criticism that Chinese schools had produced too few graduates who could apply what they learned to the real world. Teachers who had espoused the progressive theory of "pragmatism" in the 1920s had no interest in turning out unemployed poets. Still, they held fast to the view that education should address the aspirations of the individual and the advancement of learning, rather than the needs of society and an instrumental approach to knowledge. This faith placed them at odds with the conservatives on educational policy in general and scientific education in particular.

The liberal position was spelled out in the pages of *Tu-li p'ing-lun*, a journal established by intellectuals in Peiping in 1932, to discuss ways of dealing with the nation's crisis. The chairman of the Tsinghua history department, Tsiang T'ing-fu, normally a defender of the government and critic of student protesters, was among the first to take issue with Ch'en Kuo-fu's proposal for educational reform. Tsiang charged that Ch'en was reviving the heresy of the self-strengtheners, who had tried to graft modern technical skills onto an unchanged traditional society. It was the refusal of Ch'ing officials to promote all-around modernization, to change institutions and ideas along with techniques, that had caused China to fall so far behind Japan and left it, nearly a century later, still vulnerable to attack. China's problem, Tsiang explained, was not a shortage of technical manpower, but the fact that the political, social, and economic system gave the Chinese no opportunity to use their skills. As for the widespread interest in humanities and politics, these were the proper concerns of all civilized men, not some fabrication of the educators. A broad, liberal curriculum was good because it helped people understand and manage their social and political affairs. If the schools stopped teaching these subjects, Chinese society would become even more confused and disordered. "The purpose of education," Tsiang argued, is not to train young people for roles drawn up by social architects, but "to educate the whole màn."[2]

Fu Ssu-nien, a leading figure of the May Fourth Movement and, after 1928, director of the Institute of History and Philology of the Academia Sinica, offered a similar critique. Fu accepted Ch'en's charge that China's schools were plagued by "gentry" attitudes: the excessive reliance on written materials, the inability of students and teachers to use their hands and feet, and the reluctance to engage in practical tasks.[3] He also agreed that the solution was to place greater emphasis on vocational training. But Fu rejected what he saw as a dangerous trend toward the mechanical learning of preset tasks – the creation of what he called "apprentice schools" – and insisted that education must foster a deeper understanding of general principles and the methods used to apply them:

Those primary and middle schools that offer chemistry should not concentrate only on making soap or those that offer physics train only electricians. Middle schools should equip their graduates to apply the knowledge learned in chemistry, not just to make soda, but to work in a chemical factory [of any type], or to apply their knowledge of plants and animals to work on [any] farm, ...

Physics should teach [students] to be able to relate the information in their books to [understanding] the electric light, the soap bubble, the changes in the heavens, the uses of heat, and all the things we meet in our environment. Botany should teach them to be able to classify and recognize all of the plants in our gardens. All work must follow step by step concrete experimentation. Textbooks are only for reference. The basis for training lies in moving our hands and feet.[4]

Fu Ssu-nien opposed Ch'en's proposals on the grounds that they would turn China's universities into technical training schools. It was alright, Fu conceded, for elementary and middle schools to perform the tasks of conveying knowledge and nurturing skills. But the purpose of higher education should be to promote scholarship, rather than simply pass on accepted truths. The job of the professor was to discover new knowledge and teach his or her students to do the same.[5] Fu laid the blame on the Ministry of Education. There was nothing wrong with China's teachers, students, or schools that they could not solve better and more quickly if the government would simply get off their backs.[6]

Few members of the intellectual establishment came to Ch'en Kuo-fu's defense.[7] One exception was the dean of Tsinghua University, Ku

---

[2] Chiang T'ing-fu, "Ch'en Kuo-fu hsien-sheng ti chiao-yü cheng-ts'e" [Educational policy of Mr. Ch'en Kuo-fu], *TLPL* 4 (12 June 1932):6–8.

[3] Meng Chen [Fu Ssu-nien], "Chiao-yü peng-k'uei chih yüan-yin" [The reason for the collapse of education], *TLPL* 9 (17 July 1932):2–6.

[4] Meng Chen [Fu Ssu-nien], "Chiao-yü kai-ke chung chi-ko chü-t'i shih-chien" [Some concrete issues in educational reform], *TLPL* 10 (24 July 1932):6–7.

[5] Meng Chen [Fu Ssu-nien], "Kai-ke kao-teng chiao-yü chung chi-ko wen-t'i," 2–3.

[6] Ibid., 8–9.

Yü-hsiu. Ku, who had been trained in engineering (Sc.D., M.I.T., 1928), shared the view that education in the sciences should serve immediate, practical goals. His 1933 article, "What Kind of Science Do We Need?" was a reply to Ch'en's liberal critics. Ku attacked current educational practices, which served only a small minority of students who had the talent and opportunity to pursue academic careers, while ignoring the majority, who could be trained to perform the more mundane tasks of applying existing knowledge to local needs. The age of great scientific discoveries was past, Ku argued; what was left were the particulars. Since only a few individuals of enormous genius could make a contribution to science, since China was too poor to explore these peaks of knowledge, and since there was no point in rediscovering what was already known, Chinese young people should be taught to lower their sights and be satisfied with working out those few details that might be useful in China. Chinese scientists trained abroad and aspiring to international prominence had misguided their students into believing that they too could join the priesthood of research scholars. Instead, students should be told that their horizons in science were limited and their responsibility was to serve the nation. The main point, Ku concluded, is "to help young people understand that not all students of science can go on to do research."[8]

The initial response to Ch'en Kuo-fu had come from Tsiang T'ing-fu and Fu Ssu-nien, historians whose "useless" discipline was threatened by Ch'en's proposals. But Ku's attack on scientific teaching and research prompted a rebuttal along narrowly technical lines. Ku's critics pointed out that science is not a body of facts inherited from the past, but a process of discovery, so that scientists of every age can and should seek new knowledge. Pure and applied research go hand in hand, each contributing to the other; there is no better training for an engineer or an industrial manager than immersion in the research process. To be content to copy others would condemn China to a perpetual state of backwardness and promised no greater success than fighting today's enemies with yesterday's weapons, or relying on the League of Nations to remove Japanese armies from China. It is clear from this exchange that the two sides differed on the very purpose of science: Ku and the conservatives wanted to train technicians who could apply existing knowledge to accelerate the process of economic growth in the absence of social or cultural change, while the liberals saw science as a process through which

[7] For other criticisms of Ch'en, see for example: Shu Yung [Jen Hung-chün], "Tang-hua chiao-yü shih k'o-neng ti ma?" [Is a party-ized education possible?], *TLPL* 3 (5 June 1932):12–15; and Shu Yung [Jen Hung-chün], "Tsai-lun tang-hua chiao-yü" [More on party-ized education], *TLPL* 8 (10 July 1932):10–13.

[8] Ku Yü-hsiu, "Wo-men hsü-yao tsen-yang ti k'o-hsüeh?" [What kind of science do we need?], *TLPL* 33 (1 January 1933):12–15.

young people might grow intellectually and morally and contribute in the long run to a better China. Wu Hsien, one of the few practicing scientists who spoke out on this issue, shared the liberal bias. Given the low level of China's development, Wu explained, there was little prospect that Chinese scientists would make important discoveries, but for students in China as elsewhere, "promoting the research atmosphere and teaching research methods are more important, by far, than the results of the research itself."[9]

Educational reforms, 1933–1935

The influence of the liberals on Kuomintang education policy was fleeting, and in the end utilitarianism prevailed. Ts'ai Yüan-p'ei, hero of the May Fourth Movement, failed in an attempt to give scholars and teachers a larger role in China's schools. After the formation of the Ministry of Education in 1928, policies emanating from that body came to reflect the will of the Kuomintang right. The liberals resisted until 1933, when Nanking appointed as minister Wang Shih-chieh, a man with strong ties to the Peking educational establishment, who was nonetheless willing to carry out the Kuomintang program. "The shifting of emphasis to knowledge of science capable of practical application, " Wang explained to his colleagues in the north,

is in great part an answer to the growing realization that, whatever intrinsic qualities knowledge in itself may possess, the present society demands new tools and a new type of man, and that knowledge should be harnessed to produce results in connection with the economic development of the country.[10]

The reforms introduced by Wang Shih-chieh attacked secondary school science, which most observers considered one of the weakest links in Chinese education. Part of the problem was the dominance of academic subjects, which bore little relation to the future of most students or the needs of the country. In 1934, 70 percent of all secondary students were in middle schools, 20 percent in normal schools, and only 10 percent in vocational schools, yet there were few opportunities for students to move up the academic ladder and great demand for teachers and people with

---

[9] Responses to Ku: Sun I, "Tu Ku Yü-hsiu 'Wo-men hsü-yao tsen-yang ti k'o-hsüeh' hou" [After reading Ku Yu-hsiu's "What kind of science do we need?"], *TLPL* 36 (22 January 1933):14–18; and " 'Wo-men hsü-yao tsen-yang ti k'o-hsüeh' ti t'ao-lun" [Discussion of "What kind of science do we need?"], *TLPL* 38 (19 February 1933):17–18. Wu Hsien: Wu Hsien, "Kuan-yü k'o-hsüeh yen-chiu chih wo-chien" [My views on scientific research], *TLPL* 101 (20 May 1934):17.

[10] Wang Shih-chieh, "Education," 1031–2. Other details: Allen Linden, "Politics and Higher Education in China," 220–4.

practical skills. To improve technical training, MOE guidelines called for a shift of resources within the secondary sector from academic and normal to vocational schools, primarily in the fields of agriculture, industry, and commerce. Even as preparation for higher education, moreover, the middle schools did a poor job, especially in the sciences. Beginning in 1932, therefore, Nanking issued mandatory middle school curricula that raised the requirements in English, math, and science (Appendix, Table A.6). The ministry, in cooperation with the Academy of Sciences, produced sets of equipment for physics, chemistry, and biology that were made available to middle schools at half their cost.[11]

These measures produced mixed results. Spending on vocational education never reached the ministry's target of 35 percent of all secondary expenditures, but its share did rise from less than 10 percent in 1931 to more than 14 percent in 1936. The number of vocational schools increased at about the same pace. Meanwhile, science instruction in the mainstream secondary schools continued to exhibit many of its old, unseemly traits. A 1933 survey of over half the country's 200 higher middle schools found that most offered two years of chemistry and physics, at least one year of biology, and several different math courses. A second survey conducted in 1936 provided greater detail and a less flattering picture of many of these same schools. More than 60 percent of lower and 20 percent of upper middle schools omitted the required chemistry lab, and most of those that did have labs failed to meet the standard. Most lower schools offered demonstration teaching only; upper schools generally had only one set of equipment for each two or three students. Even more serious was the ministry's failure to improve the morale of middle school science teachers, whose attitudes appear to have changed little from the depressing state of the 1920s. In one sample of the 173 middle school chemistry teachers surveyed in 1936, most were men in their mid-twenties or thirties who had graduated from a normal school (if lower school teacher) or university (upper school teacher) with a major or minor in chemistry, and in some cases had received advanced

---

[11] Many observers: Wang Shih-chieh, "Education," 1048; League of Nations' Mission, *Reorganization of Education in China*, 108–9; Wang Chih-chia, "Wo-kuo k'o-hsüeh chiao-yü chin-hou ying chü chih fang-chen" [Plans for the future of our science education], *KH* 24, no. 5 (May 1940):348; Chu Yüan-shan, "Sheng-ch'an chu-i chih li-k'o chiao-shou" [Science teaching for production], *CYTC* 9, no. 1 (1917), bk 21:11,403–9; Chia Kuan-jen, "Chung-teng hsüeh-hsiao li-hua-hsüeh chiao-shou-fa kai-liang i-chien-shu" [Proposals for improving science teaching methods in middle schools], *CYTC* 10, no. 10 (1918), bk 26:14,309–12. Distribution of students, 1934: Wang Shih-chieh, "Education," 1048. MOE guidelines: *Chung-kuo chiao-yü nien-chien, ti-erh-tz'u* [The second China education yearbook], ed. Chiao-yü-pu [Ministry of Education] (Shanghai: Shang-wu yin-shu-kuan, 1948), 1,023–4. Curriculum: *China Year Book* (1933), 531–2. Equipment: Wang Shih-chieh, "Education," 1052; *China Year Book* (1938), 315.

training abroad. Few had prepared for careers in education, however, and most hoped to find jobs in research, engineering, or other scientific professions. Teaching was in all cases a last resort. Their average classroom load was twelve to eighteen hours, and often as many as twenty-four to thirty hours per week. Since most teachers left their posts at the first opportunity, only half had as many as six years of experience and few more than ten years experience.[12]

Given the lack of commitment on the part of the faculty, the graduates of China's middle schools were often ill prepared for higher education. Beginning in 1932, the ministry administered standard primary and secondary school graduation exams in all subjects. The results of these exams showed that students were most competent in Chinese, history, and literature, and least competent in science and math. Entrance exams given by Chinese colleges and universities in the 1930s revealed the same imbalance. University professors testified that they could assume no previous study of science on the part of incoming students and generally had to start from the beginning.[13]

Nanking's policies toward higher education also favored the pure and applied sciences. Colleges and universities were reorganized, money allocated, and increasingly strong measures adopted to force reluctant students into these fields. The basic law on higher education, promulgated in 1929, required that each university include at least one college of science, engineering, agriculture, or medicine. To correct the imbalance in college enrollments, in 1933 the ministry ruled that institutions at this level must maintain a proportion between students admitted to the various schools and departments. When these modest efforts failed to produce the desired change, in 1935 a ceiling of thirty new admissions was placed on each department of arts, humanities, and social studies. During the same period, regulations were issued to give preference in the allocation of government scholarships for study abroad to candidates who selected "practical" courses.[14]

These measures produced results. The percentage of college and university students studying science and technology nearly doubled between 1931 and 1936, while those in arts and humanities declined by a

[12] Vocational expenditures and schools: *Chung-kuo chiao-yü nien-chien* (1948), 1,428. 1933 survey: Jen Hung-chün, "I-ke kuan-yü li-k'o chiao-k'o-shu ti tiao-ch'a" [An investigation of science textbooks], *KH* 17, no. 12 (December 1933):2029–31. 1936 survey: Tai An-pang, "Chung-kuo hua-hsüeh chiao-yü ti hsien-chuang," 89–107.

[13] MOE exams: Wang Shih-chieh, "Education," 1051–2. Entrance exams: Wang Chih-chia, "Wo-kuo k'o-hsüeh chiao-yü," 348. University professors: L. G. Morgan, *Teaching of Science to the Chinese*, 89.

[14] Basic law: Wang Shih-chieh, "Education," 1032. MOE regulations: ibid., 1032, 1040. Overseas scholarships: Y. C. Wang, *Chinese Intellectuals and the West*, 128; Kuo Tze-hsiung, "Higher Education in China," 46–7.

Table 9.1 *Student enrollments in higher education, by subject, 1928–1936*

Subject	1928–9 Number	%	1931–2 Number	%	1936–7 Number	%
Sciences						
Pure science	1,910	7.6	3,930	8.9	5,485	13.2
Engineering	2,777	11.1	4,084	9.3	6,989	16.8
Agriculture	1,085	4.3	1,413	3.2	2,590	6.2
Medicine	977	3.9	1,800	4.1	3,395	8.2
Total	6,749	27.0	11,227	25.5	18,459	44.4
Arts						
Letters	5,464	21.8	10,066	22.8	8,364	20.1
Law/politics	9,466	37.9	16,487	37.2	8,253	19.8
Education	1,661	6.6	4,231	9.6	3,292	7.9
Commerce	1,695	6.8	2,156	4.9	3,243	7.8
Total	18,286	73.0	32,940	74.5	23,152	55.6
Total	25,035	100.0	44,167	100.0	41,611	100.0

Source: *Chung-kuo chiao-yü nien-chien* (1948), 525–6. See Appendix, Table A.7 for complete statistics, 1928–57.

corresponding amount. Most of the gains were in pure science and engineering, while agriculture and medicine continued to lag (Table 9.1 and Appendix, Table A.7). The composition of graduating classes confirmed this shift: in 1930, graduates of colleges of liberal arts (17,000) outnumbered those in the pure and applied sciences (8,000) by more than two to one, but by 1937 the number of each was almost exactly the same (15,200). Beginning in 1933, a majority of students approved by the ministry for study abroad indicated the intention to major in some technical field (Appendix, Tabel A.3).[15]

Although Nanking's control over missionary education was less direct, the 1929 regulation requiring private schools to register with the government gave an unmistakable message to the churchmen and resulted in parallel changes in mission school curricula and enrollments. Science courses in mission colleges and universities increased from 35 percent of all semester hours taught in 1925 to 41 percent in 1933. Students in these schools also gravitated to the sciences: In 1925 only 30 percent of course enrollments had been in the sciences; by 1933, 44 percent of all under-

[15] Graduates, 1930 and 1937: E-tu Zen Sun, "The Growth of the Academic Community, 1912–1949," 391. Students approved for study abroad, 1933: Kuo Tze-hsiung, "Higher Education in China," 46–7.

Table 9.2. *Chemistry students, Yenching University, 1926–1934*

Year	University enrollment	Course enrollments chemistry (fall)	Undergraduate chemistry majors	Graduate students in chemistry
1926–7	542	165	19	3
1927–8		216	26	7
1928–9		250	31	7
1929–30		254	48	12
1932–3	783	300	55	20
1933–4	779		66	18

Sources: "Statement Presented by The College of Natural Sciences of Yenching University (October 1929)," RG 1, ser. 601, box 41, fld. 341, RAC, 38; *Yenching Natural Sciences News* 1 (January 1933), 15; and Earl Cressy, "Christian Colleges in China: Statistics," 18–21.

graduates who had declared a major picked some branch of science, with chemistry accounting for more than half the total. Almost half of all undergraduates at Nanking University chose to major in chemistry, physics, or biology, and enrollments in the chemistry department at Yenching, on both undergraduate and graduate levels, increased in similar fashion (Table 9.2).[16]

The only detailed breakdown of students by college and department is for the academic year 1931–2 (Table 9.3). These figures show that among the pure sciences the largest number of students opted for chemistry, followed by math and physics; only a handful chose the life or earth sciences, fields that could have contributed to China's economy, but were consistently undervalued by Nanking. The attraction of students to chemistry and relative disinterest in biology is indicative of the broader preference for industry over agriculture, a bias shared by teachers, school administrators, and bureaucrats. Among engineering students, civil, electrical, and mechanical were the preferred specialties, and chemical ranked last – probably a fair reflection of the job market.

Statistics compiled during this period also shed light on the place of women in higher scientific education (Appendix, Table A.8). Throughout the 1930s and 1940s, women remained a minority in all fields, although their share of enrollments increased over time. Pure science and medicine were among the areas where women fared best, engineering and agriculture where they made the fewest inroads. By the mid-1930s, a few

---

[16] Mission school statistics: Earl Cressy, "Christian Colleges in China: Statistics," *Bulletin* 33 (1933–4):18–25. Nanking: Earl Cressy, "Christian Higher Education in China, A Study for the Year, 1925–26," 86.

Table 9.3. *Numbers of students in colleges and universities,
by department and college, 1931*

College	Number	%
Colleges of science		
Chemistry	1,062	24.2
Math	676	15.3
Physics	640	14.5
Biology	414	9.4
Geography	151	3.4
Geology	85	1.9
Psychology	59	1.3
Astronomy	36	0.8
Other	825	18.8
N.A.	459	10.4
Total	4,407	100.0
Colleges of engineering		
Civil	1,246	38.2
Electrical	690	21.1
Mechanical	500	15.3
Mining	282	8.6
Chemical	176	5.4
Architecture	61	1.9
Other	196	6.0
N.A.	116	3.5
Total	3,267	100.0

Source: Chung-kuo chiao-yü nien-chien (1934), 4:54–5.

women began to appear in the upper ranks of Chinese scientists. Their impact on the natural sciences remained negligible, however, as compared to other fields such as literature and the arts.

Science in higher education

In the universities, as in middle schools, Nanking found it easier to change the size and shape of the bottles than their content. The most complete survey of higher science education in China during this period was conducted in the fall of 1933 by W. E. Tisdale for the Rockefeller Foundation.[17] Tisdale, like other foreign observers, may have been

[17] Unless otherwise noted, following details are from W. E. Tisdale, "Report of Visit to Scientific Institutions in China."

unduly critical of work that fell below the standard he had known in the West. But since he advised the foundation to shift its support from missionary to Chinese institutions, it is likely that his views, although harsh, were well intentioned. In any case, he has left behind the most complete account of science programs in the top twelve public and fifteen private universities of Republican China (Table 9.4).

With a few notable exceptions, Tisdale judged the laboratories, libraries, and other science facilities of these schools to be inadequate. There was not a single commercial firm in China that manufactured scientific apparatus. Imported equipment was expensive, scarce, and generally slow in arriving. Some common reagents were available through foreign supply houses in the port cities, but these were usually too expensive for student use, and more exotic reagents had to be ordered from abroad at considerable cost and delay. A few universities tried to bridge the gap by fashioning homemade substitutes, although such efforts were generally unsuccessful because the machines and tools to make the equipment were also lacking. No amount of ingenuity could fill the library shelves. Some universities received the most important current publications, but none had adequate reference materials or back issues of the major journals needed for teaching and research.

Tisdale was equally critical of the teachers and their methods. Teaching was almost solely by lecture, with students taking a minimum of twenty hours and often as many as thirty or forty hours each week. The principal medium of instruction was English, which may have been necessary given the dearth of Chinese texts, but the language proficiency of most students and teachers was too low, even in the mission colleges, to permit the exchange of much information. Tisdale found that most faculty members conducted no research. Foreign instructors were generally missionaries first and science teachers second. Their motivation and training had not prepared them for the laboratory. The situation of Chinese professors was more complex. Many had completed higher degrees in foreign universities and been exposed to advanced research. Tisdale questioned how well their previous study in China had prepared them for this experience, however, and how much they had derived from it. After completing their education abroad, he explained, the new graduates

return to the milieu of a Chinese university – a milieu as foreign to them as they are to the institution in which they had their training. Under the best conditions abroad they have had excellent opportunities to react to their environment, but they have had no opportunity to make their environment react to them. They have learned to follow the inspiration and leadership of their associates, but they have gained no experience in inspiring others or in directing the activities of others.

Table 9.4. *Summary of W. E. Tisdale's findings on scientific education in Chinese colleges and universities, 1933*

	Natural sciences				Engineering					Agriculture
	Phys	Chem	Biol	Math	Civ	Mech	Elec	Chem	Min	
Public Universities										
Chekiang	4	4 R	4 R	4 R	gd	gd	gd	gd		new
Chiao-t'ung	2	2		2	ex	ex	ex			
Hunan	*	*			pr	pr	pr		pr	
Central	4 R	4 R	4 R	4 R	ex	gd	gd	gd		pr
Peiping	*	*	*	*	pr					pr
Peiping Normal	2	2	2	2						
Peiyang	*	*		2	fr	fr	new		fr	
Peking	2	2	2	4						
Shansi	2	2	2	2	pr	pr				
Sun Yat-sen	2	2	4 T	4	pr		pr			pr
Tsinghua	4 R	4 R	4 R	4	gd	gd	gd			new
Wuhan	2	2	4 R	2	new	new	new			new

Private universities

Aurora	2	2	2 T	2			
Catholic	2	2	2 R	2			
Chung-hua	*	*	*	*			
Futan	*	*	*	*	pr		
Ginling	2	2	2	2			
Hangchow	*	2	*	*	pr		
Hua-chung	2	2 R	2 T	2			
Lingnan	4 R	4 R	4 T	4	new		gd
Nankai	*	*	2	4		pr	gd
Nanking	2	4 R	4 R	2		pr	
Shanghai	2	2	4 T	2			
Shantung Christian	2	2	2	4			
Soochow	2	2	4 T	2			
St. John's	2	2	4 T	2	fr		
Yenching	4 R	4 R	4 R T	2			

Notes: * = equipment and staff inadequate for two-year course
2 = equipment and staff adequate for two-year general course
4 = equipment and staff adequate for four-year major course
R = experimental research activity by faculty
T = research activity by faculty limited to taxonomy
pr = poor   fr = fair   gd = good   ex = excellent

Source: W. E. Tisdale, "Report of Visit to Scientific Institutions in China (September – December 1933)," 7.

In all, Tisdale could identify fewer than two dozen faculty, Chinese and foreign, actively engaged in research.[18]

It should not be supposed that Tisdale's views were unique or limited to foreign critics with unreasonably high standards. Teachers in the best Chinese universities agreed that students read too many books, particularly foreign books, and listened to too many lectures, many of which were taken from the same books, while doing too little research or work of a practical nature. "If we want to modernize Chinese education," observed the historian Fu Ssu-nien, "we must train people to use their hands and feet and make the burning of books and burying of scholars our political platform."[19]

Others noted the lack of research and connectedly of graduate programs in Chinese universities. Yenching was the first to award the master of science degree, beginning in 1922. Tsinghua established a research institute in Chinese studies in 1925, and was the first public university to grant a master's – to Fei Hsiao-t'ung in sociology, in 1934. The Tsinghua graduate program was expanded in 1929 with the creation of a research academy that housed institutes for each of the major scientific disciplines. Since most jobs in advanced teaching and research went to graduates of foreign universities, however, the academy attracted few students and granted only twenty-seven degrees before the war. The institute of chemistry enrolled eleven students and produced two notable graduates: Ma Tsu-sheng (Ph.D., Chicago, 1938) and Chang Tsing-lien (Ph.D., Berlin, 1936). Proponents of university expansion pointed out that graduate programs would generate the data, theories, and literature needed to reflect the peculiarities of China, keep the better students and teachers at home, and stimulate the development of education generally. After 1934, six public and five private universities were authorized to grant master's degrees, although the first doctorate from a Chinese institution (National Taiwan University) was not awarded until 1961. In most cases, higher education in China ended with the bachelor's degree.[20]

[18] Hours per week: League of Nations' Mission, *Reorganization of Education in China*, 159–60. Quotation: W. E. Tisdale, "Report of Visit," 5. Faculty members: ibid., 4–6, Appendix.

[19] Books and lectures: Chiang T'ing-fu, "Tui ta-hsüeh hsin-sheng kung-hsien chi-tien i-chien" [A few ideas on the contributions of new university students], *TLPL* 69 (24 September 1933):8–9. Quotation: Meng Chen [Fu Ssu-nien], "Chiao-yü peng-k'uei chih yüan-yin," 4.

[20] Lack of research and graduate programs: League of Nations' Mission, *Reorganization of Education in China*, 171; W. E. Tisdale, "Report of Visit," 16; and Ts'ai Yüan-p'ei, "Lun ta-hsüeh ying-she ko k'o yen-chiu-so chih li-yu" [Reasons that universities should establish research institutes], *TFTC* 32, no. 1 (January 1935):13–14. Yenching: William Pruviance Fenn, *Christian Higher Education in Changing China, 1880–1950*, 81. Other mission graduate programs: Earl Cressy, "Christian Higher Education in

Tisdale found only six universities that offered a full four-year course of study in the sciences. Three of these were public institutions: Central, Tsinghua, and Chekiang. Three were Christian: Yenching, Nanking, and Lingnan. Four universities, all national, had substantial programs in civil, mechanical, and electrical engineering. The best was Chiao-t'ung, followed by Tsinghua, Central, and Chekiang. Only the last two had departments of chemical engineering, which Tisdale described as "well provided for." On the other hand, he found only two good colleges of agriculture, both attached to mission schools: Nanking and Lingnan. Even these programs were spotty, however, and Tisdale concluded that there was no school in China that could teach veterinary medicine, animal breeding, entomology, or forestry, and no government institution with an agricultural college of any real quality.[21]

The leading Chinese universities

A closer look at the leading universities of Nationalist China reveals the bargain struck between the scientists and the state. Nanking provided the money, much of which helped establish new or strengthen old programs in the applied sciences: agriculture, medicine, and engineering. After 1928, public institutions at every level were better equipped to meet national goals. Meanwhile, professional scientists avoided the debates on education, preferring to concentrate their efforts in the laboratory and classroom, the kingdom in which they reigned supreme. And most used their positions to enforce a curriculum that was not utilitarian, as the ministry preferred, but adhered to the high academic standards they had been exposed to in the West. Increased government support tempered by tolerance of or inability to control the scientists, combined to produce rapid advances in the quantity and quality of higher science education. It is striking how much China's schools improved in such a short space of time. By the mid 1930s, the leading Chinese universities supplanted their missionary rivals as the foremost centers of scientific education, a point that is best illustrated by a close look at the big three: Tsinghua, Central, and Chekiang.

Characteristic of the mood that accompanied the Kuomintang

China," 142; Earl Cressy, "Christian Colleges in China," *Bulletin* 30 (1932–3):24. Tsinghua: *Ch'ing-hua ta-hsüeh hsiao-shih kao*, 113–14, 202; Y. C. Wang, *Chinese Intellectuals and the West*, 124. Taiwan: ibid. 1934 regulations: Wang Shih-chieh, "Education," 1037.

21 A 1933 survey of twenty leading Chinese universities revealed that all had introductory courses in chemistry and physics, but many had no biology. See: Jen Hung-chün, "I ke kuan-yü li-k'o chiao-k'o-shu ti tiao-ch'a," 2029.

takeover was the transformation of Tsinghua from its previous role as prep school for America-bound Boxer Fellows to a full-fledged national university. Tsinghua was reorganized in 1929 into colleges of arts, science, and law. Engineering was added in 1932. Entrance was by examination, the competition stiff. All students were required to take math and science in their freshman year, after which they could specialize in a chosen field. Like all students in Nationalist China, they were required to perform an independent research project in their senior year.[22]

The college of science included six departments: math, physics, chemistry, biology, psychology, and earth sciences. Prior to 1937, the college of engineering offered degrees in civil, mechanical, and electrical. During the 1930s, enrollment in the sciences was around three hundred and in engineering four hundred. The hallmark of the Tsinghua program was its emphasis on research. Following the American model, the college of sciences taught primarily theory and experimental techniques, in order to prepare graduates for teaching and scholarship, while neglecting practical applications. By 1934, there were fifty-eight laboratories, most with modern equipment and generous supplies. Science and engineering maintained the highest standards, both to start and to finish the program. During the years 1929 to 1937, only half the students who entered these colleges made it to graduation. Attrition rates in physics and chemistry were even higher.[23]

Chemistry was the largest of the science departments with eighty students and a dozen full and assistant professors, including several of China's most prominent chemists, all younger men who returned to China in the late 1920s or 1930s: Sah Pen-t'ieh (Ph.D., Wisconsin, 1926), Kao Ch'ung-hsi (Ph.D., Wisconsin, 1926), Huang Tzu-ch'ing (Sc.D., M.I.T., 1935), Li Yün-hua (Ph.D., Columbia, 1927), and Chang Ta-yü (Ph.D., Dresden, 1933). The chairman, Chang Tzu-kao, was older (b. 1886), less educated (M.S., M.I.T., 1915), and primarily an administrator rather than a scholar. The Tsinghua curriculum, modeled on the University of Wisconsin, combined a core of required courses in each branch of chemistry with a wide variety of electives. All courses included lab work. The standard ratio was three hours of lab for each two hours of lecture. A new chemistry building, opened in 1933, contained fifteen laboratories and, by 1936, 100,000 *yüan* (U.S.$29,700) worth of equipment. The library held over seven hundred reference books and subscribed to more than forty journals in this field. The university published two journals, *Ch'ing-hua hsüeh-pao* [Tsinghua

---

[22] *Ch'ing-hua ta-hsüeh hsiao-shih kao*, 111–12, 121–5, 144.
[23] Ibid., 124–8, 143, 181–3, 197–200.

Journal] and *Li-k'o pao-kao* [Science Report], which contained current research in the sciences, much of it by the Tsinghua faculty.[24]

National Central University (NCU), the successor institution to National Southeastern, located in Nanking, emerged during this period as a leading center for the study of science. In 1931, NCU enrolled over 1,800 students, of whom 183 men and 33 women were in the college of sciences. The university produced fifteen to twenty graduates per year in chemistry, physics, and math, 5 in botany, zoology, geology, and experimental psychology. According to the dean of the science faculty, students in physics, math, and biology had a real interest in science for its own sake, whereas those in chemistry tended to be concerned with commercial applications.[25]

Most faculty members were engaged in some form of research. Important work in chemistry began with the arrival in 1927 of Tseng Chao-lun. Tseng worked with younger colleagues, including Chang I-tsun and Wang Pao-jen, graduates of Southeastern, and Edith Chu Ju-hwa (Ph.D., Michigan, 1936). During his four years at NCU, Tseng published twenty articles, many of which appeared in the University's own *Science Report* [*Kuo-li chung-yang ta-hsüeh k'o-hsüeh yen-chiu lu*], which was launched in 1930 as one of the first journals in China to publish research in the physical sciences. After Tseng's departure in 1931, work in organic chemistry was continued by two new recruits: Kao Tsi-yu (Ph.D., Illinois, 1931) and Yüan Han-ch'ing (Ph.D., Illinois, 1932).[26]

As the leading university in the captial, NCU responded quickly to the Kuomintang call for more practical and applied work. Prior to 1927, none of the government universities included a college of medicine, and medical schools under Chinese control had improved little since the visit by the Rockefeller commission in 1914. One of the first acts of the Nationalist Ministry of Education was to establish a medical college within NCU. Initially, a joint program was arranged with St. John's University, whereby Central provided basic lab courses, while St. John's taught the clinical work at an affiliated Shanghai hospital. In the fall of 1929, this temporary union was dissolved, and NCU launched its own four-year medical college. When Knud Faber of the League of Nations Health Organization toured China in 1931, he described it as the best of the four national medical schools then in existence.[27]

[24] Ibid., 144, 183, 199–204.
[25] L. G. Morgan, *Teaching of Science to the Chinese*, 93–5.
[26] Chang I-tsun, "Chung-kuo ti hua-hsüeh," 4.
[27] Roger S. Greene, "Proposal for Aid to the Medical School of the Central National University of China," 6 March 1929, RG 1, ser. 601, box 24, RAC. Knud Faber: Knud Faber, *Report on Medical Schools in China*, (Geneva: League of Nations Health Organization, 1931), 17.

The NCU engineering program made similar strides. Since the establishment of Peiyang University at the end of the previous century, engineering had been taught almost exclusively in schools under Chinese administration and hardly at all in the mission colleges. By the 1930s, the best programs – Chiao-t'ung, Tsinghua, Central, and Chekiang – were quite good. But most work was in civil or mechanical engineering, little or none in chemical. During the 1920s, a few technical schools offered classes on the manufacture of acids and bases, brewing, tanning, and other chemical industries. Finally, in 1928, National Central University established the first department of chemical engineering. In 1933, the chemistry department, with twenty students, was still the smallest in the college, which included 230 students in civil, mechanical, and electrical engineering.[28]

The NCU college of agriculture, which traced its roots to a department of agriculture established in 1918 under the Nanking Higher Normal School, offered regular degrees and carried on research in taxonomy, crop breeding, and other branches of agricultural science. After 1927, colleges of agriculture were established at several other national universities, many of which did work on the development and application of new crop strains. There remains some question as to the quality of this work, however. In contrast to engineering, which had always been under Chinese control, agriculture was almost exclusively the work of the mission schools. In 1933, Tisdale found the agricultural college at NCU, purportedly the best of those under Chinese administration, in a sorry state, concluding that it was "not sufficiently developed to be rated."[29]

The last of the top government schools, National Chekiang University, was founded in Hangchow in 1928. Chekiang housed colleges of arts and sciences, engineering, and agriculture. Most students took at least one year of laboratory science. By 1933, four-year majors were offered in physics, chemistry, and biology. Among the science departments, chemistry had the largest enrollment and the only program of faculty research. The chairman, Lih Kun-hou (Ph.D., Munich, 1932), had excellent equipment for his work on the absorption spectra of iron compounds. C. C. P'an (M.I.T., 1916) did research on the manufacture of paper from silk residues. The school of engineering included

---

28 Engineering: Chang Hung-yüan, "San-shih nien-lai Chung-kuo chih hua-hsüeh kung-ch'eng" [Chinese chemical engineering during the past thirty years], *San-shih-nien-lai chih Chung-kuo kung-ch'eng* [Chinese engineering during the past thirty years], ed. Chou K'ai-ch'ing, 2 vols. (Taipei: Wen-hua shu-chü, 1967), 1:1–2; W. E. Tisdale, "Report of Visit," 12.

29 Agriculture: ibid., 18; P. W. Tsou, "Letter Addressed to the Committee of the China Foundation...." 27 December 1926, CMB, sec. II, box 38, RAC; and T. H. Shen, "First Attempts to Transform Chinese Agriculture," 213, 221–2.

departments of civil, mechanical, electrical, and chemical, the last of which was headed by Li Shoo-hen (Ph.D., Illinois, 1925), who worked on the combustion of coal. The department maintained large, well-equipped laboratories for advanced courses on qualitative and quantitative analysis, physical chemistry and mechanics, and complete plants for instruction in papermaking, tanning, and dyeing. The college of agriculture was quite poor — Yoshi Kuno suggested that its graduates be given only one and one-half years of credit toward an American degree — and still in the process of reorganization, when Tisdale visited in 1933. New recruits to the faculty included the first two men in China who specialized in agricultural chemistry: H. O. Lou (Ph.D., Ohio State, 1924) and S. L. Whang (Ph.D. Cornell, 1932). There were over one hundred students in the first two years, but no upperclassmen. Plans for the department existed only on paper.[30]

The most dramatic improvement in scientific education and research occurred at National Peking University. As late as 1931, a Ministry of Education inspection team found the school's facilities woefully lacking, the faculty absent or preoccupied with tasks other than teaching, and the courses regularly attended by less than a third of the students, most of whom were only hanging around to pick up their diplomas. The appointment of a new president, Chiang Monlin, backed by a matching grant of 200,000 *yüan* per year from the China Foundation, heralded the first serious entry of *Pei-ta* into the sciences. Beginning in the fall of 1931, laboratory facilities were expanded, new students recruited at both the undergraduate and graduate levels, and several scientists brought to Peking from universities in the south, among them two from Nanking: Leo Shoo-tze, who served as dean of the Peking science faculty, and Tseng Chao-lun, who became chairman of the chemistry department. During the 1930s, biology, physics, and other sciences continued to lag, but the chemistry program made rapid strides. In 1933, there were 170 students in chemistry courses, including 10 seniors writing research theses, mainly on tanning. Leo taught industrial chemistry and gave instruction on tanning, electric furnace extraction of tungsten and beryllium (two minerals China held in abundance), and the fermentation of molasses to produce alcohol. Tseng continued his work on organic chemistry, bringing several younger men to Peking to join this work.[31]

The leading Chinese institution of higher learning in southern China was National Sun Yat-sen University, which was founded in Canton in 1924. When Tisdale visited the university in 1933, he found that basic

[30] W. E. Tisdale, "Report of Visit," 1–6. Yoshi Kuno, *Educational Institutions in the Orient*, 63.

[31] MOE team: *Pei-ching ta-hsüeh hsiao-shih*, 195–6. Chiang Monlin reforms: ibid., 198–207. Other details: W. E. Tisdale, "Report of Visit," 20–1.

chemistry courses were offered in the first two years and industrial chemistry, including tanning, ceramics, soap making, and metallurgy, in the upper grades, although the laboratories were "ill-equipped" for teaching, much less research. The chairman of the chemistry department, Whang Siar-hong (Ph.D., Berlin, 1930), had served as a chemist in the local arsenal. German, rather than the ubiquitous English, was the required foreign language in this department, but few students had language skills adequate to the task. Instruction was primarily by lecture, supplemented by the professor's notes, without the benefit of a textbook. There was no research under way in any branch of the sciences.[32]

Conclusion

Among the many criticisms made of the Nationalists in the Nanking Decade, none seems better deserved than the charge that they mistreated the intellectuals, driving many to opposition or despair, and in the end lost the mind of China. But their treatment of the scientists and the effect on scientific education do not fit this pattern. Two factors encouraged cooperation between the scientists and the state. First, the Ministry of Education was interested less in high-sounding ideas and more in practical results. This fact distinguishes the Nationalists from the educational leaders of the May Fourth era, who talked a great deal about "science," but did little to promote the study of nature or the application of its lessons to China's needs. The liberals had all the right slogans to describe a culture that might, in an ideal world, have been conducive to science. But in practice, they and their students kept to literature and humanities, while the study of nature languished. The Nationalists reversed this process, force-feeding students and resources into science and engineering, while limiting access to the arts. This practice was not always popular, but it produced results.

The emphasis on practicality was a new, or at least revived, justification for the study of science. During the nineteenth century, Chinese officials had introduced training in subjects that served the goals of self-strengthening, and many missionaries taught science as part of medicine. The big push for science in the twentieth century came from the Rockefeller program of medical education and research. But utilitarianism was not always evident in the schools under Chinese control. Prior to 1928, no Chinese university had a medical school, and in a country where 90 percent of the people lived on the land, no Chinese university had

---

[32] *National Sun Yat-sen University: A Short History* (n.p., 1936); and W. E. Tisdale, "Report of Visit," 23–5.

a respectable college of agriculture. With the notable exception of engineering, the sciences entered Chinese (as distinct from missionary) education as a self-contained academic activity with few links to the surrounding environment. Nanking concentrated its attack on the separation between theory and practice. Medicine, engineering, and to a lesser extent agriculture occupied a larger place in the Chinese university after 1928. Students and money flowed to these fields. College seniors were required to do research theses, and the available evidence suggests that many worked on topics of an applied sort. More than before, education addressed national needs.

A second reason for the success of scientific education under the Nationalists was that the Ministry rejected the harsher measures proposed by hard-liners like Ch'en Kuo-fu and left the scientists and science teachers more or less alone. This tolerance, or indifference, may have been due to the fact that few educational administrators knew what the scientists were doing or tried to find out. Conversely, scientists of the second generation stayed clear of politics, and out of sight their academic preoccupations attracted little attention. In any case, the scientists had the classrooms to themselves, went about their business, and taught the courses by their own lights.

In sum, the case of education suggests a surprising compatibility between political authority and social change. On one hand, the government adopted policies that favored utilitarian goals. On the other hand, Chinese society had been fragmented to a degree that allowed technical experts to carry out these policies in ways they saw fit. This tacit understanding, first worked out in education, would be more severely tested by the conduct of scientific research, which received greater attention as the conflict with Japan mounted.

10

Scientific research: the balance threatened

During most of the Nanking Decade, the compromise struck in scientific education also applied to research. The Nationalists, who preferred a utilitarian approach to science, created agencies for industrial and agricultural research and urged scholars in the universities and national academies to pursue projects that would produce useful results. Scientists of the second generation preferred to continue the pure or basic research they had begun as graduate students in the West, in order to win recognition in the international scientific community and, they believed, serve China better in the long run. For the most part, Nanking was willing to tolerate this diversity, with the result that research of all types flourished during the 1930s.

From the beginning this was an uneasy compromise, however, and in the end it did not work. The conflict was brought to the surface by the approach of war. The more China was threatened by Japan, the more Nanking pressured the scholars to shift attention to utilitarian ends, and the more the scholars felt a violation of their professional space. As the decade drew to a close, spokesmen for the Kuomintang on one side and the scientists on the other clashed openly over the course of research and the place of science in the new China.

Chemical research in Chinese universities

From what has been said thus far, it should come as no surprise that the most and best scientific research in China during the Nanking Decade was performed by Chinese scientists working in institutions under Chinese control. The reader might also anticipate that unlike their predecessors in the mission colleges, many of whom explored problems of an applied sort, Chinese scientists of the second generation concentrated on basic research, addressed to the fundamental problems of their disciplines. What might be surprising is that the leading centers of this research were the universities. Given the prominence of the Academia

Table 10.1. *Institutional affiliation of authors in the Journal of the Chinese Chemical Society, 1933–1937*

Institution	Number of articles	%
Chinese universities		
Tsinghua	64	
Peking	44	
Central	19	
Nankai	6	
Other	14	
Total	147	66
Mission universities		
Yenching	22	
Other	16	
Total	38	17
Research institutions		
Academia Sinica	13	
Henry Lester Institute	5	
National Geological Survey	4	
National Academy of Peiping	3	
Other	12	
Total	37	17
Total	222	100

Sinica during this period and the Soviet-style practice of post-1949 China, whereby research and education have been performed in separate institutions, one might expect that the habit of insulating these activities has deep roots. During the 1920s and 1930s, however, the Chinese followed the American model, in which research was dispersed among laboratories that were at the same time expected to educate the next generation of scientists and laymen alike.

The supremacy of the universities is demonstrated by the example of chemistry. Contents of the *Journal of the Chinese Chemical Society*, the leading organ in this field during the years 1933 to 1937, show that university-based scholars published the largest number of articles and that members of just four institutions – Tsinghua, Peking, Yenching, and Central – accounted for over half the total (Table 10.1).

It follows that the most prolific research chemist of the Nanking Decade, Sah Pen-t'ieh, was a professor at Tsinghua University. Sah's chief contribution lay in the identification of organic compounds, which was a matter of considerable significance for the development of chemistry not only in China, but around the world. Organic compounds are complex substances, often difficult to distinguish and isolate. The key step in the process of organic analysis in use in the 1930s was the reaction of an "unknown" compound with a known reagent to produce a "derivative," which could be recognized by the characteristic shape of its crystals and the temperature at which they melt. This assumed, of course, that chemists had established the melting points of substances of known composition, which could serve as reference points for the identification of new unknowns. Creating a reliable chart of melting points was one of the more mundane, but important tasks facing organic chemistry in its formative years. This work required more patience and skill than theoretical sophistication or elaborate equipment and was, therefore, well suited to the state of Chinese science and the strength of its young scholars. During the 1930s, the Tsinghua lab turned out dozens of articles by Sah and his students, whose senior (B.S.) theses were to synthesize and establish the melting points of given compounds. Results of the studies were first published in China, after which they spread quickly to the United States and Europe, where scholars were anxious to make use of Sah's data. The most popular American textbook on this subject, first published in 1935 and reprinted in several later editions, cited more studies by Sah Pen-t'ieh than any other scholar.[1]

Although his principal activity was in basic research, Sah also dabbled in problems of an applied sort. During his postdoctoral studies at Yale, he had investigated the vitamin content of Chinese litchi nuts, based on the biological assay of laboratory rats. After returning to China, he continued this work, concentrating on the vitamin C content of fruits and vegetables in the Peiping market and the dietary needs of the local people. Later, he turned to the synthesis of the blood coagulant vitamin K, interest in which increased sharply after the outbreak of war, and the female sex hormone estrone, which has various medical applications. Yet in all these cases, it was the fundamental problems of chemistry rather than their applications that commanded Sah's attention. In this sense, the environment acted not so much to dictate the objects of Sah's work, as to place limits on its design and success.[2]

The case of vitamin K, for example, demonstrates the dependency of

---

[1] For reports on this research, see articles by Sah Pen-t'ieh et al., JCCS and SRNTU, 1932–40. Textbook: Ralph L. Shriner and Reynold C. Fuson, The Systematic Identification of Organic Compounds: A Laboratory Manual (New York: Wiley, 1935, 1940, 1948). Chapter 8 of the 1948 edition cites 261 articles on the preparation of derivatives of which the largest number, 33, were by Sah. Sah's work is also abundantly cited in Ernest H. Huntress and Samuel P. Mulliken, Identification of Pure Organic Compounds (New York: Wiley & Sons, 1941).

[2] Work at Yale: Arthur Smith and Peter Sah, PSEBM 25 (1928):63–4. Vitamin C: Sah et al., JCCS 2 (1934):73–83, 184–91. Estrone: Sah, JCCS 13 (1946):77–118.

chemical research on the development of allied industries that were notably lacking in China. Vitamin K is a naturally occurring substance (in fact, a group of substances, K-1, K-2, etc.) discovered in 1935 and first synthesized in 1939. In Europe this synthesis was performed from a starting material, 2-methyl naphthalene, produced by coal tar distilleries that supplied a wide variety of materials to the manufacturers of drugs, dyes, and other products in Europe's burgeoning organic chemical industry. In China proper, however, there was only one experimental coal tar distillery, attached to the Ching-hsing Mining Bureau, and the narrow range of its products did not include specialized compounds of this type. Most of Sah's work was devoted to finding a way to make vitamin K from more common materials, such as naphthalene, that were available in China.[3]

Working under inferior conditions, it was not surprising that Chinese scientists should be less productive than their colleagues in the West. Since they were joined in a common race for peer recognition and personal satisfaction, however, many Chinese felt frustrated by this inequity, a reaction illustrated by Sah's experience with vitamin C. The discovery and synthesis of vitamin C were achieved during the period 1928–33 by European chemists, most notably Albert Szent-Gyorgyi, who won the Nobel prize for this accomplishment in 1937. The synthesis of a new compound from known starting materials was one of the most important and exciting challenges in organic chemistry, and it was natural for Sah, who had been working on vitamin C in the Chinese diet, to try his hand at this task. During the mid-1930s, he proposed several methods and sought to interest the European masters in his work, but when the syntheses were finally performed, they took no account of publications from China.[4] Sah's reaction to this slight reveals the pique felt by a proud, competitive scientist working in the backwaters of this global enterprise:

These experimental results [achieved in Europe] are apparently identical in principle with the theoretical synthesis proposed by the author [Sah]. While we wish to extend our hearty congratulations to the Swiss chemists for their excellent performance, regrets must be expressed that due to lack of materials and expenses, the Chinese laboratories can only wish to compete in experimental work with our foreign friends, especially at the present time when the general atmosphere in this country is totally political and far from being scientific.[5]

Tseng Chao-lun, another of China's leading university-based chemists, also devoted the bulk of his time to basic research, including study of the

---

[3] Sah's research on vitamin K: *Recueils des Travaux Chimiques*, 59 (1940):461–70; *JCCS* 13 (1946):119–54.
[4] Sah, *SRNTU* 2 (1933):167–90.
[5] Sah, *JCCS* 3 (1934):289.

Grignard reaction, preparation of organic and inorganic reagents, qualitative analyses of organic chemicals, and mathematical calculations in the field of physical chemistry. None of this work was particularly distinguished, however, or won the kind of recognition enjoyed by some of Tseng's peers. More interesting are the studies done by Tseng and one of his female students, Edith Chu Ju-hwa, that contributed more to China than to chemistry, namely, their research on the seasoning monosodium glutamate, or MSG.

Monosodium glutamate is, as the name implies, the sodium salt of glutamic acid, an amino acid discovered in Germany in 1866. The precise structure of the salt remained in question until it was established by Tseng and Chu in 1931, using a simple, elegant procedure to determine the location at which sodium attaches to the dibasic acid. Subsequent papers explored two key steps in the manufacturing process. The starting material for the synthesis of this compound was a protein, glutenin, part of the "crude gluten" left behind after starch had been removed from wheat flour. After separation, the gluten was mixed with hydrochloric acid to form glutamic acid hydrochloride, which was then reacted to produce the salt. Tseng and Chu showed that the yield of the hydrochloride depended not on the purity of the acid, but on the dryness of the gluten and the time of refluxing. At this stage, the mixture was black with impurities, which were removed by warming in the presence of active charcoal, filtering, and concentrating the solution under a vacuum until the white hydrochloride crystals appeared. The researchers next compared the efficacy of various charcoals and confirmed the superiority of those of vegetable origin. In what might be considered a pioneering study of consumer preference, they prepared and tasted eighteen derivatives of glutamic acid to see if any was as good as the original. None matched MSG, a conclusion that seems to be shared by diners throughout the world.[6]

After completing this research, Chu Ju-hwa went on to the University of Michigan where she received her doctorate in 1936. She returned to Peking and, following the Japanese invasion, moved with Tseng and other members of the Pei-ta faculty to Kunming, where she worked on the synthesis of vitamin K from locally available starting materials, an effort they hoped would help in treating the wounded during the war. Chu left China in 1944 to teach and continue her research in the United

---

[6] History and manufacture of MSG: Tseng Chao-lun and Hu Mei, *JCCS* 3 (1935):154–72; and Wang Shih-mo, "Tan-pai-chih chung chih ku-suan yü ku-suan-na t'iao-wei-fen." Structure of MSG: *SQNUP* 3 (1931):1–5. More precisely, the compounds in question are the dextrorotatory stereoisomers, d-glutamic acid and monosodium d-glutamate. Hydrolysis: *SQNUP* 3 (1932):53–68. Charcoal: *SQNUP* 3 (1932):69–74; *JCCS* 1 (1933):35–36. Taste: *NCUSR* 1 (1931):11–18; *JCCS* 1 (1933):188–98.

States. She never returned, but remains the most important female Chinese chemist of the prerevolutionary era.[7]

Another feature of Tseng Chao-lun's work was his skill in fashioning new devices and putting old pieces together in new ways. Throughout this period, Chinese chemists were dependent on foreign supply of almost all equipment and reagents. This meant at best long delays in carrying out planned experiments and at worst no experiments at all. During the 1930s, Tseng and a colleague, Hu Mei, reported the development of new or improved copper ignition tubes, glassblowing lamps, and apparatuses for sublimation and the determination of melting points.[8]

Whereas most research in China was on organic compounds, there were a few scholars who worked in inorganic chemistry. The first of these, Kao Ch'ung-hsi, wrote his dissertation on the behavior of the rare earth metals, selenium and tellurium, at Wisconsin in 1926, after which he returned to join the faculty at Tsinghua University. More prominent was Kao Ch'ung-hsi's prize student, Chang Tsing-lien, who received his doctorate from Berlin in 1936 for the study of heavy water (deuterium oxide). Chang returned to Tsinghua in 1937, bringing with him the equipment he needed to continue this research, and then transported it to Kunming where he worked throughout the war. His several publications on heavy water received worldwide attention and are generally considered the most significant chemical research performed in wartime China.[9]

Equally rare were physical chemists, who study the basic structures of molecules and the mechanisms by which they react. The lone star in this field was Sun Ch'eng-o. Sun took his doctorate in 1933 at Wisconsin where he met Albert Sherman, a colleague of Henry Eyring, who had developed methods for calculating the activation energies of chemical reactions, an early application of quantum mechanics and one of the pillars of modern chemical kinetics. In his first published work, Sun applied these methods to determine the activation energies of reactions involving organic compounds. One of these papers, published with Eyring, gained wide recognition. Sun returned to China in 1934 to take a position in the department of chemistry at National Peking University. During the next three years, he continued work on activation energies and resumed his dissertation research on the measurement of dipole

[7] Vitamin K: Chu Ju-hwa and Shen Zoe-ing, *JCCS* 10 (1943):119–23.

[8] Improved apparatus: *JCCS* 1 (1933):37–43, 143–82; *SQNUP* 4 (1934):250, 283–332.

[9] Work by Kao and Chang on rare metals: *JCCS* 1 (1933):116–19; 2 (1934):6–12. Heavy water: Chang Tsing-lien, *Science* 100 (1944):29–30; Yen Chih-hsien, "Chung-kuo hua-hsüeh-chia tui-yü wu-chi hua-hsüeh ti kung-hsien" [Contributions of Chinese chemists to inorganic chemistry], *HHSC* 8, nos. 8–9 (1953):260–3. Recognition of Chang: Alice H. Kimball, comp. *Bibliography of Research on Heavy Hydrogen Compounds* (New York: McGraw-Hill, 1949), 47–8.

moments. The dipole moment is a quantitative measure of electromagnetic polarity between the two "ends" of a molecule, which can be used to determine molecular structure, a matter of particular interest to organic chemists. To carry out these experiments Sun used several electronic devices manufactured abroad, but made his own telephone receiver, dielectric cell, and storage batteries.[10]

One of the few local industries to attract the attention of university-based scholars was tanning. By the late nineteenth century, factories abounded in and around Peking and Tientsin for the tanning and dyeing of skins and hides taken from the herd animals that graze the steppes of northern China and central Asia. Traditionally, this process relied on a vegetable tannin such as nutgall, of which China was the world's leading producer and exporter. Later, synthetic tanning agents including formaldehyde and chromium compounds were used, generally in conjuction with vegetable tannins. China's first modern tannery was established in Tientsin in 1898, and by the 1910s there were more than thirty such plants in Peking alone. Beginning in the 1920s, American professors at Yenching University offered courses on the chemistry of tanning. They were joined in 1932 by Ts'ai Liu-sheng (Ph.D., Chicago, 1932), who published several papers on the chemistry of aldehyde tannage, including the reaction between formaldehyde and collagen, or hide protein, and the subsequent adsorption of hydrogen chloride by the formaldehyde-protein complex. In the fall of 1932, Leo Shoo-tze, who had come north with Tseng Chao-lun, established a laboratory for teaching courses on tanning at Peking University. Leo had done his doctorate on chromic acid (Columbia, 1919) and while in Peking concentrated on the chemistry of chromium tannage. His untimely death in 1935 halted teaching and research on one of the most promising of China's native chemical industries.[11]

Chemical research in the academies

Universities remained the most productive centers of scholarship throughout the 1930s, but the Nationalists also created two national

---

[10] Important study: Eyring, Gershinowitz, and Sun, *JCP* 3 (1935):786-96. Recognition of this work: Samuel Glasstone, Keith J. Laidler, and Henry Eyring, *The Theory of Rate Processes* (New York: McGraw-Hill, 1941), 85-152. Other work by Sun: *JACS* 56 (1934):1096-1101; *JCP* 3 (1935):49-55. Activation energies: *JCCS* 3 (1935):1-5, 293-5; 4 (1936):1-5, 98-102, 340-3; 5 (1937):1-2. Dipole moments: *JCCS* 4 (1936):473-6; 5 (1937):22-4, 39-41, 236-8.

[11] Tientsin and Peking factories: Li Ch'iao-p'ing, *Chung-kuo hua-hsüeh shih*, 451-2. Yenching: *Yenching Natural Science News*, 1 (January 1933):14, in RG 1, ser. 601, box 41, fld. 342, RAC. Articles by Ts'ai et al.: *JCCS* 2 (1934):87-107, 193-7, 291-7; 3 (1935):16-21, 296-300. Leo Shoo-tze: Tseng Chao-lun and Mai Shu-huai, *SQNUP* 2 (1937):17-41.

academies that were devoted to scientific research. The National Central Research Academy, or Academia Sinica, established in Nanking in 1928, was the premier institution in this field. Under the academy were thirteen research institutes – for physics, chemistry, engineering, geology, astronomy, meteorology, psychology, history and philology, social sciences, zoology, botany, mathematics, and medicine. The institutes of physics, chemistry, and engineering were located in Shanghai, near the only reliable supply of electricity, gas, and water and near those industries that could most profit from their work. The other institutes were in Nanking, along with the academy's administrative offices. Before the war, the Academia Sinica received as much as one-third of all government expenditures for pure and applied research and won additional grants from the American and British Boxer indemnity funds. It was by far the best-funded, most prominent research organ in Republican China.[12]

At the outset, most of the academy's limited resources were used to purchase fixed assets – land, buildings, books, and equipment – to provide the infrastructure for what was viewed as a long-range undertaking. Remaining funds supported a small staff that worked on problems of an applied sort, whereas basic or pure research was deferred to a better-endowed future. The chemistry institute concentrated on analysis of native products, such as drugs, plant fibers and oils, minerals, petroleum, and salt. The first director, Wang Chin, was long on administrative background and connections in Nanking, but short on laboratory experience. By the mid-1930s, facilities had improved and the staff expanded under the leadership of a new director, Chuang Chang-kong, who oversaw the transition to pure research.[13]

Chuang Chang-kong's chief contribution was to the study of sterols and sex hormones. During his stay in the Gottingen laboratory of Adolph Windaus, Chuang succeeded, where Windaus and others had failed, in establishing the structure of ergosterol. Back in China, he continued work on the synthesis of this and other compounds, published thirteen articles in the most respected German chemical journal, and won recognition in the world of organic chemists. During the same period, he also studied Chinese medicinal herbs, establishing the structure of two new plant alkaloids. Among other members of this institute, Woo Sho-chow [Wu Hsüeh-chou] investigated the ultraviolet absorption spectra of polyatomic molecules, and Chou Tsan-quo extracted and analyzed alkaloids from Chinese medicinal herbs, studied the behavior of san-

[12] In 1935, total annual government expenditures on research were estimated at 3.5 to 4 million *yüan*, and the annual budget of the Academia Sinica was 1.2 million *yüan*, of which about 120 thousand *yüan* went to each institute. Beginning in 1930, the China Foundation and, after 1937, the Sino-British Boxer Fund also provided support. See: Ting Wen-chiang, "Chung-yang yen-chiu-yüan ti shih-ming," 5–8.

[13] *Academia Sinica (1928–1948)*, (n.p., n.d.); and *The Academia Sinica and Its National Research Institutes* (Nanking: Academia Sinica, 1931).

tonins, and discovered derivatives useful as anthelmintic drugs. Most of the chemistry institute's resources were devoted to basic research, although a few staff members did studies of an applied sort, most notably on glass and on the use of native alunite for the manufacture of aluminum and fertilizer.[14]

The smaller National Academy of Peiping was created in 1929, in large part to mollify scholars in the old capital, who felt cheated by the shift of resources and attention to the south. The Peiping Academy had nine institutes – for physics, chemistry, radiology, zoology, botany, pharmacology, physiology, geology, and history. Initially, it was to receive an allocation of 50 thousand *yüan* per month, but actual payments reached only 30 thousand; the various Boxer indemnity funds helped pay the rest. During the 1930s, the staff of the academy numbered over two hundred, and several of its institutes, including chemistry, published the results of work on pure and applied problems. After 1937, when most of these operations moved to Kunming, the size of the staff and scope of its activity were greatly reduced, focusing almost entirely on local applications, such as the manufacture of natural dyes, gasoline substitutes and engine oils, syntheses of compounds related to vitamin K, and analyses of Chinese drugs.[15]

Applied chemical research

The universities and research academies of the Nanking era focused most of their resources on pure or basic research, whereas other institutions established under both public and private auspices did work of an

[14] For a description of research performed in the academy's various institutes, see: *Academia Sinica (1928–1948)*; and *The Academia Sinica and Its National Research Institutes*. Chuang Chang-kong's life and work are described in: Liu Kuang-ting, "Chu-ming yu-chi hua-hsüeh-chia Chuang Ch'ang-kung chien-chuan," 31–3. Chuang's most important papers were on the structure of ergosterol, *Annalen der Chemie*, no. 500 (1933):270–80, and the synthesis of polyalicyclic alpha-ketones, *Berichte deutsch Chemisch Gesellschaft* 69B (1936):1494–1505. The work on herbals was published in the *Transactions of the Science Society of China* 7 (1932):187–215, and in *Berichte* 72B (1939):519–25. Other articles on sterols and sex hormones appeared in *Berichte*, 1935–41.

[15] For details on the history and work of the Peiping Academy, see: T'ao Ying-hui, "Ts'ai Yüan-p'ei yü chung-yang yen-chiu-yüan (i-chiu-erh-ch'i – i-chiu-ssu-ling)" [Ts'ai Yüan-p'ei and the Academia Sinica, 1927–1940], *Chung-yang yen-chiu-yüan chin-tai-shih yen-chiu-so chi-k'an* [Bulletin of the Institute of Modern History, Academia Sinica] 7 (June 1978):15–19; *Kuo-li Pei-p'ing yen-chiu-yüan yüan-wu hui-pao* [Bulletin of the National Academy of Peiping] 1 (1930):2; "Kuo-li Pei-p'ing yen-chiu-yüan shih-chou-nien chi-nien" [Commemoration of the tenth anniversary of the National Academy of Peiping], *KH* 24, no. 2 (1940):144–6; *Kuo-li Pei-p'ing yen-chiu-yüan kung-tso pao-kao* [Work report of the National Academy of Peiping] (Kunming: Pei-p'ing yen-chiu-yüan, 1939, 1942); *Pei-p'ing yen-chiu-yüan* [The National Academy of Peiping] (Nanking (?): Hsing-cheng-yüan, hsin-wen-chü, 1948); and *China Year Book*. (1937–43), 412; (1937–44), 271; (1937–45), 357–9.

applied sort. In the field of chemistry, we have already mentioned the Golden Sea Research Institute, the pioneer laboratory for industrial research established in 1922 by salt and soda magnate, Fan Hsü-tung. During the 1930s, Golden Sea produced useful studies, several of which appeared in the institute's own journals. The fermentation lab developed modern techniques for selecting and cultivating yeasts and ferments used in the manufacture of wine, vinegar, sugar, soy sauce, and dyes. Other researchers tested native sources of bauxite and alunite and studied methods for refining these ores to make aluminum, or explored the production of fertilizer from mineral ores, seaweed, and compost. Golden Sea scholars traveled throughout China to inspect factories and advise on the introduction of new techniques. During the war, the institute moved to Szechwan, where its focus shifted to analysis of subterranean brines, the extraction of salts, and the manufacture of soda. After 1949 Golden Sea was incorporated into the new order: the bacteriology lab became part of the Chinese Academy of Sciences, and the remainder of the institute came under control of the Ministry of Heavy Industry.[16]

The Nationalists, determined to hasten economic development, were not content to leave industrial research in private hands. Their main entry in this field was the National Bureau of Industrial Research (NBIR), established in Nanking in 1930 under the Ministry of Industry. The NBIR was divided into two departments, chemical and mechanical, the first of which included laboratories for chemical analysis, fermentation, ceramics, paper, oils and fats, and special research. Under the direction of Ku Yü-tsuan, holder of a doctorate in industrial management (Cornell, 1931), the bureau's principal tasks were to analyze raw materials, test finished products, and develop better methods of manufacture. Results of this work were published in the bureau's Chinese-language journal, *Kung-yeh chung-hsin* [Industrial Center], which appeared monthly beginning in 1932.[17]

[16] For details on the Golden Sea Research Institute, see citations in Chapter 6, note 27, and *Report of the China Foundation for the Promotion of Education and Culture*, Ninth Report (1934), 67–8; Tenth Report (1935), 53–4; and Eleventh Report (1936), 48–9. The major pubications of the institute included: *Huang-hai-she yen-chiu tiao-ch'a pao-kao* [Research and investigation reports of the Golden Sea Institute], 1932–7; *Huang-hai fa-hsiao yü chün-hsueh* [Golden Sea fermentation and bacteriology], 1939–51; *Huang-hai hua-kung hui-pao* [Golden Sea chemical industry report], 1930s.
[17] For an overview of the NBIR, see: Li Erh-k'ang, "Erh-shih-san nien-tu chung-yang kung-yeh shih-yen-so hua-hsüeh-tsu kung-tso chih hui-ku" [Look back at the work of the chemical department of the National Bureau of Industrial Research during 1934], *KYCH* 4, no. 1 (1935):6–9; Ku Yü-chen and Fan Ching-p'ing, "Hua-hsüeh fen-hsi shih kung-tso kai-k'uang" [Survey of the work of the chemical analysis laboratory], *KYCH* 4, no. 1 (1935):20–3; and Ku Yü-tsuan, "I-nien-pan i-lai chih chung-yang kung-yeh shih-yen-so" [The National Bureau of Industrial Research during the past year and a half], *KYCH* 5, no. 1 (1936):1–50.

Another government organ that did work in applied chemistry was the Sinyuan Fuels Research Laboratory, founded in 1931 by a grant from the family of the industrialist Chin Hsin-yüan, and lodged under the National Geological Survey in Nanking. The Sinyuan Lab was fully equipped for research on a variety of problems related to fuels, results of which were published in its own journal, *Jan-liao chuan-pao* [Fuels Report]. During its first two years, this lab produced studies of two types: microscopic analyses to establish a typology of Chinese coals, and chemical and physical studies of coal, including low-temperature distillation to produce coke and coke byproducts. These three institutions contributed to two of the most successful efforts in applied chemical research from this period – the production of gasoline and other liquid fuels, and the fermentation of wines and sauces.[18]

GASOLINE

By the mid-1930s, the Chinese viewed with concern the mounting threat of war and their country's dependence on imported gasoline and other petroleum products. Since the beginning of the century, they had attempted to extract and refine petroleum from fields in Shensi Province, but the results·were disappointing. By the 1930s, domestic production of gasoline ˙accounted for less than 1 percent of total consumption, which was between 30 and 40 million gallons per year. Imports of kerosene were even higher, over 100 million gallons annually, but this posed a problem more for the country's balance of payments than its security. Nanking's chief worry was fuel for trucks, tanks, and planes. Faced with a similar challenge, Europeans had begun to develop synthetic fuels and gasoline substitutes, but the results of this work were secret, so the Chinese had to rediscover or reinvent solutions on their own.[19]

Although lacking petroleum, China was richly endowed with coal, and it was widely known that gasoline could be extracted from coal tar, a byproduct in the low-temperature distillation of coal to produce coke. In fact, attempts to manufacture synthetic gasoline had already begun at the Ching-hsing Mining Bureau. Ching-hsing, located near Shihchiachuang in Hopeh Province, was the site of a major coal field, opened up by the

18 Hsieh Chia-jung and Chin K'ai-ying, "Jan-liao yen-chiu yü Chung-kuo ti jan-liao wen-t'i" [Fuels research and China's fuel problem], *KH* 17, no. 10 (October 1933):1717–24. The Sinyuan Lab was established by a grant from Soh-tsu King (Chin Shao-chi) in honor of his father, Chin Hsin-yüan: *The National Geological Survey of China, 1916–1931*, 2. Biography of King: *Who's Who in China*, 4th ed. (Shanghai: China Weekly Review, 1931):91–2.
19 Import figures: Li Erh-k'ang, "Wo-kuo hua-hsüeh kung-yeh kai-k'uang chi ch'i fa-chan t'u-ching" [The situation and path of development of our country's chemical industry], *KYCH* 6, nos. 9–12 (1937):339.

Coking plant, Ching-hsing Mining Bureau, Hopeh. (Kung-yeh chung-hsin, *1935.*)

Germans at the turn of the century and taken over by the Chinese at the end of World War I. In 1923, the Mining Bureau, which operated the largest coking facility in China proper, set up a coal tar extraction plant to manufacture synthetic gasoline, production of which reached two thousand barrels per year. Ching-hsing, the only coal tar facility in China proper, attracted the attention of scholars from throughout the country, who studied methods for making synthetic fuels, analyzed the products, and assessed the economics of this enterprise. These studies demonstrated that whereas the quality of Ching-hsing gasoline was excellent, the limited output and high cost made it uncompetitive against foreign imports. Research on coal distillation continued, before the war in East China and after 1937 in the coal fields of Szechwan, but never yielded a satisfactory solution.[20]

Scholars at the NBIR and Sinyuan laboratories also studied methods of producing liquid fuel from solid fossil sources. As one proceeds from hard through soft coal or shale to liquid fuels and from coal tar through

[20] Yen Yen-ts'un, "Ching-hsing lien-chiao-ch'ang fu-ch'an chih ch'i-yu" [Byproduct gasoline of the Ching-hsing coking plant], *HHKY* 10, no. 1 (January 1935):69–80. Ching-hsing production figures: Li Erh-k'ang, "Wo-kuo hua-hsüeh kung-yeh," 339.

*Experimental coal gas engine, National Bureau of Industrial
Research, Nanking. (Kung-yeh chung-hsin, 1932.)*

the light oil and gasoline fractions, there is a corresponding increase in
the ratio of hydrogen to carbon. This means that coal can be converted
to gasoline by either of two methods: *low-temperature distillation*, a
process that drives gases from the coal, concentrating carbon in the coke
and hydrogen in the tar, followed by further distillation of the tar to
separate the light oil fractions that make up gasoline; or *hydrogenation*,
the addition of hydrogen to coal under high temperature and pressure to
yield low-boiling liquids. Distillation can also be applied to shale with
similar results. During the 1930s, Chinese researchers conducted
laboratory experiments on coal distillation, surveyed the shale deposits
of Szechwan, Shensi, and Kwangtung, and prepared studies on the
feasibility of these processes. Their findings were uniformly disappointing: the techniques were complicated, the equipment unavailable, and the
required investment prohibitively high. The record of Japanese manufacture of synthetic fuels in Manchuria, which produced substantial
volumes but at enormous cost, confirmed the impracticality of coal and
shale technologies.[21]

---

[21] Research on coal distillation: Hsiao Chih-chien, *Bulletin of the Geological Survey of
China* 21 (1933):13; Li Erh-k'ang, *KYCH* 4, no. 7 (1935):365–72. Hydrogenation:
Hsiao Chih-chien and Lo Ch'ing-lung, *HHKC* 4 (1937):248–54; Tu Chang-ming,
KYCH 5, no. 2 (1936):63–5. Shale: Pin Kuo, *Bulletin of the Geological Survey of China*
24 (1934):15; Ku Yü-chen and Cheng Su-ming, *KYCH* 3, no. 10 (1934):313–18.

Chinese researchers achieved greater success using vegetable oil. The extraction of oil from soybeans and other oil-bearing seeds to serve as both food and fuel was a major industry in late traditional China. In the early twentieth century, foreign investors introduced power-driven presses and a modern industry developed, centered in southern Manchuria and largely under Japanese control. Chinese in Shanghai, Canton, and other cities copied these methods, and by 1931 there were more than one hundred modern bean mills in China proper. Meanwhile, substandard oils produced in the hinterland were shipped to factories in and around Hankow, where they were refined by the addition of lime or caustic soda to precipitate the impurities. Again, the new method was pioneered by foreigners, and Chinese followed their lead. By the early 1930s, modern plants to press vegetable oils and purify them by chemical techniques were an established part of Chinese industry.[22]

Vegetable oils, like many organic materials, contain a mixture of long-chained molecules of hydrogen and carbon that can be "cracked" under high temperature and pressure to yield larger numbers of shorter chains, six to twelve carbons in length, of the type found in gasoline. Crude petroleum is cracked to increase the gasoline fraction. The same process can be applied to vegetable oils with similar results. In 1934, researchers in China began to investigate the extraction, cracking, and refining of oils from cotton, rape, soybean, sesame, peanut, and tung seeds. These studies showed that the best results came from liquid-phase cracking using an aluminum chloride catalyst. Yields as high as 43 percent gasoline were reported from cottonseed oil. By the end of 1935, successful laboratory production of gasoline from vegetable oil was an established fact.[23]

The problem with this method was not a shortage of equipment or skilled manpower, but the cost of raw materials. During the 1930s, Chinese agriculture was depressed and the price of oil-bearing seeds correspondingly high. The Sinyuan Lab set up an experimental plant in Nanking to produce gasoline from vegetable oil by the methods described. This operation was a technical success, but its products could not compete with cheaper imported gasoline. This situation changed when China's supply lines were severed by the outbreak of war. In 1938, scientists from the Sinyuan plant were dispatched to Chungking to set up the Tung Li Oil Works, the first and largest of many synthetic gasoline factories established during the war. Research on this method continued

[22] Li Ch'iao-p'ing, *Chung-kuo hua-hsüeh shih*, 413–15.
[23] For research on the cracking of vegetable oil, Ku Yü-chen and Cheng Su-ming, *KYCH* 4, no. 1 (1935):64–9; Cheng Su-ming, *KYCH* 4, no. 8 (1935):412–15; Ku Yü-chen, et al., *KYCH* 4, no. 12 (1935):466–79; Pin Kuo, *KYCH* 5, no. 2 (1936):66–70. Cottonseed – 43%: Lo Tsung-shih and Ts'ai Liu-sheng, *JCCS* 4 (1936):157–71.

at the West China Academy of Sciences in Szechwan, but no additional breakthroughs were recorded.[24]

One way to reduce dependency on petroleum imports was to manufacture synthetic gasoline from alternate sources, such as coal or vegetable oil. Another way was to make the gasoline substitute, ethyl alcohol (ethanol). By the 1930s, the "gasohol" industry was well established in Europe, where the laws of several countries required the addition of 20 to 30 percent ethanol in all gasoline sold. The Chinese saw the potential for stretching their own petroleum by similar means. The first problem was to manufacture the alcohol, imports of which had risen to over six million gallons in 1929. To overcome this bottleneck, Nanking joined with private investors to set up several jointly owned factories in Shanghai and other cities. By 1935, China was essentially self-sufficient in industrial alcohol.

The manufacture of ethanol was only the beginning, however. It had to be shown that gasohol would work. Mixtures of gasoline and anhydrous (water free, or pure) alcohol are miscible even at very low temperatures and can be used in an internal combustion engine without complication. But mixtures of gasoline and common industrial ethanol, which is only 95 percent pure, tend to separate at lower temperatures and become ineffective for this purpose. The manufacture of anhydrous alcohol and its storage in moisture-proof containers were too expensive and difficult for the Chinese to consider. In 1934, researchers at the NBIR found another solution: the addition of blending agents to lower the "critical solution temperature" below which gasoline and alcohol separate. Experiments showed that a mixture of 70 percent gasoline, 25 percent normal ethanol (95% pure) and 5 percent amyl or butyl alcohol would work in an internal combustion engine, even if the mixture absorbed some additional water. These blending agents could be produced locally and at reasonable expense. Although the high cost of Chinese gasohol made it an unattractive alternative before the war, after 1937 production of alcohol for use as a gasoline substitute became one of the major industries of Nationalist China.[25]

WINES AND SAUCES

A second line of research was designed to improve the performance of the fermentation industries. The fermenting of grain to produce wine is an

[24] Tung Li: Shen Chin-t'ai, "Hua-hsüeh kung-ch'eng-hsüeh" [Chemical engineering] *Chung-hua min-kuo k'o-hsüeh chih* [Record of science in the Republic of China], ed. Li Hsi-mou (Taipei: Chung-hua wen-hua ch'u-pan shih-yeh wei-yüan-hui, 1955), 1–6.
[25] Alcohol import and production figures: Li Erh-k'ang "Wo-kuo hua-hsüeh kung-yeh," 340. Research: Ku Yü-chen, et al., *KYCH* 3, no. 3 (1934):112–17; Ku Yü-chen, *KYCH* 5, no. 4 (1936):161–8; Hsieh Kuang-ch'ü, *KYCH* 5, no. 12 (1936):565–77.

ancient art in China, where legend attributed its invention to the mythical rulers of antiquity. For more than eight hundred years, this industry had flourished in privately owned factories on the Lower Yangtze. Alcoholic beverages produced in these wineries were of two types: *huang-chiu* [yellow wine], made by the fermentation of rice; and stronger distilled spirits, such as *kao-liang*, made from sorghum. Next to wine, the most important traditional fermentation product was soy sauce, followed by bean paste and vinegar.[26]

The fermentation industries are distinguished by their reliance on microorganisms that cause chemical changes in proteins and carbohydrates, forming products with special characteristics, such as taste or physiological effect. The chemical changes are of three types: saccharification, the breaking down of complex carbohydrates to simple sugars; fermentation itself, the conversion of sugar to alcohol and carbon dioxide; and the decomposition of proteins to their constituent amino acids. These reactions are caused by enzymes produced by fungi, especially yeasts and molds, or in fewer cases by bacteria. The procedures and equipment employed in this industry are often quite complex, but at the heart of the process lies the work of the microbes. Since there are some one hundred thousand species of fungi, selection, cultivation, and application of the right yeast or mold are matters of grave importance to wine and sauce makers. Contamination by an errant bug can turn alcohol to vinegar and spoil the whole batch.

Before the introduction of modern science, the Chinese had no notion of the chemistry of fermentation or the role played by microorganisms. They attributed the changes observed in this realm, as in others, to the grand forces of the universe: Rice and millet were said to contain the elemental force of *yang*, leavening agents the *yin*, and the meeting of *yin* and *yang* produced heat, which they took to be the cause of fermentation. Theory aside, accumulated experience had taught Chinese fermenters to follow practices that assured good results. The "leavens," or media on which the fungi grew, were prepared from specially selected grains, flours and herbs, following elaborate prescriptions. The leavening-room, well seeded by the spores of previous occupants, was carefully swept, cleaned, and sealed with mud. Temperature was controlled by the opening and closing of windows, the periodic turning of the leaven cakes, and by restricting this activity to prescribed seasons, following climatic and other local conditions. Makers of Shaohsing wine, the most famous of the yellow variety, steeped their rice in an acidic "sour liquor," which retarded growth of bacteria and favored reproduction of yeast, similar to the role played by lactic acid in modern alcohol production. The

[26] Li Ch'iao-p'ing, *The Chemical Arts of Old China*, 180–4.

T'ien-kung k'ai-wu, a seventeenth-century guide to industrial technology, explains why the rules of yeast making had to be followed with such care:

If, at the wine-yeast maker's home, the culturing of moulds is not properly done, or not regularly tended to, or if the implements are not properly cleaned, then a few pellets of the badly prepared yeast can spoil whole bushels of the grain. For reliable yeast, therefore, the wine maker must procure it from those who have good reputations and are well-known as leaven manufacturers.[27]

Despite these precautions, all was not well with Chinese wine and sauce makers of the 1930s. European importers had introduced an inexpensive beverage – beer – that enjoyed broad appeal, and several more potent spirits for the new generation of Chinese with cosmopolitan tastes. The use of scientific methods assured foreign manufacturers better and cheaper products. To add insult to injury, these methods included the "amylo process," invented by the French microbiologist Leon Calmette and based on the fungus *Mucor*, which he had extracted from a Chinese wine leaven! Owing to the general decline of the Chinese economy and the increase in foreign competition, production of Shaohsing wine fell by 27 percent during the years 1933 to 1937. The soy sauce industry came under similar pressure, due to the rise in the price of wheat used in making the soy sauce leaven and the competition from Japanese producers, who introduced several new techniques to increase the quality and cut the cost of their sauces.[28]

Both the NBIR and the Golden Sea Institute placed high priority on bringing Chinese fermenters into the twentieth century. The chemistry department of the NBIR made a systematic study of existing wine and soy sauce factories, evaluating techniques and gathering samples of leavens. Back in the laboratory, yeasts and molds were extracted from these leavens, cultivated in pure strains, identified, and tested for their ability to effect the desired reactions. This research demonstrated that the microbes used in the manufacture of even the finest Chinese wines and sauces dominated more by their reproductive powers than their efficacy as fermenting agents. In the absence of scientific methods and modern quality controls, the heartiest and most fertile fungi drove out their biologically weaker, but from the point of view of wine and sauce makers better, rivals. Use of inefficient agents reduced yields of the desired

---

[27] *Yin–yang*: ibid., 3, citing a Ming edition of the *History of Wine* [*Chiu-shih*]. Other details: ibid., 175–9, 197–8. Quotation: Sung Ying-hsing, *T'ien-kung k'ai-wu: Chinese Technology in the Seventeenth Century*, trans. E-tu Zen-Sun and Shiou-chuan Sun (University Park: Pennsylvania State University Press, 1966), 290.

[28] Amylo process: Ch'en T'ao-sheng, *KYCH* 3, no. 2 (1934):64–7. Shaohsing production: Li Erh-k'ang, "Wo-kuo hua-hsüeh kung-yeh," 344. Soy sauce: Ch'en T'ao-sheng, et al., *KYCH* 3, no. 10 (1934):324–32; Li Ch'iao-p'ing, *Chemical Arts of Old China*, 178–9.

product – in the case of *kao-liang* to just 44 percent – and increased the time required for fermentation. Since the leaven alone could account for one-third of the total cost of production, it was important to find and use the most effective microbes.

Researchers at the NBIR identified those species that produced the highest yields, cultivated pure strains, and disseminated samples to factories willing to try them. The equipment and techniques used in traditional factories also came under scrutiny, particularly as they affected environmental conditions. Failure to control temperature meant that yeasts and molds could be bred only in certain seasons, and during cold periods the rate of fermentation slowed significantly. Research demonstrated that some of these problems could be overcome by the use of high-pressure boilers and heating devices. Finally, the high cost of beans and wheat prompted a study of ways to produce soy sauce from less expensive substitutes. In many cases, the use of improved strains and techniques reduced the time required for fermentaton by as much as 90 percent, producing higher yields of wine and sauce with no reduction in quality, even from the cheapest starting materials. By the late 1930s, research and development in this field was an established success.[29]

IMPLEMENTATION

The mission of the NBIR was to improve the technical performance of Chinese industry. Researchers in the chemical department focused on two areas, fuels and fermentation products, and in each case their work was a success. But an assessment of the bureau's performance must also take account of the manner in which it brought its insights and innovations to bear on the factory floor. In this regard, each of these industries posed distinct problems and opportunities.

Gasoline was a new product, introduced from abroad, important to national security and therefore to the government. Spokesmen for the NBIR and its chemical department argued in favor of developing synthetic gasoline and gasoline substitutes on the grounds that they were vital to the nation's defense. Their appeal was addressed to higher levels of the bureaucracy, and they depended on government support for facilities to perform the research and legislative measures, such as requiring the sale of gasohol, that would favor the new industry. They received some but not all of what they asked for. The government supported the study

---

[29] For annual summaries of the work of the NBIR, its chemical department, and fermentation laboratory, see: Li Erh-k'ang, *KYCH* 3, no. 2 (1934):61–3; Li Erh-k'ang, "Erh-shih-san nien-tu," 6–9; Chin P'ei-sung, *KYCH* 4, no. 1 (1935):46–51; and Ku Yü-tsuan, "I-nien-pan i-lai," 7–11. 44% yield: *KYCH* 3, no. 2 (1934):64. Cost of leaven: *KYCH* 3, no. 8 (1934):252–3.

of alternate fuels, invested in the manufacture of ethanol, and built an experimental plant to make gasoline from vegetable oil. Prior to 1937, however, Nanking refused to take those steps that would raise the price of gas at the pump or attempt any of the very expensive projects designed to extract liquid from solid fossil fuels. History demonstrated the wisdom of this middle course: Gasoline and kerosene prices were kept down during the 1930s, a time when Chinese industry and transportation needed cheap fuel. Then after the outbreak of war, alcohol and vegetable oil factories were brought quickly on line, and China avoided the enormously expensive and ultimately unsuccessful production of gasoline from coal or shale.

Wine and soy sauce were traditional Chinese products manufactured in widely dispersed, privately owned factories of various sizes. In this case, the NBIR had to demonstrate that its suggestions would lead to more effective, profitable operations. The bureau adopted an aggressive program of on-site investigation, research, and extension. Teams were sent to the most successful factories to study manufacturing techniques and collect raw material samples. Back in the laboratory, leavening agents were studied and the best of them bred and distributed to wine and sauce producers. The bureau tested materials and products submitted by Chinese manufacturers, received visitors to its laboratories and experimental plants, answered questions in person or by mail, and provided summer training programs for university students and factory technicians. These services were provided gratis, and they were used. In 1930, the chemical laboratory received twelve requests for analysis of materials or products from outside parties. By 1934, the number of outside requests had grown to 113, surpassing assignments generated from within the lab itself. A six-week training program on the theory and practice of fermentation offered in the fall of 1935 prompted more than one hundred applications from technicians in soy sauce factories of the Lower Yangtze region. To accommodate the large number, a second session was held the following year.[30]

The National Research Council and the debate over research

The compromise between the scientists and the state that made possible simultaneous advances in pure and applied research held through the first half of the 1930s. It was undone by the coming of war. The rise of the Japanese threat prompted Nanking to demand a more utilitarian

[30] NBIR extension program: Ku Yü-tsuan, "I-nien-pan i-lai," 10–11. Inspection teams: Ch'en T'ao-sheng, *KYCH* 3, no. 5 (1934):168–75. Chemical analysis: Ku Yü-chen and Fan Ching-p'ing, "Hua-hsüeh fen-hsi-shih," 20–3.

approach to science, and the scientists responded by defending their sphere of autonomy and their commitment to basic research. This debate over research policy became entwined with the early history of the National Research Council (NRC), an official advisory body set up to help China chart a direction for the nation's science. The following account describes the debate, the development of the NRC, and the relationship between them.

Chemical research of the Nanking Decade was performed in two separate channels. The first of these, for pure or basic studies, was lodged in colleges, universities, and research academies, was performed by professional scientists, and constituted a local expression of the global scientific enterprise. Researchers in this channel were trained abroad. Most held doctorates from European or American universities. With few exceptions, they built upon research begun as graduate students in foreign universities and, if successful, published the results in European and American journals or in Chinese journals that used Western languages and were addressed to an international audience. In most respects their work differed little from what one would expect from scholars operating on the periphery of the same disciplines anywhere else. They were playing the game of science as well as their knowledge, wits, and local facilities permitted.

In almost all cases, work of this type belongs to what has been called "normal" science, the routine accretion of data to confirm and enhance the currently accepted model of nature, rather than challenge or seek to replace the reigning "paradigm" itself. Sah Pen-t'ieh's analysis of organic compounds and Chuang Chang-kong's synthesis of sterols, two of the most widely recognized contributions by Chinese chemists of this period, were both "normal" additions to the tree of knowledge. The one Chinese who moved beyond the accepted paradigm to suggest a new model for explaining chemical phenomena, Wu Hsien, was a scientist of exceptional scope and brilliance, who also benefitted from working at Peking Union Medical College, a world-class research center that was only coincidentally located in China.

The second channel, for applied research, also borrowed heavily on knowledge and skills from abroad, but its center of gravity was in China. Researchers in this channel labored in government or corporate laboratories, focused on immediate, practical applications of proven techniques, and published their results in Chinese for a local audience. Few of these men and women had been abroad. None that I have been able to identify held an advanced degree from a foreign university. Sometimes university-based scientists ventured into the applied fields, but even in these instances their research was reported in English-language academic journals with the apparent purpose of reaping professional glory. I have found no example of lateral mobility – the movement of scholars from one channel to the other – or of joint meetings, consultations, or other bridges between these separate worlds.

During most of the Nanking Decade, this division of labor seems to have satisfied all sides. Kuomintang leaders, particularly conservatives who favored scholarly service to the state, wanted applied research, which they got from the ministerial institutes, and more technicians, which the colleges and universities were producing faster than the job market could absorb them. In exchange, the conservatives were willing to turn a blind eye to the academic scientists, most of whom were spending public monies on activities that were of little use in China – or anywhere else. Official tolerance waned, however, as the Japanese threat mounted. By 1935, Nanking began to demand greater devotion on the part of scientists to the needs of the state. This triggered a debate over the course of scientific research and the role of science in China's revival.

One of the earliest and sharpest attacks came in 1935 from Richard Li [Li Erh-k'ang], head of the chemical department of the National Bureau of Industrial Research, on those scholars who, Li maintained,

emphasize pure research and personal reputation, neglect practical engineering and forget the needs of society. It must be recognized that at this time of national crisis, everyone is burying his head [in pure research] and making not the least contribution to the resurrection of China. They are working to win the Nobel Prize or other symbols of vanity. But to win a prize or establish a great name in science cannot restore the country. Tagore won a prize [Nobel prize, literature, 1913], but India is still a subject nation. Einstein has the reputation of a great scientist, but he has no country to go home to, no place to rest his feet. We must all recognize that to fulfill their responsibility, technical personnel must sacrifice personal fame, work to achieve practical results, and contribute to the nation.[31]

Criticisms of this type irritated many scholars, heightening tensions and pushing the issue onto Nanking's agenda.[32]

During the last two years of the Nanking era, the debate on scientific research became the focus of deliberations in the National Research Council, the nation's highest advisory body on scientific and scholarly affairs, whose history testifies to the growing conflict between the scientists and the state.[33] Under the "organic law" of the Academia Sinica promulgated in 1928, the task of advising the government was assigned to the "members" of the academy, an elite group of scholars to be chosen from throughout the nation on the basis of academic merit.

[31] Quotation: Li Erh-k'ang, "Yen-chiu hua-hsüeh-che ying-yu chih jen-shih" [What knowledge chemical researchers ought to have], KYCH 4, no. 8 (1935):407.

[32] For a contemporary view of this issue by a prominent scholar, see: Wu Hsien, "Kuan-yü k'o-hsüeh yen-chiu chih wo-chien," 15–17.

[33] Unless otherwise noted, this history of the NRC is based on T'ao Ying-hui, "Ts'ai Yüan-p'ei yü chung-yang yen-chiu-yüan," 1–50.

The selection of these members was postponed, however, while the academy was preoccupied with the task of its own formation. The first group was finally chosen in 1947.

Meanwhile, in 1934, Ting Wen-chiang, who had just been appointed secretary-general of the academy, judged that the time was ripe to create an advisory body of some kind. He had two reasons for haste. First, whereas most of the institutes were up and running, the academy was not performing its advisory and coordinative role, and this at a time when the purpose and direction of China's scientific research were coming under dispute. Second, and more pressing, was the fate of the academy itself. The health of the academy's president, Ts'ai Yüan-p'ei, was declining, and the method for selecting his successor unsettled. Under the regulations then in force, candidates for the presidency should be nominated by the members of the Academia Sinica and final selection made by the government. But since no members had been chosen, if a vacancy should occur, the selection of a successor might fall to the government and the scholarly community would lose its voice entirely. For these reasons, Ting set about finding a way to assure China's scholars the role they had been promised, but not delivered, in the policy-making structure.

Ting's solution was to change the academy's law to provide for a new National Research Council (NRC) that would play the part originally assigned to the members of "directing, coordinating, and encouraging" research activities throughout China. Sources that describe these events fail to explain why Ting chose this circuitous route and only hint at the problems he met along the way. Apparently concerned that some in the government would oppose his plan, he sought the support of Chu Chia-hua, a member of the powerful Kuomintang Central Political Commission (CPC), which oversaw the work of the government's nonmilitary agencies. Chu differed with Ting on the scope of the NRC membership, but eventually supported the proposal, which was approved by the CPC in May 1935.[34]

The method of selection and eventual makeup of the first council demonstrate that it fairly represented the nation's scholarly community and maintained its independence from political control. The president of the Academia Sinica and heads of its institutes, which numbered ten in 1935, were ex officio members. An additional thirty members, including at least one but not more than three from each discipline represented in

---

[34] Chu: Chu Chia-hua, "Ting Wen-chiang yü chung-yang yen-chiu-yüan" [Ting Wen-chiang and the Academia Sinica], *Chu Chia-hua hsien-sheng yen-lun-chi* [Dissertations of Dr. Chu Chia-hua], ed. Wang Yee-chun and Sun Pin (Taipei: Institute of Modern History, Academia Sinica, 1977), 748–9.

the academy, were selected in a three-stage process: Faculty of the national universities and independent colleges submitted nominations of men and women who had made significant contributions to their fields of learning; an election committee composed of the president of the Academia Sinica and the presidents of the national universities elected members by ballot from this list of nominees; and the national government made the final selection, in effect ratifying the decisions made by the election committee. All members served for a term of five years. Selection of the first two councils, in 1935 and 1940, followed these procedures without apparent political interference. The results of both elections show that the scholarly community was represented by its most illustrious figures. In chemistry, for example, the first council included Chuang Chang-kong, head of the institute, and elected members Wu Hsien, Hou Te-pang, and Chou Tsan-quo, all of whom belong on any short list of leading chemists.[35]

In January 1935, at the time he was seeking support for this scheme, Ting Wen-chiang offered his own views on the critical questions of what kinds of research Chinese scholars should do, and whether and how the government should try to influence the research agenda. Writing in *Tung-fang tsa-chih* [Eastern Miscellany], a periodical read by many intellectuals and policymakers, Ting made three points. First, with regard to scholars and researchers within the Academia Sinica itself, "The most important, practical task is to make scientific studies of our resources and methods of manufacture in order to solve the problems of industry." Second, also within the Academy, scholars should be given the opportunity to pursue their own pure or basic research. Ting justified this concession on utilitarian grounds, arguing that the industrialization of the West proves the interdependence of knowledge of both types: "At their root, 'pure' and 'applied' are one." Finally, Ting adopted a radical libertarian stand on scholarship *outside* the Academia Sinica. This was the realm of freedom, he maintained, where science and the scientist were best left alone:

Should the Central Research Academy, as the nation's "highest" research institution, try to control all scientific research? No! No! The country can control all [other] things, but it cannot control scientific research, because science knows no "authority" and accepts the control of no "authority." Therefore, no country – not even the Soviet Union – has an agency to control scientific research. True, in China, which has a shortage of money and trained people, the government's own research institutes should hold discussions to avoid inadvertent duplication and secure planned cooperation.... The Academy might occasionally use its position to coordinate and exchange information with institutions in [China], but it

---

[35] NRC membership: Lin Wen-chao, "Chung-yang yen-chiu-yüan kai-shu" [Summary of the Academia Sinica], *CKKCSL* 6, no. 2 (1985):21–2, 26.

cannot obstruct the development or mechanically control the research of others.[36]

Ting Wen-chiang's position – to encourage applied research while leaving the power of decision in the hands of the scholars themselves – was tested by the approach of war. In the spring of 1936, Tokyo announced a new "positive diplomacy" aimed against the mainland of Asia, and the Japanese army began to press beyond Manchuria into China proper. As an observer in the U.S. Embassy noted, "practically all major developments of the six months [January to June 1936] were factors working for the disintegration of China." In Peiping and other major cities, Chinese students poured into the streets to protest Japanese aggression and their government's weak response. At the same time, the Communists began to call for a united front of all Chinese against the common enemy. In this situation, Nanking faced a choice between leading the popular movement for national resistance or risk being swept aside by it. The mounting crisis raised pressures on Chinese in all walks of life to come to the defense of the nation.[37]

In April, the newly created National Research Council met to discuss and make recommendations on the role of research in a China whose future was now in peril. President Ts'ai Yüan-p'ei's address, subsequently adopted as the council's "Work Outline," set forth the agreed consensus. In the main, this document confirmed the points made a year earlier by Ting Wen-chiang. The priority, all agreed, should be on applied research. Given the current crisis, scientists should "pay greatest attention" to the "problems of our nation's resources and products," and scholars within the academy should "place their technical abilities at the service of the nation" and "respond to the demands of the government." At the same time, there should be room for some basic research. Pure and applied science are interdependent, Ts'ai reminded his listeners. "If we emphasize one and neglect the other, in the end we may fail everywhere." Finally, even with war clouds darkening the horizon, Ts'ai Yüan-p'ei and the members of the National Research Council, like Ting Wen-chiang before them, refused to restrict the freedom of scholars working outside the Academia Sinica:

In the West, the so-called principle of "academic freedom" holds that the interests and views of the researcher should decide the direction [of his research] and that he must not be controlled by other people. Academic freedom is the basis for the advance of learning, and within the bounds of reason, this [principle] is fully respected. . . . It is the opinion of the Central Research Academy that we

---

[36] Ting Wen-chiang, "Chung-yang yen-chiu-yüan ti shih-ming," 5–8.
[37] Quotation: *Foreign Relations of the United States, 1936*, 4:231, cited in Dorothy Borg, *The United States and the Far Eastern Crisis of 1933–1938* (Cambridge: Harvard University Press, 1964), 182. Other details: ibid., 176–83.

ought to devote every effort to the spirit of free scientific research, and that's that.[38]

This ringing defense of academic freedom by the majordomo of the May Fourth era and representatives of the scholarly community could not stem the rising tide of pressure on scientists to produce useful results. The four-point directive issued by this session of the NRC established narrow guidelines for research within the academy and sought to influence, insofar as possible, research outside the academy as well. Under this directive, scholars were to "pay particular attention to the pressing, practical problems of country and society." The academy and other scholarly organizations in China were to consider this standard in distributing research funds. The results of their research were to be reported on a timely basis to the NRC. And government agencies were invited to discuss their research requirements with the academy and its appropriate institutes.[39]

This statement by the National Research Council marked the first attempt by China to establish a national policy for scientific research, the thrust of which was that science should serve the state. The Executive Yuan, China's highest policy-making body, instructed research organizations throughout the country to carry out the four points, after which Ts'ai Yüan-p'ei wrote these organizations, asking them to submit lists of projects that could produce, within two to three years, the most useful results on the nation's most pressing problems. Once these lists had been assembled and reviewed, the academy would assign tasks to the appropriate agencies. These were the first steps in a program to coordinate and direct the work of scientists throughout China.[40]

The scope of the Academy's authority was unclear, however, and its influence on the scholarly community uncertain. After the directive was announced, there followed a divisive battle over its implementation, particularly as it affected scholars outside the academy. Ku Yü-hsiu, the Tsinghua dean who had defended Ch'en Kuo-fu's educational reforms, spent much of his time trying to persuade reticent scholars to accept official direction. Ku was a tough bargainer, threatening, for example,

---

[38] "Kuo-li chung-yang yen-chiu-yüan chin-hsing kung-tso ta-kang" [Outline for the performance of work by the NCRA], *Kuo-li chung-yang yen-chiu-yüan p'ing-i-hui ti erh-tz'u pao-kao shu*, [Second report of the National Research Council of the National Central Research Academy], 83–7. This is the report by Ts'ai Yüan-p'ei to the second annual meeting of the first National Research Council, on April 16, 1936. The passages cited here have been translated from direct quotations of this speech in T'ao Ying-hui, "Ts'ai Yüan-p'ei yü chung-yang yen-chiu-yüan," 21–7.

[39] Four-point directive: Ku Yü-hsiu, "K'o-hsüeh yen-chiu yü kuo-chia hsü-yao" [Scientific research and the nation's needs], *TLPL* 210 (19 July 1936):5.

[40] Ts'ai Yüan-p'ei: Chou T'ien-tu, *Ts'ai Yüan-p'ei chuan* [Biography of Ts'ai Yüan-p'ei] (Peking: Jen-min ch'u-pan-she, 1984), 373.

that if the academy's chemistry institute did not take up practical research problems, such as developing synthetic fuels, then a new unit should be created to take its place. Taking a less combative approach, Tseng Chao-lun, one of the few scientists with an acute sense of political reality, warned his colleagues against separating themselves from society, refusing to apply their knowledge and skills on behalf of the nation, and harboring the arrogance of a "special class."[41]

But some scholars resisted. "Because of the severity of the national crisis," the physicist Sah Pen-t'ung countered, "some people in China think we should not do pure scientific research that has no immediate applications. This point of view is wrong." These dissenters continued to argue, in opposition to the call for an applied agenda, that the advancement of science depends on an unfettered quest for truth by scholars free of political control, that pure and applied research are inseparable and the ability to produce practical results dependent on an understanding of basic principles, and that in the long run the nation will be revived in the same measure as its scholarship flourishes. Not to be outdone in expressions of patriotism, some hastened to wrap their scholarship in the flag:

If the Chinese nation is to stand up in the world and win the respect of others, then we must make a contribution to the advancement of science. And if we want to make this contribution, then we must allow the free development of science in China.[42]

Conclusion

The revival of the Chinese state and the return to China of a generation of trained scholars combined during the Nanking Decade to produce impressive results in scientific research. As late as 1920, China had no laboratory research of any kind. By 1937, Chinese chemists, working in China without foreign assistance, had produced hundreds of articles

[41] Ku: Ku Yü-hsiu, "K'o-hsüeh yen-chiu yü kuo-chia hsü-yao," 5–8; and Ku Yü-hsiu, "Ch'i k'o-hsüeh t'uan-t'i lien-ho nien-hui ti i-i ho shih-ming" [Significance and mission of the joint annual meeting of seven science organizations], *TLPL* 215 (23 August 1936):8–10. Tseng: Tseng Chao-lun, "Kuo-nan ch'i-chien k'o-hsüeh-chieh t'ung-jen ying fu ti tse-jen" [Responsibilities that must be borne by our scientific colleagues during the period of national crisis], *KH* 20, no. 4 (1936):255–6.

[42] Sah: Sah Pen-t'ung, "Ch'un-ts'ui k'o-hsüeh yü shih-yung k'o-hsüeh" [Pure science and practical science], *TLPL* 236 (30 May 1937):14–17. Patriotic quote: P'eng Kuang-ch'in, "Lun k'o-hsüeh yen-chiu chih t'ung-chih" [Discussing control of scientific research], *TLPL* 214 (16 August 1936):7–9. See also: P'eng Kuang-ch'in, "K'o-hsüeh ti ying-yung" [Application of science], *TLPL* 199 (3 May 1936):11–13; and Ch'en Hsün-tz'u, "So-wang yü Chung-kuo k'o-hsüeh-chia-che" [Expectations of Chinese scientists], *KH* 20, no. 10 (October 1936):884–90.

that merited publication in the world's leading scholarly journals, and Chinese industries profited from the studies of China's own labs. This transformation was due in part to political development: the rise of a new order committed to increasing the country's wealth and power through the application of modern techniques. It was also in part the product of new social forces: the arrival of a second generation of scientists, prepared to do research on the standard they had learned in the West. The concurrence of these political and social changes produced conflict, but it also yielded products in pure and applied research of a type China had not previously enjoyed.

Still, some fields were conspicuous by their absence or lack of development. Chinese chemists investigated problems of medicine, nutrition, and public health; they sought ways to improve established industries, such as tanning, fermentation, and seasonings, and create new ones, most notably liquid fuels. They did little or nothing, however, in the areas of agriculture, engineering, or the basic chemical industries. Given the position of farming in China, the neglect of agricultural chemistry is shocking. Prior to 1949, there was not one agricultural college of any real quality under Chinese (as distinct from missionary) administration. The National Agricultural Research Bureau, established in 1932, helped develop new crop strains, but did nothing to promote chemistry. Surveys of the literature reveal little research on the chemistry of water, soils, fertilizers, crops, or other similar topics. One sample of 442 chemical articles published between 1929 and 1949 includes only one item on agricultural chemistry. Chemical engineering, the study of industrial processes, was also neglected. Before 1949, most research in this field was coauthored by Chinese and foreigners and appeared in foreign journals. Only two books (plus one translation) and a handful of articles were published in Chinese. Finally, the role of the basic chemical industries, acids and sodas, was not on China's research agenda. Throughout this period, we find only two articles on the manufacture of sulfuric acid, the most important of all industrial chemicals, by state-of-the-art techniques.[43]

As the Nanking Decade drew to a close, the compromise between Chinese scientists and the state began to come undone. The conflict was

---

[43] National Agricultural Research Bureau: T. H. Shen, "First Attempts to Transform Chinese Agriculture," 221–9. Soil and fertilizer chemistry: Han Tsu-k'ang and Chang Yüan-lang, "Chung-kuo hua-hsüeh-chia tui-yü fen-hsi hua-hsüeh ti kung-hsien" [Contributions of Chinese chemists to analytical chemistry], HHSC 8, no. 10 (1953):340–5. List of articles, 1929–49: Li Ch'iao-p'ing, Chung-kuo hua-hsüeh shih, 295–354. Chemical engineering: Ku Yü-chen, "Chung-kuo hua-hsüeh-chia tui-yü hua-hsüeh kung-ch'eng ti kung-hsien" [Contributions of Chinese chemists to chemical engineering], HHSC 8, no. 10 (1953):333–9. Sulfuric acid: Yen Chih-hsien, "Chung-kuo hua-hsüeh-chia tui-yü wu-chi hua-hsüeh ti kung-hsien," 296–300.

in part over authority: the power of the state versus professional autonomy. It was also over the role of science: whether it was an instrument for immediate, practical change or a search for knowledge whose results would be useful, but in their own time. Ultimately, it was over freedom: whether the loyalty of an individual Chinese should be to the group or to himself or herself, and whether that group should be defined as the Chinese nation or the community of scientists. This conflict was not resolved; it could not be resolved. But it could grow sharper, and it did – after the outbreak of the war and the retreat of scholars and bureaucrats to a remote area of China, where the very existence of the nation hung in the balance.

11

The chemical industry and the limits of growth

An examination of China's chemical industry during the Nanking Decade reveals little tension between Chinese state and society. This may seem surprising in view of the growing rift between the bureaucracy and the scholarly community, described in the preceding chapters. And it contradicts some recent scholarship that presents the Nationalists as exploiting and oppressing, rather than representing the interests of China's capitalist class. Still, the record of the chemical sector shows that public and private efforts combined to introduce current technologies and efficient production that succeeded in displacing foreign imports and controlling the Chinese market. In industry, as in education and research, the Nanking Decade witnessed the flowering of Chinese chemistry.

New technologies

Technological change in Nationalist China occurred primarily through the importation of machines and methods from the West. Most technical transfers were carried out by Chinese engineers and technicians, working for private Chinese corporations that produced for the civilian sector. Chinese government enterprises played a secondary role, mainly in the manufacture of armaments and their chemical inputs. Foreign-owned companies and joint ventures contributed little. A few foreign engineers helped to contruct the largest plants, and foreigners were employed in the government-owned arsenals. But Chinese built and operated most modern chemical factories, large and small. Finally, local chemists and chemical researchers made few significant innovations. Rather, Chinese entrepreneurs purchased and Chinese engineers installed state-of-the-art machinery from the West. The most important initiatives in the chemical sector were the electrolysis of salt, the manufacture of sulfuric acid, nitrogenous fertilizer, and ethyl alcohol, and the refining of kerosene and other petroleum products.

ELECTROLYSIS OF SALT

The electrolysis of salt, first developed in the United States in the 1890s, assumed growing importance after the turn of the century. In this process, a strong electric current creates opposite charges on electrodes submerged in cells filled with sodium chloride (salt) solution. Positively charged sodium ions gather at the negative pole, combining with water to form sodium hydroxide (caustic soda) and hydrogen gas. Negatively charged chlorine ions gather at the positive pole to form chlorine gas, which can be used to make liquid chlorine, bleaching powder (chloride of lime), or hydrochloric acid. The method was first tried in China during World War I by Chinese investors working with a German technician, but the attempt failed due to the high cost of electricity, the major expense in operating factories of this type.[1]

A second, more successful effort was undertaken in the late 1920s by Wu Yün-ch'u, the Shanghai chemical magnate whose monosodium glutamate factory required large quantities of hydrochloric acid. The simplest and cheapest way to make hydrochloric acid is by the reaction between sulfuric acid and salt. At this time, however, the Chinese produced no sulfuric acid for the commercial market, so Wu opted for the more direct, if unproven (in China) method of electrolysis. The T'ien-yüan Electrochemical Factory was established in 1928, with 200 thousand *yüan* raised from stockholders in Wu's T'ien-ch'u MSG company and an exemption from tax on the principal raw material, refined salt. In the fall of that year, Wu learned that a French firm located in Haiphong had gone bankrupt and was planning to sell its electrolysis plant. He went to Haiphong to inspect the equipment, found it in good condition and bought the entire lot, including 120 electrolytic cells, for one-third the market price. These were Allen-Moore diaphragm cells, which were somewhat less efficient than the competing mercury cells, but were cheaper and required less energy, the principal considerations for Chinese manufacturers at this time. The equipment was shipped to Shanghai, a site chosen because of its proximity to the T'ien-ch'u plant that would consume most of the acid and the need for a reliable supply of electricity, unavailable in most other parts of China. The salt came from nearby Chekiang. Like Wu's other plants, T'ien-yüan was an immediate success. By 1937, its capital investment had increased to 1 million *yüan*, 300 cells were in constant operation, and annual production of hydrochloric acid reached 1,500 tons. The plant also made caustic soda

---

[1] Electrolysis: D. W. F. Hardie, *Electrolytic Manufacture of Chemicals from Salt* (London: Oxford University Press, 1959), 60–5. World War I venture: Wu Ch'eng-lo, "Chung-kuo hua-hsüeh kung-yeh wen-t'i," 8.

and bleaching powder, which were sold to soap, paper, and dye factories in the Shanghai area.[2]

Once established, this new technology played an expanding role in China, as in the West. The electrolysis of salt is typical of many chemical industries that make several distinct, even unrelated, products by a single, flexible method. The advantage of such industries is that the outputs can be adjusted to meet changing market conditions. Thus, the T'ien-yüan factory was created to manufacture hydrochloric acid, but its byproducts – bleaching powder, caustic soda, and liquid chlorine – assumed greater importance with the passage of time. Caustic soda, in particular, was vital to the development of chemical and allied industries throughout the world, and as the demand for this product increased, electrolysis plants were reprogrammed to make more of it.

The demand for caustic soda rose sharply during the late nineteenth century, when high-speed presses began to consume mountains of pulp paper, made by soaking wood fibers in caustic soda, and textile factories began to treat cotton cloth by the Mercer process, which also uses this chemical. Originally, caustic soda was made by reacting soda ash with lime, water, and sodium carbonate. This simple and efficient technique was first used in China during World War I to causticize natural soda. Large-scale production began in 1932, when the Yungli company set up a factory on the Tangku site that used the same method. With an output of four thousand tons per year, Yungli was the leading producer of caustic soda in China. At about the same time, the Chinese recognized the possibility of making this product by electrolysis. Besides T'ien-yüan, the Kwangtung Provincial Electrochemical Factory, a government enterprise, opened in 1935, using Vorce-type diaphragm cells designed and built in the United States. The daily output of this plant reached six tons of caustic soda, along with similar amounts of bleaching powder and hydrochloric acid. A second, private firm, the United Chemical Works, was also established in Canton in 1935 by Chinese students who had returned from the United States. The opening of this plant was delayed by the shortage of electricity, however, and there is no evidence that the problem was solved before the outbreak of war in 1937.[3]

[2] T'ien-yüan: Ch'en Chen and Yao Lo, eds., Chung-kuo chin-tai kung-yeh-shih tsu-liao, 522–6; Li Ch'iao-p'ing, Chung-kuo hua-hsüeh shih, 2:374, 416; and Yü Jen-chün, "Wo-kuo shih-yen tien-chieh kung-yeh chih ching-chi kuan" [Economic outlook for our country's salt electrolysis industry], HHKY 19, nos. 1–2 (April 1947):4–6. Electrolytic cells: W. L. Badger and E. M. Baker, Inorganic Chemical Technology (New York: McGraw-Hill, 1941), 150–70. Salt: China Industrial Handbooks, 659. Investment and production figures: Shina Kōgyō, 150. For biography of Wu Yün-ch'u, see: HHSC 8, no. 11 (November 1953).

[3] History of caustic soda: L. F. Haber, Chemical Industry during the Nineteenth Century, 95–8. Natural soda: Kuo Pen-lan, "Chang-chia-k'ou chih liang ta hua-hsüeh kung-yeh," 137–9. Yungli: Li Ch'iao-p'ing, Chung-kuo hua-hsüeh shih, 374. T'ien-yüan and

Wu Yün-ch'u, the founder of China's electrolysis and other chemical industries, exhibited many of the traits of an industrial pioneer. He learned business from the bottom up, recognized opportunities and pursued them boldly, and remained directly involved in the management of all his factories. In addition to the enterprises already discussed, Wu went on to build a nitric acid factory that made use of surplus hydrogen gas from the electrolysis plant, and a chemical porcelain factory, a spin-off of an earlier in-house operation. Like Fan Hsü-tung, he formed his own research and development unit, the Chinese Industrial Chemical Research Institute. Wu was a man of public spirit, a founder of the Chinese Society of Chemical Industry and at least two charities. During the war, he moved his factories to the rear areas to help carry on the struggle against Japan. After 1949, Wu remained in China and supported industrial development under the Communists. He died in Shanghai in 1953.[4]

MANUFACTURE OF SULFURIC ACID

The manufacture of sulfuric acid, the most important of industrial chemicals, also illustrates the rapid changes of the Nanking Decade. Prior to 1933, the Chinese produced acid in government arsenals, where they introduced some improvements. The Hanyang Arsenal, for example, installed Gay-Lussac and Glover towers, which served to recover nitrogen oxides, increase the efficiency of the lead chambers, and clean up the air. On the other hand, the Tehchow Arsenal, which was more representative of this industry, had no such refinements and simply burnt sulfur and niter, feeding the gases directly into lead chambers, in a small and primitive system. Sulfuric acid was combined with potassium or sodium nitrate to produce nitric acid and both acids used to make guncotton.[5]

One foreign concern attempted to manufacture acid in China, albeit with limited success. Major Brothers, later called the Kiangsu Chemical

---

Kwangtung caustic soda factories: ibid., 374, 416; *Shina kōgyō*, 68–70, 150; "Kuo-nei kung-yeh," *HHKC* 3, no. 1 (March 1936):74; Wu Ch'eng-lo, "Chung-kuo hua-hsüeh kung-yeh wen-t'i," 8; Yü Jen-chün, "Wo-kuo shih-yen tien-chieh kung-yeh," 4–9. United Chemical Works: Donald C. Dunham, "The Influence of American Education in South China," (15 February 1935), 61–2, in Records of the U.S. Department of State Relating to the Internal Affairs of China, 1930–1939, Department of State Decimal File 893, (Scholarly Resources Inc., 1985, microfilm).

[4] Obituary of Wu Yün-ch'u: *HHSC* 8, no. 11 (November 1953).

[5] Hanyang arsenal: Ch'en T'ao-sheng, "Hu Han hua-hsüeh kung-yeh k'ao-ch'a-chi," 234–40. Tehchow arsenal: Wei T'ing-ying, "Chi-nan ... hua-hsüeh kung-yeh k'ao-ch'a-chi," 212. Towers: L. F. Haber, *Chemical Industry during the Nineteenth Century*, 102–8; and F. Sherwood Taylor, *A History of Industrial Chemistry* (New York: Abelard-Schuman, 1957), 189–90.

Works, was founded in 1901 by British investors in Shanghai to supply local factories with sulfuric, hydrochloric, and nitric acids. Owing to the high cost and unreliable supply of raw materials, however, this enterprise foundered and was sold in 1922 to another group of foreigners, who experienced similar difficulties. As late as 1933, the total output of Major Brothers was less than 50 tons of sulfuric acid per year. In 1936, after the introduction of new equipment and imported sulfur, production increased to 2,700 tons, but by this time, the Chinese had entered the scene.[6]

The first private Chinese venture and eventually the largest producer of sulfuric acid in China was the K'ai-ch'eng Acid Company. Founded in Shanghai in 1931 with an initial investment of 500 thousand *yüan*, K'ai-ch'eng suffered from the high cost and low quality of Chinese sulfur-bearing ores and was unable to begin production until 1933. This factory had three lead chambers, featured Gay-Lussac and Glover towers, and burned pyrites from neighboring Chekiang Province. Although total plant capacity was 4,500 tons per year, production peaked at around 3,000 tons. Over half the output was sold to arsenals; the rest went to factories producing textiles, dyes, enamelware, soda water, plastics, and soap.[7]

In 1933, the same year K'ai-ch'eng began operation in Shanghai, the largest plant in northern China, the Li-chung Acid Company, opened in Tientsin. Li-chung was founded with an initial investment of 200 thousand *yüan* by a group of men with connections to several provincial governments in this region. The plant, which burned pyrites from Honan and Shensi in lead chambers, was designed by engineers from the Nankai University Institute of Applied Chemistry in Tientsin. It began operation in 1934 and reached an annual output of 1,200 tons, all of which was sold in northern China.[8]

These private concerns employed lead chambers, which had been used in China since the turn of the century. The introduction of new technology was the work of government factories that supplied sulfuric acid for the manufacture of gunpowder and high explosives. The leaders in this field were the provincial governments of Kwangtung and Kwangsi. Canton, capital of Kwangtung, had long served as China's gateway to Southeast Asia and more recently as the marketplace for goods and ideas

---

[6] Major Bros.: Wu Ch'eng-lo, "Chung-kuo hua-hsüeh kung-yeh wen-t'i," 5; *China Industrial Handbooks*, 655–7; and *Shina kōgyō*, 123–4, 152.

[7] K'ai-ch'eng: Wu Ch'eng-lo, "Chung-kuo hua-hsüeh kung-yeh wen-t'i," 5; and *Shina kōgyō*, 148. Technology: Ting Szu-hsien, "Chien-she kuo-fang hua-hsüeh kung-yeh chih wo-chien" [My views on establishing a chemical industry of national defense], *HHKC* 2, no. 1 (June 1935):82. Pyrites: *China Industrial Handbooks*, 658–9. Market for acid: ibid., 659–61. Production: "Kuo-nei kung-yeh," 75.

[8] Li-chung: *Shina kōgyō*, 143–4.

introduced to this region from the West. Within China, this was one area that most prized innovation and the entrepreneurial spirit. It is ironic, therefore, that in contrast to the other major centers, Shanghai and Tientsin, the chemical industries of the south came under official control. In 1927, the government of Kwangsi, using 600 thousand *yüan* from the army budget and with the assistance of German technicians, built lead chambers at Wuchow to supply acid for the provincial powder plant. This operation failed and was reorganized as the Liang-Kwang Provincial Sulfuric Acid Factory, with an injection of funds from Kwangtung Province and a Chinese director. Meanwhile, in 1933, the government of Kwangtung established a second enterprise, the Kwangtung Provincial Sulfuric Acid Factory, in Canton. This plant was built by the Chemical Construction Company of New York and managed by B. F. Wong, a graduate of the University of California (B.S., 1923), who had previously served as professor of chemistry at Canton's National Chungshan University. By 1936, the combined annual capacity of the two southern plants was 7,500 tons, although actual production was limited by inadequate demand. The entire output went to supply provincial gunpowder and fertilizer factories. It was expected that the production of acid would increase when the provincial ammonium sulfate factory, then under construction, was completed.[9]

The Kwangtung plant was built around the "contact" process, which had been developed in Germany at the end of the nineteenth century to supply the synthetic dye industry with acids of greater concentration than those produced by the chamber method alone. The name of this process derives from the fact that the key step, oxidation of sulfur dioxide to trioxide, is effected by bringing the gas into contact with a metallic catalyst, first made of platinum, later vanadium. Commercial production of sulfuric acid by this method began around 1890 and assumed growing importance, primarily in Germany, after the turn of the century. The contact process places heavy demands on the manufacturer, for the metallic catalyst is easily contaminated and requires the use of pure sulfur and careful controls. It also offers advantages over the chamber method: The plant occupies less space and requires no towers or other special buildings, there is no need to construct and maintain massive lead chambers, and the product is more concentrated and therefore easier and more economical to transport. The first factory in China to make sulfuric acid by the contact process was established in 1927 at the government

[9] Liang-Kwang: ibid., 155. Kwangtung: "Kuo-nei kung-yeh," 74; and Donald C. Dunham, "The Influence of American Education in South China," 61. Both plants: Wu Ch'eng-lo, "Chung-kuo hua-hsüeh kung-yeh wen-t'i," 5; Fan Ching-p'ing, "Kuo-ch'an liu-suan chih fen-hsi" [Analysis of sulfuric acid produced in China], *KYCH* 5, no. 12 (December 1936):578–82; Shen Tseng-tso, "Liu-suan kung-yeh chi ch'i hsin ch'ü-shih" [The sulfuric acid industry and its new direction], *KYCH* 12, no. 1 (October 1948):10.

arsenal in Liaoning Province. During the next decade, the arsenals at Tehchow, Chengtu, Taiyuan, and Kung-hsien also adopted this technique. The Kwangtung Acid Factory introduced American equipment, using a vanadium catalyst, and burned imported sulfur, also from the United States, because of the limited quantity and poor quality of local supplies. The first contact plant for commercial production was built in 1936 as a part of the Yungli ammonium sulfate complex, the last and greatest of China's prewar chemical enterprises.[10]

MANUFACTURE OF NITROGENOUS FERTILIZERS

In addition to acids and sodas, among the most important industrial chemicals are the nitrogen-bearing compounds. Nitrogenous fertilizers, such as ammonium sulfate, supply this element to the soil, making possible the continuous high yields of modern agriculture. But nitrogen has also been used in less constructive ways, for it is the prime component of nitric acid and, hence, the high explosives: nitroglycerin, nitrocellulose, and trinitrotoluene (TNT). Although it is among the most abundant elements in nature, nitrogen must be "fixed," or combined with other elements, before it can be used for other purposes. The artificial fixing of nitrogen in the form of synthetic ammonia was one of the major industrial breakthroughs occasioned by World War I. During the 1920s, all the major Western powers erected ammonia factories and began to manufacture nitrogenous fertilizer in great abundance, knowing that these facilities could be converted to produce explosives, if war should recur. By 1931, world production of ammonium sulfate neared two million tons, but consumption was just over half that amount. The great strategic importance of fixed nitrogen led to the expansion of this industry beyond all economic reason.[11]

Although China's military planners recognized the strategic benefits of nitrogen fixation, it was the needs of agriculture that brought this matter to public attention. After the mid-1920s, China's imports of grain and flour mounted, exceeding 2.6 million tons and a cost of $62 million,

[10] History of contact method: L. F. Haber, *Chemical Industry during the Nineteenth Century*, 102–8; F. Sherwood Taylor, *History of Industrial Chemistry*, 424–6; and Aaron J. Ihde, *The Development of Modern Chemistry* (New York: Harper & Row, 1964), 675. Arsenals: Li Ch'iao-p'ing, *Chung-kuo hua-hsüeh shih*, 370; and Wu Ch'eng-lo, "Chung-kuo hua-hsüeh kung-yeh wen-t'i," 5. Kwangtung: Ch'en Te-yüan, "Liu-suan yü hsiao-suan chih-tsao fang-fa chih ko-ming" [Revolution in the method of manufacture of sulfuric and nitric acids], *HHKY* 12, no. 1 (January 1937):7; and "Kuo-nei kung-yeh," 74.

[11] Aaron Ihde, *Development of Modern Chemistry*, 678–80. In 1931, world ammonium sulfate production was 1.9 million and consumption 1.1 million metric tons. See: Hsü Shan-hsiang, "Chung-kuo jen-tsao fei-liao chih hsien-tsai chi chiang-lai" [The present and future of China's artificial fertilizer], *HHKY* 8, no. 1 (January 1933):16.

about one-eighth of all imports, in 1931. One way to stem the drain of cash for the purchase of food was to increase yields of rice and wheat, a strategy that called for greater inputs of chemical fertilizers. Japan, with a similar type of agriculture, one-fifth the population, and one-tenth the arable land, consumed more than five times as much ammonium sulfate as China. Until this time, chemical fertilizers had been used primarily in southern China, where the proximity to foreign markets and the high price of agricultural goods encouraged high rates of investment. The spread of this practice to northern China and greater application of fertilizer throughout the country seemed to some Chinese observers a wise strategy.[12]

Given that China needed more fertilizer, there remained the question of where it should come from. During the late 1920s, China's imports of ammonium sulfate increased by 50 percent per year, reaching 190 thousand tons and a cost of more than $6 million in 1930. Some Chinese pointed to the world glut of this chemical and argued that China should remain a buyer. But the global trend was in the opposite direction. Japan had joined the race with Western producers by opening new ammonium sulfate plants, and the Soviet First Five-year Plan also called for heavy investment in this industry. To many Chinese, the argument seemed compelling: In the short run, manufacturing ammonium sulfate in China would raise agricultural yields and stem the outflow of cash for grain and fertilizer; and in case of war, these plants could be converted to produce nitric acid and explosives.[13]

In 1931, the Ministry of Industry began to explore the problem of building nitrogen fixation and ammonium sulfate plants. After talks with British and German firms broke down, Nanking turned to Fan Hsü-tung, a man with good political connections and a successful record for manufacturing chemicals on a large scale. In 1934, Fan accepted a government loan of 5.5 million *yüan* at 7 percent for five years and changed the name of his corporation to the Yungli Chemical Industry Company to reflect its broader scope. He purchased 275 acres of land at Hsiehchiatien on the Yangtze River near Nanking as the site for the new plant, a spot chosen because of the need to import large, bulky

[12] Grain and flour imports: ibid., 2. Japan: ibid., 7, 19. *Shina kōgyō*, 166–7, shows that during the period 1926–35, of all ammonium sulfate in China, 69% was consumed in southern China, 25% in central China, and 6% in northern China. Hsü Shan-hsiang was one of those who advocated the manufacture and greater application of ammonium sulfate in China.

[13] Ammonium sulfate imports: *Shina kōgyō*, 169; and Hsü Shan-hsiang, "Chung-kuo jen-tsao fei liao," 4. Japan and Soviet Union: ibid., 19. Compelling argument: ibid.; and Ho Shang-p'ing, "Tan-chih hua-hsüeh fei-liao yü wo-kuo nung-yeh" [Nitrogenous chemical fertilizer and our country's agriculture], *HHKY* 8, no. 1 (January 1933): 21–30.

Yungli Ammonium Sulfate Factory, Hsiehchiatien, Kiangsu. (Hua-hsüeh kung-ch'eng, 1937.)

Yungli Nitric Acid Factory, Hsiehchiatien, Kiangsu. (Hua-hsüeh kung-ch'eng, *1937.)*

equipment, including a 100-ton ammonia synthesizer. Meanwhile, Hou Te-pang was dispatched to the United States, where he bought equipment and signed a contract with the American Cyanamid Company to supervise construction. By the end of 1936, four separate factories were in place: one each for synthetic ammonia, sulfuric acid, nitric acid, and ammonium sulfate. Production began in February of the following year.[14]

In the Haber process employed at Hsiehchiatien, pure hydrogen and nitrogen gases were combined under high temperature and pressure to form ammonia. The hydrogen was produced in a water gas generator, the nitrogen by heating coke. The ammonia, condensed to liquid form, was combined with sulfuric acid, made by the contact method, to

[14] Hsiehchiatien: "Yung-li liu-suan-ya-ch'ang ch'eng-kung chih i-i" [Significance of the success of the Yungli Ammonium Sulfate Factory], *HHKC* 4, no. 2 (June 1937):111–12; "Yung-li hua-hsüeh kung-yeh kung-szu liu-suan-ya-ch'ang ch'eng-li ching-kuo chi ch'i kai-k'uang" [The process of establishment and current status of the Yungli Chemical Industry Company Ammonium Sulfate Factory], ibid., 183–95; Wu Ch'eng-lo, "Chung-kuo hua-hsüeh kung-yeh wen-t'i," 9; Wu Ch'eng-lo, "San-shih-nien-lai Chung-kuo chih suan-chien hua-hsüeh kung-ch'eng" [China's acid and soda chemical engineering during the past thirty years], *San-shih nien-lai chih Chung-kuo kung-ch'eng* [Chinese engineering during the past thirty years], ed. Chou K'ai-ch'ing, 2 vols. (Taiwan: Hua-wen shu-chü, n.d.), 1:15–16; and Jung-chao Liu, *China's Fertilizer Economy* (Chicago: Aldine, 1970), 10, 129.

produce ammonium sulfate. Enormous machines moved great mounds of raw materials in a continuous process, all on a par with the international standards for this industry. The Yungli plant had a capacity of 150 tons of ammonium sulfate per day, or 54,000 tons per year, an amount equal to one-third of China's current annual imports.

The Hsiehchiatien complex was the most modern and complete chemical facility in China and demonstrates the high level achieved by this industry prior to World War II. Key pieces of equipment were imported, but much of the rest was manufactured in the company's own foundry and workshops. Only three foreign advisers took part in construction of the plant; none was needed to supervise its operation. The entire work force was Chinese, including more than eighty college graduates, twenty of whom had received higher engineering degrees from foreign universities, giving this factory one of the greatest concentrations of trained manpower in China. The enterprise was financed by low-interest government and private loans guaranteed by Yungli's assets at Tangku, which kept the cost of construction low and enabled the company to obtain the newest and best equipment. Separate factories were planned for sodium nitrate, ammonium nitrate, and nitrochalk, while Yungli researchers worked on refining sulfur and coke.

Hsiehchiatien, by far the largest synthetic ammonia facility in China proper, was preceded by two smaller plants of a similar type in Shanghai and Canton. The first of these was the T'ien-li Nitrogen Gas Company, established in Shanghai in 1936 by Wu Yün-ch'u with an initial investment of one million *yüan* and second-hand nitrogen fixation equipment purchased from the Du Pont Corporation. Wu entered this field because it offered a profitable outlet for the hydrogen gas produced in abundance as a byproduct of his electrolysis plant. Since it was known that Yungli would manufacture ammonium sulfate, T'ien-li specialized in nitric acid, using the Ostwald process and equipment purchased from France. This factory produced nearly two thousand tons of nitric acid per year, along with lesser amounts of liquid ammonia and ammonium nitrate. The Yungli complex also included a nitric acid plant that used the same technology, albeit on a smaller scale.[15]

Further south, the Kwangtung Provincial Fertilizer Factory opened in 1935 to meet that region's growing demand for chemical nutrients. This plant made its own ammonia, which was combined with sulfuric acid from the provincial acid factory to produce forty tons of ammonium sulfate per day. Two other departments manufactured phosphorus and

[15] T'ien-li: Wu Ch'eng-lo, "Chung-kuo hua-hsüeh kung-yeh wen-t'i," 9; "Kuo-nei kung-yeh," 74; and Hsi Yeh, "T'ien-li tan-ch'i-ch'ang kai-k'uang" [Status of the T'ien-li Nitrogen Gas Factory], HHSC 6, no. 2 (February 1951):10. Production figures: *Shina kōgyō*, 153.

potassium fertilizers, which were consumed in south China in lesser amounts but became increasingly important as the repeated application of nitrogen left the soil depleted of these other minerals.[16]

Ethyl alcohol, or ethanol, one of the most common organic solvents, is used in the manufacture of countless products and can serve as a gasoline substitute. It is produced by the fermentation of sugar, which may be obtained directly from molasses, the waste product of a sugar refinery, or indirectly from starches, such as grain, corn, or potatoes, that have been ground up and saccharified by any of several processes. In either case, the concentration of alcohol obtained from fermentation is around 10 percent. Distillation removes excess water and raises the product to the standard industrial grades of 95 percent or above. The techniques of this industry were introduced into China by Europeans, who monopolized the manufacture of ethanol in that country prior to 1920.

The first Chinese enterprise in this field was the P'u-i Sugar Factory of Tsinan, which in 1922 began to make alcohol from molasses at the rate of twelve hundred gallons per day. This factory closed following the Tsinan Incident of 1927, and within two years imports of ethanol exceeded six million gallons. With recognition of the importance of this industry, the China Alcohol Factory was established in Shanghai in 1934 as a joint venture between the government and an overseas Chinese with large sugarcane holdings in Java. The factory had the most modern British (Blair) equipment and a daily capacity of seven thousand gallons. Another major producer was the Kwangtung Provincial Sugar Factory, whose two alcohol plants had a combined capacity of fifty-four hundred gallons per day. The Hsien-yang Alcohol Factory, founded in Shensi in 1936, employed a German technician and German (Haig) equipment to reach a daily capacity of twelve hundred gallons of anhydrous alcohol. It was the only facility in China that made a product of this purity. By 1937 there were nine major distilleries under Chinese management, with a total annual capacity of 7.2 million gallons. Much of the equipment was imported, but the technicians, with the exception of those at Hsien-yang, were all Chinese. As a result of these efforts, China's imports of ethanol fell by more than 95 percent, to less than 200 thousand gallons, in 1936.[17]

[16] Kwangtung plant: "Kuo-nei kung-yeh," 74; Wu Ch'eng-lo, "Chung-kuo hua-hsüeh kung-yeh wen-t'i," 10; Chou Ch'ang-yün, "T'u-chiang fei-liao-hsüeh" [Soil fertilization studies], Chung-hua min-kuo k'o-hsüeh chih, 22.

[17] P'u-i: Wei T'ing-ying, "Chi-nan … hua-hsüeh kung-yeh k'ao-ch'a-chi," 207; and Ch'en T'ao-sheng, "Chi-nan hua-hsüeh kung-yeh chih kai-k'uang" [Status of the Tsinan

SYNTHESIS OF ORGANIC COMPOUNDS

The Chinese had less success with other industries that rely on the synthesis of organic, or carbon-based, compounds. Whereas a few techniques for treating organic substances – fermentation, for example – have long histories in China and other parts of the world, man's whole relationship to the carbon compounds changed during the latter half of the nineteenth century, when the synthesis of chemical dyes, first in England, then Germany, opened a broad new avenue of industrial development. Before the rise of petroleum in the 1920s, the primary source of organic substances was coal tar, a black sticky substance produced as a byproduct in the distillation of coal to produce coke. The extraction of carbon compounds from coal and the assemblage from these simple starting materials of many useful products is one of the revolutions that has given substance, color and shape to the modern world.

The Chinese discovered the importance of these developments through the drug industry. In the early 1920s the International Dispensary of Shanghai, a Chinese company established in 1907, began to produce Lysol, an antiseptic derived from coal tar. The tar was obtained from the Shanghai Gas Company, a British concern founded in 1866 to supply that city with coal gas (methane) to fire its street lamps. Shanghai Gas, the first and largest company of its kind in East Asia, manufactured methane by the dry distillation of coal and marketed a variety of byproducts including tar. Just as some Chinese began to appreciate the importance of coal tar, however, hopes for this industry were destroyed by larger economic forces.[18]

The problem, far removed from drugs and dyes, was with the iron industry. The manufacture of iron requires high temperatures achieved by burning coke. Coke is made by the distillation of coal. Supply of the byproduct, coal tar, in large quantity and at low cost depends on the consumption of coke and thus on the health of the parent iron industry. At first the prospects for these industries seemed bright, for China is

chemical industry], *HHKY* 3, no. 1 (January 1925):247. Later developments: Lu Pao-yü, "San-shih-nien-lai Chung-kuo chih chiu-ching kung-yeh" [China's alcohol industry during the past thirty years], *San-shih-nien-lai Chung-kuo kung-ch'eng*, 2:1–2 (all articles in this volume are numbered individually, each beginning with page 1); and Li Ch'iao-p'ing, *Chung-kuo hua-hsüeh shih*, 400–1. Trade statistics: *China Year Book* (1935), 195; (1936), 55; (1938), 67; and Li Erh-k'ang, "Wo-kuo hua-hsüeh kung-yeh kai-k'uang chi ch'i fa-chan t'u-ching," 340.

18 Shanghai Gas: Wu Ch'eng-lo, "Chung-kuo hua-hsüeh kung-yeh wen-t'i," 10. International Dispensary: *China Industrial Handbooks*, 663–5; and Wu Ch'eng-lo, "Ts'ung Shang-hai hua-hsüeh kung-i chan-lan-hui kuan-ch'a Chung-kuo hua-hsüeh kung-yeh chih hsien-chuang" [Observing the present situation of the Chinese chemical industry on the basis of the Shanghai Chemical Handicraft Exhibition], *HHKY* 2, no. 1 (January 1924):11–12.

blessed with large deposits of iron ore and coal, which have been mined for centuries in one of the world's oldest iron-making traditions. Modern machine methods were introduced in the late nineteenth century, most notably at Hanyang, where fuel and ore from neighboring provinces were assembled to make iron on a grand scale.

After a fast start, however, the iron, coke, and hence coal tar industries in China collapsed. The boom came during World War I. Expansion of Chinese industry created a demand for construction materials just at the time when foreign supplies were diverted to Europe, resulting in a tenfold increase in the price of iron and steel in China during the decade 1912–22. Mines, refineries, smelting, and coking furnaces opened throughout the country to meet current demand and in anticipation of future growth. Then came the crash. In 1922, surplus production in China and elsewhere, a downturn in the global economy, and the return of foreign suppliers to the Far East caused a sharp decline in the price of iron. The Hanyang Iron Works, the industry leader for thirty years, closed in 1924. Throughout the 1920s and 1930s, China exported coal and iron ore, while importing over 90 percent of its finished iron and steel. The crash of 1922 marked the end of large-scale iron production under Chinese control until after 1949.[19]

No iron production meant no coke, and thus no ready supply of cheap tar on which to build an organic chemical industry. The Chinese did keep one coke and tar plant in operation, and this remained a cause for hope throughout the Republican period. The Ching-hsing Mining Bureau, located near Shihchiachuang in Hopeh Province, was the site of a major bituminous coal mine opened by German engineers at the turn of the century and taken over by the Chinese at the close of World War I. Under control of the provincial government, Ching-hsing operated China's largest coking facility and only coal tar extraction plant, whose principal purpose was to experiment with the manufacture of synthetic gasoline. On the whole, however, the scope of these operations was limited and the results disappointing. In the end, the Chinese paid for their failure to develop coal tar. By the mid-1930s the cost of synthetic dyes, principally aniline and indigo, the demand for which grew along with the textile industry, exceeded $13 million per year and accounted for nearly 4 percent of China's total import bill.[20]

[19] Coal and iron: Hou Te-feng and Ts'ao Kuo-ch'üan, "San-shih-nien-lai Chung-kuo chih mei-k'uang shih-yeh" [China's coal mining industry during the past thirty years], San-shih-nien-lai chih Chung-kuo kung-ch'eng, 2:1–25; Hu Po-yüan, "San-shih-nien-lai Chung-kuo chih kang-t'ieh shih-yeh" [China's iron and steel industry during the past thirty years], ibid, 1–15. Hanyang Iron Works: Ch'en T'ao-sheng, "Hu Han hua-hsüeh kung-yeh k'ao-ch'a-chi," 247.

[20] Ching-hsing: Yen Yen-ts'un, "Ching-hsing lien-chiao-ch'ang fu-ch'an chih ch'i-yu," 69–80; China Year Book (1933), 522; (1935), 51; and Data Papers on China, 1931 (Shanghai: China Institute of Pacific Relations, 1931), 36–8. Coke and tar production:

PETROLEUM REFINING

The failure to establish viable coke and coal tar production might have been forgiven, for by the time China's iron industry collapsed a new source of carbon compounds, petroleum, was supplying the organic chemical industry in the West. Yet petroleum was no more plentiful in China than coal tar. The Chinese remained dependent on the importation of petroleum products, first kerosene, then gasoline.

The first uniquely Western product to win an important share of the Chinese domestic market was kerosene, which displaced native vegetable oil as the preferred fuel in lamps during the 1890s. Kerosene imports rose to a peak of nearly 300 million gallons in the late 1920s, before declining to one-third that amount in 1935. The reason for the decline was the high price of this fuel, inflated further by high tariffs, which caused many consumers to revert to cheaper vegetable oil, and gave rise to an industry, centered in Canton, to distill kerosene from less expensive diesel fuel. By 1930, more than 160 small distilleries had been created, using a simple technology based on a locally made still. Production of kerosene by this method reached 32 million gallons in 1933, prompting foreign suppliers to raise the price of diesel fuel and lower that of kerosene. The local reprocessing industry survived until 1937, mainly because the Kwangtung provincial government paid one-quarter of its costs.[21]

Whereas the imports of kerosene decreased, those of another liquid fuel, gasoline, increased, from 29 million gallons in 1929 to 55 million in 1937. The expansion of roads, cars, trucks, and buses created greater demand, but the retail price of gasoline remained unchanged. By the end of the Nanking Decade, the volume of gasoline imports was still less than half that of kerosene, but the strategic and economic significance of gasoline was much greater. Throughout the 1930s, liquid fuels accounted for 8 to 9 percent of China's total import bill.[22]

The Chinese remained dependent on imported petroleum, because they failed to find any at home. The first modern oil wells in China were sunk in 1907 by Japanese experts invited to explore fields near Yenchang, in northern Shensi Province. In 1914 the Standard Oil Company won a concession to investigate this region more fully, but the Americans found little oil and soon pulled out. Their judgment was correct: Production at

Li Erh-k'ang, "Wo-kuo hua-hsüeh kung-yeh kai-k'uang," 339. Trade statistics: ibid., 338–9; and *China Year Book*, various years.
[21] Kerosene imports: *China Year Book* (1933), 168; (1936), 50, 56; (1938), 68. Tariff: Frank Kai-ming Su and Alvin Barber, "China's Tariff Autonomy, Fact or Myth," *Far Eastern Survey* 5, no. 12 (June 1936):121. Canton industry: Liu Ta-chün [D. K. Lieu], *Chung-kuo kung-yeh tiao-ch'a pao-kao* [Report on a survey of Chinese industry], 3 vols. (Shanghai: Ching-chi t'ung-chi yen-chiu-so, 1937), 1:123–5.
[22] *China Year Book* (1936), 56; (1938), 68.

Yenchang peaked at around two thousand barrels in 1916, before declining to a mere trickle during the 1930s. Refining was done in a few simple distillation pots. Yenchang, moreover, was the most productive source of crude petroleum in prewar China, whose peak annual output was around twenty-six hundred barrels, less than 1 percent of total consumption. The failure to find adequate sources of petroleum and to refine gasoline and other liquid fuels, major burdens throughout the Nanking Decade, grew heavier with the approach of war.[23]

An overview of the chemical industry

It is easier to describe particular factories or technologies than draw an overview of the Chinese chemical industry and its place in China's economy as a whole. Statistics from the warlord period are few and unreliable, and even the Nanking Decade is poorly recorded. The best data are for 1933, when D. K. Lieu (Liu Ta-chün) and the China Institute for Economic and Statistical Research conducted surveys of industry in Shanghai and throughout the nation. Lieu's figures can be supplemented by a more detailed assessment of the chemical sector, carried out by the Japanese-operated South Manchurian Railway Company in 1937. Finally, data on foreign investments are provided by Hou Chi-ming and C. F. Remer. Taken together, these studies provide a fair picture of the size, scope, and position of China's chemical industries in the middle 1930s.

Lieu's nationwide survey covered Chinese-owned factories in seventeen provinces of China proper – excluding Manchuria, which was then under Japanese control, and the western portions of the country, which had little industry in any case. Within this area Lieu identified 2,435 factories that employed power-driven machinery and a minimum of thirty workers. Half were in Shanghai. The average size was small, although perhaps comparable to that of factories in other countries at a similar level of development: U.S.$44,000 (167,000 yüan) in capital, 205 workers, 180 horsepower of motive power. Textiles and foodstuffs accounted for more than three-fourths of total output by value. Over 95 percent of the factories were privately owned; one-quarter were joint-stock companies.[24]

Included in Lieu's survey were 148 chemical plants with a combined

---

[23] Petroleum in China: *China Year Book* (1921–22), 159; (1933), 522; (1935), 49–52; Li Ch'iao-p'ing, *Chung-kuo hua-hsüeh shih*, 388; Cheng Yu-kwei, *Foreign Trade and Industrial Development of China*, 264. Petroleum import figures: Li Erh-k'ang, "Wo-kuo hua-hsüeh kung-yeh kai-k'uang," 339.

[24] Liu Ta-chün, *Chung-kuo kung-yeh tiao-ch'a pao-kao*, 1:9; 2:64, 160, 192, 291, 410, 422, 428. For calculation of exchange rates between U.S.$ and Ch$, see: Cheng

output of 50 million *yüan*. This figure represents only 4.5 percent of the output of all factories in the survey, but was sufficient to place chemicals third, behind textiles and foodstuffs, on the list of Chinese manufactures. In terms of average size (178,000 *yüan* in capital, 186 workers) and location (57 percent in Shanghai), the chemical plants were similar to all factories, although their power consumption was low, only 61 horse-power (about one-third the average). Like Chinese industry generally, the chemical sector produced mostly consumer goods: four-fifths of the factories in this group made matches, soap and candles, enamelware, cosmetics, medicines, and plastics. In 1932, the ratio of profit to investment for the chemical group and the total sample was the same, 45 percent. Yet over half the chemical factories were organized as joint-stock companies, which was more than twice the average and suggests a particular confidence in this sector.[25]

One reason for such confidence is indicated by Lieu's second study, this one limited to factories in the Shanghai area and the years 1931–3. The Chinese economy fared poorly in the early 1930s, a period the Shanghai municipal government described as a "disastrous business depression." Yet Lieu found that chemical production was one of the fastest-growing sectors of the economy. The number of Shanghai chemical plants using power-driven machinery and employing at least thirty workers jumped from twenty-eight in 1931 to seventy-eight in 1933. Thirteen new factories were opened in five new lines: acids, calcium and magnesium carbonate, oxygen and acetylene, dyes, and plastics. During this period, the capitalization and value of output of this sector more than doubled, power consumption increased by 80 percent, and seventeen hundred workers were added to the work force.[26]

Several of China's largest and most modern chemical plants, established after 1933, were omitted in Lieu's survey, but included in a study conducted in 1937 by the research branch of the South Manchurian Railway Company. These data (Table 11.1) included two dozen producers of acids and sodas, along with all the factories discussed, except the massive Yungli ammonium sulfate plant, which began operation only at the very end of the Nanking Decade.[27]

Yu-kwei, *Foreign Trade and Industrial Development of China*, xi, 262–3. For discussion of these issues, see: Albert Feuerwerker, *Economic Trends in the Republic of China*, 17–18, 36.

[25] Liu Ta-chün, *Chung-kuo kung-yeh tiao-ch'a pao-kao*, 2:48, 160, 192, 267, 401. The ratio of profit was calculated by subtracting expenses from sales and dividing the remainder by capital investment. See: ibid., 2:48, 64, 401, 428.

[26] Survey: D. K. Lieu, *The Growth and Industrialization of Shanghai* (Shanghai: China Institute of Pacific Relations, 1936), 77–8. "Depression": Lloyd Eastman, *The Abortive Revolution: China under Nationalist Rule, 1927–37* (Cambridge: Harvard University Press, 1974), 187.

[27] *Shina kōgyō*.

Table 11.1. *Chinese manufacturers of sodas and acids*

Name	Location	Year began operation	Capital, *yüan*	Production in 1936, metric tons (unused capacity)
Soda ash				
T'ung-i Soda Factory	Szechwan	1918	50,000	482
Chia-yü Soda Factory	Szechwan	1920	50,000	374
Yungli Soda Company	Tangku	1924	2,000,000	51,270
Pohai Chemical Works	Tangku	1926	600,000	4,976
Total			2,700,000	57,102
Caustic soda				
Yungli Soda Company	Tangku	1924	2,000,000	4,976
¨ao T'ien-li Factory	Tientsin	1925	· 15,000	603
T'ien-yüan Electrochemical Company	Shanghai	1928	600,000	1,508
Northwest Industrial Company	Taiyuan	1935	400,000	489 (1,508)
Kwangtung Provincial Electrochemical Factory	Canton	1935	1,400,000	(2,111)
Total			4,415,000	7,576 (3,619)
Sodium sulfide				
Pohai Chemical Works	Tangku	1926	600,000	2,413
Chao-hsin Chemical Works	Shanghai	1931	200,000	1,508
Ho-chi Chemical Works	Tangku	1932	200,000	3,076
Tao-i Soda Works	Tsinan	1933	10,000	205
T'ung-shêng Chemical Works	Taiyuan	1934	23,000	362
Shen-yü Soda Works	Hangchow	1934	3,000	603
Ta-ch'ing Soda Works	Tientsin	1936	25,000	1,206
Wei-hsin Chemical Industry Society	Tsingtao	1936	500,000	2,413
Total			1,561,000	11,786
Sodium silicate				
Lao T'ien-li Factory	Tientsin	1925	15,000	1,086
Pohai Chemical Works	Tangku	1926	600,000	4,041

Table 11.1 (*cont.*)

Name	Location	Year began operation	Capital, *yüan*	Production in 1936, metric tons (unused capacity)
Hsing Hua Soda Works	Tientsin	1929	45,000	1,267
K'ai-yüan Soda Company	Shanghai	1930	40,000	2,087
Total			700,000	8,481
Sulfuric acid				
Liang-Kuang Provincial Sulfuric Acid Factory	Wuchow	1927	560,000	(2,606)
Tê-li Three Acid Factory	Tangshan	1931	50,000	(483)
Li-chung Acid Manufacturing Company	Tientsin	1933	200,000	905
K'ai-ch'êng Acid Manufacturing Company	Shanghai	1933	750,000	4,463
Kwangtung Provincial Sulfuric Acid Factory	Canton	1933	—	(4,500)
Northwest Industrial Company	Taiyuan	1934	55,000	72
Chi-ch'êng Three Acid Factory	Sian	1934	50,000	121
Total			1,665,000	5,561 (7,589)
Nitric acid				
T'ien-li Nitrogen Gas Company	Shanghai	1936	1,000,000	1,942
Northwest Industrial Company	Taiyuan	1934	55,000	68
Chi-ch'êng Three Acid Factory	Sian	1934	50,000	36
Total			1,105,000	2,046
Hydrochloric acid				
Pohai Chemical Works	Tangku	1926	600,000	1,206
T'ien-yüan Electrochemical Company	Shanghai	1928	600,000	1,508

Table 11.1 (cont.)

Name	Location	Year began operation	Capital, yüan	Production in 1936, metric tons (unused capacity)
Kwangtung Provincial Electrochemical Factory	Canton	1935	1,400,000	(1,810)
Total			2,600,000	2,714 (1,810)

Sources: For Yungli Soda: Ou-yang I, "P'ing-tung hua-hsüeh kung-yeh k'ao-ch'a-chi" [Record of an investigation of the chemical industries east of Peiping], HHKY 5, no.1 (February 1930): 77–110. For the Kwangtung factories: "Kuo-nei kung-yeh" [Chinese industry], HHKC 3, no.1 (March 1936): 74. For all others: Shina kōgyō, tables on pp. 81–4, 125, 131, 137.

Compared with Chinese producers, foreigners played a relatively small role in China's chemical industries, particularly as regards the introduction of new techniques. Measured in terms of capital, the foreign presence appears quite large. In 1936, foreign investment in manufacturing in China, excluding Manchuria but including Hong Kong, was $332 million. More than half of this amount was in textiles; the second largest category, chemicals, accounted for $75 million, or 22 percent of the total. (Of this amount, $63 million was British.) In contrast, the total capitalization of factories included in Lieu's survey was only $107 million (Ch$407 million), of which the chemical group accounted for less than $7 million (Ch$26 million). In sum, foreigners invested more than ten times as much in the manufacture of chemical products in China as did the Chinese themselves.[28]

Production figures tell a different story. Liu Ta-chung and Yeh Kung-chia have shown that in 1933 Chinese firms accounted for two-thirds of the gross value of factory output in all of China and for 78 percent of the output in China proper (excluding Manchuria). Figures for the chemical sector show even greater dominance by the Chinese: 71 percent in all of China, 82 percent in China proper. In both cases, the

[28] Foreign investments: Hou Chi-ming, Foreign Investment and Economic Development in China, 1840–1937 (Cambridge: Harvard University Press, 1965), 13, 16, 81; and C. F. Remer, Foreign Investments in China (New York: Macmillan, 1933). Chinese investment: Liu Ta-chün, Chung-kuo kung-yeh tiao-ch'a pao-kao, 2:48, 64.

foreigners invested more capital, but the Chinese produced more goods. Albert Feuerwerker has argued that output is a more accurate measurement of the Chinese–foreign balance, because most Chinese factories engaged in light manufacturing, where labor could be substituted for capital. Although this may be true, it is interesting that Chinese chemical factories accounted for a greater share of production in their sector than did Chinese manufacturers as a whole, even though the chemical producers had a higher ratio of capital to labor than that for all Chinese factories. In other words, China's chemical sector competed more effectively against foreign producers in China than Chinese industry generally, and did so with less capacity for labor substitution.[29]

Most foreign manufacturers, like the Chinese themselves, made consumer goods. The first foreign firm to manufacture a chemical product in China, J. Llewellyn and Company, established in Shanghai in 1853, made drugs and cosmetics. During the 1880s foreigners set up plants to make matches, glass, paper, and soap. Major Brothers, founded in 1901, manufactured acids, albeit with limited success. This was the only foreign-owned producer of basic chemicals in China proper. As Robert Dernberger has pointed out, given the limited development of the Chinese economy, there was little market for producer goods in the chemical or any other sector. The notable exception was Manchuria, where the Japanese carried out a massive program of industrialization, which the author has described in a separate study.[30]

An assessment of the chemical industry

The development of China's chemical industries followed a now familiar pattern. To begin, it was the lure of profits earned by foreign importers that inspired Chinese to try to manufacture chemicals – salt, soda, MSG, acids, and ammonium sulfate – on their own. Once established, chemical factories in China spawned links to other industries. The manufacture of refined salt provided the raw materials Fan Hsü-tung needed to make soda. Wu Yün-ch'u moved from one opportunity to the next: The

[29] Liu Ta-chung and Yeh Kung-chia, *Economy of Chinese Mainland*, 142–3, 426–8. Albert Feuerwerker, *Economic Trends in the Republic of China*, 35–6. Comparison of chemical and all factories: Liu Ta-chün, *Chung-kuo kung-yeh tiao-ch'a pao-kao*, 2:48, 64, 267, 291.

[30] Foreign manufacturers: Hou Chi-ming, *Foreign Investment and Economic Development in China*, 85–8. Foreign manufacture of consumer goods: Robert Dernberger, "The Role of the Foreigner in China's Economic Development, 1840–1949," *China's Modern Economy in Historical Perspective*, ed. Dwight H. Perkins (Stanford: Stanford University Press, 1975), 37–9. For details on the Japanese chemical industries of Manchuria, see: James Reardon-Anderson, "China's Modern Chemical Industry, 1860–1949," *OSIRIS*, 2d ser., 2 (1986):215–19.

manufacture of MSG created a demand for hydrochloric acid; the electrolysis plant that made this acid generated a surplus of hydrogen gas, which was converted to ammonia; the ammonia was used to make nitric acid; and so on. The interlocking complex of acid, soda, gunpowder, and fertilizer factories in Kwangtung and Kwangsi demonstrates the habit of chemical manufacturers to "take in each other's laundry." Unfortunate, but characteristic of Republican China was the failure to extend such links to the agricultural sector. Attempts to manufacture chemical fertilizer came too late to bridge the widening gap between city and countryside.

Capital for the chemical industry came primarily from the private sector. Both Fan Hsü-tung and Wu Yün-ch'u invested profits from their first successful enterprises to help finance later ventures. Commercial banks and individual investors bought stock in these corporations. Provincial governments operated arsenals, a few key plants in areas not served by private producers, such as the Ching-hsing coal tar facility, and one large complex in southern China. The central government made subsidized loans and direct investments to establish those industries, such as alcohol and ammonium sulfate, where private enterprise had not done the job and self-sufficiency was considered a matter of national interest. For the most part, however, the manufacture of chemicals in China before World War II was the work of Chinese capitalists.

In addition to playing the role of investor of last resort, the governments of Republican China encouraged domestic chemical producers in several ways. Beginning in 1917, salt for industrial use and, later, soda products themselves were exempted from tax. Monopolies on the manufacture of certain chemicals within designated areas protected some factories against competition from within China and abroad. Tariff schedules issued after 1929, when the Nanking government revised its "unequal treaties" with the foreign powers and regained tariff autonomy, raised the rates on most chemical products manufactured in China. Rebates on railway freight charges for Chinese chemicals favored domestic products over imports, particularly in the inland cities. Comparative price data show that in the ports of entry, Tientsin and Shanghai, Yungli soda was only slightly cheaper than the Brunner-Mond brand; but further inland in Tsinan, Taiyuan, and Hankow, it was much cheaper.[31]

---

[31] Tax exemptions and monopolies: Ou-yang I, "P'ing-tung hua-hsüeh kung-yeh k'ao-ch'a-chi," 97–9; Fan Jui, "Yung-li chih-chien kung-szu ta-shih-chi," 255; and Ch'en Chen and Yao Lo, eds., *Chung-kuo chin-tai kung-yeh-shih tsu-liao*, 514. Shipping rebates: ibid., 517. For a complaint on the low tariff on soda, which was fixed under the "unequal treaties" at 5% *ad valorum*, see Li Ch'iao-p'ing, "Kuan-shui yü hua-hsüeh kung-yeh" [Customs duties and the chemical industry], *HHKY* 6, no. 1 (January 1931):14. The first independent tariff schedule, fixed by the Chinese themselves under

Chinese chemical manufacturers had little difficulty obtaining the necessary equipment and skilled manpower. Generally, the Chinese were able to buy what they needed, individual machines or whole plants, from Europe and the United States. In some cases they bought used equipment at a fraction of the regular price. Foreign advisers and technicians played an important role in the construction and operation of China's arsenals, but other chemical plants, public and private, were manned almost exclusively by Chinese. The founders of several companies – Yungli, Pohai, Li-chung, T'ien-ch'u, and others – included men with training or experience in chemistry. More engineers were recruited by and trained in these plants. In at least one company, Yungli Soda, these men discovered the secret of the most current techniques and built a successful enterprise, despite the lack of assistance from the international cartel that controlled this industry.[32]

The raw materials used in China's chemical factories present a mixed picture. The absence of coal tar held back development of the organic industries. Sulfur, the key component of sulfuric acid, and nitrates, used to make nitric acid, were also a problem. Since the Sung dynasty (960–1279 A. D.), the Chinese had been distilling pyrite (iron disulfide) to obtain sulfur for the manufacture of gunpowder and other products. The same historic methods continued in use in the Republican period, but the quality of Chinese ores was generally low, and the facilities for mining, refining, and transporting were inadequate to bring a steady supply of high-grade material to the factory gate. As a result, many acid factories relied on imported sulfur and saltpeter. Between the wars, Chinese production of sulfur, most of which was from Shansi and Hunan provinces, fell to as low as two thousand tons per year, whereas imports exceeded five thousand tons. The dependence on foreign nitrates was almost as great. On the other hand, China enjoyed an abundance of inexpensive, high-quality salt, coal, limestone, and other important materials. The high quality and low price of their refined salt, for example, enabled the Chinese to manufacture and export soda to Japan, a country whose other industries were far in advance of those in China.

---

revised treaties and issued on 1 January 1933, raised duties to a maximum of 50% on items that competed with domestic manufactures, while lowering them on other items that aided home industries. See: H. D. Fong, "China's Industrialization: A Statistical Survey," *Data Papers on China* (Shanghai: China Institute of Pacific Relations, 1931), 40–1. Thomas Rawski, *China's Transition to Industrialism* (Ann Arbor: University of Michigan Press, 1980), 21, reports that under this schedule, tariffs on nitric acid rose by 54%, on hydrochloric acid by 50%, and on sulfur black by 328%. Under pressure from Japan, a new schedule was issued in July 1934, lowering rates on textiles, but raising those on chemicals and pharmaceuticals by an additional 14–100%. See: Frank Su and Alvin Barber, "China's Tariff Autonomy," 115–22.

32 Recruitment and training: Ch'en Hsin-wen, "Chi Yung-li hua-hsüeh kung-yeh kung-szu p'ei-yang jen-ts'ai ti tso-fa," 28–34.

On balance, the supply of raw materials constituted a net plus for Chinese chemical manufacturers.[33]

Because of these advantages, China's chemical industry made impressive strides during the years before World War II. In spite of them, further growth was limited by the small size of the Chinese market. Comparsion of production and trade figures demonstrates that in almost every case the principal constraint on this industry was insufficient demand (Table 11.2 and Table 11.3). Imports of acids and sodas increased steadily during the first part of the century to supply the growing number of factories for the bleaching, sizing, and dyeing of textiles and the manufacture of soap, paper, glass, and other products. Yungli Soda was the first to challenge foreigners for a share of the Chinese market. After 1930, when most of the factories described had entered production, imports in the major categories of industrial chemicals declined sharply. As these statistics show, the development of China's chemical industry followed the pattern of foreign penetration to create demand for new products, followed by successful import substitution on the part of local manufacturers.[34]

By 1936, Chinese production exceeded the imports in every category, except caustic soda. The most striking successes occurred where they counted most, in sulfuric acid and soda ash. Yet both these products serve to illustrate the chief limitation on this industry, which was the underlying weakness of the domestic market. China's largest private acid factory, K'ai-ch'eng, had a capacity of fifteen tons per day, but actual production reached only nine tons. The Kwangtung sulfuric acid plant had a similar capacity and was similarly underutilized. The problem was not the quality of the Chinese product, which was good, nor the price, which in the case of K'ai-ch'eng declined substantially during the first years of operation. Rather, it was the inadequacy of demand: China had few or none of those industries, such as iron and steel, coal tar, chemical fertilizer, and synthetic dyes, that consume large quantities of acid; other industries used sulfuric acid, but in modest amounts. The same constraints held down production of soda ash. Thanks in part to con-

---

[33] Sung sulfur and use of pyrites: Zhang Yunming, "Ancient Chinese Sulfur Manufacturing Processes," *ISIS* 77 (1986):487–97. Sulfur and saltpeter: Ting Szu-hsien, "Chien-she kuo-fang hua-hsüeh kung-yeh chih wo-chien," 76–7; Ch'en T'ao-sheng, "Hu Han hua-hsüeh kung-yeh k'ao-ch'a-chi," 234; Wu Ch'eng-lo, "Chung-kuo hua-hsüeh kung-yeh wen-t'i," 5–8; and *China Industrial Handbooks*, 658–9. Import figures: Maritime Customs of China. *Returns of Trade and Trade Reports* (1906–1919); *Foreign Trade of China* (1920–1931); and *The Trade of China* (1932–1938) (Shanghai: Inspector General of Customs, indicated years). Chinese soda exports: *Shina kōgyō*, 17–18.

[34] Thomas Rawski, *China's Transition to Industrialism*, 16, points out that between 1919 and 1936 Chinese factory cloth production increased eight itmes, creating a huge demand for chemicals used in dyeing and bleaching.

Table 11.2. *Chinese imports and production of soda*

Year	Production (Yungli) Soda ash	Imports Soda ash	Caustic soda	Silicate and sulfide	All soda[a]
1905–1909	—	—	—	—	13,755[b]
1910–1915	—	—	—	—	26,836[b]
1916	—	—	—	—	2,542
1917	—	—	—	—	13,030
1918	—	—	—	—	13,384
1919	—	—	—	—	53,047
1920	—	—	—	—	44,490
1921	—	—	—	—	24,650
1922	—	—	—	—	55,689
1923	—	37,290	7,372	—	53,132
1924	3,300	44,787	7,540	7,805	62,098
1925	1,800	50,161	8,687	7,821	72,563
1926	4,500	47,211	10,345	8,750	75,973
1927	14,000	52,782	9,030	10,145	79,398
1928	15,000	51,020	13,860	9,770	84,871
1929	18,000	56,368	11,503	9,758	85,914
1930	18,000	64,941	13,219	6,332	92,761
1931	23,000	46,360	12,096	6,080	72,588
1932	32,000	29,260	9,780	7,262	48,621
1933	34,000	23,767	16,560	7,137	51,759
1934	37,000	29,363	19,390	6,797	58,967
1935	45,000	27,143	19,471	4,210	54,696
1936	51,000	25,049	18,037	3,683	50,277
		Production (unused capacity)[c]			
1936	57,102	—	7,576 (3,619)	20,267	101,713[d] (3,619)

[a] All sodas comprise, for imports: ash, bicarbonate, caustic, silicate, sulfide, and Chile saltpeter; for production: ash, caustic, silicate, sulfide, and natural soda.
[b] Annual averages.
[c] Figures from Table 11.1 totals.
[d] Production of natural sodas in 1936 was 16,768 tons: *Shina kogyo*, 81.
Sources: For import figures: Maritime Customs of China, Returns of Trade and Trade Reports (1906–1919): Foreign Trade of China (1920–1931); and The Trade of China (1932–1938) (Shanghai: Inspector General of Customs, indicated years). For Yungli soda production figures: *Shina kōgyō*, 15–16.

Table 11.3. *Chinese imports and production of acid*

Year	Imports			
	Sulfuric	Nitric	Hydrochloric	All acids[a]
1905–1915	1,420[b]	—	—	—
1924	2,859	1,456	782	5,932
1925	3,102	1,294	1,251	6,664
1926	3,811	1,723	1,692	8,260
1927	6,122	2,490	2,047	11,583
1928	5,996	2,289	2,795	12,344
1929	3,826	1,151	3,507	9,668
1930	3,083	1,263	3,473	9,109
1931	4,049	1,631	2,112	9,924
1932	3,001	1,520	1,249	7,050
1933	3,342	2,361	1,697	9,299
1934	1,317	2,962	2,040	8,762
1935	764	1,971	2,347	7,192
1936	580	342	2,440	5,890
	Production (unused capacity)[c]			
1936	5,561 (7,589)	2,046	2,714 (1,810)	10,701 (9,399)

[a] All acids, besides sulfuric, nitric, and hydrochloric, for imports includes various acids for which separate figures are not available; for production, acetic acid (380 tons).
[b] Annual average.
[c] Figures from Table 11.1 totals.
Sources: For import figures: as in Table 11.2.

cessions granted by the Chinese government, the price of Yungli soda was lower than the British variety in every one of the seven cities where they competed. The Chinese also exported some soda, primarily to Japan, although beyond their shores they faced international chemical combines of enormous size and sophistication that dominated most markets. At home, Yungli soda was accepted; the business was a success. Yet no one in China moved to create a second modern factory for the manufacture of soda ash. The reason was lack of demand. Growth of the chemical sector depended primarily on the expansion of Chinese industry as a whole, and the horizons of both were limited.[35]

[35] Sulfuric acid: Ting Szu-hsien, "Chien-she kuo-fang hua-hsüeh kung-yeh chih wo-chien," 82; Donald Dunham, "The Influence of American Education in South China," 61. Soda

Only in the case of caustic soda did the Chinese fail to produce more than they imported, and here, too, the fault lay partly with the local market. Most caustic soda was made, in China as elsewhere, by the electrolysis of salt. Like many successful chemical technologies, electrolysis is a flexible method that can be used to make different products in response to shifting market demands. In the early years, electrolysis plants made caustic soda and disposed of the unwanted chlorine by converting it to bleaching powder. Later, when chlorine replaced powder as the preferred bleach in the industrial West, most plants were reprogrammed to make liquid chlorine. The Chinese failed to recognize the importance of liquid chlorine, which was used only in a few water treatment plants; the chlorine from their electrolysis plants went to make bleaching powder and hydrochloric acid, the markets for which were small and shrinking. Meanwhile, foreign manufacturers diversified their operations, earned a growing share of profits from the sale of chlorine, and dropped the price of caustic soda, which competed successfully against Chinese brands.[36]

These findings provide little support for the major interpretations that have been offered to explain the record of industrial development in Republican China. Two schools describe China's development as slow, but offer different explanations for this torpor. Marxists and neo-Marxists blame the international environment: Western manufacturers, aided by low tariffs and other privileges exacted by the use of force, undersold local producers, thus preventing, or at least delaying, the rise of native industry. In this view, successive Chinese governments were either too weak to defend China or so perfidious that they could be bought off. In either case, foreigners called the shots. A second interpretation, which applies more narrowly to the Nanking Decade, is based on the empirical work of Lloyd Eastman, Parks Coble, and others. They also consider Chinese development retarded, but blame this outcome on Nanking. Rather than defending the interests of Chinese capitalists against the lower classes, as the Marxists have charged, the Nationalists exploited commercial and industrial enterprises, siphoned wealth away from productive endeavors, and made it impossible for industrial or any other kind of development to occur. What the imperialists failed to do from without, these scholars argue, the Nationalists achieved from within. In either case, the cause was political and the loser was China. A third opinion, put forward by Thomas Rawski, emphasizes economic rather than political factors and comes to a happier

prices: *Shina kōgyō*, 114-15. In 1933, exports of Yungli soda reached seven thousand tons, 80% of which went to Japan. See: ibid., 17-18.

[36] Yü Jen-chuñ, "Wo-kuo shih-yen tien-chieh kung-yeh," 4-9. Comparative price data show that in all categories British sodas, which accounted for 70-80% of imports, were more expensive than Chinese, but that the margin was less for caustic soda than for soda ash. See: *Shina kōgyō*, 90-94, 114-15.

conclusion. Rawski holds that Chinese industry grew steadily throughout the first half of the twentieth century, without concern for who was in power. The driving force behind this growth was the accumulation of economic factors. Neither foreign influence nor the succession of Chinese governments had much to do with it.[37]

My study does not try to judge the pace of China's development or explain its causes. But the details it provides on one sector of industry and the conclusions it draws from them depart from all three accounts of Republican economic history. In contrast to Eastman, Coble, and Rawski, I find that the Nationalists played a significant, positive role in promoting the development of the chemical industries. Central and provincial governments built and operated arsenals and other factories and introduced new technologies for the manufacture of several basic and intermediate chemicals. Nanking provided special incentives – high tariffs, low taxes, monopolies, rebates, and loans – that made private investment attractive and successful. More broadly, the Nationalists created a sense of security, stability, and direction that encouraged Chinese to stay in China and bet on its future. The growth of China's chemical industries was not, as Rawski would have it, cumulative and smooth. Rather, it leapt forward after 1927, and the cause was at least in part political. Nor was the role of the government a negative one, following Eastman and Coble. Surely the Nationalists did not do all they could, and much of what they did was bad, but focusing on the chemical sector alone, the evidence shows they gave more help than harm.

Despite government support and private effort, China's most successful chemical industries soon reached their limits. Contrary to the assertions of the Marxists, however, these limits were imposed less by the international environment than by the Chinese market itself. Foreign importers created the demand for industrial chemicals in China. Foreign suppliers made available at reasonable prices the machinery and techniques to produce these chemicals. And when the Chinese proved capable of manufacturing chemicals, they were quick to capture the bulk of their own market. Neither the Chinese government nor the foreign imperialists choked the chemical industry in China. It succeeded within the confines of its own economy, but not beyond.

Conclusion

However one scores and explains the development of Chinese industry in this period, there remains the question of how this development affected

[37] Lloyd Eastman, *The Abortive Revolution*, 226–43; Parks M. Coble, Jr., *The Shanghai Capitalists and the Nationalist Government, 1927–1937* (Cambridge: Harvard University Press, 1980), 1–12; and Thomas Rawski, *China's Transition to Industrialism*, passim.

the study and practice of science. On the positive side of the ledger, industry provided the bulk of the Nationalists' tax base and thus the wherewithal to support scientific education and research. Industrialization transformed a corner of China's landscape, giving young people a better feel for the texture and pace of the modern world. Some chemical industries, such as seasonings, tanning, fermentation, and fuels, generated problems that could be taken into the laboratories of China's universities and research institutes. Most important, industry provided jobs for men and women at all levels of technical training and expertise. The good news was that the industry gave the Chinese a view of the peaks of modern science and engineering, posing questions, generating answers, and opening new opportunities for China's best and brightest.

The bad news was that the development of this sector was limited and the opportunities few. Our survey of China's chemical industries shows that modern factories were few in number and most employed no more than one skilled technician. Meanwhile, each spring, China's universities graduated hundreds of chemistry majors. What were these young people to do? The answer they hoped for lay in rapid economic growth, foreign trade, travel, technology. The answer they got was the Japanese invasion, retreat to the hinterland, and war.

PART IV
The war
1937—1945

During the War of Resistance (1937–45), the cycle of political and social change came full circle – with important implications for Chinese science and scientists. Early in the twentieth century, the decline of the old order had removed the chief obstacle to the introduction of science, and new forces emerged to promote this cause. By the 1930s, as the opportunities created by the chaos of the interregnum were exhausted, the revival of political authority and return to China of a new generation of trained scholars established a balance that carried the development of science forward. The war upset that balance, destroyed the basis for an autonomous scientific profession in China, and shifted power to the state.

Three factors account for this change. First, the Japanese occupation of northern and eastern China drove hundreds of thousands of Chinese, including much of the intellectual and technical elite, out of the coastal cities and into the hinterland, breaking their links to the outside world. Isolated from their prewar base, urban intellectuals were less able to defend their identity, values, and standards against the resurgent power of the state. Second, the challenge of survival in the interior forced scientists, engineers, and other émigrés to adapt their knowledge and skill to local circumstances, to innovate, improvise, and invent, rather than adopt ready-made formulas from abroad. This increased the leverage of the experts in the debate over science policy, but it also strengthened the utilitarian character of this enterprise, reducing the value of and commitment to learning for its own sake. Finally, the pressures of war caused the government to intervene more forcefully in all areas of national life, including science and technology. Those responsible for the fate of China sought to mobilize society behind economic production and national defense and considered other goals, such as the quest for knowledge, a costly, expendable distraction. Throughout the war, these features – isolation, innovation, and government intervention – shaped the development of science throughout Resistance China.

The isolation of science from the cities and the coast changed the whole nature of this enterprise. Before the war, Chinese scientists and engineers had done little that was not directly connected to life in China's modernizing cities and to the continuous traffic of people, products, and ideas between these cities and the West. Then, in a matter of months, Japanese forces occupied the areas that had been home to factories, universities, research academies, and other institutions where science and technology were studied, developed, and applied. Rejecting surrender, the Nationalists retreated inland to the mountainous "Great Rear" – the provinces of Szechwan, Kweichow, and Yunnan, the cities of Kunming,

National Health Administration laboratory, Nanking, damaged by Japanese bombing, 1937. (Courtesy of the Rockefeller Archive Center.)

Chengtu, and the wartime capital of Chungking – taking with them not only soldiers and bureaucrats, but industries and industrialists, schools and scholars, research laboratories and research scientists, students, teachers, intellectuals, patriots, scoundrels, and inbetweeners of every stripe. The personalities, institutions, and traditions of the Nanking Decade were uprooted from their native habitat and moved to this colder clime.

Some of the urban émigrés took refuge in areas under Communist control, which were even further removed from the comfort and security of the coast. In this case, migration had a contrary effect – rather than removing the Communists from the cities, it brought them into contact with the forces of modernization and change. Most members of the Chinese Communist Party (CCP) had spent the Nanking Decade fleeing Nationalist armies, hiding out on barren mountaintops, or escaping destruction on the Long March (1934–5). Japan's invasion interrupted the civil war and gave the "Red Bandits" new life. The establishment in

1937 of the United Front, a temporary alliance between rival Chinese parties designed to strengthen the nation's resistance, provided the Communists access to capital, technology, and skills from those parts of China that had previously been denied them. During the war, tens of thousands of refugees from Peking, Shanghai, Nanking and Tientsin, students, scholars, intellectuals, factory managers, and technicians, most bringing new skills, a few carrying tools, machines, and other supplies, came to the Communist capital of Yenan, beyond the bend of the Yellow River in the remote northwest. This influx of educated, skilled, urban activists introduced modern science and technology to Communist China.

As they retreated inland, Chinese scientists and engineers discovered the need to shift from adoption and adaptation to innovation and device. Chinese science of the late nineteenth and early twentieth centuries had its share of heros: the discoverers of new truths, inventors of new techniques, makers of fashion. But on the whole, the record of prewar China was marked by the wholesale adoption of prepackaged goods from the West. The war changed that. Historians have long noted the impact of the "Yenan way" on the development of Chinese Communism: the attention to practical applications, simple technologies, and popular culture, the molding of Marxist theory to fit the reality of peasant life. What the present account shows is the degree to which war inspired similar trends throughout Resistance China. After 1937, scientists and engineers in both Nationalist and Communist areas discovered the need to adopt local remedies and adapt established techniques, to innovate, improvise, and invent. Unable to rely on old solutions from outside China, they were forced to create new solutions from within.

Finally, the war prompted political authorities, both Nationalist and Communist, to extend their control over a wider range of activities, completing a trend begun in the mid-1930s. Government intervention was designed to promote the application of knowledge to the immediate needs of production and defense. After 1937, the Nationalists created hundreds of state-owned factories, reshaped education and research to serve practical needs, and mobilized the country for war. Many scientists resisted this encroachment on what they considered their own turf. The ensuing struggle between scholars and bureaucrats in Nationalist China ended in a draw.

Under similar circumstances and for similar reasons, Communist authorities tightened their grip on science in the territory under their control. Locked in a desperate battle for survival and whipsawed by conflicts within the leadership of the party itself, Yenan demanded of its intellectuals strict adherence to a narrow orthodoxy, whose prime virtue was utility. Many scholars and students from the urban areas favored a program that included basic theoretical knowledge, if not for its own sake, then at least as an investment in the reconstruction of China after the war. But Mao Tse-tung and party leaders campaigned to "Rectify" the behavior of misguided cadres, insisting, without room for compromise, on immediate, useful results. In contrast to Nationalist China, where this conflict ended in stalemate, Communist authorities brooked no dissent. The impact of these events on Chinese science endures.

12

Science in Nationalist China: the wartime experience

The isolation of Chinese scientists and engineers from their prewar base, innovation in bringing modern techniques to bear on the hinterland, intervention of government authority into the conduct of science, and resistance by scientists and other scholars to this pressure – all these factors were evident in the wartime experience of Nationalist China. The government built factories and extended its control over the economy. Emigré engineers applied their knowledge and skills in novel ways. Scholars taught in schools and did research in institutes that had removed to the interior. Chungking urged scientists and engineers to concentrate on the immediate requirements of production and defense. But cooperation between state and society faltered, and the conflict over the purpose and direction of science grew more intense.

The chemical industry in wartime China

The migration of Nationalist China from its base on the Lower Yangtze to the remote southwest brought with it the expansion of government into all areas of the economy. From 1937 to 1940, 639 privately owned factories, including more than 117 thousand tons of machinery and equipment, were hauled from Shanghai, Nanking, and Wuhan to Szechwan and other western provinces, where they were reassembled for wartime service. The government provided loans and investments to support the operation of established factories and help start new ones. At the same time, the National Resources Commission (NRC), a branch of the Ministry of Economic Affairs (MEA), took a direct role in setting up and managing more than one hundred major enterprises in every branch of industry. By 1944, the MEA had registered nearly five thousand factories, public and private, in the Nationalist areas.[1]

---

[1] Factories moved to Szechwan: Cheng Yu-kwei, *Foreign Trade and Industrial Development of China*, 108. Public sector: *The China Handbook*, ed. Chinese Ministry of Information, Chungking (New York: Macmillan), for the following years: (1937–43),

Among the largest and most important wartime industries were for the manufacture of chemicals. One survey of China's industrial technologies conducted in 1944 describes advances made during the war in the manufacture of acids, sodas, porcelain, bricks, glass, plastics, sugar, paper, drugs, dyes, alcohol, petroleum, and other chemical products.[2] Next to electric power, the largest number of NRC factories was devoted to chemicals, including eight alcohol, two synthetic oil, and five acid and soda plants. Chemical factories also received the lion's share of government aid to the private sector: more than $17 million in loans and $19 million in investments, or one-third of the total outlay. The largest number of factories registered by the MEA, over one-quarter of the total, was devoted to chemicals. Starting from the base year, 1938 (100), the MEA index of industrial production for 1944 stood at 352, a figure exceeded by every chemical product: soda ash, 394; caustic soda, 671; bleaching powder, 550; sulfuric acid, 457; hydrochloric acid, 433; alcohol, 2,464; and gasoline (including synthetic), 83,154.[3]

The record of the Szechwan salt and soda industry demonstrates the Nationalists' success in weaving together modern imported and established local techniques. For centuries, the natives of western Szechwan had drilled for the rich brine that lies beneath its rugged mountains. Salt from these wells was originally used as a flavoring or preservative, but the growth of the Szechwan paper industry in the early 1900s created a demand for soda and other minerals extracted from the brine. The needs of eastern Szechwan, about forty-five hundred tons per year, were filled by imports of high-grade soda, one-third of which came from the Yungli factory, two-thirds from Brunner-Mond. Manufacturers in western Szechwan, beyond the reach of supplies from the coast, introduced modern techniques for processing the local salt. Wells of the P'eng-shan region yielded sodium sulfate, which was used to make soda by the Leblanc method, once the premier process in this industry but abandoned in most parts of the world by the end of the nineteenth century. The first Szechwan factory, established in 1912, turned out a product containing 50 percent sodium carbonate, making this region self-sufficient in soda ash.[4]

Following the Japanese invasion in 1937, Fan Hsü-tung, Hou Te-pang, and other Yungli engineers withdrew to Szechwan to set up a salt refinery

---

433–6; (1937–44), 277–8; (1937–45), 364–6. Private sector: ibid. (1937–43), 436–40; (1937–44), 289. MEA-registered factories: ibid. (1937–45), 363–4.

[2] Hao Ching-sheng, "K'ang-chan ch'i-nien lai chih k'o-hsüeh" [Science during the last seven years of the War of Resistance], *Chung-kuo chan-shih hsüeh-shu* [Chinese wartime scholarship], ed. Sun Pen-wen (Shanghai: Cheng-chung shu-chü, 1946), 182–9.

[3] NRC factories, government aid, MEA register: see note 1. MEA index: *The China Handbook* (1937–44), 278; (1937–45), 369.

[4] *Shina kōgyō*, 70–8.

at Tzuliuching and a soda factory at Wutungchiao. Their initial intent was to use the Solvay process, which had worked well in Tangku. However, this process converts a relatively low percentage (70 percent) of salt to soda and requires very pure sodium chloride, which could be extracted from well brine only at enormous cost, so Fan and Hou traveled to Germany in search of ways to overcome these problems. The Germans had the solution, but because of their close ties to Japan, refused to transfer patents to China. Undaunted, Hou proceeded to New York, where he directed research on-site and by long-distance in Hong Kong, Shanghai, and Szechwan. Finally, in 1943, successful trials were completed using the "Hou united soda method," which combined the Solvay and synthetic ammonia (Haber) methods in continuous operation to make pure soda from carbon dioxide, ammonia, and salt. The virtue of this process was that it converted all of the salt to useful products – sodium carbonate (soda) and ammonium chloride (fertilizer). At the time, however, the Japanese blockade of China prevented delivery of the necessary equipment. Yungli and other factories fell back on the more primitive, but reliable, Leblanc method, using machinery built on the spot. The Wutungchiao plant produced 180 tons, and many smaller factories 30–50 tons per month. By 1944, annual production of soda ash in the Nationalist areas exceeded 5,600 tons.[5] Dorothy Needham, who visited the Szechwan salt wells near the end of the war, paints a vivid portrait of the hybrid technology developed there:

The salt wells . . . go down about 2,500 feet; water is poured in, and the brine is raised by a long cylindrical bamboo bucket with a valve at the bottom. In the old-fashioned method, going back two thousand years, the bamboo cylinder is raised, on its bamboo rope, by a very large horizontal drum (some 6 yards diam.) driven by about 6 buffaloes. In the newer setups there is a steel drum and an electric winch. The boring is done by means of an iron tool of special

---

[5] Inapplicable at the time of its invention and resisted for years thereafter, the Hou soda method eventually gained acceptance. Having chosen to remain in China after the revolution, Hou was sent by the government in November 1949 to inspect chemical factories at Dairen in southern Manchuria. Noting the proximity of Solvay and ammonia plants, he proposed to combine them, using the method that bears his name. Peking approved the plan and work began, but was interrupted in the mid-1950s by Soviet advisers who opposed the use of ammonium chloride as fertilizer, thus dismissing a major justification for this technique. During the Great Leap Forward, which was in part a rebellion against Soviet tutelage, Hou was appointed vice minister of chemical industry, after which development of the combined soda method resumed. In 1964, a national scientific inspection committee proclaimed the project a success, and it became a model for the manufacture of soda throughout China.

Hou Te-pang: Sung Tzu-ch'eng, "Hou Te-pang ch'eng-kung chih lu," 34–7; Li Chih-ch'uan and Ch'en Hsin-wen, "'Hou-shih chien-fa' ti tan-sheng ho fa-chan" [Birth and development of the "Hou soda method"], HHTP 8 (1982):475–8. Leblanc method in China: Wu Ch'eng-lo, "San-shih-nien-lai Chung-kuo chih suan-chien hua-hsüeh kung-ch'eng," 1; and Li Ch'iao-p'ing, Chung-kuo hua-hsüeh shih, 373. Soda ash production: China Handbook (1937–43), 445; (1937–45), 372.

Bamboo piping, salt wells, Tzuliuching, Szechwan. (Joseph and Dorothy Needham, Science Outpost.)

shape which pounds up and down in the borehole, being raised and lowered by a dozen men jumping on and off a beam. A slight rotary motion is given by hand to the cable carrying the tool. A wooden casing has to be made, reaching down to the level of the brine, in order to prevent surface soil drainage into the wells. All the constructional part of the machinery is made of wood (maximum size of poles about 8″ diameter) bound together with bamboo rope, and wedges driven in to tighten [the members] . . . .

Mr. Wu Ching showed us round the Chiuta Salt Factory. Here the brine is evaporated by charcoal fires, and after concentration is sprayed over great fences of bamboo and twigs, some 30 ft. high, out in the open air. KCl, boric acid, iodides and bromides are separated and purified. After removal of the potassium, acid is added, and the boric crystallizes out; or calcium borate is decomposed by addition of sodium carbonate, giving sodium borate and lime.

On the following day, we visited a factory which uses natural gas [drawn from the same wells] for evaporating the brine. Gas burning at the top of iron tubes some 5″ in diam., mostly methane. There is enough gas for about 600 of these burners. The gas is collected into a sort of reservoir from which bamboo pipes branch out. It is said that each burner can be turned off independently, but according to the workers themselves, they used water to quench a flame and had no means of turning [it] off and on.[6]

Dorothy Needham, diary entry of 16 January 1945, in Joseph Needham and Dorothy Needham, *Science Outpost: Papers of the Sino–British Science Co-operation Office, 1942–46* (London: Pilot Press, 1948), 241–2. For illustrations of these techniques, see ibid., figures 16 and 17. For a fuller description of this industry, see: Li Jung, "An

Production of basic industrial chemicals, sodas and acids, in wartime China reached only a fraction of their prewar peaks. Much more successful was the manufacture of liquid fuels. Government leadership, exploitation of local resources, and technical innovations by refugee engineers combined to make this one of the bright spots of the Nationalists' wartime industry. On the eve of the war, Nanking had grown increasingly concerned about its dependence on foreign oil and stepped up exploration for new sources. Investigation of the Shensi fields produced no new finds, but in 1935 news surfaced of oil in Yümen County, Kansu Province, deep in the interior and far from the centers of Chinese industry and commerce on the coast. An American geologist hired to study the site issued a pessimistic report, and interest in Kansu waned. With the outbreak of war and the retreat inland, however, a Chinese team was sent to investigate the region. Based on this team's more optimistic findings, technical personnel were shifted from Shensi to Kansu, and drilling began in 1939. The following year, they struck oil.[7]

Owing to the difficulty of wartime transportation, the Kansu Petroleum Bureau, which was set up to manage drilling and refining at this site, could not obtain high-quality foreign equipment. All facilities connected to the Kansu enterprise were built and operated by the Chinese themselves, using local resources. One refinery located in the Yümen fields first used shell stills, relatively primitive devices later replaced by more modern pipe stills, to produce straight distillates. A second refinery, located east of the fields, included a semicracking unit. By 1944 the output of these refineries reached 6.4 million gallons of petroleum products – mostly gasoline, some kerosene and diesel – far above the prewar peak, but still short of Chungking's needs. Further exploration by the National Resources Commission in Szechwan and other provinces located some natural gas, but no oil.[8]

The supply of crude petroleum and gasoline made from it were never sufficient to power Nationalist tanks, trucks, and planes, much less civilian transport. Buses, the chief mode of intercity transit, were converted to burn charcoal in producer-gas generators that worked adequately on level terrain, but were too weak to power the worn-out, overloaded buses over the mountains. The search for alternative fuels

Account of the Salt Industry at Tzu-liu-ching," *ISIS* 39, no. 118 (November 1948):228–34; and Li Ch'iao-p'ing, *Chemical Arts of Old China*, 61–5.

[7] Chinese petroleum industry: Li Ch'iao-p'ing, *Chung-kuo hua-hsüeh shih*, 381, 390; *China Handbook* (1937–43), 483–4; Shen Chin-t'ai, "Hua-hsüeh kung-ch'eng-hsüeh," 6; Shen Chin-t'ai, "Chung-kuo hua-hsüeh-shih kai-shu" [Summary of the history of chemisty in China], *Chung-kuo k'o-hsüeh-shih lun-chi*, 40–2.

[8] Output, 1944: Cheng Yu-kwei, *Foreign Trade and Industrial Development of China*, 264. By comparison, peak petroleum production in China before the war was 2,600 barrels, or 82,000 gallons, reached in 1916. See: *China Year Book* (1921–2), 159.

continued. One source, coal or shale, that had been studied without notable success before 1937, proved equally wanting thereafter. Several coking facilities were set up in the rear areas, and researchers at the West China Academy of Sciences near Chungking tested samples of Szechwan coal to determine potential yields. The NRC operated a coal distillation plant at Kienwei in western Szechwan to make synthetic gasoline and other products from the rich deposits of bituminous coal found in that region. Kienwei gasoline, high in octane, was suitable for use as aviation fuel, but was never produced in large volume.[9]

More amenable than fossil fuels were the hydrocarbons found in living plants. The extraction of gasoline and other fuels from vegetable oil had begun in the early 1930s, at the National Bureau of Industrial Research and the Sinyuan Fuels Laboratory in Nanking. In 1937, Sinyuan set up an experimental plant to produce gasoline and kerosene by heating vegetable oil in the presence of an aluminum chloride catalyst. Based on the success of these experiments, the NRC sent technicians from this lab to establish a factory in Chungking. Production at this plant, named the Tung-Li Oil Works, began in the spring of 1940, using tung oil extracted from the seeds of the tung tree, which is native to western China.[10]

Initially the Tung-li factory attempted thermal cracking of oil in locally made pipe stills. By the spring of 1940, production reached 500 gallons of gasoline per day, but the operation was only a qualified success. Thermal cracking requires sophisticated equipment made of special alloys capable of withstanding high temperatures and pressures. Materials of this type were unavailable in wartime China. Corrosion, damage, and breakdown of machinery repeatedly interrupted work. The product contained excessive amounts of fatty acids and had a low octane rating. To overcome these problems, Tung-li and other factories in the areas under Nationalist control shifted to saponification, a chemical process in which the tung oil was mixed with lime to form a calcium soap that could be distilled to produce a crude petroleum substitute. This technique yielded high-grade gasoline equal in volume to 20 percent of the raw tung oil.[11]

By 1942 the Tung-li works was turning out 4 thousand gallons of gaso-

9 Buses: Joseph and Dorothy Needham, *Science Outpost*, 51. West China Academy: Tseng Chao-lun, "Erh-shih-nien-lai Chung-kuo hua-hsüeh chih chin-chan," 1523; and Li Lo-yüan, "Low Temperature Carbonization of Szechwan Coal" (in English), *HHKC* 6, nos. 3–4 (December 1939):33–7. Kienwei: *China Handbook* (1937–44), 278.

10 For a sample of research on the synthesis of gasoline from vegetable oil, see articles in *KYCH* 4, no. 1 (January 1935):64–9; 4, no. 8 (August 1935):412–15; 4, no. 12 (December 1935):466–79; 5, no. 2 (February 1936):66–70. Tung-li Oil Works: Shen Chin-t'ai, "Hua-hsüeh kung-ch'eng-hsüeh," 5; and Shen Chin-t'ai, "Chung-kuo hua-hsüeh-shih kai-shu," 39.

11 Ibid., 39; Ch'en T'i-jung, "San-shih-nien-lai Chung-kuo chih lien-yu kung-yeh" [China's oil refining industry during the past thirty years], *San-shih-nien-lai chih Chung-kuo kung-ch'eng*, 2:1–4.

Experimental cracking still, National Bureau of Industrial Research, Chengtu, Szechwan. (Joseph Needham, Chinese Science.)

Table 12.1. *Production of synthetic gasoline from vegetable oil, Nationalist China, 1940–1944 (1,000 gallons)*

Factories	1940	1941	1942	1943	1944
Government	14 (1)	38 (3)	204 (12)	441 (16)	408 (13)
Private	8 (3)	72 (12)	74 (17)	201 (9)	157 (13)
Total	22 (4)	110 (15)	278 (29)	642 (25)	565 (26)

Note: Number in parentheses is the number of factories.
Source: Ministry of Economic Affairs, cited in the *The China Handbook* (1937–45), 370.

line per month, a record that attracted others into this field. Chungking alone had thirteen vegetable oil refineries, both government and private, producing liquid fuels with locally manufactured equipment and having a combined monthly output of more than 18 thousand gallons of gasoline, 13 thousand gallons of kerosene, and 80 thousand gallons of diesel. This industry spread to other parts of Szechwan and production expanded (Table 12.1). In a similar but separate effort, one Foochow University graduate built an apparatus out of old steel gasoline drums, bamboo piping, and derelict boilers from the Foochow naval yard to crack the turpentinelike oil obtained from distilled pinewood roots to make

Power alcohol factory, Szechwan. (Joseph and Dorothy Needham, Science Outpost.)

gasoline that was used to run buses in Fukien. Innovations of this type helped keep the traffic of wartime China rolling.[12]

Still more successful than the manufacture of synthetic fuels was the production of a gasoline substitute, alcohol. During the 1930s, the Nationalists had experimented with the use of ethanol as a gasoline substitute. After 1937, Chungking invested heavily in alcohol plants as a means of stretching China's limited fuel supply. The first distillation towers built by Chinese alone were erected in Szechwan in August 1938 by technicians from the China Alcohol Factory. By 1944 there were fifty-one alcohol factories (twelve government, thirty-nine private) in Szechwan, most with an annual output of 250 thousand gallons and a few three times this large (Table 12.2). Over one hundred factories of various sizes were scattered among the other provinces under Nationalist control. The principal raw material used in these factories was molasses or a strong spirit (*shao-chiu* or *kan-chiu*) made from it. By the end of the war, their combined monthly production neared one million gallons.[13]

World War II had mixed effects on China's chemical industries. The war dealt a severe blow to those industries such as acids, sodas, and fertilizers, where most progress had been made before the war and where the prospects for continued development were brightest. Annual produc-

[12] Gasoline: Ibid., 4. Fukien: Joseph and Dorothy Needham, *Science Outpost*, 227.
[13] Lu Pao-yü, "San-shih-nien-lai Chung-kuo chih chiu-ching kung-yeh," 2–3.

*Hydrochloric acid factory using traditional methods, Szechwan.
(Joseph Needham, Chinese Science.)*

tion of acids and sodas in Nationalist China during the war was around
one-eighth of that achieved in China proper, excluding Manchuria and
Taiwan, before 1937, a record that paralleled the decline of total industrial production from the prewar to wartime periods (Table 12.3).[14] Some
chemical engineers in the Nationalist areas admitted that due to the
shortage of spare parts and high cost of materials, it was often cheaper to
fly chemical products in over the Himalayan Hump from India than
make them locally. China's wartime factories, they argued, should be
considered "experimental, for training engineers and technicians," rather
than true productive enterprises.[15] On the other hand, the war spurred
the manufacture of products such as synthetic fuels and alcohol that had
been lacking or made in smaller quantities during the prewar era.

The retreat inland separated Nationalist manufacturers from the imported machines, blueprints, and advice that had stimulated technical

[14] Peak industrial production of Nationalist China during the war was less than one-eighth
the prewar level of China proper. Similarly, wartime production of acids and sodas, 7–9
thousand tons, was about one-eighth the prewar peak of 60 thousand tons. See: Cheng
Yu-kwei, *Foreign Trade and Industrial Development of China*, 109, 264–5.
[15] Joseph and Dorothy Needham, *Science Outpost*, 237.

Table 12.2. *Production of ethanol, Nationalist China, 1940–1944*
(1,000 gallons)

Factories	1940	1941	1942	1943	1944
Government	1,144 (6)	1,653 (22)	3,517 (31)	4,551 (44)	4,470 (43)
Private	2,715 (37)	4,504 (159)	5,835 (118)	6,164 (131)	6,624 (127)
Total	3,859 (43)	6,157 (181)	9,352 (149)	10,715 (175)	11,094 (170)

Note: Number in parentheses is the number of factories.
Source: Ministry of Economic Affairs, cited in *The China Handbook* (1937–45), 370.

Table 12.3. *Production of chemicals, Nationalist China, 1940–1944*
(metric tons)

Product	1940	1941	1942	1943	1944
Sulfuric acid	561	411	598	596	712
Nitric acid	10	12	19	28	24
Hydrochloric acid	317	197	279	226	549
Soda ash	1,290	1,598	2,209	4,232	5,676
Caustic soda	397	607	741	2,282	2,768
Sodium sulfide	21	60	45	68	77
Bleaching powder	103	468	696	646	963

Source: Ministry of Economic Affairs, cited in *The China Handbook* (1937–45), 371–2.

change and economic growth during the Nanking Decade. This isolation prompted innovation in the design and construction of factories of all types. "One may see," reported Joseph Needham, who traveled throughout Nationalist China during the last three years of the war,

a coal carbonization plant, in which all the piping, scrubbing towers, and metal parts have been constructed out of old gasoline drums. Or one may find a steel rolling mill operated by a salvaged river-steamer engine, and an excellent blast furnace made with steel plates from sunken river steamers.[16]

Enterprises of this type, which would have been dwarfed by the industrial complexes of coastal China, loomed large in the barren hinterland and bore evidence to the success of Nationalist engineers amidst the depravation of war.

[16] Ibid., 54.

Scientific education in wartime China

The same forces that affected industry – isolation, innovation, and government intervention – also shaped education in the Great Rear. Chinese students and scholars joined the retreat to the hinterland, where they carried on their studies through a combination of artifice and will. Chungking, preoccupied with the problems of survival, took steps to ensure that the education of these refugees served the needs of production and defense. The students responded favorably, flooding departments and colleges of engineering created for this purpose. But in the end, China's young engineers, ample in number, were poorly prepared for work in the real world. One reason for this failure was that many Chinese educators remained hesitant to embrace the utilitarian cause.

In 1938, as China's defenses crumbled, most of the country's leading universities, their faculty, staff, students, and as much of the libraries, laboratories, and other facilities as they could carry with them, followed the Nationalist retreat inland. The most famous of the wartime universities, National Southwest Associated [*Hsi-nan lien-ho ta-hsüeh*, commonly called *Lien-ta*], was set up in Kunming by students and teachers from Peking, Tsinghua, and Nankai universities, who had marched more than 1,300 miles to reach the relative security of mountainous Yunnan. Other schools moved to Sian, Chungking, and Chengtu. Altogether, some sixty-two colleges and universities made the trek west. Their combined enrollment grew from thirty thousand in 1938 to seventy thousand by the end of the war.[17] Everything in these schools – the books, paper and pencils, clothes, food, drugs, and other basic necessities – was in short supply. Life in the Great Rear was a constant struggle for survival; education and research had to compete with the business of staying alive.[18]

At *Lien-ta*, colleges of arts, sciences, commerce, engineering, and education were pieced together from the member institutions. Enrollment peaked at around three thousand students, most of whom were in engineering and commerce, the fields where the prospects for employment were brightest. The faculty numbered 350. This favorable student – teacher ratio might have enabled scholars to specialize and devote more time to research. In fact, almost all their energy was consumed by the mundane tasks of living and working under perilous conditions. Joseph Needham, who visited *Lien-ta* regularly, described the difficult conditions faced by scholars and students, even late in the war, when the worst days were behind them:

[17] Enrollments: Ch'en Li-fu, *Chinese Education during the War (1937–42)* (Chungking: Ministry of Education, 1942), 3, 31; *China Handbook (1937–45)*, 399.
[18] *China Handbook (1937–43)*, 369.

All the departments are housed in "hutments" built of mud brick, and roofed very simply with tiles or tin sheets, though some have curving roofs in the great tradition of Chinese architecture. Inside, the floors are beaten earth, with a little cement, and extreme ingenuity has been used in fitting up laboratories for research and teaching under these conditions. For example, since no gas is available, all the heating has to be done with electricity. When the supply of element wire for heaters (home-made out of clay) ran out some time ago, work was at a standstill until it was found that gun lathe shavings from one of the Yunnan arsenals would do very well. When microscope slides could not be had, windowpanes broken by air raids were cut up, and the unobtainable coverslips were replaced by local mica. Many other instances could be given of Chinese ingenuity and initiative. In many cases there are no air raid shelters, and the population scatters to the hills if a raid looks like being serious. When the siren goes, all the most valuable apparatus of the Associated University is lowered into large petrol drums built into the floors of each mud-brick building, to guard against anything but a direct hit. Even in its humble buildings, the University has been bombed several times, and many of the rows of huts destroyed.[19]

The *Lien-ta* college of science included departments of math, physics, chemistry, biology, and earth sciences. Enrollment was around three hundred. The faculty numbered as many as 100, with 8 to 14 full and assistant professors in each department. The purpose of the college and design of its curricula were essentially the same as those of the constituent universities before the war, except for the addition of courses on the application of science to wartime needs. What changed was the potential for implementing this program. Laboratories and experimental facilities, specimens and reagents, library reference works and current periodicals were all in short supply. In contrast to the industrial research institutes and schools of engineering, which received the most and best of everything, the college of science, whose purpose was primarily academic, had the lowest priority and received only the most meager supplies.[20]

The *Lien-ta* department of chemistry included all of the leading figures from Tsinghua, Peking, and Nankai universities, except Sah Pen-t'ieh, who chose to remain in Peking. The basic curriculum was the same as before the war, except for the addition of a few courses in pharmacological chemistry and some industrial applications, such as dyeing and fermentation. The number of Tsinghua chemistry courses that included laboratory sessions dropped, however, from twenty before the war to five in Kunming, and the surviving labs offered only six or seven experiments each semester. Laboratories were limited in number and size and had no running water, gas, or electricity. Early in the war, reagents and other materials were smuggled from Shanghai via Haiphong

---

[19] Quotation: Joseph Needham, *Chinese Science* (London: Pilot Press, 1945), 13–14. Other details: *Ch'ing-hua ta-hsüeh hsiao-shih kao*, 289–316.
[20] Ibid., 336–9; W. E. Tisdale, "Report of a Visit to Scientific Institutions in China," 26–8.

Chemical glassware made in wartime Nationalist China. (Joseph Needham, Chinese Science.)

to Kunming, but after 1940, when this route was closed, teachers and students had to get by with what they could find or make for themselves. Imagination and self-reliance filled some gaps: Alcohol lamps replaced gas burners; well water was distilled; some common reagents, such as hydrogen peroxide, silver nitrate, hydrochloric and nitric acids, were made on the spot. One reporter who visited Kunming in 1939 found the "simplest household implements, pots, kettles, meatgrinders, were used for chemical experiments." But precision instruments were lacking, and the selection of experiments was dictated by the availability of supplies. Under these conditions, laboratory standards and the experimental skills of the students plummeted.[21]

The college of engineering, which had the largest enrollment and best facilities of the five *Lien-ta* colleges, included departments of mechanical,

[21] The chemistry faculty included from Tsinghua, Chang Tzu-kao, Huang Tzu-ch'ing, Kao Ch'ung-hsi, Chang Ta-yü, Su Guoc-jen, and Chang Ch'ing-lien; from Pei-ta, Tseng Chao-lun, Chu Ju-hwa, Ch'ien Shih-liang, Sun Ch'eng-o, and Liu Yün-fu; and from Nankai, Yang Shih-hsien and Ch'iu Tzung-fou. Faculty and curriculum: *Ch'ing-hua ta-hsüeh hsiao-shih kao*, 345–7. Smuggling: John Israel, "Galaxy of Stars: The Lianda Faculty. IV. The College of Natural Sciences," (unpublished paper), 20–2. Reporter, 1939: Hubert Freyn, *Chinese Education in the War* (Shanghai: Kelly & Walsh, 1940), 38.

electrical, civil, and chemical engineering, the last of which grew out of work begun at Nankai University in the 1930s. For upperclassmen the chemical engineering department offered twenty specialized courses, taught by the department chairman, Ch'en K'o-chung, and other Nankai faculty. In cooperation with the National Resources Commission, *Lien-ta* engineers worked on road construction, water conservancy, and problems of chemical engineering, such as the manufacture of coke and dyes.[22]

The example of *Lien-ta* illustrates the sweeping changes that occurred in educational priorities during the war, when the government intervened to favor the applied fields, particularly engineering, at the expense of all others, including the pure sciences. Ch'en Li-fu, who served as minister of education from 1940 until the end of the war, continued his efforts to enforce greater discipline on China's campuses and reshape the curriculum of higher education to serve utilitarian goals. During his tenure, more than fifty subjects with a direct bearing on industrial and military problems were introduced into China's colleges and universities. In the case of chemistry, many departments shifted attention to tanning, dyeing, refining of fuels, and the manufacture of ceramics, alcohol, paper, and other products. In all the schools under Nationalist control, the application of knowledge to achieve immediate results was accorded the highest priority.[23]

Students were quick to accept this utilitarian approach. Most university students, because they wanted to help save the nation, saw (correctly) where the best job opportunities lay, or both, elected to study engineering, followed by humanities and the arts; agriculture, medicine, and natural science were far down the list of preferred majors. In 1943, of the nearly six thousand candidates who took the entrance exam for National Central University, 31 percent chose engineering, 25 percent law, 12 percent each agriculture and education, 10 percent letters, and only 5 percent each medicine and science. Some observers expressed concern that young people were being seduced into fields that promised the most immediate returns – a "dangerous phenomenon," in the view of Academia Sinica president, Chu Chia-hua, that threatened to undermine the foundation of basic science – but to no avail. By 1944, less than 9 percent of students in schools of higher learning chose pure science – a

---

[22] *Ch'ing-hua ta-hsüeh hsiao-shih kao*, 358–63, 370.

[23] Role of Ch'en Li-fu: E-tu Zen Sun, "The Growth of the Academic Community," 416; and Wilma Fairbank, *America's Cultural Experiment in China, 1942–1949* (Washington, DC: GPO, 1976), 122. Curriculum: Ch'en Li-fu, *Chinese Education*, 3. Fifty subjects: *China Handbook* (1937–43), 384. Chemistry courses: *Ch'ing-hua ta-hsüeh hsiao-shih kao*, 346; and William Band, *Science in the Christian Universities at Chengtu, China* (New York: Associated Boards for Christian Colleges in China, 1945), 6.

Table 12.4. *Enrollment in Chinese higher education, Spring 1944*

Subject	National	Provincial	Private	Total	%
Arts					
Liberal arts	3,492	638	3,816	7,946	11.4
Social studies	7,580	306	6,972	14,858	21.2
Commerce	2,932	1,310	4,851	9,093	13.0
Education	873	242	1,121	2,236	3.2
Total	14,877	2,496	16,760	34,133	48.8
Sciences					
Natural sciences	2,945	183	2,728	5,856	8.4
Engineering	10,284	586	2,079	12,949	18.5
Medicine	2,891	963	1,640	5,494	7.9
Agriculture	3,582	735	1,003	5,320	7.6
Total	19,702	2,467	7,450	29,619	42.3
Teacher's course	5,724	483	—	6,207	8.9
Grand total	40,303	5,446	24,210	69,959	100.0

Source: The China Handbook (1937–45), 330.

figure below the level of 1928, when the Nationalists first began their program to strengthen and expand education in the sciences (Table 12.4)![24]

Whereas the ministry directed funds to the applied fields and the students followed these priorities, the fate of Chinese education remained in the hands of the teachers and school administrators, who were not always faithful to Ch'en Li-fu's utilitarian approach. One report on the state of engineering education in the Nationalist areas in 1942 describes the wide gap between these programs and the society they were supposed to serve. According to this report, the typical engineering graduate invariably discovered that "what he has studied is of no use, and what is of use he has not studied," and the people who were supposed to benefit from expert services found that the young technicians "cannot put up

[24] NCU entrants, 1943: Hao Ching-sheng, "K'ang-chan ch'i-nien lai chih k'o-hsüeh," 195. Students seduced: ibid. Chu Chia-hua: Chu Chia-hua, "Ch'ing-nien yü k'o-hsüeh" [Youth and science], *Chu Chia-hua hsien-sheng yen-lun-chi* [Dissertations of Dr. Chu Chia-hua], ed. Wang Yee-chun and Sun Pin (Taipei: Institute of Modern History, Academia Sinica, 1977), 58. Enrollments, 1928: *China Year Book* (1933), 537. Enrollments, 1944: *China Handbook* (1937–45), 330. Hubert Freyn, *Chinese Education in the War*, 43, states that in 1932 one-third of the students at National Central University were studying science and technology, whereas in 1939 the fraction rose to two-thirds.

with suffering, are unwilling to take responsibility and unable to deal with practical problems." The author blamed this shortcoming on the character of the education itself. China's engineering curricula were copied from European and American models and based on advanced technologies that had nothing to do with China. Meanwhile, Chinese scholars had failed to investigate the situation in their own country. The engineering colleges had done little to promote work-study programs or internships, students were allowed to while away their summer vacations rather than gain useful hands-on experience, and university research facilities had failed to draw in people and problems from the real world. Those charged with training China's future engineers tended to look down on the challenges of applied science as being of little scholarly value, and the educational bureaucrats preferred to concentrate resources on areas that promised more immediate rewards. The result, this study concluded, was that Chinese engineering education "falls between two stools," and, lacking clear purpose, has become "education for its own sake."[25]

This assessment was confirmed by other reports on the performance of China's young engineering graduates. By far the largest employer in this field was the National Resources Commission, which operated more than one hundred major industrial enterprises and oversaw the government's massive program to plan and direct wartime and postwar industrial development. During the war, the NRC employed twelve thousand administrative and technical personnel, of whom 80 percent were under thirty-five years of age and five thousand were university graduates. The commission's system of recruitment was highly professional: Candidates were judged on their performance in school and on a uniform competitive examination, and hiring decisions were based on scholarly and technical credentials, rather than personal or political connections. But the result, according to one recent study, was that NRC factories became bloated with young inexperienced university graduates who were unable or unwilling to contribute to the tasks of production. In 1945, members of the American War Production Mission to China reported that these novices were "cluttering up" NRC plants, which in some cases could be run just as well by intelligent and experienced technicians. The problem was too much theory and too little practice. "The greatest need," as Chinese chemical engineers told Dorothy Needham, "is for men of the shop foremen type; without good men of this kind, the products cannot always reach a high quality."[26]

[25] Ssu Hao, "Mu-ch'ien Chung-kuo kung-ch'eng chiao-yü chu wen-t'i" [Various questions on present Chinese engineering education], *Hsüeh-hsi sheng-huo* [Study life] 3, no. 2 (20 July 1942):92–5.
[26] William C. Kirby, "Technocratic Organization and Technological Development in

In sum, Ch'en Li-fu and the Ministry of Education invested heavily in the applied sciences, particularly engineering, with the result that the number of engineering programs, the size of their enrollments, and the quality of their facilities all advanced relative to the pure sciences and other fields of study. Many graduates of these programs got what they had hoped for – jobs in industry, in many cases government-owned industries. But the evidence suggests that most were ill prepared for the factory floor and that at least part of the blame lies with the nature of the education itself. One reason these universities failed to prepare graduates for the real world was that many of the teachers remained in the ivory tower of academia – a fact also demonstrated by their continued preference for basic or theoretical research.

Scientific research in wartime China

Just as he sought to promote more useful education, Ch'en Li-fu also tried to channel support to research that would serve the immediate needs of production and defense. Some scholars followed this lead, shifting their attention to problems of industry, agriculture, and public health. But others continued research begun during the previous decade, in China or abroad, that addressed basic questions about the workings of nature, although bearing little connection to the practical problems of China at war.

The exile affected research in each branch of science differently. Biologists and geologists lacked books and supplies, but were compensated by the opportunity to collect specimens of previously unknown plants and animals or study unexplored topographies. Taxonomic and descriptive work, which required no special equipment, made significant advances during the war. Mathematicians, astronomers, and physicists were handicapped by the lack of current literature and access to colleagues abroad, but scholars in these fields had time to think, and many used this time productively, in most cases to pursue work along theoretical and computational lines. All of the most important wartime research in physics – by S. T. Ma, T. Y. Wu, and T. S. Chang on statistical and quantum mechanics, and Chou P'ei-yüan on fluid dynamics – was theoretical rather than experimental; among the students in Kunming, China's most famous scientists, future Nobel prizewinners T. D. Lee and C. N. Yang, began their training in theoretical physics.[27]

China: The Nationalist Experience and Legacy, 1928–1953," *Science and Technology in Post-Mao China*, ed. Denis Fred Simon and Merle Goldman (Cambridge: Harvard University Press, 1989), 30–1. Needham: Joseph and Dorothy Needham, *Science Outpost*, 237.

[27] The most complete account of scientific research in Nationalist China during the war is Joseph and Dorothy Needham, *Science Outpost*. For additional information on botany,

Dr. Chu Ju-hwa, Institute of Chemistry, Peiping Academy, Heilongtan, near Kunming. (Joseph Needham, Chinese Science.)

In the experimental sciences, such as chemistry, research was more
difficult, and in some cases impossible. Many chemists turned to stra-
tegically important, but scientifically banal problems of an applied sort,
such as testing local products or developing substitutes for foreign imports. Tseng Chao-lun conducted research on glycol lubricants, Chu
Ju-hwa on antimalarial drugs, while others worked on the manufacture
of glass, paper, textiles, fuels, and the like. A few took up purely theoretical problems that did not depend on laboratory devices. Sun Ch'eng-o
extended his previous study of mathematical models by calculating
ionic potentials, the numerical relationship between the valence and
radius of an ion, which can be used to predict its physical properties.
The only significant experimental research by a university-based chemist
was Chang Tsing-lien's work on heavy water. Chang, who began this
research as a graduate student in Germany, returned to China in 1937,
bringing with him a supply of heavy water and the necessary instruments.
During the war, he conducted a series of studies in his *Lien-ta* laboratory

see: C. Y. Chang, "Botanical Work in China during the War, 1937–43," *Acta Brevia
Sinensia*, no. 6 (April 1944):3–6. On astronomy: Y. C. Chang, "The Research Activity
of the National Institute of Astronomy," *Acta Brevia Sinensia*, no. 2 (March 1943):7–8.
On mathematics and other sciences: *Acta Brevia Sinensia*, no. 3 (May 1943):31–3. On
physics: C. Y. Fan, "Advance of Physics in War-time China," *Acta Brevia Sinensia*, no. 8
(December 1944):3–5. For details on research at *Lien-ta*, see: *Ch'ing-hua ta-hsüeh
hsiao-shih kao*, 339–52, 376.

that established several previously unknown properties of heavy water and won widespread recognition both in China and abroad.[28]

The Academia Sinica also moved inland in 1938. Its institutes were dispersed among several sites, the one for chemistry in Kunming. Woo Sho-chow, who served as director of the Institute of Chemistry during most of the war, continued his work on ultraviolet absorption spectra, switching form gases to liquids, because equipment for the former was unavailable. Most other research at this institute focused on problems having direct industrial and military applications: the synthesis of sulfanilamides (used in sulfa drugs) and santonins (an anthelmintic), analysis of salt ores and brines, manufacture of fertilizers, paper, and iron, purification of castor oil, and low-temperature distillation of coal.[29] The National Academy of Peiping also settled in Kunming, where its Institute of Chemistry carried out research designed to aid the war effort: the distillation of wood, manufacture of synthetic gasoline, synthesis of drugs, purification of vegetable oils, and extraction of vegetable dyes. Even those problems that had a theoretical component — such as the synthesis of vitamin K, or the chemical analysis of herbal medicine — were justified by their possible applications.[30]

Besides Kunming, the other major center of wartime research was Beipei, a small town outside Chungking, which was home to eighteen scientific and educational institutions, including the National Bureau of Industrial Research (NBIR). The NBIR, under the direction of Dr. Ku Yü-tsuan, maintained seventeen laboratories for research on problems related to industry. The fermentation division, under Dr. Chin P'ei-sung, produced acetone, alcohol, butyl alcohol, and other solvents; the pure chemicals division, headed by Dr. Li Erh-kang, made acids, alkalies, and salts; and the motor fuel division, directed by Dr. Ku Yü-chen, worked on the cracking of vegetable oils to produce synthetic gasoline.[31]

[28] Sun Ch'eng-o and ionic potentials: *JACS* 50 (1928):2855−72; *JCCS* 5 (1937):148−53, 195−203; 7 (1940):62−75; 10 (1943):19−21, 77−9; 11 (1944):118−19, 121−4; 12 (1945):24−8. Analysis of Sun's work: Hsü Hsien-kung, "Chung-kuo hua-hsüeh-chia tui-yü yu-chi hua-hsüeh ti kung-hsien" [Contributions of Chinese chemists to organic chemistry], *HHSC* 9, no. 6 (1954):230−3. Chang Tsing-lien: Tsing-lien Chang, "Recent Researches on Heavy Water," *Science* 100 (14 July 1944):29−30; and Yen Chih-hsien, "Chung-kuo hua-hsüeh-chia tui-yü wu-chi hua-hsüeh ti kung-hsien," 260−3. Recognition of Chang's work: Alice H. Kimball, comp. *Bibliography of Research on Heavy Hydrogen Compounds*, 47−8. A brief survey of work in both pure and applied chemical research is Tseng Chao-lun, "Progress of Chemical Research in China," *Acta Brevia Sinensia*, no. 3 (May 1943):15−18. Other details: Chang I-tsun, "Chung-kuo ti hua-hsüeh," 9−12; and Joseph and Dorothy Needham, *Science Outpost*, 80−2.
[29] Academia Sinica during the War: Yeh Ch'i-sun, "Work of the Academia Sinica, 1937−42," *Quarterly Bulletin of China Bibliography* 3, no. 1 (1943):7−20; and *China Handbook* (1937−44), 263; (1937−45), 351.
[30] *Pei-p'ing yen-chiu-yüan*, 12−13. *China Handbook* (1937−43), 412; (1937−44), 271; (1937−45), 357−9.
[31] Joseph and Dorothy Needham, *Science Outpost*, 101.

Study of tung oil for manufacture of plastic, Associated University
[Lien-ta], Kunming. (Joseph Needham, Chinese Science.)

The politics of science

Evidence from the fields of education and research reveals a persistent
conflict over the role of science in wartime China. Spokesmen for the
state favored a narrow definition of learning that stressed its application
to practical needs. The scholars, or at least some of them, held out for a
broader, deeper, more long-term view of science as search for knowledge, unconstrained by the test of utility. This conflict sharpened with
the passage of time and remained unresolved at war's end.

Early in the war, as the spirit of national resistance ran high, there
were hopeful signs of cooperation among the parties engaged in science

and technology. "Although this has been a most difficult period," wrote one observer in 1939, "it has been a good opportunity for the unification of chemical research and chemical industry." New industries had sprung up to meet the demands of war, new methods were devised and resources found to make up for lost imports, and everyone pitched in to save the nation:

In some cases, scholars have taken control of factories, research departments have been set up within scholarly organizations, and scholarly organizations have stepped in to solve problems on behalf of the factories.... In the present circumstances, chemical research and chemical industry have become one.

At the same time, this observer foresaw the potential for conflict between the narrow agenda of Chinese industry and the broader needs of science. The question will inevitably arise, he warned, of "how to promote research on chemical industry without compromising the freedom to do [other] chemical research."[32]

As the war continued and patience on all sides wore thin, this augury proved correct. Spokesmen for the pragmatic or statist approach argued eloquently for scientists to support the immediate needs of war. Lu Yü-tao, editor of the journal *K'o-hsüeh*, took up the government's cause by urging the scientists to come out of their ivory towers and join in solving the problems confronting China. Lu charged that many scientists considered their research a "sacred, lofty" affair that should not be "vulgarized" by the demands of the marketplace. In fact, he countered, much current research is excessively theoretical and serves no useful purpose, and those who perform it refuse to recognize their responsibility to society as a whole. The more deeply the scientists enter into these arcane specialties, Lu pointed out, "the heavier becomes the odor of their pedantry." Chungking calls on its scientists to do applied studies, while the scientists reply that the government should support basic research. "In the end, the two parties remain on opposite banks, shouting their war cries. They are like two cogwheels that have not been engaged."[33]

The chief force behind the movement for applied learning was the conservative wing of the Kuomintang, but early in the war, when hopes for the united front between Nationalists and Communists were high, leftists in Chungking also promoted the view that science should serve the state.

---

[32] Chou Fa-ch'i, "Hua-hsüeh yen-chiu yü hua-hsüeh kung-yeh" [Chemical research and chemical industry], *HHKC* 4, nos. 3–4 (December 1939):66–7.

[33] Lu Yü-tao, "Hsien-shih-hsing ti k'o-hsüeh yen-chiu" [Practical scientific research], *Wen-hua hsien-feng* [Cultural pioneer] 1, no. 9 (27 October 1942):11–14; and Lu Yü-tao, "K'ang-chan ch'i-nien lai chih k'o-hsüeh chieh" [The scientific community during the last seven years of the War of Resistance], *Chung-kuo chan-shih hsüeh-shu* [Chinese wartime scholarship], ed. Sun Pen-wen (Shanghai: Cheng-chung shu-chü, 1946), 166–80. "Odor of pedantry:" ibid., 180. "Opposite banks": ibid., 166.

Articles in the Communist-sponsored, Chungking-based journal *Ch'ün-chung* [The Masses], criticized those scholars who favored the teaching of basic science while resisting utilitarian studies. Since the outbreak of the war, one author was happy to report, many scientists had come around to the defense of their country. Now, all sides should promote this trend, and those who continue to resist should be made to see the error of their ways.[34] Although the language of these articles was blunted in deference to the united front, much of this analysis was indelibly Marxist. Yüan Han-ch'ing, an important figure in Nationalist science, who had taught chemistry at two leading universities during the 1930s and headed an institute in Lanchow during the war, argued that the progress of Chinese science depended on the prior development of China's industry and the liberation of that country from imperialist control:

Science becomes a necessary instrument only after the development of industry. . . . Once there is the need, once there is a path, then naturally people will be willing to receive training in scientific research. . . . Our country is a semi-colony oppressed by imperialism. In this social setting, the immature shoots of our science cannot see the light, receive the rain, or breathe the air. . . . If we want to promote scientific research, we must first liberate the nation and struggle for the industrialization of the society.[35]

On the other side were scholars and scientists who favored an approach rooted in pure theoretical learning. These men and women cannot be charged with lacking concern for China. Most of them had left the relative comfort and security of homes on the coast and in some cases had foregone offers of employment abroad in order to join the resistance, hardships and all, in the Great Rear. They defended a program that included academic learning, because they believed it was in the long-term interests of China. The center of opposition was Kunming, which boasted a large concentration of scholars and intellectuals, a favorable location at the terminal for flights over the Hump, and the protection of local warlord Lung Yün against political interference from Chungking. The Kunming intellectual community shared a faith in the right of scholars to set their own agenda. "The basic goal of the university," explained political scientist Ch'ien Tuan-sheng, "is the quest for knowledge, it is not utility. If university education is able, at the same

---

[34] For leftist views, see: Tzu Nien, "Fa-hui 'Wu-ssu' yün-tung so t'i-ch'ang ti k'o-hsüeh ching-shen" [Develop the scientific spirit advocated by the "May Fourth" movement], *Ch'ün-chung* 2, no. 24/25 (15 May 1939):797–8; and Wu Tsao-hsi, "Erh-ch'i k'ang-chan ti k'o-hsüeh yün-tung" [Scientific movement in the second stage of the War of Resistance], *Ch'ün-chung* 3, no. 13 (27 August 1939):343–7.

[35] Yüan Han-ch'ing, "Lun t'i-kao k'o-hsüeh ti yen-chiu" [On raising up scientific research], *Hsin min-tsu* 1, no. 12 (15 May 1938):5–6. Yüan's wartime experience: Joseph and Dorothy Needham, *Science Outpost*, 136.

time, to produce something useful, that is an ancillary function and not its original goal."[36]

Several of Ch'ien's colleagues extended this principle to the study of science. In 1944, Hao Ching-sheng, a popular writer on scientific and educational affairs, praised the many effective applications of science made during the war, but pointed out that these successes rested on education and reseach from the prewar era, without which none of the later achievements would have been possible. In the course of the next decade, Hao observed, China would need 2.5 million technical cadres, all of whom must be trained in scientific theory. During the war, the country had paid in blood to raise itself to the ranks of the Allied powers. It could remain in this company after the war only if Chinese science kept pace with international standards. The long-term needs of the nation depended on an investment in basic education and research, a commitment Chungking refused to honor. "From the independent academies above to the research institutes of the universities below," Hao lamented, "all are in economic straits and cannot develop. This is a situation we must deplore."[37]

Many scientists voiced similar misgivings about Chungking's neglect of pure research. Hu Hsien-su, the most prominent botanist of his generation, explained that many modern advances that "seem to arise from an almost mystical science, are in fact the result of constant research and experimentation by ordinary scientists, burrowing away in their laboratories, wracking their brains." In Hu's opinion, this activity, far from being alien to society, deserved greater understanding and more generous support.[38] Chu Chia-hua, the geologist who became acting president of the Academia Sinica after the death of Ts'ai Yüan-p'ei in 1940, criticized Chungking for placing too much emphasis on short-term results and appealed for a more balanced program of pure and applied studies. The experience of the Soviet Union, Chu argued, shows that successful applications depend on knowledge derived from basic research.[39] In 1944, the National Research Council followed Chu's lead by calling for more emphasis on basic research, improvement of the facilities of universities and research institutes, greater efforts to extend the knowledge and fruits of Chinese science beyond the state to the people, and closer cooperation with scientists overseas.[40]

[36] Ch'ien Tuan-sheng quotation of 1940, in unpublished ms. by John Israel, cited in E-tu Zen Sun, "The Growth of the Academic Community," 416.
[37] Hao Ching-sheng, "K'ang-chan ch'i-nien lai chih k'o-hsüeh," 189, 194–5, 198.
[38] Hu Hsien-su, "K'o-hsüeh yü chien-kuo" [Science and national reconstruction], Wen-hua hsien-feng 2, no 21 (10 October 1943):5–6.
[39] See selected speeches by Chu Chia-hua, 1943 and 1944, in Chu Chia-hua hsien-sheng yen-lun-chi, 43–5, 58, 85–7.
[40] "Chung-shih chung-yang yen-chiu-yüan p'ing-i-hui chih chien-i" [Emphasizing the

As the struggle wore on, critics of Ch'en Li-fu charged that the Nationalists were using science to strengthen their military power, while maintaining the dominance of traditional values and a dictatorial hold on society. In October 1943, Chungking launched a movement to promote the "science of national defense." Among those who took issue with this effort were spokesmen on the left, who had favored utilitarianism early in the war when it served the alliance between Nationalists and Communists, but reversed their position after the united front broke down. One article in the leftist journal *Ch'ün-chung*, although agreeing that science should help defend China, pointed out that the current effort failed to combine technological progress with mass mobilization or extend the benefits of science to the general public. The author pointed out that the same people who defended the "science of national defense" also favored traditional philosophy, while rejecting the role of science in remaking Chinese culture – a clear reference to the Ch'en brothers and the Kuomintang right. The proper role of science, he concluded, was to enrich the masses and transform their culture, not simply to strengthen the state.[41]

Opinions of this type were not limited to the left, however. The editors of the centrist *Wen-hua hsien-feng* [Cultural Pioneer], who supported the movement for a "science of national defense" and other elements of the pragmatic agenda, argued that science should also contribute to the formation of a new "national culture." In their view, earlier defeats at the hands of imperialism had demoralized the Chinese, causing them to abandon traditional learning and embrace indiscriminate Westernization, with the result that "our nation's scholarship and culture remain the tail wagged by foreign scholarship and culture." Only when China develops its own science and other forms of learning, they concluded, can the Chinese regain their independence and pride: "If a country does not respect scholarship and culture, then it cannot advance, become independent and strong. If a people cannot absorb the world's culture, cannot invent or create, and make no contribution to the world, then they are lost."[42]

Participants and observers offered different explanations for the causes underlying this dispute. Lu Yü-tao considered it the natural outcome of differences in the training and experience of scientists and bureau-

proposal of the National Research Council of the Academia Sinica], *Wen-hua hsien-feng* 3, no. 12 (21 March 1944):2.

41 Ting Ssu, "K'o-hsüeh ching-shen, k'o-hsüeh t'ai-tu" [Scientific spirit, scientific attitude], *Ch'ün-chung* 9, no. 12 (June 1944):498–504.

42 This view is expressed in response to the proposals of the Advisory Committee of the Academia Sinica, already mentioned: *Wen-hua hsien-feng* 3, no. 12 (21 March 1944):2. The quotation is from the companion editorial, "Hsüeh-shu yü chien-kuo" [Scholarship and national reconstruction], ibid.

crats: "Those charged with practical tasks tend to see things too closely, while scholars tend to see things at too long range, and the near and the far do not connect."[43] Leftist critic Wu Tsao-hsi gave an historical interpretation: Chinese students and scholars had failed to overcome the "traditional poisons of bureaucratic and literary education," and scientific and technical educators had not kept up with the times.[44] Joseph Needham rejected the popular tendency to blame the impracticality of scholars on their "age-old aloofness" or "disinclination for manual work," concluding that it was primarily a matter of failed policy. In Britain, Needham explained, scientists cooperated with the government because they knew that their wartime service was temporary and that they would be allowed to return to basic research after the war. But China offered no such guarantee, Chinese scientists feared being drawn into a form of permanent servitude to the state, and many responded by refusing to cooperate at all.[45]

The commitment of Chinese scientists to professional autonomy and basic learning remained intact. One might expect the years of isolation and hardship to erode values that had been inculcated in the course of studying in Europe or America and working in urban China during the prewar decade, but there were countercurrents. First, Chinese scholarly associations that had been inactive during the period 1937–41, when many scholars were scattered or in transit, were restored after 1942 as the military situation improved, communities stabilized, and scientific research resumed. The revival of conferences, publications, and other activities helped rekindle a sense of common purpose and professionalism and keep alive the faith in science as a discrete intellectual enterprise.[46]

Also important was the reopening of contact with foreign scholars. During the first five years of the war, Chinese scientists were almost completely cut off from their colleagues in the West. Most of the books and equipment carried on the retreat inland were damaged or lost; those that arrived intact were used up or became outdated. Some supplies reached Kunming and Chungking from Hong Kong, before that city fell to the Japanese at the end of 1941, but during this period the Chinese neither benefited from nor contributed to the common wealth of global science. They were removed from the main pit of scientific excavation and felt, in Joseph Needham's rather melodramatic phrase, that they had been "buried alive."[47]

Beginning in 1942, when they entered the Pacific War, the British and

[43] Lu Yü-tao, "K'ang-chan ch'i-nien lai chih k'o-hsüeh chieh," 180.
[44] Wu Tsao-hsi, "Erh-ch'i k'ang-chan ti k'o-hsüeh," 345.
[45] Joseph and Dorothy Needham, *Science Outpost*, 266–7.
[46] Lu Yü-tao, "K'ang-chan ch'i-nien lai chih k'o-hsüeh chich," 167–70.
[47] Joseph and Dorothy Needham, *Science Outpost*, 17.

American governments reopened the lifeline to China and restored contact between Chinese scientists and the world outside. At first, the only line of supply was the airlift over the Hump from India, a treasure reserved for higher military and diplomatic priorities, leaving little room for books, journals, and instruments. Until 1944, the Allies sent microfilms of current scientific and technical journals that were distributed through a network of reading rooms located in universities and research institutes throughout the Nationalist areas. Chinese scholars found it difficult to use microfilm, a technology then in its infancy, but the access it provided to current scholarship was a great psychological boost. Actual books, journals, and other more substantial supplies began to arrive during the last year of the war. Scholarship also moved in the opposite direction, as the Allies supported the translation of scholarly articles produced in China and helped find publishers in the West. A small number of Chinese scholars and students, many in the sciences, were invited to study in Britain and the United States. All of these activities helped restore the sense of participation in the world scientific community and strengthen sagging professional spirits.[48]

Conclusion

China's scientists, like Chinese in other walks of life, responded to the Japanese invasion by leaving the comfort of homes and workplaces in cities along the coast to join the hard scrabble of the Resistance. Adaptation to life in the Great Rear forced the scientists to become more self-reliant and innovative and to recognize the need to apply their knowledge to practical ends. The Nationalist government advanced this trend by asserting greater control over all aspects of science and technology, urging men and women in these fields to serve the national cause, and allocating funds to support this purpose. But many scientists retained a sense of their professional autonomy and a commitment to learning that transcended the demands of the moment. By war's end, the conflict between science and the state remained unresolved.

[48] The American program is described in Wilma Fairbank, *America's Cultural Experiment in China*, 46–65. The British program is in Joseph and Dorothy Needham, *Science Outpost*, 16–26, 56–75.

13

Science in Communist China I: innovations in industry

The War of Resistance drove the Nationalists into the hinterland, isolating them from the cities, the coast, and the foreign connections that held the key to development of science during the Nanking Decade. It had the opposite effect on the Communists, however, bringing them for the first time into contact with these same forces of modernization. The Communists, outlaws in Nationalist China, had taken to the hills after 1927 to escape the government's "bandit suppression" campaigns and missed the dramatic changes of the 1930s. Before the war, Chinese Marxists thought of science mainly as a slogan in the battle to shape China's New Culture, a banner waved in polemical debates that raged inside and outside the party. But Japan's invasion brought the Communists respite from civil war, an influx of refugees from the urban areas, and new resources that could be applied to the manufacture of goods within the Red base areas. With these changes, the Communists discovered science as a way of understanding their material surroundings, an instrument for creating wealth and power, a means of waging revolution and war.

Science and Chinese communism, 1921–1939

Before the war, "science" appeared in the lexicon of Chinese Marxism as a philosophical abstraction that bore little connection to the study of nature. During the first two decades after its founding in 1921, the Chinese Communist Party (CCP) was dominated by men and women whose principal experience had been as labor organizers, propagandists, military officers, or rural reformers. Few who took part in the Long March or the events preceding it had a vocation separate from politics. No figure of any significance within the party had pursued advanced studies in science. Reared in the charged atmosphere of the May Fourth

era, Chinese followers of Marx saw in science not a method for
understanding the physical environment or an instrument for transforming nature, but a model for erecting a new culture and a blueprint for
saving China. They were drawn to Marxism, in part, because it offered a
"scientific" explanation of history and charted a certain course for the
future. Science in their view was proof that the revolution would succeed.

This enthusiasm for the scientistic value of Marxism was consistent
with part of the Marxist tradition, but not the whole of it. Chinese of the
1920s and early 1930s were ignorant of the extensive writings by
European Marxists on problems of nature. Marx, Engels, and Lenin were
serious, if amateur, students of science and maintained that their theories
should apply with equal precision to natural and social objects alike.
Indeed, a basic tenet of their faith was its omnipotence: The claim that
Marxism could explain human history was inextricably linked to its
mastery of nature. Engels wrote extensively on the "dialectics of nature,"
and Lenin, unsettled by the philosophical implications of relativity and
quantum mechanics, engaged in long, heated debates on the relationship
between Marxist theory and modern physics. These controversies
continued during the 1920s, pitting the "mechanism" of Nikolai
Bukharin against the "idealism" of Abram Deborin. The nature of
matter, the forces that move it, the method of human cognition, and the
models that describe natural phenomena – all these topics were intensely
scrutinized and tested against the laws of dialectical materialism.[1]

Chinese Marxists picked up the thread of this discussion in the
mid-1930s, when Russian texts on the subject were first translated into
Chinese. The publication in Shanghai in 1934 of the book *Where is
Philosophy Going?* [*Che-hsüeh tao ho-ch'u ch'ü*], by the Marxist
philosopher and ex-CCP member Yeh Ch'ing, sparked a lively intramural debate. In this work, Yeh set forth the view that dialectical
materialism constituted the final stage in the evolution of human
thought, ending the history of philosophy and establishing a permanent
orthodoxy that would remain unchanged for all times. He credited the
influence of natural science for raising Marxist theory to this height and
called dialectical materialism "the science of pure thought" – sufficient in
itself to explain all the observable facts of man and nature. Yeh staked

---

[1] The classical works applying Marxist theory to the study of nature are: Engels's
Anti-Duhring (1877), *Ludwig Feuerbach and the End of Classical German philosophy*
(1886), and *Dialectics of Nature*, written betweei. 1873 and 1883, but not published
until 1925; and Lenin's *Materialism and Empirio-Criticism* (1908) and the posthumous
Philosophical Notebooks (1925–29). For a fuller discussion of Engels's writings and the
subsequent debates in the Soviet Union, see: Loren R. Graham, *Science and Philosophy in
the Soviet Union*, (New York: Knopf, 1972), 3–68; and David Joravsky, *Soviet Marxism
and Natural Science, 1917–1932* (New York: Columbia University Press, 1961).

out the extreme position that Marxists had found the one, correct, final explanation for all things. Scientists and philosophers would have to adjust their thinking to this fact.[2]

Yeh Ch'ing was roundly attacked by Communist party philosophers in Shanghai, led by Ai Szu-ch'i and including among others, Ch'en Po-ta, who later served as secretary to Mao Tse-tung and editor of Mao's writings. The crux of their criticism was that by equating dialectical materialism with science, Yeh was overlooking the more important relationship between Marxist theory and human history. Ai Szu-ch'i charged that Yeh was a "Deborinist" (by this time, Deborin had fallen out of favor in Russia), meaning that he reduced philosophy to science and removed it from its social context. Philosophy, according to Ai, could not be replaced by mathematical formulas, nor was it fixed, unchanging, and dead. Rather it must be constantly updated to reflect the evolving experience of man. Ai's chief work, *Philosophy of the Masses* [*Ta-chung che-hsüeh*] (1936), offered to replace Yeh's mechanical theories with a popular version of dialectical materialism, illustrated with examples from daily life, on the whole a more open, flexible, living version of the faith. After moving to Yenan in 1937, Ai and other members of this group continued their study of Marxism, including its relationship to science, while translating Engels's works on the dialectics of nature. Among those who joined their discussions was Mao Tse-tung, who was himself developing an interpretation of Marxism in many ways similar to that of Ai Szu-ch'i.[3]

While these debates were going on in Shanghai, in Yenan Mao for the first time took up the serious study of Marxism, work that resulted in his principal theoretical writings, "On Practice," "On Contradiction," and "On Dialectial Materialism," all of which were developed during the years 1936 to 1938. Drawing on the same (translated) Russian texts that were available to scholars in Shanghai, Mao too discovered the

---

[2] This discussion of the debate between Yeh Ch'ing and Ai Szu-ch'i is based on O. Briere, "L'effort de la philosophie Marxiste en Chine," *Bulletin de l'Université l'Aurore*, serie 3, tome 8, no. 3 (1947):318–21, 330; O. Briere, "Les Courants Philosophiques en Chine depuis 50 ans (1898–1950)," *Bulletin de l'Université l'Aurore*, serie 3, tome 10. no. 40 (1949):578, 609–15; and Vsevolod Holubnychy, "Mao Tse-tung's Materialistic Dialectics," *China Quarterly* 19 (July–September 1964):9–10. See also: Frederic Wakeman, Jr., *History and Will*, 222–6.

[3] Debate of the 1930s: O. Briere, "L'Effort de la philosophie Marxiste en Chine," 327–31. Yenan period: *Shen-Kan-Ning pien-ch'ü tzu-jan pien-cheng-fa yen-chiu tsu-liao* [Research materials on the dialectics of nature in the Shen-Kan-Ning Border Region], ed. Shen-hsi sheng kao-teng yüan-hsiao tzu-jan pien-cheng-fa yen-chiu hui, Yen-an ta-hsüeh fen-hui [Shensi Provincial Higher Schools Dialectics of Nature Research Society, Yenan University Branch] (Sian: Shen-hsi jen-min ch'u-pan-she, 1984), 41–95, 298–311. This source is cited hereinafter as *PCF*. In addition to Ai Szu-ch'i, the intellectuals who supported this activity in Yenan include Mao Tun, Chou Yang, Kuo Mo-jo, Hsü T'e-li, Yü Kuang-yüan, Ch'en K'ang-pai, Ko Po-nien, Kao Shih-ch'i, and Tung Ch'un-ts'ai.

relationship between science and philosophy. In "Dialectical Materialism," he drew parallels between the history of science and class struggle,
and in "On Contradiction" he described the dialectic – or "contradiction" – as the elemental force in nature. Comments later attributed to
Mao extend this explanation to the behavior of the atom, a realm Lenin
had also claimed for the dialectic, thereby affirming the dominion of
Marxist theory over the material environment. On the whole, however,
Mao was not much interested in theoretical explanations of natural
phenomena.The main point about his philosophical writings is not what
they say about the internal models or "paradigms" of nature (or anything
else), but about the external relationship between ideas and their
surroundings, or what Mao called the "unity of theory and practice."[4]

This concept, the "unity of theory and practice," can be traced to
Marx and Engels, but Mao made it the center of his philosophy, and
herein lies the chief implication of his writings for science. Science
provided for Mao a model for understanding and dealing with the world.
Just as the scientist subjects his theories to the cold light of experimentation, so each person must test his or her ideas in the real world of
practice. Whether in material production, class struggle, or scientific
research, practice alone determines whether or not a particular idea
corresponds to the laws of the external world and thus whether or not it
will succeed.[5] This emphasis on practice separated Mao from Deborin
and Yeh Ch'ing, both of whom embraced the notion that dialectical
materialism was the final answer, sufficient to explain matter, motion,
and the myriad of products that follow. For if ideas had to be proven
through practice, as Mao contended, then theory could never rest.
"Marxism-Leninism has in no way summed up all knowledge of truth,"
he observed in his 1938 treatise, "On Practice," "but is ceaselessly
opening up, through practice, the road to the knowledge of truth."[6]

[4] In addition to the works of Russian Marxists of the 1920s, Mao read Engels's
Anti-Dühring and Ludwig Feuerbach, Marx's Theses on Feuerbach, and Lenin's
Materialism and Empirio-criticism and Philosophical Notebooks. It is uncertain whether
or not he had access to Engels's Dialectics of Nature. See: Vsevolod Holubnychy, "Mao
Tse-tung's Materialistic Dialectics," 11–12. "Dialectical Materialism": Mao Tse-tung,
"On Dialectical Materialism," trans. Stuart R. Schram, The Political Thought of Mao
Tse-tung, rev. and enl. ed. (New York: Praeger, 1969), 184. "On Contradictions": Mao
Tse-tung, "On Contradictions," Selected Works of Mao Tse-tung, 4 vols. (Peking:
Foreign Languages Press, 1967), 1:313. Behavior of atom: See, for example, Mao's
comments made in the mid-1960s on the theories of the Japanese physicist Sakata
Shiyouchi, cited in Frederic Wakeman, History and Will, 227. Note: The application of
dialectical materialism to scientific phenomena is one of the things that has distinguished
Chinese from Soviet Marxism. Even during the periods of most intense politicization,
such as the Cultural Revolution, the Chinese have made few attempts to relate Marxist
categories to scientific theory. By contrast, Soviet scientists and theoreticians have had a
perennial interest in this problem.
[5] Mao Tse-tung, "On Practice," Selected Works, 1:296–7.
[6] Mao Tse-tung, "On Practice," cited in Frederic Wakeman, History and Will, 234.

In sum, Ai Szu-ch'i and other party philosophers in Shanghai and Mao Tse-tung in Yenan reached similar conclusions about the relationship between philosophy and science. They departed from earlier Marxists by depreciating the role of dialectical materialism in explaining natural phenomena. Rather, they emphasized the importance of testing ideas in the laboratory of practice. On the surface at least, this makes Mao's thought more compatible with the spirit of modern science than, for example, those Leninists who have argued with little regard for the facts that nature must obey the dialectic. It has also had profound implications for the conduct of science in Communist China.

During the 1930s, when these ideas were worked out, science was for the Communists a distant shadow, a slogan in polemical debates and philosophical tracts, hardly a tool for changing the world. Yet all this would change, and change quickly, for in 1939 came the challenge to build modern industries in the Red base areas. In a twinkling, Yenan discovered science not as a category of thought, but as an instrument of power; now Mao's theories about science could be tested in practice.

Industrial development in the Communist base areas

What focused the attention of Chinese Communist policymakers on modern science and technology was the movement launched in the base areas in 1939, to achieve self-reliance in all aspects of production and defense. Before the war, the Communists had enjoyed access to only the simplest handicraft enterprises. After the formation of the United Front in 1937, they received limited financial assistance from the Nationalists and were allowed to import from outside the border regions items they could not produce within. With the breakdown of the United Front and the blockade of the Red areas in 1939, however, the Communists were forced to rely entirely on their own resources, manufacture a wide range of goods for military and civilian use, and develop for this purpose their own cadre of scientific and technical personnel.

The headquarters of the Chinese Communist Party during World War II was the Shensi-Kansu-Ninghsia (Shen-Kan-Ning) Border Region, whose capital, Yenan, served as the command post for a loosely organized movement scattered across northern and eastern China. Tucked into the remote and impoverished northwest, Shen-Kan-Ning covered a territory the size of Indiana, with 1.5 million people, more than 4 million sheep, and fewer resources than almost any other part of China. Before the arrival of the Communists at the end of 1935, this area was without industry, save for a few rug factories and one "modern" enterprise, the Shensi oil fields. In fact, the development of other parts of

China during the early twentieth century may have had a negative impact on this unhappy region: Its traditional handicrafts, spinning and weaving, were undermined by the importation of yarn and cloth, and mining of salt, the area's leading export, declined as salt producers on the coast captured a larger share of the market.[7] "We have heard much of the terrible bitterness of this road," wrote one traveler who passed through this area in 1920,

and certainly what with the ravages of brigands and the natural infertility of the soil, the few inhabitants were poor to the verge of starvation. Yenan seems to be the centre of the most desolate area, by far the poorest region I have traversed in China outside the actual deserts.[8]

Ironically, this impoverished niche served the Communists well. The proximity of Red forces to Japanese lines in northern China hastened the formation of the United Front that gave the Communists a brief respite from civil war and access to goods from the richer Nationalist areas. Even after the collapse of CCP-KMT cooperation, the Japanese, who held other Communist bases hostage to periodic raids, saw no reason to invade beyond the Yellow River, leaving Shen-Kan-Ning to its own devices. Communists in east China enjoyed access to a richer and more highly skilled society, but the vulnerability of these bases to Japanese attack constrained their ability to erect stable institutions or carry out long-term projects. It was west of the Yellow River, in a relatively secure stretch of time and space, that the Communist experiment with industrial and technical development achieved its earliest and greatest success.

Most factories in the Shen-Kan-Ning Border Region were created between 1939 and 1941. This was the high tide of industrialization, after the shift to self-reliance and before the effects of overspending and inflation forced Yenan to cut back expenditures in all areas. A close examination of one sector of Communist industry, chemicals, shows how this process worked. Unlike other products – textiles, for example, which relied on a mixture of public and private enterprises, handicraft and producer cooperatives – chemicals were made in capital-intensive, publicly owned factories, several of which were managed directly by the Communist Eighth Route Army. This industry should not be confused with its counterparts in Japanese-occupied and Nationalist China, where the levels of technology and capital were much higher. Still, in Shen-Kan-Ning as elsewhere, the manufacture of chemicals required a concentration of skilled technicians, elaborate equipment, and capital, all

---

[7] Peter Schran, *Guerrilla Economy: The Development of the Shensi-Kansu-Ninghsia Border Region, 1937–1945* (Albany: State University of New York Press, 1976), 16, 32, 138; Andrew Watson, *Mao Zedong and the Political Economy of the Border Region* (Cambridge, Eng.: Cambridge University Press, 1980):7.

[8] Eric Teichman, *Travels of a Consular Officer in North-West China* (Cambridge, Eng.: Cambridge University Press, 1921), 62–3, cited in ibid., 5.

Table 13.1. *Publicly owned factories in the Shen-Kan-Ning Border Region, December 1942*

Type	Factories	Workers	Capital (*yüan*)	*Yüan* per worker
Chemicals (medicine, soap, leather, pottery, petrol, etc.)	12	674	17,030,000	25.3
Textiles	7	1,427	26,900,000	18.9
Tools	9	237	3,662,792	15.5
Printing	3	379	5,200,000	13.7
Paper	12	437	4,100,000	9.4
Coal and charcoal	12	432	1,777,070	4.1
Bedding, shoes	8	405	1,001,100	2.5
Totals	63	3,991	59,670,962	15.0

Source: Mao Tse-tung, "Financial and Economic Problems," in Andrew Watson, *Mao Zedong,* 157. N.B.: Mao gives a total of sixty-two factories, but figures actually total sixty-three.

of which had to be introduced from outside the border region and drawn together under central control. Several chemical products – explosives, petroleum, drugs, paper, and ink - had a high strategic or political importance; even simple consumer products, like leather, soap and candles, required inputs that only the state could provide.

Because of their importance, government investment in the chemical industries continued unabated during the economic crisis of 1942–3. The output of chemicals continued to grow, while investment and production in other sectors, including textiles, declined. The total number of state-run factories plunged from a peak of ninety-seven in 1941 to sixty-two in the consolidation of 1942, before climbing back to eighty-two in 1943. By contrast, the number of chemical producers remained more or less constant, with a dozen paper mills and ten to twelve factories making leather, drugs, soap, matches, and petroleum products, all of which showed steady increases in productivity. A snapshot of the publicly owned factories in Shen-Kan-Ning at the end of 1942 demonstrates the importance of this sector (Table 13.1).[9]

[9] Decline of textile production in 1943: Mark Selden, *The Yenan Way in Revolutionary China* (Cambridge: Harvard University Press, 1971), 256–8. Numbers of factories: Kao Tzu-li, "Wei kung-yeh-p'in ti ch'üan-mien tzu-chi erh fen-tou" [Struggle for complete self-sufficiency in industrial products], *K'ang-Jih chan-cheng shih-ch'i chieh-fang-ch'ü k'o-hsüeh chi-shu fa-chan shih tsu-liao* [Historical materials on the development of science and technology in the liberated areas during the War of Resistance], ed. Wu

The manufacture of gunpowder and explosives

The single most important chemical industry in Communist China was for the manufacture of gunpowder and higher explosives. Arsenals of surprising sophistication were erected in at least three north China bases, using techniques and equipment introduced from outside the border regions and adapting traditional technologies from within. The success of this effort enabled the Communists to carry on the battle against better-equipped Japanese and later Nationalist forces.

The largest Communist arsenal was located at Ch'a-fang, in An-sai County, north of Yenan. An earlier and more primitive operation, which had moved with the Red Army on the Long March, was first lodged at Wu-ch'i-chen in northern Shensi, where in 1936 Edgar Snow found over one hundred workers making and repairing a variety of weapons under the direction of technicians from the Mukden and Hanyang arsenals and from factories in Shanghai, Canton, Tientsin, and Peking. In 1938, this activity came to rest in Ch'a-fang, where factories were set up to manufacture machinery, weapons, and explosives.[10]

The transformation of the Ch'a-fang Arsenal from primitive repair shop to industrial enterprise was the work of Shen Hung, a self-made industrialist and engineer who was subsequently rewarded for his efforts with the titles of "labor hero" and "father of industry" of the Shen-Kan-Ning region. Shen, born in 1905 to a poor family in Chekiang Province, left home as a teenager to work in Shanghai, where he first encountered power-driven machinery and began reading about modern science and technology. In 1931, Shen and several friends pooled their cash to open a machine shop, which grew to employ more than thirty workers. After the Japanese occupation, Shen moved his operation up the river to Hankow, then for unexplained reasons chose to go north to Yenan, rather than follow the main stream of retreat west to Chungking. He arrived in Yenan in February 1938 with seven workers, a small collection of Japanese and English books on machine building, and ten pieces of equipment, including lathes, drills, and presses. These were added to the meager facilities of the existing arsenal, and the whole operation moved to Ch'a-fang, where Shen was appointed chief engineer.[11]

Heng, 5 vols. (Peking: Chung-kuo hsüeh-shu ch'u-pan-she, 1983–5), 2:8–12. This source is cited hereinafter as *KJCC*.

10 Edgar Snow, *Red Star Over China* (New York: Grove Press, 1961), 271.

11 Shen biography: "Mo-fan kung-ch'eng-shih Shen Hung t'ung-chih" [Model engineer, comrade Shen Hung], *CFJP*, 10 May 1944, in *KJCC*, 2:57–60; Huang Hai-lin, "Tzu-li keng-sheng hsieh-tso fu-wu – i Shen-Kan-Ning Ch'a-fang ping-kung-ch'ang chi-ch'i-pu" [Self-reliance, cooperation, and service – remembering the machine department of the Shen-Kan-Ning Ch'a-fang Arsenal], *KJCC*, 2:223. Shen fared well in post-1949 China, where he rose to become vice minister of machine building and a prominent figure in the

Attached to the Ch'a-fang Arsenal was a chemical plant, located at nearby Tzu-fang-kou. The director of this plant, Ch'ien Chih-tao, represents a second stream of Communist technical expertise, more bourgeois and scholarly than the skilled workers and self-made engineers represented by Shen Hung. Ch'ien had graduated from the chemistry department of Chekiang University, where he served for one year as a teaching assistant before going on to do research at the Central Industrial Bureau in Nanking and later at another institute in Taiyuan. After reading an article in a Communist newspaper about the problems the Red Army was having in coping with poison gas, he felt moved to join the resistance against Japan. In the spring of 1938, Ch'ien arrived in Yenan, where he was appointed director of the newly established chemical branch of the Ch'a-fang Arsenal. His staff included several former Chinese university chemistry students, among them Wang Shih-chen, a woman who joined the factory in 1942 to direct the work of chemical analysis, testing, and training younger recruits.[12]

Engineers at the Ch'a-fang Arsenal consulted Japanese textbooks, scholars at the Yenan Natural Science Institute, and experienced workers to design and construct equipment used to make weapons and ammunition. Power at this plant was supplied by two truck engines, from a 1933 Soviet Zis and 1929 Chevrolet, that were fueled with charcoal gas and hooked up to nineteen pulleys driven by a single overhead shaft. In 1944, American journalist Gunther Stein visited the arsenal, where he found 330 men at work in

a dozen low buildings with the strangest assortment of machinery I ever saw in use: ancient lathes, planing, drilling, rolling and stamping machines made in China, the United States, England, and Germany which were bought second-, third-, or probably tenth-hand in Sian before the Kuomintang blockade; and simple new machines of various kinds made in the arsenal itself or in one of the Border Region's new engineering workshops.

The main plant made machines that were used in the chemical factory to manufacture sulfuric, nitric, and hydrochloric acids and glycerine, and from these and other raw materials the explosives black gunpowder, smokeless gunpowder (nitrocellulose), nitroglycerine, and TNT. The

---

People's Republic: Donald W. Klein and Anne B. Clark, *A Biographical Dictionary of Chinese Communism, 1921–1965*, 2 vols. (Cambridge: Harvard University Press, 1971), 2:756–7.

[12] Tzu-fang-kou plant: "Shen-Kan-Ning pien-ch'ü ping-kung fa-chan chien-shih" [Brief history of the Shen-Kan-Ning Border Region arsenals], *KJCC*, 1:170. Ch'ien biography: "Mo-fan kung-ch'eng-shih Ch'ien Chih-tao t'ung-chih ch'uang-li pien-ch'ü chi-pen hua-hsüeh kung-yeh" [Model engineer comrade Ch'ien Chih-tao establishes the border region's basic chemical industry], *CFJP*, 16 May 1944, in *KJCC*, 2:61–3; Huang Hai-lin, "Tzu-li keng-sheng hsieh-tso fu-wu," 223–4. Wang Shih-chen: Hua Shou-chün, "Tzu-fang-kou hua-hsüch-ch'ang" [Tzu fang-kou Chemical Factory], *KJCC*, 5:230–1.

chemical factory also made alcohol, acetone, potassium chlorate, paper, and ink, the last two for the manufacture of border region currency.[13]

Sulfuric acid was made by the lead chamber method, the somewhat dated but tried-and-true technique used in China since the beginning of the century. The machine shop built the presses needed to press block lead imported from outside the border regions into plates, an oven for the reaction between sulfur and steam, and a rotary fan to throw steam onto the sulfur. This equipment yielded a dilute (2 percent) acid, which was further concentrated by the "descending pan method." Porcelain pans were laid out in steplike fashion along the sloping chimney of a furnace cut into the hillside. Dilute acid was poured into the uppermost pan and allowed to trickle down from pan to pan. By the time it reached the bottom, most of the water had evaporated, leaving a more concentrated product. Sulfuric acid was reacted with sodium nitrate to make nitric acid.[14]

All the raw materials used in the arsenal were obtained from inside the border regions, although many had to be transported some distances and across enemy lines. Most of the sulfur was from mines – three in 1940, three dozen by 1945 – in Ho-ch'ü County in the northwest corner of Shansi. The pyrites were refined by a traditional technique in which the ore was heated, causing the sulfur to separate and, upon cooling, form solid crystals of pebble size. Nitrates were extracted by washing and heating the rich "niter-earth" scraped up by peasants from around privies, village walls, and slaughterhouses. According to one report, by 1945 the Shansi-Suiyuan Border Region produced 333 metric tons (550,000 catties) of sulfur and over 240 tons (400,000 catties) of nitrogen, all of which went to supply arsenals throughout the base areas.[15]

Sulfuric and nitric acid were used to treat glycerine for the manufacture of higher explosives. The glycerine was made by the "calcium soap"

13   Power source: Harrison Forman, *Report from Red China* (New York: Holt, 1945), 79. Quotation: Gunther Stein, *The Challenge of Red China* (New York: McGraw-Hill, 1945), 174. Products: Tung Wen-li, "Chieh-fang-ch'ü kung-ch'ang chi-shu kung-tso tien-li" [A note on the technical work of the factories in the liberated areas], *KJCC*, 1:160–1; "Shen-Kan-Ning pien-ch'ü ping-kung fa-chan," *KJCC*, 1:170; Mao Yüan-yao, "Hui-i tsai Yen-an ts'an-chia chün-shih kung-yeh ti p'ien-tuan" [Note on remembering joining military industry in Yenan], *KJCC*, 1:178; Huang Hai-lin, "Tzu-li keng-sheng hsieh-tso fu-wu," 2:227–9. Sources of information at Ch'a-fang: ibid., 229.
14   Lead chambers: ibid., 227. Sulfuric and nitric acids: Tung Wen-li, "Chieh-fang-ch'ü kung-ch'ang," 1:160.
15   Transportation of niter: Chang Te-yao, "Chin-Ch'a-Chi k'ang-Jih ken-chü-ti chün-shih kung-yeh chien-she kung-tso hui-i" [Recollections of military industry construction work in the Chin-Ch'a-Chi Anti-Japanese Base], *KJCC*, 1:192. Ho-ch'ü: *CFJP*, 3 May 1946, p. 2. Sulfur: Chang Te-yao, "Chin-Ch'a-Chi k'ang-Jih ken-chü-ti," 1:195. Niter: *CFJP*, 12 February 1943, p. 2; 3 May 1946, p. 2; 24 October 1946, p. 4. Shansi-Suiyuan production, 1945: *CFJP*, 3 May 1946, p. 2. N.B.: See Tung Wen-li, "Chieh-fang-ch'ü kung-ch'ang," 1:160, for details of acid manufacture in Chin-Ch'a-Chi.

method, in which sesame oil was heated in the presence of lime to remove the fat, leaving glycerine in solution. The glycerine was then mixed with sulfuric and nitric acids and heated in a small porcelain container to produce nitroglycerine. The Ch'a-fang machine shop provided the necessary equipment, including a centrifugal drying machine, press, and cutter. At first, these operations were marred by explosions, gas poisoning, and other hazards. To guard workers against such dangers, reaction pans were submerged in water to prevent combustion and chimneys were erected to draw off the gases. By the end of the war, the production of nitroglycerine was a going concern.[16]

A separate, similarly successful operation was established east of the Yellow River in the Shansi-Chahar-Hopeh (Chin-Ch'a-Chi) Military Region, which was exposed to the threat of Japanese attack.[17] Early in the war, Communist forces in this area had to rely on captured arms and ammunition. In the spring of 1939, following the imposition of the Kuomintang blockade, the Central Committee issued a call for greater self-sufficiency, and Chin-Ch'a-Chi responded by establishing a military industrial department to manufacture weapons and ordnance, including gunpowder, high explosives, and the chemicals used to make them. Factories were set up in western Hopeh, the area least accessible to the Japanese. At first, these plants made black gunpowder, for which components – charcoal, sulfur, and saltpeter – were available locally and could be made or refined by traditional techniques. A second explosive was made by mixing potassium chlorate (apparently scraped from the heads of matches) with captured TNT.

In 1942, the Chin-Ch'a-Chi Arsenal widened the range and increased the explosive power of its products. A military industrial research laboratory was set up with a staff of twenty technicians, most of whom had studied at or graduated from universities in Peking or other parts of Hopeh Province. This lab began by developing an ammonium nitrate explosive. At that time, the Japanese Army, hoping to win support from the peasants of the north China plain, was selling or distributing the

---

[16] Glycerine: CFJP, 17 January 1945, p. 4. Machinery: Huang Hai-lin, "Tzu-li keng-sheng hsieh-tso fu-wu," 2:227. Accidents: Tung Wen-li, "Chieh-fang-ch'ü kung-ch'ang," 1:160–1.

[17] For details on the Chin-Ch'a-Chi arsenal and lab see: ibid.; Chang Te-yao, "Chin-Ch'a-Chi k'ang-Jih ken-chü-ti," 1:182–97; Chang Te-yao, "K'ang-Jih chan-cheng shih-ch'i Chin-Ch'a-Chi ken-chü-ti ti cha-yao sheng-ch'an" [Production of explosives in the Shansi-Chahar-Hopeh Base Area during the War of Resistance], KJCC, 4:296–304; Chang Chen, "Chin-Ch'a-Chi ken-chü-ti kung-yeh chi-shu yen-chiu-shih ti kai-mao chi ch'eng-chiu" [Description and accomplishments of the Industrial Technology Research Laboratory of the Shansi-Chahar-Hopeh Base Area], KJCC, 5:252–62; Wu Yüan-ku, "Chin-Ch'a-Chi k'ang-Jih ken-chü-ti ti ti-lei yü ti-lei-chan" [Mines and mine warfare of the Shansi-Chahar-Hopeh Anti-Japanese Base Area], KJCC, 5:273–85.

chemical fertilizer ammonium sulfate. Ammonium sulfate, when reacted with a solution of niter-containing soil (potassium or sodium nitrate), yielded ammonium nitrate, which could be combined with nitrated cotton to form an "ammonium nitrate mixed explosive." Soon, the Japanese discovered what was being done with the fertilizer and cut off supply. The Chin-Ch'a-Chi Arsenal then began making its own ammonium sulfate by reacting ammonium gas, extrated from the dry distillation of animal bones, with sulfuric acid. The resultant explosive, which was much more powerful than black gunpowder, remained one of the most important weapons of Communist forces during the War of Resistance and the civil war that followed.

Finally, in 1943 the Chin-Ch'a-Chi lab succeeded in making nitroglycerine and other high explosives. The glycerine was produced by the calcium soap method, similar to that employed in Shen-Kan-Ning. The fabrication of sulfuric acid took somewhat greater imagination. Lacking the means to press lead plates required to make chambers, technicians in Chin-Ch'a-Chi developed a batch technique, using large acid-resistant water cisterns, a type of local pottery reminiscent of the carboys used in eighteenth-century Europe before the invention of the lead chamber. The dilute acid was concentrated by a laborious method that required ladling the acid from the cistern. Nitric acid was made by heating sulfuric acid and saltpeter in one cristern and trapping and condensing the steam in another.

Further east, Communist technicians in the Chiao-tung liberated area of central Shantung also succeeded in making higher explosives. Early in the war, Communist arsenals in Shantung made black gunpowder, using sulfur imported from neighboring Shansi. When this supply was cut off by the Japanese blockade, the Chiao-tung region opened its own pyrite mines and extracted sulfur by use of a wood-burning furnace made according to instructions in the *T'ien-kung k'ai-wu*, the seventeenth-century guide to industrial technology. Meanwhile, resistance forces in Shantung launched their own research unit, which later became known as the Chiao-tung Industrial Research Laboratory. In 1941, five men sharing two chemistry books set up a primitive operation that began to manufacture ink, soap, and dry cells. Two years later, when asked by the Chiao-tung military supply department to produce explosives, these amateur chemists consulted their texts and developed methods, similar to those already described, for making sulfuric and nitric acids, glycerine and nitroglycerine. The manufacture of these explosives requires solvents – alcohol, ethyl ether, and acetone. Alcohol was made in a distillation tower erected by a student who came to Shantung from the Southwest Associated University in Kunming, and the ether by reacting alcohol with sulfuric acid. But highly nitrated guncotton of the type used to make smokeless gunpowder demands a more powerful solvent, so technicians from this arsenal explored the nearby K'un-yü Mountains to cut trees from which they extracted wood vinegar and from the vinegar, acetone. By the end of the war, there were nine arsenals in the Chiao-tung region, employing more than ten thousand workers and producing over thirteen tons (20 thousand catties) of explosives per month. The Chiao-tung Industrial Research Laboratory survived into the late 1940s, when it reportedly employed a staff of 320 workers, including 48 researchers – 17 Chinese, 28 Japanese, 2 Italians, and 1 German – educated at the university level or above, and maintained a library of forty thousand volumes, most in foreign languages. Among the more prominent alumni of this lab were one minister of chemical industry, one minister of coal, and several other high officials of the People's Republic.[18]

The arsenal of the east China Huai-nan base was less successful than those in the north. Most ammunition used in this area was made by repacking explosives extracted from captured or unexploded enemy shells. The Huai-nan arsenal made some black gunpowder by mixing local saltpeter, red arsenic (arsenic disulfide) obtained from pharmacies, and soot scraped from cooking pots. Members of the Communist New Fourth Army sometimes captured or purchased "yellow gunpowder" (probably ammonium nitrate), but the cost was high and the supply uncertain. On occasion, scavengers were sent to Shanghai to buy nitric acid or old cinema film, which contained nitrocellulose that could be used as a substitute for TNT. But all efforts to produce higher and more reliable explosives failed. After the end of the war, when the Nationalists reoccupied the Lower Yangtze, the New Fourth Army was driven northward, where it linked up with the Communist forces in Shantung, who had established an extensive network of arsenals. Only then, in 1947, did the New Fourth Army begin to develop the capacity to make higher explosives.[19]

Other chemical industries

In addition to the arsenals, the Communists established factories that made a wide range of products for both military and civilian use. Among

[18] Wang Hsü-chiu, "Kang-Jih chan-cheng ho chieh-fang chan-cheng shih-ch'i Chiao-tung kung-yeh yen-chiu-shih ti k'o-yen kung-tso" [The scientific research work of the Chiao-tung Industrial Research Laboratory during the War of Resistance and the War of Liberation], *KJCC*, 4:268–80. The minister of chemical industry was Ch'in Yu-chai, alias Ch'in Chung-ta; minister of coal, Hsü Chin-ch'iang. See: ibid., 280.

[19] Wang Hsin-min, "Hua-tung chieh-fang-ch'ü ti chün-shih kung-yeh" [Military industry in the East China Liberated Area], *KJCC*, 5:286–307. Wu Yün-to, a technician and "labor hero" who worked in the arsenals of the New Fourth Army, provides insight into

the chemical industries, none was more important than the manufacture of soap. Before the Communists arrived in northern Shensi, the cleansing agent used by peasants in this region was natural niter, extracted from the salt lakes of Inner Mongolia, which contained about 50 percent washing soda (sodium carbonate). After the fall of Wuhan in October 1938, when the CCP called upon the border regions to meet a greater share of their own needs, one group of students who had come to Yenan with previous training in chemistry decided to try their hands at making soap. In this group were Hua Shou-chün and Wang Shih-chen, both of whom later made important contributions to the chemical industries of the base areas. They experimented with methods of making caustic soda from limewater and local niter and heating it in the presence of various animal and vegetable fats, finally settling on castor oil. When their class left for the front in the spring of 1939, they took with them their little factory for making "Resistance Soap."[20]

The New China Chemical Factory, established in 1939 at Ch'iao-erh-kou, a few miles east of Yenan, made soap on a much larger scale. At first, New China causticized natural soda with lime and mixed the caustic with sheep or beef fat. Despite the factory's meager facilities – its laboratory had no balance, test tubes, glass bottles, or other special devices – the chief researcher, Tsinghua University graduate Tung Wen-li, and his three assistants, all middle school graduates, introduced several improvements. Within two years, New China had greatly increased the output of soap and developed such new products as chalk, toothpaste, sodium bicarbonate, saltcake, alcohol, ink, candles, and Turkey red oil.[21]

By early 1942, as the demand for chemical products outpaced production, Yenan made additional allocations to this sector, even while cutting back in other areas. Singled out for special treatment, New China responded. Its output of soap more than doubled, from 200 thousand bars in 1941 to 480 thousand in 1943. More to the point, productivity rose sharply: According to one report, after May 1943 the number of chemical workers declined by 27 percent, whereas chemical products increased by 55 percent. Soda output jumped from eleven to twenty-two

the difficulties faced by Communist industry in central China. See: Wu Yün-to, *Son of the Working Class: An Autobiography of Wu Yün-to* (Peking: Foreign Languages Press, 1956), 106, 134, 138, 160–2.

[20] Hua Shou-chün, "K'ang-Jih fei-tsao" [Resist Japan soap], *KJCC*, 4:323–6.

[21] Research facilities and staff: Tung Wen-li, "Chieh-fang-ch'ü kung-ch'ang," 1:159; Chiang Hsiang, "Hsin Hua hua-hsüeh-ch'ang ch'an-p'in tseng-chia, ping tsai shih-yen chih-tsao hsin ch'an-p'in" [Production increases and new research and development products of the New China Chemical Factory], *KJCC*, 2:188–90; *CFJP*, 1 June 1941, p. 2; 30 May 1942, p. 4. Other products: *CFJP*, 28 May 1941, p. 2; 8 February 1942, p. 4; Tung Wen-li, "Chieh-fang-ch'ü kung-ch'ang," 1:158; Fan Mu-han, "Shen-Kan-Ning pien-ch'ü ti kung-yeh chien-she" [Industrial development in the Shen-Kan-Ning Border Region], *KJCC*, 5:176.

Washing pulp for papermaking in the Shen-Kan-Ning Border Region. (Joseph and Dorothy Needham, Science Outpost.)

catties and soap from 113 to 270 bars per man-day of labor.[22]

Also vital to the border region economy were liquid fuels and other petroleum products obtained from wells and refineries in Yenchang County, east of Yenan. The Yenchang fields, opened by foreign companies at the beginning of the century, were operated with little success by the Chinese themselves during the 1920s and 1930s. When the Communists seized this area in 1937, production had fallen to a mere trickle.[23]

The man most responsible for revitalizing this enterprise was Ch'en Chen-hsia, a former Shanghai factory worker, labor activist, and technician with the China Merchant Navigation Company, who arrived in Yenan in 1938 and was asked to carry out a study of the oil fields.[24]

---

[22] 1942 allocations: *CFJP*, 8 February 1942, p. 4. Cutbacks in other sectors: Mark Selden, *Yenan Way*, 255–8; Andrew Watson, *Mao Zedong and the Political Economy*, 17. Soap production, in 1941: *CFJP*, 8 February 1942, p. 2. In 1943: Yung Ying Hsü, *A Survey of the Shensi-Kansu-Ninghsia Border Region*, 2 vols. (New York: Institute of Pacific Relations, 1945), 1:95. Productivity, 1943: Kao Tzu-li, "Wei kung-yeh-p'in," 2:13. Other evidence of increased productivity: *CFJP*, 8 October 1943, p. 4.

[23] Yenchang, 1936: Edgar Snow, *Red Star Over China*, 268–9.

[24] Ch'en Chen-hsia: Yung Ying Hsü, *Survey*, 1:97; "Mo-fan ch'ang-chang Ch'en Chen-hsia t'ung-chih ch'uang-chien mei-yu kung-yeh ch'eng-chi hao" [The good achievements of model factory manager comrade Ch'en Chen-hsia in establishing the kerosene industry], *CFJP*, 18 May 1944, in *KJCC*, 3:130.

When Ch'en began this project in 1939, ten to twenty wells, located at Yenchang, Yung-p'ing, and Yen-wu-kou and operated by steam-driven pumps, were producing four to five tons (six–nine thousand catties) of crude oil per month. The oil was bottled at the wellhead and brought to Yenchang by pack animals, where two simple pipe stills converted it to 22 barrels of gasoline, 150 barrels of kerosene, and 200 pounds of paraffin per month. The Yenchang refinery had a staff of around one hundred men. The common workers were from the local area and had ten to twenty years of experience; managers and technicians were veterans of the Long March or intellectuals who had come to the border region after the outbreak of the war. The low grade of Shensi crude, poor refining equipment, and shortage of technical personnel all contributed to a disappointing product: less than 11 percent gasoline, as compared to more than 25 percent from contemporary American refineries.[25]

Although the refining equipment was outdated and in bad repair, the chief limiting factor was inadequate supplies, so the initial emphasis was on sinking new wells. In 1941, after several failures, the explorers hit a gusher at Ch'i-li-ts'un, west of Yenchang. Together with wells added in 1942, production of crude oil climbed to over five hundred tons per month, more than one hundred times the level of 1939.[26] Even after constructing a new furnace, however, output of the refinery never surpassed 120 tons per month, leaving a substantial surplus of unprocessed crude. Some of this was stored in slate-lined tanks. The rest was sold locally or exported to the Nationalist areas when possible. As a result of these developments, petroleum production remained high during 1943 and 1944, before declining again in 1945.[27] Most of the gasoline fueled Yenan's small fleet of trucks, the paraffin went to make candles, and kerosene and other byproducts were exported to Chungking.[28] American journalist Harrison Forman described the threadbare Yenchang fields as he found them in 1944:

[25] Yen-ch'ang and Yung-p'ing oil fields: Liu Ting, Ch'ien Wei-jen, and Fu Chiang, "Yen-ch'ang Yung-p'ing yu-k'uang tiao-ch'a pao-kao chi ch'u-pu i-chien-shu" [Investigation report and preliminary ideas on the Yenchang and Yung-p'ing oil fields], KJCC, 2:121–6; Ch'en Chen-hsia and Hu Hua-ch'in, "Yen-ch'ang shih-yu-ch'ang mu-ch'ien ti kai-liang ho chin-hou ti fa-chan chi-hua" [Present improvements and plans for future development of the Yenchang Petroleum Factory] (Report of 1939), KJCC, 3:152–6; CFJP, 12 June 1941, p. 2; 7 and 8 October 1941, p. 4. Yenchang staff: Hsü Ch'ang-yü, "Yen-ch'ang shih-yu-ch'ang hui-i p'ien-tuan" [Recollections of the Yenchang Petroleum Factory], KJCC, 5:213–14. Comparison of Yenchang and American refining: Liu Ting, et al., "Yen-ch'ang," 2:123.

[26] Ch'i-li-ts'un: Fan Mu-han, "Shen-Kan-Ning pien-ch'ü ti kung-yeh chien-she," 5:170.

[27] Crude oil and refining, 1939–44: Hsü Ch'ang-yü, "Yen-ch'ang shih-yu-ch'ang," 5:215–27.

[28] Export of kerosene: Lyman P. Van Slyke, ed., The Chinese Communist Movement: A Report of the United States War Department, July 1945 (Stanford: Stanford University Press, 1968), 163.

The few steel drilling bits left by Socony are almost at the limit of their usefulness. The worn-out pistons in the suction pump have been covered with cloth and leather to prevent leakage. For lack of steel, sheets of slate, quarried from the river bank, are used for lining the oil-storage tanks. As a matter of fact, improvisation and substitution are the rule here, also. The steam engine is the one installed by the Japanese engineer in 1906; the unit brought in by Socony is unusable for lack of spare parts. The kerosene and crude oil are transported from the oil fields in bamboo-and-lime casks loaded on mules and donkeys. The candles made from the byproducts are of a fairly high quality, much better than any obtainable in Chungking or other places in the Big Rear. The gasoline is mostly kept, for running the twenty-odd dilapidated trucks which operate in Yenan and its immediate vicinity.[29]

The Communists also experimented with alternative sources of fuel. An attempt to distill coal tar met with little success. But engineers in the Chin-Ch'a-Chi Border Region reportedly made synthetic gasoline by cracking vegetable oil and distilled ethanol as a gasoline substitute, although there is no record of how much fuel was produced or how well it worked.[30]

By the early 1940s, factories in Shen-Kan-Ning were making a wide range of other chemical products. The Chen Hua Paper Factory, established in An-sai County in 1938 by a returned student from Germany, Liu Hsien-i, at first produced only a few hundred reams per year, a small fraction of the border region's annual requirement of around eight thousand reams. This operation was transformed in 1941 when the chief technician, Hua Shou-chün, developed a method for extracting fibers from *ma-lan* grass, which grows abundantly in the Yenan region, by washing the grass in limewater, boiling, and rinsing. With this breakthrough, production increased rapidly, to over five thousand reams per year. The leading Chinese Communist newspaper, *Liberation Daily* [*Chieh-fang jih-pao*], and other familiar Yenan publications were printed on *ma-lan* paper. Unfortunately, there was no bleaching powder to whiten the paper, and production reached only 80 percent of demand by 1943. The shortage was filled by importing traditional bamboo paper for printing books and other uses.[31]

Methods for the manufacture of glass were worked out by Lin Hua,

29 Harrison Forman, *Report from Red China*, 82–3.
30 Coal tar: *CFJP*, 10 January 1945, p. 4. Chin-Ch'a-Chi: Kao Ai-t'ing, "Hui-i Chin-Ch'a-Chi pien-ch'ü kung-k'uang-chü ti ch'ou-chien ho fa-chan" [Remembering the establishment and development of the Industrial and Mining Bureau of the Chin-Ch'a-Chi Border Region], *KJCC*, 2:247–8.
31 Chen-hua: I Jüi-chen, "Ma-lan-ts'ao tsao-chih shih-yen ti ch'eng-kung chieh-chüeh-le pien-ch'ü ti chih-chang k'un-nan" [Success of the experiments in making paper from *ma-lan* grass solved the paper problem of the border regions], *KJCC*, 1:163–6; Chiang Hsiang, "Chen-Hua tsao-chih-ch'ang ts'an-kuan chi" [Notes on visit to the Chen Hua Paper Factory], *KJCC*, 2:178–81; *CFJP*, 20 May 1941, p. 2; 12 November 1941, p. 3. Production figures: Kao Tzu-li, "Wei kung-yeh-p'in," 2:12. Shortage: Yung Ying Hsü, *Survey*, 1:95.

graduate of a Chinese university chemistry department, and Tung Wen-li, a chemical engineer at the New China Chemical Factory. Relying on the advice of local workers and their own research, these two men substituted available materials – soda, quartz, sand, lime, and magnesium or graphite oxide – for the aluminum oxide and industrial soda commonly used in modern glassmaking operations and built their own ovens. The result was the border region's first glass factory, whose 1942 output reached 140 thousand syringes and 40 thousand vaccine ampules, along with some glass laboratory equipment.[32]

The story of tanning shows how technicians in the Communist areas developed alternate sources for chemicals needed to keep an established industry alive. Traditional tanners in northern Shensi treated sheepskins with a mixture of salt, sodium sulfate, and millet flour to produce a crude but serviceable leather. In early 1939, technicians dispatched to these handicraft factories introduced more modern techniques, which depended on some chemicals – blue vitriol (chrome alum), tannin, sulfuric acid, and dyes – that could be imported from Sian; they learned to make other materials – in particular, the leather softener, Turkey red oil – on their own. After the imposition of the blockade cut off the supply of blue vitriol and tannin, a leather factory engineer took workers into the mountains to gather bark and nutgall, which are rich in tannin, the substance that turns raw hides into leather. By 1943, these factories were making their own tannin and using it to produce a high-quality leather from cow and sheep skins.[33]

By 1941, a drug factory set up by the Eighth Route Army was employing a staff of over one hundred workers to make more than one hundred different types of Western drugs. Two years later, when this factory moved to Chang-erh-ts'un, research improved, more than ten new drugs entered production, and a pharmaceutical school was added to train druggists and other experts. A second drug factory, Kuang-Hua, established in 1939, made Chinese medicines.[34]

The drug industry received a boost when technicians at the New China

32 Glass: Ma Hai-p'ing, "Shen-Kan-Ning pien-ch'ü k'o-hsüeh chi-shu ho tzu-jan pien-cheng-fa yen-chiu kai-k'uang" [Situation of research on science, technology, and the dialectics of nature in the Shen-Kan-Ning Border Region], *KJCC*, 1:110; Lin Hua, "Shen-Kan-Ning pien-ch'ü po-li t'ao-ts'u ho nai-huo ts'ai-liao kung-yeh ti ch'uang-chien" [Creation of the glass, porcelain, and fire-resistant materials industry in the Shensi-Kansu-Ninghsia Border Region], *KJCC*, 4:305–22; *CFJP*, 31 July 1942, p. 4.

33 Chang Po, "Wo so chih-tao ti Yen-an Hsing-Hua p'i-ke-ch'ang" [The Yenan Hsing-Hua Leather Factory that I knew], *KJCC*, 3:301–10; Chi Tse, "Chih-ke-ch'ang tzu-chi chih-tsao tan-ning" [The leather factories themselves make tannin], *CFJP*, 15 December 1943, in *KJCC*, 4:103–4.

34 ERA Drug factory: *CFJP*, 10 June 1941, p. 2; "Pa-lu-chün chih-yao-ch'ang chih-ch'eng yao-p'in ssu-shih chung" [Forty types of drugs manufactured by the Eighth Route Army Drug Factory], *KJCC*, 2:195; Li Wei-chen, "K'ang-Jih chan-cheng shih-ch'i wo chün yao-hsüeh shih-yeh ti fa-chan" [Development of our army's pharmaceutical industry during the War of Resistance], *KJCC*, 5:318–38. List of drugs: Lyman Van Slyke,

Chemical Factory developed a method for making sodium bicarbonate, an important item in the Western medical dispensary that was difficult to obtain in the interior of China. Prior to 1941, the New China factory had converted natural block soda from the northern province of Suiyuan to caustic soda, using the lime method, to make soap. The supply of block soda dried up, however, prompting a shift to natural crystalline soda from Mongolia. Engineer Tung Wen-li's study showed that crystalline soda was not only richer in sodium carbonate, but also contained sodium bicarbonate, which was absent from the block variety. Using primitive devices, Tung was able to extract a very pure (97 percent) bicarbonate, which went to supply Yenan's hospitals and dispensaries.[35]

Even more remarkable than the medicine shops of Shen-Kan-Ning was the New China Drug Factory, created in the spring of 1944 in the Chiao-tung liberated region of eastern Shantung. This factory was an outgrowth of the work of two elementary school graduates who had joined the Eighth Route Army and, because of their education, had been placed in charge of drug procurement. They began by studying Japanese books on chemistry and later learned from a former (female) university medical student that it was possible to manufacture certain drugs by combining simpler substances in the laboratory. "If we had been able to buy the drugs," explained factory manager, Liu Jang, "we would never have thought of these methods of making them." Still, by the end of the war the facilities of this factory were quite limited and the staff was made up of peasants with no more than primary school education. Reinforcements were supplied from the Chiao-tung Industrial Research Laboratory, including four researchers, one of whom had graduated from a college of pharmacology in Manchuria and one from a research lab in Tsingtao. Thereafter, technicians at the New China Drug Factory were able to manufacture over one hundred different drugs, including caffeine, ephedrine, sulfonamide, glucose, soda, chloroform, and bromides extracted from seawater. By the fall of 1946, the factory employed 280 workers and occupied 254 buildings, covering an entire village located in northern Shantung between Yentai and Lai-yang.[36]

Matches were the most elusive prize sought by chemical manufacturers of the liberated areas. Before 1939, all of the matches used in Shensi were imported from Sian. The lack of flammable materials and machinery to

---

Chinese Communist Movement, 163; Yung Ying Hsü, *Survey*, 2:99. Kuang-Hua: ibid., 100.

[35] *CFJP*, 1 June 1941, p. 2; 30 May 1942, p. 4.

[36] New China: Li P'u, "Chieh-fang-ch'ü ti hua-hsüeh kung-ch'ang" [Chemical industry in the liberated areas], *Ch'ün-chung* 13, no. 2 (27 October 1946):26–8; "Shan-tung Tzu-po Hsin-Hua yao-ch'ang ch'ang-shih" [History of the Shantung Tzu-po New China Drug Factory], *KJCC*, 4:367–77. New China in 1945: Leng Tzu-sheng, "Hsin-Hua yao-ch'ang ti yu-nien shih-tai hui-i p'ien-tuan" [Memories of my youth in the New China Drug Factory], *KJCC*, 3:321–8.

make matches delayed production until 1943, when technicians at the Border Region Match Factory succeeded in making yellow phosphorus by treating bone ashes with sulfuric acid. The quality of the matches was good, but the quantity limited until the spring of 1944, when the introduction of a tipping holder, which held the sticks while they were dipped into chemicals, raised output to three hundred cartons per day. Still, this represented only about 15 percent of current imports, which cost the border regions over 600 million *yüan* annually. At the end of 1944, two chemical technicians, Ch'ü Po-ch'uan and Ch'eng Shu-jen, made a breakthrough in the manufacture of potassium chlorate, a better and safer combustible than phosphorus, that led to a much improved match. By the end of the war, Yenan offered training programs for match factory technicians, aimed at spreading this industry to other parts of the liberated areas.[37]

Conclusion

The year 1939 marked a turning point in the Chinese Communist approach to science. Before that date, the Communists had no program for applying knowledge of nature in a useful way. Science entered their lexicon only as an abstraction in a philosophical debate of someone else's making. After 1939, the Communists erected a wide range of industries, several of which required a deeper understanding of nature and the methods of transforming it. Thus began their serious interest in science and its potential for creating wealth and power.

The success of Yenan's modest program of industrialization depended on the same factors evident in the Nationalist areas. First was a government that played a direct role in establishing factories and intervened more forcefully in the economy as a whole. Second were technical experts from the coastal urban areas, who introduced into the border regions methods they had learned in prewar China and devised ways of building on local technologies and resources. As in the case of the Nationalists, this combination of aggressive government leadership and innovation by émigré engineers yielded progress in development hiterto unknown in these parts of China. Yet in neither case did cooperation carry over to education and research. In Yenan, as the next chapter shows, the conflict was sharper than that in Kuomintang China, the scars deeper and more lasting.

[37] Match manufacture, 1941–4: *CFJP*, 29 April 1944, p. 1; Tung Wen-li, "Chieh-fang-ch'ü kung-ch'ang," 1:159–60; Fan Mu-han, "Shen-Kan-Ning pien-ch'ü," 5:177–8. Tipping holder: Yung Ying Hsü, *Survey*, 2:98. Potassium chlorate: *CFJP*, 7 January 1945, p. 4; "Pien-ch'ü huo-ch'ai-ch'ang shih-chih lü-suan-chia ch'eng-kung" [Success in experimental production of potassium chlorate by the Border Region Match Factory], *CFJP*, 26 June 1945, in *KJCC*, 5:94. Training program: *CFJP*, 17 September 1945, p. 2.

14

Science in Communist China II: scientists versus the state

In addition to bringing modern knowledge to the remote northwest and sparking innovations in the application of new techniques, the war also triggered the intervention by Communist authorities into the study and practice of science, much as occurred in Nationalist China. In both cases, the motive of the state was to mobilize society behind the struggle for survival. In both, this meant shifting resources and attention from pure or basic learning to practical application. In both, the politicians met with resistance from the scholars, who kept their faith in a deeper, more long-term view of science. What distinguishes the two cases was the degree of enthusiasm with which each party sought to impose its will and the extent to which it succeeded. The Nationalists fought with kid gloves, allocating funds and setting priorities, but leaving the scholars who resisted them more or less alone. The Communists fought with brass knuckles, imposing a strict orthodoxy, demanding obedience, and crushing those who refused to comply. In the short run, the Communists created a more unified, purposeful movement that succeeded in seizing power. In the long run, these measures would have a devastating effect on the development of science in China.

Science and the state in Communist China

The conflict between political and technical elites in Communist-ruled areas of China derived in part from the contrasting biographies of these two groups. Both were drawn from the same social stratum, but they arrived in Yenan by different routes. The older Communist party leaders had risked everything to challenge what they saw as corrupt and ineffective authority. They took to the hills when life in the cities became dangerous, teetered on the brink of extinction during the Long March, and finally found refuge in one of the harshest corners of China. The younger, better-educated intellectuals also took risks and faced hardships in making their way to Yenan, but they arrived later, and brought with them more deeply rooted ties to urban comfort, established authority, and vested careers. The party, needing the intellectuals to perform the tasks of production and war, set up schools to train technicians and other experts. The intellectuals, viewing the CCP as an instrument for achieving their twin goals of saving China and advancing careers, flocked to these schools and the jobs offered their graduates. Yet from the beginning, there was a tension between the party leaders, who favored the practical training of technicians to perform the pressing tasks at hand, and the scholars, who considered it their duty to create a richer, deeper culture that would equip China's youth for the larger needs of the nation and its future. The conflict was similar to that between scientists and the state in Nationalist China. The difference lay in the outcome: The Communists resolved it by a decisive, even devastating exercise of force.

The influx of refugees from occupied China injected new life into the Communist movement. According to one estimate, by the end of 1940, 100 thousand immigrants, about half of them students, teachers, writers, and intellectuals, had arrived in the Shen-Kan-Ning Border Region. Although the flow slowed somewhat after the imposition of the Nationalist blockade, another 86 thousand came between 1941 and 1944.[1] Many of these young people had studied in China's best schools and universities, where they had begun to prepare for modern, urban careers. They were outraged by Japan's aggression, critical of Nanking's weak response, and drawn to Yenan as a Mecca of national resistance. For them the resistance was a cause worth fighting for, but only one stop on a longer journey. They saw no conflict among the goals of liberating China from foreign occupation, rebuilding the country after the war, and pursuing their own professional ambitions.

To attract these young people, absorb them into the border regions and prepare them for the tasks of war, the party established schools of various types. When the Communists arrived in northern China in 1935, the Shen-Kan-Ning area had 120 primary schools with an enrollment of 2,000. The literacy rate was 10 percent. There were more sorcerers than students in this region! By 1940 the number of primary schools had increased to 1,341, primary enrollment to 43,625, and new middle and normal schools had been set up to accommodate more than 1,000 elementary graduates.[2] In 1942, Yenan, which visitors described as a "student city," boasted more than twenty full-time institutions to train

---

[1] Peter Schran, Guerrilla Economy, 99.
[2] Wu Heng, "Shen-Kan-Ning pien-ch'ü ti k'o-hsüeh p'u-chi kung-tso" [Scientific popularization work in the Shen-Kan-Ning Border Region], KJCC, 5:156–8.

cadres for military, administrative, and political tasks or for more specialized work in propaganda, ideology, labor, transportation, the arts, medicine, pharmacology, and science and technology. By the end of the war, 100 thousand students reportedly passed through the largest of these schools, the Anti-Japanese University.[3]

Urban émigrés contributed valuable skills and youthful enthusiasm to the Communist cause. Many also carried liabilities that placed them in conflict with party authorities. The new arrivals came from comfortable bourgeois backgrounds that proved a poor preparation for the hardship of life in the countryside. They were drawn to Yenan by a spirit of nationalism and lacked a partisan preference for the Communist party, knowledge of Marxism, or commitment to class struggle and social revolution. They thought of themselves as modern-day "men of talent," who had a responsibility to attack hypocrisy and abuse of power, wherever they found it. Finally, they viewed the pursuit of knowledge as a human right, an end in itself, an experience to be shared by men and women of all nations, *not* a special privilege or instrument of the state. These several features pitted the urban intellectuals against Communist party authorities, in science as in other fields.

Previous studies have focused on the writers and artists in this group. Many members of the left-wing Shanghai literary scene who arrived in Yenan in the late 1930s made important contributions to Communist culture and propaganda, but were at the same time ill prepared for strict party discipline and the hard life of the hinterland. Reared in the relatively free, antiauthoritarian atmosphere of Shanghai, these men and women considered it both a duty and a privilege to describe and criticize the world around them. They viewed themselves as members of an artistic elite, held to high standards, and followed cosmopolitan styles and trends. Under their influence, literary and artistic activities in Yenan changed from the simple populism that had characterized Communist culture of the 1930s toward a quest for the best, emulation of foreign models, and adherence to professional ideals that sometimes separated the artists from the masses. Critics charged these intellectuals, not unfairly, with "raising standards behind closed doors."[4]

A similar shift occurred in science and technology, where domination by semieducated generalists and practitioners gave way to professional teachers and scholars who believed in the value of a broad, theoretical education – even if it meant postponing some rewards to a later date. Leaders of the scientific community were committed as much to science

---

[3] Jane L. Price, *Cadres, Commanders, and Commissars: The Training of the Chinese Communist Leadership, 1920–45* (Boulder: Westview Press, 1976), 137–64.
[4] Ellen R. Judd, "Prelude to the 'Yan'an Talks': Problems in Transforming a Literary Intelligentsia," *Modern China*, 11, no. 3 (July 1985):380–95.

as to China, as much to China as to communism. They thought of the present and the future, of the border regions and the nation as a whole. The growing tension between the scientists, students, and teachers, on one hand, and CCP leaders on the other was evident in the record of the Natural Science Institute, the Natural Science Research Society, and the coverage accorded to science in the Yenan press. In each case, the party's original motive, to spur production, was bent by the scholars in the direction of deeper, more basic and theoretical learning.

Yenan's scientific institutions, 1939–1941

The Natural Science Institute (NSI) was founded in Yenan in 1939, for the purpose of training technicians who could step directly into the factories that had been created to realize the goals of self-reliance. The initiative came from the CCP's Finance and Economics Department, whose deputy director, Li Fu-ch'un, served as the institute's first president. Li's speech at the "Natural Science Forum" in December 1939, explained that the purpose of the institute was to attract scientific and technical personnel from other parts of China, to use them to train more technicians, and to channel this resource into production. The whole enterprise was designed to produce immediate, practical results.[5]

From the outset, the school operated in tandem with experimental factories that served to train cadres and conduct research into new technologies. A machine shop was set up in 1940, followed in 1941 by a chemical factory that was later expanded to include plants for the manufacture of alcohol, glass, soda, and soap. The NSI worked closely with other government enterprises, such as the New China Chemical Factory, to study manufacturing techniques. Attached to the biology department was the Kuang-Hua Farm for training students in agriculture.[6]

The Natural Science Institute was first located in hilltop caves at Tu-fu-ch'uan-k'ou, near Ch'i-li-pu, outside the south gate of Yenan.

---

[5] Establishment and goals of NSI: Hu Chi-ch'üan, "Hui-i Yen-an tzu-jan k'o-hsüeh-yüan ti hsüeh-hsi sheng-huo" [Remembering the study life at the Yenan Natural Science Institute], KJCC, 1:137; Ch'ü Po-ch'uan and Wei Chih, "Yen-an tzu-jan k'o-hsüeh-yüan" [The Yenan Natural Science Institute], KJCC, 4:242–4; and Hua Shou-chün, "Yung tzu-jan pien-cheng-fa chih-tao k'o-hsüeh chi-shu ti yen-chiu" [Use the dialectics of nature to direct scientific and technical research], PCF, 305–6.

[6] Chemical factory: Hsü Ming-hsiu, "Chien-k'u ch'uang-yeh, wei chiao-hsüeh ho pien-ch'ü chien-she fu-wu' [The difficult enterprise, service to education and the construction of the border regions], KJCC, 1:126–9; Ch'ung Shih, "Tzu-jan k'o-hsüeh ti yen-chiu kung-tso" [Natural science research work], PCF, 101. Soda and soap: CFJP, 30 May 1942, p. 4. Glass: Ma Hai-p'ing, "Shen-Kan-Ning pien-ch'ü k'o-hsüeh chi-shu ho tzu-jan pien-cheng-fa yen-chiu kai-k'uang," 1:110; CFJP, 31 July 1942, p. 4. Kuang-hua: Hu Chi-ch'üan, "Hui-i Yen-an tzu-jan k'o-hsüeh-yüan," 1:140.

Chemistry student in the Shen-Kan-Ning Border Region. (Joseph and Dorothy Needham, Science Outpost.*)*

It boasted Shen-Kan-Ning's only science building, which included a chemical laboratory equipped for qualitative and quantitative analysis and other experiments. Glass instruments and reagents were obtained through the CCP office in Chungking from agents in Hong Kong. Laboratory equipment was manufactured in the institute's own machine shop. Facilities were in fact limited, consisting chiefly of what could be scavenged or improvised from bits of glass and metal. Paper, ink, and other materials were scarce and of poor quality. Students passed the few available textbooks from hand to hand. Looking back, one participant recalled the school's crude, but spirited operation:

We did not have any modern scientific apparatuses or equipment, let alone spacious and well-lit classrooms. The conditions of school life were even harder. Classes were conducted and meals taken in open air. Bricks and tree stumps were our stools, and our knees served as desks. Walls were smeared with soot and used as blackboards. When the ground was levelled, a raised portion was left to serve as platform. The students had no paper to write on, and so they did their exercises on the ground. They had no pens to write with, and so they tied nibs onto goose feathers or twigs. Both the faculty members and the students slept in caves, each of which was shared by three people in the case of faculty members and by eight, ten, or more than ten people in the case of students. They all slept in earth beds, which were so crowded that they found it difficult even to turn

Outdoor geometry class, Shen-Kan-Ning Border Region. (Joseph Needham, Chinese Science.)

about. Some students use to say jokingly, "When we want to turn in our sleep, we have to cry, 'one, two, three,' so that all turn together."[7]

The man responsible for guiding the NSI through its formative years was Hsü T'e-li, who replaced Li Fu-ch'un as president at the end of 1940 and emerged as the foremost spokesman for a professional approach to the study of science. Born near Changsha, Hunan, in 1877, Hsü was a generation older than most Communist leaders and ranked respectfully among the Party's "five elders." After a brief, unsuccessful classical education, he spent his young adulthood studying and teaching modern Western subjects, primarily math and science, a career that took him to Japan (1910), France (1919–24), and after joining the CCP, to Moscow (1929–31). As a teacher in Changsha, his students included several future party leaders, including Mao Tse-tung. In 1932, Hsü moved to the Communist base in Kiangsi, where he was named president of the Lenin Normal School, and in 1935 completed the Long March to the north. By 1940, when he was picked to head the Natural Science Institute, "Old

---

[7] Tu-fu-ch'uan-k'ou: Hu Chi-ch'üan, "Hui-i Yen-an tzu-jan k'o-hsüeh-yüan," 1:137. Laboratory: Hsü Ming-hsiu, "Chien-k'u ch'uang-yeh," 1:132; and Hu Chi-ch'üan, "Hui-i Yen-an tzu-jan k'o-hsüeh-yüan," 1:142. Quotation: Ts'ao Ch'ing-yüan, "Inherit and Exemplify the Work Style of the Days of Yenan – an Account of Interview Given by Veteran Comrade Hsü T'e-li to Students of the Peking Institute of Technology," *Chung-kuo ch'ing-nien*, no. 3 (1 February 1961), in *Survey of the China Mainland Press* (cited hereinafter as *SCMP*), no. 253, 20–1.

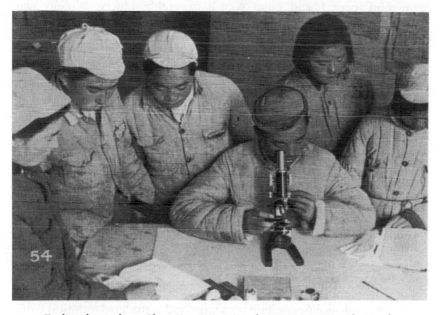

Embryology class, Shen-Kan-Ning Border Region. (Joseph Needham, Chinese Science.)

Hsü" was among the most venerable party officials. His age, his record as a teacher, and his personal association with Mao helped insulate him from attack in the bitter struggle that was to come.[8]

In addition to Hsü T'e-li, other top members of the institute's administration included vice president Ch'en K'ang-pai, and dean of studies Ch'ü Po-ch'uan. The faculty, with the exception of Hsü and one other former university professor, were young, averaging around thirty years of age, and had attended universities in China or abroad. Math was taught by Sun Hung-ju, physics by Yen P'ei-lin. The chemistry faculty, the largest and generally regarded as the best, included three returned students from Germany – Ch'en K'ang-pai, Liu Hsien-i, and Ch'ü Po-ch'uan (Ph.D., Dresden, 1938) – and others who had graduated from universities in China – Li Su, Tung Wen-li, Hua Shou-chün, and Wang Shih-chen. Several members of this group made important contributions to Yenan's chemical industries: Lin Hua (glass), Liu Hsien-i and Hua

[8] Hsü biography: Donald Klein and Anne Clark, *Biographical Dictionary*, 363–6; Ch'eng Fang-p'ing, "Hsü T'e-li t'ung-chih yü tzu-jan k'o-hsüeh chiao-yü" [Comrade Hsü T'e-li and natural science education], *CKKCSL* 3, no. 4 (December 1982):4–9. A full-length biography of Hsü is: Chou Shih-chao, *Wo-men ti shih-piao* [Our model teacher] (Peking, 1958). Hsü's role in NSI: Wu Heng and Yen P'ei-lin, "Huai-nien Hsü-lao, wo-men ti hao lao-shih" [Remembering old Hsü, our great teacher], *KJCC*, 1:90–1; Hsü Ming-hsiu, "Chien-k'u ch'uang-yeh," 1:132.

Shou-chün (paper), and Tung Wen-li (several products). Biology was headed by Lo T'ien-yü. Mining, the weakest department, had only two instructors, Chang Ch'ao-chün and Wu Heng.[9]

Students were recruited by examination from schools in the border region and from among young intellectuals who had come to Yenan from other parts of China and were assigned to the NSI by the party's organization department. Enrollment in the first class of the university section began in 1940 with twenty to thirty students, all middle school graduates, and peaked at eighty in 1942. There was also a preparatory course for around 150 students who had previously completed primary school. Hsü T'e-li remembered the students in the university section as "advanced intellectual youths who had come from different parts of the country," many from "rather well-off families." Although precise data on the background of these students is lacking, they were probably similar to those at the party's Central Research Institute, where a substantial majority were men and women in their twenties from educated urban families, who had never held a job or joined a political party before coming to Yenan.[10]

Like their colleagues in other parts of China, most NSI students wanted to study politics or military affairs and considered technical work a dead end.[11] Several directives and editorials from this period criticized the prejudice against technical work, which was widespread among Communist cadres.[12] Overcoming it was one of the chief reasons for establishing NSI in the first place. Personal testimonies to this bias were

[9] Administration: Ch'ung Shih, "Tzu-jan k'o-hsüeh," 100; Hua Shou-chün, "Yung tzu-jan pien-cheng-fa," 306. Faculty: Wu Heng and Yen P'ei-lin, "Huai-nien Hsü-lao," 1:90–1; Yü Kuang-yüan, "K'o-hsüeh ti kuang-hui tsai Yen-an shan-yao" [The brilliance of science sparkles in Yenan], KJCC, 1:120–4; Hu Chi-ch'üan, "Hui-i Yen-an tzu-jan k'o-hsüeh-yüan," 1:138–41. List of NSI faculty, 1939: Ch'ü Po-ch'uan and Wei Chih, "Yen-an tzu-jan k'o-hsüeh-yüan," 4:243.

[10] NSI students, recruitment and enrollment: Hu Chi-ch'üan, "Hui-i Yen-an tzu-jan k'o-hsüeh-yüan," 1:137, 140; Ch'ung Shih, "Tzu-jan k'o-hsüeh," 101; Ch'ü Po-ch'uan and Wei Chih, "Yen-an tzu-jan k'o-hsüeh-yüan," 4:246–8. List of twenty-four students in the first class of 1940: P'eng Erh-ning, "Yen-an tzu-jan k'o-hsüeh-yüan – wo-ti mu-hsiao" [The Yenan Natural Science Institute – my alma mater], KJCC, 3:335. Hsü quotation: Ts'ao Ch'ing-yüan, SCMP 253:22. CRI students: Peter Seybolt, "Terror and Conformity: Counterespionage Campaigns, Rectification, and Mass Movements, 1942–1943," Modern China 12, no. 1 (January 1986):47–8.

[11] Cadres devalued technical work: "Lun ching-chi yü chi-shu kung-tso" [On economic and technical work], CFJP, 2 June 1941, in KJCC, 1:50–3; Hu Chi-ch'üan, "Hui-i Yen-an tzu-jan k'o-hsüeh-yüan," 1:137.

[12] On May 1, 1941, the Central Committee issued a "Decision" designed to correct the views of cadres who undervalued technical work: "Chung-kung chung-yang kuan-yü tang-yüan ts'an-chia ching-chi ho chi-shu kung-tso ti chüeh-ting" [Decision by the CCP Central Committee on party members joining economic and technical work], KJCC, 1:32–3. For editorials on this "Decision" in CFJP, 2 June 1941, and K'ang-chan jih-pao [War of Resistance daily] published in the Chin-Ch'a-Chi Border Region, and Hsin-Hua jih-pao [New China daily] in Chungking, see: KJCC, 1:50–3; and KJCC, 4:44–7.

given by four pharmacy students, who admitted, in self-study sessions conducted during the Rectification Movement in late 1942, their aversion to technical work:

In the past, I thought that those who did medical work were all politically backward. After I was sent to study pharmacy at the medical college, every time I came to Yenan to attend a meeting, people would ask me where I was [studying], and I was unwilling to say that I was studying at the medical college.

Because the conditions for studying technical subjects in Yenan are lacking, I was afraid that after graduation I would not be as good as technical cadres from the outside. And because of this, I was not secure in my studies.

When I was sent to study pharmacy, I agreed to do so, because studying natural science would raise my cultural level, and when the situation changed, I could change fields.

My objective in coming to Yenan to study was not to prepare for work in Yenan, but because I intended to go on to some glorious job in the rear areas or in the base areas behind Japanese lines. Studying technical subjects offered no future.[13]

Under Hsü's direction, the organization and curriculum of NSI followed the pattern of higher education common in prewar China. Classes, which began in the fall of 1940, were divided into three levels: the university proper for upper middle school graduates, a preparatory or middle school course for lower middle school graduates, and an extension course [pu-hsi-pan] for primary school graduates. The university was divided into four departments: physics, chemistry, mining, and biology. Because the previous scientific training of most students was minimal, the first semester, in the fall of 1940, was devoted to a review of basic math and science, including algebra, geometry, trigonometry, and introductory chemistry. When Hsü T'e-li took over at the end of 1940, the length of the course was increased – the middle school from three and one-half to five years, the university from two to three years – and the curriculum reformed to increase the teaching of basic science in the middle school and first year of college, postponing practical application to the last two years. In the spring of 1941, for example, the curriculum of the physics department included calculus, thermodynamics, electromagnetics, and optics. Political studies occupied only one day each week, regular subjects the other five. Most courses relied on standard university-level English-language texts, supplemented by lectures, which were generally Chinese interpretations of the same. Students whose English was lacking had to rely on lecture notes and, in general, found the experience difficult.[14]

The Natural Science Research Society (NSRS) was established at the

---

[13] CFJP, 24 December 1942, p. 2.
[14] Curriculum: Wu Heng and Yen P'ei-lin, "Huai-nien Hsü-lao," 1:90–1; Hsü Ming-hsiu, "Chien-k'u ch'uang-yeh," 1:128; Hu Chi-ch'üan, "Hui-i Yen-an tzu-jan k'o-hsüeh-yüan," 1:137–9. Hsü's reforms: CFJP, 24 and 25 September 1941, p. 3.

same time as the Natural Science Institute, for the same reasons and with similar results. Again, the initiative came in late 1939 from economic planner Li Fu-ch'un for the purpose of promoting production. The founding proclamation of February 1940 called on the society to popularize science through public education and the dissemination of written materials, promote scientific research and the exchange of information, apply dialectical materialism to the study of nature, and help scientists in the border regions unite with their colleagues in other parts of China. Perhaps the most important function of the NSRS was to identify from among the mass of bodies passing through Yenan individuals with special expertise and to arrange for their transfer to jobs where they could use their skills. By the end of 1941, the society had 330 members, representing the physical sciences (110), engineering (120), medicine (55), and agriculture (45).[15]

Despite the utilitarian motive behind its creation, the NSRS was led by scholars and educators – Wu Heng, Yü Kuang-yüan, Ch'ü Po-ch'uan, Hsü T'e-li, and Wu Yü-chang, another of the party's "five elders," who served as the society's first president – who defended the independence of the scientific community against strict political control. The society's first annual conference in August 1941 heard reports on research in progress, called for the publication of a scientific journal and the establishment of formal training programs and specialized societies in each of the major disciplines, and raised the qualifications for membership to include only those who had received higher scientific education, worked in some field of science or technology, or made a special contribution to the study of science. In all these respects, the Natural Science Research Society behaved like a Western-style professional association, whose purpose was more to advance the interests of its members than to place technical knowledge at the service of the state.[16]

One of the more visible activities of this society was to edit a "Science Column" [k'o-hsüeh yüan-ti] that appeared monthly in the CCP newspaper, Liberation Daily, from October 1941 to March 1943. About half the articles in this column described useful applications of scientific knowledge of the sort that might help production in the Communist base areas: methods of treating alkaline soils, manufacturing glass and soap,

[15] Founding of NSRS: Ma Hai-p'ing, "Shen-Kan-Ning pien-ch'ü," 1:111; Ho Ch'un-po, "Wu Yü-chang t'ung-chih yü tzu-jan k'o-hsüeh yen-chiu-hui" [Comrade Wu Yü-chang and the Natural Science Research Society], KJCC, 2:65; Hua Shou-chün, "Yung tzu-jan pien-cheng-fa," 303; Ch'ü Po-ch'uan and Wei Chih, "Yen-an tzu-jan k'o-hsüeh-yüan," 4:243–4. For text of Mao's speech to the first meeting of NSRS, see KJCC, 1:5. Pronouncement: "Tzu-jan k'o-hsüeh yen-chiu-hui hsüan-yen" [Pronouncement of the Natural Science Research Society], PCF, 87. Membership: Yü Kuang-yüan, "Ko-hsüeh ti kuang-hui," 1:122–3.
[16] CFJP, 4 August 1941, p. 2.

controlling insect pests, identifying and using medicinal herbs, and so forth. Others introduced basic information and theories that had little or no immediate use: the origins of the sun and earth, early biological evolution, fossils and archaeology, a comparison of human and ape intelligence, and biographical sketches of such figures as Galileo and Marie Curie. Here, as elsewhere, the scientific community sought to maintain a balanced program that included both pure and applied science.[17]

The party and the scientists: the first skirmishes, 1941

At first, during 1940 and 1941, the tension between Yenan's stated commitment to applied knowledge and the actual practice of basic academic learning was accepted, or at least tolerated, by the party center. The economic crisis brought on by the blockade was serious, and CCP leaders urgently needed better results from their factories and schools. But they also realized the need to attract intellectuals and technical experts from outside the border regions and to get the most out of those already there. The compromise, which lasted until early 1942, was to press for the greater application of knowledge to practical ends, while accepting a measure of autonomy for those scholars and intellectuals whose cooperation was essential to the program of self-reliance.

Throughout this period, Chinese Communist propaganda organs painted a rosy picture of the reception artists and intellectuals could expect in Yenan. On May 1, 1941, the Politburo issued a directive, affirming its commitment to "encourage free research, respect intellectuals, promote the movement for scientific knowledge and literature, [and] welcome scientific and artistic personnel" from outside the border regions. These themes were widely disseminated in editorials that promised scientific and artistic workers greater freedom, support, and opportunities to serve the nation, not only by raising production, but by continuing to do creative work in their specialized fields:

Do we, therefore, welcome scientists and artists only because we want them to reflect the border regions, make propaganda for the border regions, and help with the reconstruction of the border regions? No, we do not mean to limit scientific and artistic activities to propaganda and application alone. We place equal emphasis, or rather we should say we place even greater emphasis, on the achievements of science and art themselves. Although we find ourselves in the

[17] The column appeared in *CFJP*, p. 4, on the following dates: 4 October, 1 November, 13 and 27 December, 1941; 19 and 26 January, 7 February, 7 March, 3 and 30 May, 14 and 31 July, 30 August, 30 September, 30 October, 30 November, 30 December, 1942; and 30 January, 2–4 March, 1943.

wartime environment..., we ought not postpone the work of raising up science and art until after the war.[18]

Whereas the political atmosphere in Yenan remained generally favorable for the intellectuals, tensions began to surface during the latter half of 1941. The first signs came in attacks on the Natural Science Institute from cadres in industry and mining who attended the first annual meeting of the Natural Science Research Society in August 1941. The spokesman for this group was Li Ch'iang, a party veteran, telecommunications expert, and leading figure in the powerful Military Industrial Bureau, which oversaw the border region arsenal, oil fields, and other key industries. Li and his supporters charged that the institute tied up too many technical personnel whose services were needed at the production front. Given the backwardness of the border regions, they argued, there was no need for education in basic science; students should skip over the theory and get on with the practice. These critics apparently scored some points in the heated debate that followed, for one former NSI student recalls that after the conference, the curriculum of the physics department, which had previously followed the academic model of universities in the Nationalist areas, began to resemble more closely a school of engineering.[19]

The men in charge of the institute did not back down. In an article published shortly after the August meeting, Hsü T'e-li defended the school's program on the grounds that basic knowledge was a prerequisite for acquiring practical skills. If technicians specialized too early without gaining a knowledge of the underlying science, Hsü argued, they would be able to perform only routinized tasks and lack the larger vision needed to create innovations that foster long-term development. Later in the year, the second session of the Shen-Kan-Ning Border Region Consultative Council heard complaints from scientific and technical personnel that the promises that had lured them to Yenan had not been fulfilled and that the fault for this situation lay with the party center. NSI dean of studies Ch'ü Po-ch'uan charged that the sectarian policy of self-reliance within each of the three "systems" [hsi-t'ung] – party, army, and government – had led to a misallocation of resources, hampered cooperation across jurisdictional lines, and prevented solution of many

[18] Directive: "Shen-Kan-Ning pien-ch'ü shih-cheng kang-ling" [Executive program of the SKN Border Region], *KJCC*, 1:69. Quotation: "She-lun: Huan-ying k'o-hsüeh i-shu jen-ts'ai" [Editorial: welcome scientific and artistic personnel], *CFJP*, 10 June 1941, p. 1. See also editorials in *CFJP*, 7 June 1941, p. 1, and 12 June 1941, p. 1.

[19] Attack on NSI: Hu Chi-ch'üan, "Hui-i Yen-an tzu-jan k'o-hsüeh-yüan," 1:138–9. Li Ch'iang biography: Donald Klein and Anne Clark, *Biographical Dictionary*, 478–9. Role in MIB: "Shen-Kan-Ning pien-ch'ü ping-kung fa-chan chien-shih," 1:168; Huang Hai-lin, "Tzu-li keng-sheng hsieh-tso fu-wu," 2:223–4. Critics: Wu Heng and Yen P'ei-lin, "Huai-nien Hsü-lao," 1:92. Heated debate: Hu Chi-ch'üan, "Hui-i Yen-an tzu-jan k'o-hsüeh-yüan," 1:138–9.

technical problems. Ch'ü's proposal – to replace the ineffective administrative systems with a unified leadership over scientific and technical matters – posed a radical, even seditious challenge to Communist party control.[20]

By the end of 1941, those who claimed a right to professional autonomy and learning for its own sake remained in charge of scientific activities – or at least the party continued to tolerate them. The eight-point plan passed by the Consultative Council in November included most of the demands put forward by the scientists: to increase government funding for books, instruments and research; to strengthen the NSI as a "professional school" and create a separate organ for scientific research; to organize scientific societies and publish books and articles; and to send technical personnel to other areas of China or abroad for advanced study and invite scientific personnel from outside to visit the Communist areas. In early 1942, the border region government began to discuss these recommendations with scientific circles in Yenan. Cooperation between scholars and officials seemed to be on track.[21]

Perhaps the strongest evidence of official toleration for the program of basic scientific education during 1941 and early 1942 is to be found in the documents from this period that were later assigned for study in the Rectification or Cheng feng Campaign. According to Cheng feng orthodoxy, these documents were to be read as an attack on the deviation of "dogmatism" – that is, the pursuit of bookish knowledge or learning for its own sake, while neglecting the concrete reality of China – just the fault Cheng feng advocates were to find in the record of the Natural Science Institute. In fact, a careful reading shows that several of the documents supported the professional, academic values of the scientists against the utilitarianism of their critics. This is true, for example, of the "Resolution of the Central Committee of the Chinese Communist Party on the Yenan Cadre School," passed on 17 December 1941, and later taken as the seminal statement for the rectification of education in schools like the Natural Science Institute. Despite its ritual warnings against dogmatism, this resolution affirms the principles of academic quality, the pursuit of knowledge separate from politics, and the fostering of individual creativity and independence from political control, all of which were championed by NSI scientists before 1942 and sharply attacked thereafter. According to this document, cadre schools were to adopt the practice of "selecting a few choice students," to take

[20] Hsü T'e-li: Hsü T'e-li, "Tsen-ma-yang fa-chan wo-men ti tzu-jan k'o-hsüeh" [How to develop our natural sciences], CFJP, 24 and 25 September 1941, p. 3. Consultative Council: Wu Chien-chih, "Wo-men ti yao-ch'iu" [Our demands], CFJP, 10 November 1941, p. 4. Ch'ü Po-ch'uan: CFJP, 10 November 1941, p. 4.
[21] Eight-point plan: Chao I-feng, "Fa-chan pien-ch'ü k'o-hsüeh shih-yeh" [Develop science in the border regions], CFJP, 29 November 1941, p. 4. Discussion: CFJP, 27 January 1942, p. 4.

care to "examine the students' marks," and to get rid of those who failed to make the grade. In order to improve the quality of instruction, teachers should be "grouped, examined, and assigned according to the new standards." Raising the cultural level of less-educated students through basic instruction in reading, writing, and "general knowledge" of history, geography, and social, political, and natural sciences was to take precedence over the study of Marxist theory. In higher schools, such as the NSI, the study of specialized subjects, like science, should occupy 80 percent of the curriculum and political studies only 20 percent, for the Central Committee had decided that "the past error of suppressing other studies on behalf of political courses should be corrected." As for pedagogy, "the methods of research, experiment, and development of the students' minds" in order to foster "initiative and creativeness" were preferred over "doctrinaire instruction, compulsion, and obscurantism." Living conditions and equipment should be improved. The administration of the schools should follow the principle, "quality instead of size," and "all tendencies toward a monopoly by party members and conversion to the party should be corrected." Even as late as the end of February 1942, the Central Committee "Resolution on the Education of Cadres in Service," offered assurances that political education required of others would be reduced for doctors, technical specialists, writers, and artists.[22]

In sum, before the middle of 1942, the dominant opinion within the Chinese Communist scientific and technical community, as represented by the leadership of the Natural Science Institute, the Natural Science Research Society, and those spokesmen for science who gained a hearing in higher policy-making circles, favored professional autonomy and a commitment to basic knowledge, reminiscent of science as practiced in China during the Nanking Decade and as learned by Chinese students in the West. Moreover, this position was supported or at least tolerated by the party center. Although differences in material conditions were enormous, the purpose and direction of scientific education in the border regions was similar to that practiced during the 1930s in Nanking, Shanghai, and Peking.

Politics takes command, 1942–1944

The shift from professionalism to practicality was imposed on scholars and educators by external political events –namely, the intensification in

[22] Text of "Resolution on Yenan Cadre School": *Mao's China: Party Reform Documents, 1942–1944*, trans. Boyd Compton (Seattle: University of Washington Press, 1966), 74–9. "Resolution on Cadres in Service": ibid., 84.

mid-1942 of the movement for party rectification, or *Cheng feng. Cheng feng* was the first great political spasm engineered by Mao Tse-tung to reshape and extend his personal control over Chinese communism. The movement proceeded through study, criticism, and self-criticism sessions, the object of which was to assess and reform the attitudes and behavior of each participant in accordance with prescribed texts, several of which were written by Mao himself. The guiding theme of these texts was the need to "unify thought and action," by relying less on abstract knowledge of the type found in books – particularly foreign books – and more on the practical experience of the Chinese people – particularly the peasantry. The impact on inexperienced urban youth and their scholarly mentors, who favored the pursuit of basic knowledge found in laboratories, classrooms, and textbooks, was profound.

During the first few months of 1942, the campaign unfolded without clear purpose, and in this comparatively relaxed atmosphere scientific activity continued essentially unchanged. Attacks on the party leadership by malcontent writers in March and April prompted Mao's venomous reply at the Yenan Forum on Literature and Art in May, and the whole tone of the movement changed. From this point on, the themes of serving and learning from masses, of acquiring knowledge from practical experience and using it for practical ends, and of subordinating all interests to party control emerged as the unquestioned orthodoxy, whereas professional autonomy and the pursuit of knowledge that had no useful, direct application became heterodox. In the tempest of *Cheng feng*, the notions that knowledge could grow by the slow accretion of bits and pieces that had no immediate utility or that some group might be allowed to pursue such activities at public expense and without public benefit were swept aside. Among the casualties were Hsü T'e-li and the Natural Science Institute.[23]

The attack came first from within the institute, in an article in the July 23 issue of *Liberation Daily* by the chairman of the biology department, Lo T'ien-yü. Lo charged that his comrades at the NSI had placed theory above practice. They had clung to a "dogmatic" reliance on foreign textbooks, rather than going "up the mountain to collect specimens or into the factory to do work." From these books, students learned meaningless abstractions like the "size of the universe," "four or five dimensional space," and the "electron" and "atomic" theories, none of which served any useful purpose. "In some courses," Lo charged, students "simply recite the classics (chant their lecture notes)..., cling to their books and sit anxiously outside the doors of the factories and farms." Lo contrasted this shabby record with the shining example of

---

[23] Mao Tse-tung, "Talks at the Yenan Forum on Literature and Art," *Selected Works of Mao Tse-tung* (Peking: Foreign Languages Press, 1967), 3:69–94.

his own department. Here students studied local plants used to make dyes and drugs, teaching was based on an examination of real specimens rather than textbooks, and research focused on the natural resources and practical needs of the border regions. To set matters right, Lo proposed a complete reorganization of the NSI, dissolving the four existing departments (biology, physics, chemistry, earth sciences) and creating two new ones: life sciences (biology and agriculture) and physical sciences (physics, chemistry, and engineering). The former, its position and resources greatly enhanced, would continue the correct program already begun under Lo's direction. The latter, with existing departments crammed into tighter quarters, would initiate reforms along the lines of the biology example.[24]

An overwhelming majority of the institute's students and staff rejected Lo's criticisms and defended the established academic priorities. K'ang Ti, a faculty member in Lo's own department, pointed out that in the study of nature – as, in the Marxist view, in the study of society – theory and application are mutually dependent: "Without the guidance of theory, it is not possible to properly grasp politics. Similarly, in the natural sciences, without a foundation in theoretical science, it is not possible to grasp the techniques of applied science." K'ang agreed with Lo that scholars in China should not waste scarce resources on pure theoretical research, but a foundation in basic science was essential for engineers, a fact recognized even by vocational and technical schools. With such a grounding, engineers can deal with many different problems, understand their basic nature, and devise creative solutions. Without it, they are only skilled technicians, limited to the rote performance of assigned tasks and unable to respond to new and unforeseen circumstances. The NSI curriculum, K'ang explained, was designed to lay a broad theoretical base on which to build toward practical knowledge and skills. There were, of course, shortcomings and errors in the implementation of this program, but these were things for which all could share the blame. The point was for each person to examine his or her performance and set it straight. There was no reason to revamp the institute along the lines suggested by Lo T'ien-yü.[25]

Another teacher in the biology department, Lin Shan, also defended the institute, albeit from a different perspective. Lin argued that given the low level of development of science and technology in the Communist areas on the one hand and of local industry and agriculture on the other,

---

[24] Lo T'ien-yü, "Tu 'Kuan-yü Yen-an kan-pu hsüeh-hsiao ti chüeh-ting'" [On reading the "Resolution on the Yenan Cadre School"], CFJP, 23 July 1942, p. 4.
[25] K'ang Ti, "Tui Lo T'ien-yü t'ung-chih 'T'an kuan-yü Yen-an kan-pu hsüeh-hsiao ti chüeh-ting' chih shang-ch'üeh" [Discussion of comrade Lo T'ien-yü's "On reading the resolution on the Yenan Cadre School"], CFJP, 25 September 1942, p. 4, in PCF, 141–6.

the two could not be brought fruitfully together at the present time. Immediate improvement of the border region economy would come through the application of available methods and local resources, not from the classroom study of science and technology. But the long-term transformation of China required a fundamental understanding of nature and of the instruments for controlling it. If scholars in Yenan take only a short-term view and neglect the basic sciences, Lin warned,

when the day comes that we need modern science and technology to rebuild the country and have no strong and complete scientific and technological organizations and personnel, then we will lack the confidence to decide the direction of scientific and technical development and the ability to organize the strength of an even broader scientific and technical establishment in our service. It is possible that when this situation arises, the revolution will suffer a great setback.[26]

During September and October, faculty and students met regularly to discuss the issues raised by *Cheng feng* and their implications for the institute. At least two of these sessions were packed by outside participants hostile to Hsü T'e-li. On one occasion, members of the faculty were forced to admit that materials used in their courses, which had been adopted from universities in other countries or other parts of China, bore little relation to the industry and agriculture of the border regions and suffered from a "thick dogmatist coloration." Another session nearly got out of control, when it was suggested that they move from an abstract discussion of educational policy to Lo's plan for reorganization.[27]

The climax came on October 26, when comrades of the NSI formally rejected Lo's charges. They denied that the institute's curriculum had neglected the application of knowledge, and they reaffirmed the teaching of basic theories and methods, even though these were derived from foreign sources and had no immediate use. In a separate session, NSI students sided with the faculty and administration, finding that Lo's criticisms of the institute were unfounded. These meetings concluded on October 30, with the judgment that the current leadership and organization should continue essentially unchanged.[28]

Hsü T'e-li's victory was a Pyrrhic one. According to later testimony by

[26] Lin Shan, "Kuan-yü fa-chan wo-men ti tzu-jan k'o hsüeh chiao-yü yü kung-tso ti wo-chien" [My views on developing our natural science education and work], *CFJP*, 25 September 1942, p. 4.

[27] "Dogmatist coloration": "K'o-hsüeh-yüan chan-k'ai t'ao-lun 'Ju-ho yü shih-chi lien-hsi?'" [Science Institute opens discussion of "How to link up with reality?"], *CFJP*, 11 October 1942, p. 2. Out of control: "Tzu-jan k'o-hsüeh-yüan chi-hsü t'ao-lun chiao-hsüeh fang-chen" [The Natural Science Institute continues discussion of education policy], *CFJP*, 23 October 1942, p. 2.

[28] "K'o-hsüeh-yüan chiao-yü fang-chen ti t'ao-lun tsung-chieh" [Summary of discussion on the educational policy of the Science Institute], *CFJP*, 9 November 1942, reprinted in *PCE*, 147–51.

Mao Tse-tung, in October the "Natural Science Institute and all the things that went with it were handed over to the government." In any case, for the scientists and educators to approve of what they themselves had been doing was beside the point, for by this time the pressure from outside the institute could no longer be denied. On November 9, members of the industrial community, led by Li Ch'iang and including several managers and engineers from leading government-owned factories – Shen Hung (machinery), Ch'ien Chih-tao (gunpowder), Liu Hsien-i (paper), and Li Ta-chang (chemicals) – renewed the attack begun by Li Ch'iang a year before. The industrialists sided with Lo T'ien-yü, charging that the institute had imported useless education from outside the border regions and placed undue emphasis on preparing students for advanced study and future needs, while neglecting the short-term requirements of production and war: "To run a university in this way, that is by aiming high and ignoring the immediate realities, can only result in falling into the trap of dogmatism." Their conclusion was that the basic science program should be gutted and the institute transformed into a vocational school for industrial technicians.[29]

The attack on Hsü's administration was not without reason. That the long-term development of science should defer to the short-term demands of war was a reasonable and ultimately winning thesis, which Lo T'ien-yü alone among the institute's faculty seems to have understood. Students in the biology department later testified that even before 1942, Lo's program stressed on-site investigations, the collection of specimens, and crop development studies conducted in conjunction with the Kuang-Hua Farm.[30] Lo's accusation that students in other departments were studying such esoteric topics as the "size of the universe" was undoubtedly exaggerated. But the evidence suggests that the institute and its students had become dangerously alienated from the masses. This finding is suggested by a special edition of the NSI *Student Association Bulletin* [*Hsüeh-sheng-hui pi-pao*], entitled "Introduction to Organizations in the Science Institute," which was published in December 1942, apparently to show critics of the institute how politically active its students were. In fact, it had the opposite effect. All but

29 Mao: from "Economic and Financial Problems," cited in Andrew Watson, *Mao Zedong and the Political Economy*, 209. Biographical details, Li Ch'iang: "Shen-Kan-Ning pien-ch'ü ping-kung fa-chan chien-shih," 1:168; Huang Hai-lin, "Tzu-li keng-sheng hsieh-tso fu-wu," 2:223. Shen Hung: "Mo-fan kung-ch'eng-shih Shen Hung t'ung-chih," 2:57. Ch'ien Chih-tao: "Mo-fan kung-ch'eng-shih Ch'ien Chih-tao t'ung-chih," 2:61. Liu Hsien-i: Chiang Hsiang, "Chen-Hua tsao-chih-ch'ang ts'an-kuan chi," 2:178. Quotation: Li Ch'iang et al., "Tzu-jan k'o-hsüeh chiao-yü yü kung-yeh chien-she" [Natural science education and industrial reconstruction], *CFJP*, 9 November 1942, reprinted in *PCF*, 132–6.
30 Lo and biology department: P'eng Erh-ning, "Yen-an tzu-jan k'o-hsüeh-yüan," 3:336–8.

one of the organizations mentioned in this report were academic, on campus, and designed to serve student interest in literary and cultural affairs, rather than provide a bridge between students and the world outside the classroom. The single exception, an "agricultural skills group" operated by the Kuang-Hua Farm, was said to be "unknown to most students" (except, presumably, those in the biology department), a statement that supports the charge that NSI students had little to do with the peasants. If this report was meant to advertise the political sensitivity and social commitment of the NSI community, then Hsü T'e-li's critics were right.[31]

At this point, Hsü might have tried to avoid the attention of higher officials who could intervene in what had been an intramural debate. That he took the opposite course undoubtedly contributed to his undoing. Hsü chose the three hundredth anniversary of the birthday of Issac Newton, on 14 January 1943, to deliver a major speech on the nature of science and its place in China. Thumbing his nose at his *Cheng feng* critics, he admitted that on the subject of Newton, "I know only what I have read in books." What he learned from these books, moreover, was that Newton's contribution to science derived from precisely those sources that were anathema to *Cheng feng*: an interest in theory, a willingness to build on the inheritance of the past, and a broad view of time and space that extended beyond his immediate condition. Hsü branded his critics "ultra-leftists," who failed to recognize the good in others, flatly rejected the "slanted view [which] supposes that only things of immediate utility are needed," and proposed instead to "transform capitalist things into proletarian things, foreign things into things useful to China."[32]

Hsü T'e-li realized this was his swan song, for by the beginning of 1943, the larger forces of *Cheng feng* made his position untenable. The final blow came from Li Fu-ch'un, the man responsible for creating the Natural Science Institute and assigning it the task of aiding production in the first place. Even after leaving the presidency of NSI in late 1940, Li had grown concerned that the institute was failing in its original purpose, which was to train technical cadres for service in industry. In an open letter of January 1943, Li charged that many scientific and technical

---

[31] Special edition of the *Student Association Bulletin* [*Hsüeh-sheng-hui pi-pao*] was summarized in Tso Yün, "Chi k'o-hsüeh-yüan ti ch'ün-chung t'uan-t'i" [Record of the mass organizations in the Science Institute], *CFJP*, 1 February 1943, p. 4. According to the recollection of one participant, in 1940 the Kuang-Hua Farm had a staff of sixty to seventy people, including technically trained youth who had studied abroad and/or come to Yenan from the Kuomintang areas. See: "Hui-i kuang-hua nung-ch'ang" [Remembering the Kuang-Hua Farm], *KJCC*, 3:229–31.

[32] Hsü T'e-li, "Tui Niu-tun ying-yu ti jen-shih" [What we should know about Newton], *CFJP*, 14 January 1943, p. 4.

experts "still remain in their schools and other organizations and have not made practical contact with the broad production and reconstruction of the border regions." He urged these technicians to mend their ways and "join the struggle for economic development." In particular, he faulted the Natural Science Research Society, the Natural Science Institute, and the "science press" for paying undue attention to theory, while neglecting current problems, such as the expansion of arable land, the curing of livestock illnesses, and the mining and transportation of coal. Li's letter marked the end of the debate over science policy and of attempts to develop a program of basic education. Confirmation came with publication in early March of the final issues of the *Liberation Daily*'s "Science Column," which contained a long, favorable article on Lo T'ien-yü's Kuang-Hua Farm – a sign that the leftists were in complete control.[33]

Hsü T'e-li reportedly fought to defend members of the institute who were subjected to interrogation and torture during the brutal "cadre investigations" that marked the final stages of *Cheng feng*, but he and his allies within the NSI had lost all influence. In April 1943, Yenan University was reorganized to absorb the Natural Science Institute and the Lu Hsün Academy. Wu Yü-chang, principal of the university and concurrently president of the Natural Science Research Society, who had shown a preference for a detached, professional approach to science, was replaced in the former position by Chou Yang, an ally of Mao and overseer of the artistic and scholarly communities during the Rectification Campaign. From this point on, the university recruited students and teachers from among natives of the border region who had been active in military and productive affairs, whereas educated urban youths were "sent down" to live with and learn from the peasants. In October 1943, dean Ch'en K'ang-pai described the biology program and its contribution to agriculture, pharmacology, and forestry as the most successful NSI activity. Under a second reorganization carried out in May 1944, the institute was taken over by Li Ch'iang, head of the Military Industrial Bureau, who further reformed the program to serve production. Visitors to the institute in late 1944 found an abbreviated three year curriculum, with six months a year devoted to practical work-study and 20 percent of the time to productive labor. Most of the institute's students, faculty, and cadres were organized into small groups and sent to factories and farms to join the production campaign. The department of agriculture under Lo T'ien-yü concentrated on a range of applied problems: the breeding of

[33] Li's influence in NSI: Ho Ch'un-po, "Wu Yü-chang t'ung-chih," 2:66. Li letter: Li Fu-ch'un, "Li Fu-ch'un t'ung-chih kei tzu-jan k'o-hsüeh-hui ti i-feng hsin" [Letter from comrade Li Fu-ch'un to the Natural Science Society], *CFJP*, 30 January 1943, p. 4. Final issues of "Science Column": *CFJP*, 2–4 March 1943.

disease-resistant crop plants, control of insect pests, growing medicinal plants, and the development of new crops for industrial applications.[34]

The roller coaster profile of science in the border regions reflected Yenan's interest in promoting and shaping this activity for its own ends. Beginning in 1939, party leaders who understood little of science recognized that they needed its secrets to carry out the self-reliance that had been thrust upon them by the breakdown of the United Front. Faced with the need to attract scholars and intellectuals to Yenan, they accepted a permissive approach to the study and practice of science. The results were mixed. Some effort was applied directly to production, some to more basic learning – a compromise agreeable to all in the comparatively relaxed atmosphere of 1940 and 1941. What made this situation unacceptable was the economic crisis of 1941 that increased the pressure for practical results, followed by the political turn of 1942 that forced all activities in the base areas to conform to a narrow Maoist orthodoxy. One casualty of this process was science, in the sense of a systematic investigation of nature whose chief reward was understanding itself, a loss that would have long-term repercussions on the modernization of China. One benefit, on the other hand, was the countervailing increase in attention paid to China's peasants – a benefit illustrated by the party's campaign against superstition and witchcraft.

The campaign against superstition, 1944–1945

Cheng feng, although closing one door on intellectuals of the modern urban stripe, opened another to the peasants of the hinterland – the men and women who Mao believed held the key to China's future. The negative side of rectification was expressed in the decimation of the Natural Science Institute and of those nascent forces that promised to provide the knowledge and skills needed to build a new China on the model of the West. But there was also a positive side, a shift of attention

[34] Reorganizations, 1943 and 1944: Jane L. Price, *Cadres, Commanders, and Commissars*, 159–60; Hu Chi-ch'üan, "Hui-i Yen-an tzu-jan k'o-hsüeh-yüan," 1:139–40; Ch'ü Po-ch'uan and Wei Chih, "Yen-an tzu-jan k'o-hsüeh-yüan," 254–5. Price, 172, cites figures showing that most students at Yenan University in 1944 were from Shensi, Shansi, and Hopeh, from small property-owner or landlord families, had completed middle school education, and were studying administrative work. Only 49 of 1,282 Yen-ta students were in the NSI. Hsü defends NSI members: Wu Heng and Yen P'ei-lin, "Huai-nien Hsü-lao," 1:95. Li Ch'iang: Hsü Ming-hsiu, "Chien-k'u ch'uang-yeh," 1:136. Chou Yang: Donald Klein and Anne Clark, *Biographical Dictionary*, 963. Ch'en K'ang-pai: Claire and William Band, *Two Years with the Chinese Communists* (New Haven: Yale University Press, 1948), 253. Visitors, 1944: Gunther Stein, *The Challenge of Red China*, 264–7. Production campaign: Ch'ü Po-ch'uan and Wei Chih, "Yen-an k'o-hsüeh-yüan," 4:254–5. Agriculture department: I. Epstein, "Scientific Research and Education in the Border Region," *China Aid Council Bulletin* (January 1945), reprinted in Joseph and Dorothy Needham, *Science Outpost*, 201.

and resources to practical, local problems whose amelioration lay in the broad dissemination of scientific understanding. One example of the latter was the campaign to combat superstition in the Shen-Kan-Ning Border Region.

Opposition to superstition had long been part of the Communist program, which considered the rational exploitation of nature essential to the progress of the revolution. In June 1941, for example, much prior publicity was given to the arrival of an eclipse, in order to establish its natural cause and refute local mythology like the belief that "the moon is eaten by a golden toad." It was only in 1944, however, with the shift in priorities dictated by *Cheng feng*, that Yenan mounted a systematic campaign to root out and reform the witch doctors or wizards [*wu-shen*] who operated throughout the Shen-Kan-Ning region and had a deleterious effect on its economy and public health.[35]

One brand of wizardry practiced in northern China was the alleged ability to identify, control, and exorcise evil spirits, particularly those that occupy people and animals and make them sick. According to fragmentary statistics, there were over 2 thousand wizards among the 1.5 million people of the Shen-Kan-Ning region, 300 in the *San-pien* subregion of northern Shensi, which had a population of about 40 thousand, 161 in Yenan County, with a population of around 33 thousand, 30 in one subdistrict [*ch'ü*] of Ting-pien County near the Great Wall, and 3 in a community of 400 people adjacent to Yenan. These wizards were said to damage public health through the sale of useless or harmful drugs and to drain capital away from productive investment by the fees they charged and the encouragement they gave to the burning of incense, candles, paper and firecrackers, the slaughter of animals, erection of altars, and consumption of grain. One wizard "dance" reportedly cost between 1,000 and 3,800 *yüan*. (By comparsion, the price of one pound of hulled millet was around 340 *yüan*.) The San-pien subregion placed annual expenses for such activities at 200 million *yüan*, Yenan County over 33 million *yüan*, and the one subdistrict in Ting-pien County 13 million *yüan*. Further west in Kansu Province a superstitious refusal to bury dead sheep led to the spread of infectious disease and a decline in the sheep population. Rather than use available and effective drugs, peasants paid over 10,000 *yüan* to charlatans who carried on their hocus-pocus, while letting the animals die.[36]

35 Goals of border region scientists: Yü Kuang-yüan, "K'o-hsüeh ti kuang-hui," 1:123; *CFJP*, 12 June 1941, p. 1. Eclipse: *CFJP*, 10 June 1941, p. 2; 1 September 1941, p. 2; 17 September 1941, p. 3; 20 September 1941, p. 3; 22 September 1941, p. 3. Golden toad: *CFJP*, 23 September 1945, p. 2.
36 Numbers of wizards, Shen-Kan-Ning: Wu Heng, "Shen-Kan-Ning pien-ch'ü ti k'o-hsüeh p'u-chi kung-tso," 5:156. San-pien: *CFJP*, 21 August 1944, p. 2. Yenan: *CFJP*, 11 August 1944, p. 2. Ting-pien subdistrict: *CFJP*, 21 August 1944, p. 2. Community adjacent Yenan: *CFJP*, 29 April 1944, p. 1. San-pien subregion was

The cost of wizardry could also be measured by the distraction of labor on the part of the wizards and those misled by them. The Yenan press reported gatherings during the agricultural labor season of 1944 of scores and even hundreds of people at village and temple fairs to hear from and about wizards. In one community the fear of ghosts, fueled by local wizards who spread rumors and fabricated evidence by spreading sheep's blood on village doorways, caused peasants to refuse to go into the fields to harvest crops, fight fires, go out at night, or in some cases to leave their houses at all. Activities of the four hundred wizards operating in the area around Yenan were said to account for the loss of three million man-days of labor each year.[37]

To combat these ills, in the sping of 1944 Yenan launched a campaign to reduce expenditures on superstitious materials, reform wizards and return them to productive labor, and replace the widespread belief in spirits with scientific understanding. During this period, *Liberation Daily* published accounts of wizards who had admitted their errors and returned to honest society. Reformed wizards were recruited to lecture at fairs and marketplaces. In one popular exhibit designed to dramatize the message of modern hygiene, a microscope was used to display a "giant fly" that was responsible for spreading disease. Educated youth in Yenan were assigned to write new *yang-ko*, folk dances with accompanying songs, celebrating the themes of this movement. Harrison Forman, an American journalist who visited Yenan in 1944, witnessed several antiwitchcraft morality plays. In one popular drama, the witch doctor, attempting to swindle a peasant family, is unmasked by a local barefoot doctor, who shows that the heroine is not possessed by evil spirits, but pregnant! In his final soliloquy, the witch doctor admits his guilt and explains his tricks to a delighted audience.[38]

Despite the success of the 1944 campaign, which reportedly brought a

composed of Ting-pien and Yen-ch'ih counties: Andrew Watson, *Mao Zedong*, 12. Population estimated by adding 10% growth to the 1930 census figures, cited in Yung Ying Hsü, *Survey*, 1:21. Damage by wizards: *CFJP*, 23 May 1943, p. 2; 5 April 1944, p. 2; 29 April 1944, p. 2. Cost of "dance": *CFJP*, 23 May 1943, p. 2.; 11 August 1944, p. 2. Millet price: Yung Ying Hsü, *Survey*, 2:6–9, states that the price of hulled millet in June 1943 was $4,488 per picul (133 pounds) and that the general price index in Yenan grew tenfold between June 1943 and August 1944. This makes millet $340 per pound in August 1944. Annual expenses for wizardry, San-pien and Ting-pien: *CFJP*, 21 August 1944, p. 2. Yenan: *CFJP*, 11 August 1944, p. 2. Kansu: *CFJP*, 12 August 1944, p. 2; 27 December 1944, p. 2.

[37] Gatherings: *CFJP*, 5 May 1944, p. 2; 28 May 1944, p. 2. One community: *CFJP*, 12 August 1944, p. 2. Four hundred wizards: *CFJP*, 23 May 1943, p. 2.

[38] Yenan movement: "She-lun: K'ai-chan fan-tui wu-shen ti tou-cheng" [Editorial: begin the struggle against wizards], *CFJP*, 29 April 1944, p. 1. Propaganda against wizardry: *CFJP*, 29 April 1944, p. 1; 5 May 1944, p. 2; 28 May 1944, p. 2; 21 July 1944, p. 2; 11 August 1944, p. 2. Yang-ko: *CFJP*, 5 April 1944, p. 2. For a text of one such song, see: *CFJP*, 15 May 1945, p. 4. Forman: Harrison Forman, *Report from Red China*, 89–93. Other accounts of *wu-shen* activities: *CFJP*, 6 May 1944, p. 2; 28 May 1944, p. 2; 18 June 1944, p. 4; 30 June 1944, p. 2; 14 July 1944, p. 4.

decline in the sale of superstitious articles and increased enthusiasm for more rational measures, the response to a drought that struck Shen-Kan-Ning the following summer showed that wizardry continued to flourish throughout the region. In the spring of 1945, as the ground grew parched and crops wilted, the wizards came forth with offers to lead in the raising of "spirit towers" and dances to bring rain. They were followed by hordes of people, including some party cadres, who stopped work, joined the activities, and paid the sorcerers whatever they demanded.[39]

According to published reports, from April to July 1945 "spirit towers" [shen-lou-tzu] were raised and "praying for rain" [ch'i-yü] ceremonies held in twenty-one of the thirty-one villages of Ho-shui County in eastern Kansu and at other sites in northern Shensi. The cost of these activities for individual villages was reported to be in the tens of thousands of yüan, hundreds of pounds of grain, and hundreds of man-days of labor. The bill for Ho-shui County alone was more than 3 million yüan, 1.5 tons (250 piculs) of grain, and 5,000 man-days of labor, which could have been used to reclaim 76 acres (500 mou) of land and produce 3 tons (500 piculs) of grain.[40]

The Communists, like their predecessors, had a difficult time coping with popular demands for supernatural devices. Studies of southern China during the Ch'ing dynasty show that ceremonies to bring rain by the burning of incense and beating of gongs were widely supported and that government officials were expected to participate in or even lead these exercises. In the present case, Communist cadres also faced pressure to join in the raising of spirit towers and praying for rain, and many did so – in some instances willingly. Those responsible for policy in this area could not agree on how to respond. Some cadres, including the editors of Liberation Daily, favored a moderate approach: working with the peasants, even joining their ceremonies, in order to guide them toward a more scientific understanding. In this view, trying to stop the activities by fiat would provoke opposition and drive the perpetrators underground; it was better to help the masses come to their own realization that these superstitions were false. A second group considered even tactical compromises with wizardry a mistake. Rather than secure the confidence of the backward peasants as a first step toward reform, compromise encouraged the belief that superstitious activity was correct and officially sanctioned, misconceptions that would only lead to more of the same. This was particularly dangerous, the hard-liners believed, because some of those who were least clear about the difference between science and superstition were the peasant cadres themselves. The best course, in their

---

[39] Success of 1944 campaign: CFJP, 5 April 1944, p. 2; 29 April 1944, p. 2.
[40] Ho-shui: CFJP, 29 July 1945, p. 2. Shensi sites: CFJP, 8 July 1945, p. 2; 23 September 1945, p. 4.

view, was to nip the perversion in the bud by setting forth the facts and outlawing all forms of unscientific nonsense. By the end of 1945, there was no clear resolution of this dispute. Local cadres were apparently left to fight superstition and promote science as best they could.[41]

Conclusion

In both Communist and Nationalist areas of China, the war brought with it sharper conflict between scientists and the state. In both areas, political leaders placed demands on the scientists to serve the immediate practical needs of production and war, whereas the scientists insisted on their professional autonomy, the right to pursue basic education or research, and the need to balance short-term and long-term goals. The Nationalists tried to control their scholars through the allocation of funds for education and research, while leaving dissenters alone. The Communists accepted diversity within the scholarly ranks for a time, but finally intervened to remove the offenders and make sure that those who replaced them toed the line.

Communist, like Nationalist, leaders placed the interests of the state ahead of autonomous scientific development, and the war gave them the chance to enforce this preference. Several factors explain the Communists' greater success. First, they were in a more desperate situation, quicker to try desperate means, and more willing to enforce them. Second, they faced a smaller, weaker intellectual community, over whom the party had unquestioned dominion. Most of China's intellectual elite, in science as in other fields, took refuge in the Nationalist areas, where they acted together to resist external control. The smattering of young, semieducated scholars in Yenan had no such might. Finally, Mao Tse-tung had a vision and the will to impose it that was unmatched by leaders in Chungking. Nationalist politics involved the shoving, hauling, and balancing of competing interests; the Communists under Mao proceeded by campaigns of mass mobilization that swept the board clean.

Mao's program of "unifying thought and practice" proved an effective strategy for surviving in an impoverished backwater and eventually

---

[41] Ch'ing dynasty: Justus Doolittle, *Social Life of the Chinese* (London: Sampson, Low, Son and Marston, 1868), 442–6; Stephan Feuchtwang, "School-Temple and City God," *Studies in Chinese Society*, ed. Arthur P. Wolf (Stanford: Stanford University Press, 1978), 124–7. Hsiao Kung-chuan, *Rural China: Imperial Control in the Nineteenth Century* (Seattle: University of Washington Press, 1960), 225, 632 n204. Debate: "Tuan-p'ing: Kuan-yü p'o-ch'u mi-hsin" [Short comment: on rooting out superstition], *CFJP*, 19 July 1945, p. 2; "Tui 'Kuan-yü p'o-ch'u mi-hsin' ti t'ao-lun" [Discussion of "On rooting out superstition"], *CFJP*, 23 Septemper 1945, p. 4.

seizing power throughout China. Foreigners who visited Yenan at the end of the war saw none of the brutalities of *Cheng feng*, and some, such as journalist Israel Epstein, found a widespread enthusiasm for the application of knowledge to practical ends:

Production, invention and the improvisation of methods and materials cut off by the blockade are a *popular movement* in the Border Region. Teachers and students in laboratories; doctors, nurses and orderlies in hospitals; administrators, engineers and workers in factories; artisans in co-operatives, invariably talk enthusiastically and excitedly of improvement introduced by themselves or their colleagues to overcome some difficulty or other, thus enabling the wheels of the Border Region to keep on turning, and foiling outside efforts to make them stop.[42]

Less visible and less relevant at the time was the damage done to those within the Communist movement who had tried to promote a deeper and more lasting study of science. Their absence and the values that had removed them from the scene would be more deeply felt after 1949.

---

[42] I. Epstein, "Scientific Research and Education in the Border Region," in Joseph and Dorothy Needham, *Science Outpost*, 204. Italics original.

15

Conclusion

The problem of balancing political authority with the autonomy of social groups, which was crucial to the development of Chinese science during its first century, has remained so in the years that followed. There is little to say about science during the Second Civil War (1945–9): Chinese scientists and engineers returned to the cities liberatd from Japanese occupation and tried without success to restore the life and work they had known there in the 1930s. More interesting is the question of what the experience described in this book meant for science and scientists, after the establishment of the PRC in 1949. A brief glimpse at the first decade of Communist rule reveals an initial effort to balance the conflicting interests of political authority and professional expertise, followed by a vengeful reassertion of the forces unleashed in *Cheng feng*.

Chemistry and the Civil War

The final stage of China's Civil War, from the defeat of Japan in 1945 to the Communist victory in 1949, witnessed little progress in the development of chemistry or any other branch of science. Peiping, Nanking, Shanghai, and the other cities that had been home to Chinese science before the war had only begun to emerge from the occupation when Nationalist mismanagement and runaway inflation destroyed any chance for a postwar recovery. Then came the Communists, a reshuffling of authority, and the need to begin again.

In the immediate aftermath of the war, enrollments in higher education increased dramatically as refugees returned to the cities along the coast, and students, young and old, prepared for new opportunities. But with the pressure of wartime mobilization relaxed, the share of students in all branches of pure and applied science declined, while many gravitated back to the humanities, social studies, and the arts (Table 15.1).

Even students who wanted to study science found the challenge daunting, as the experience of Tsinghua University shows. Classes resumed in the fall of 1946 for 2,300 students, about twice the number at Tsinghua before the war. More than half of these were in the school of engineering, which had grown rapidly during the years in exile and now

Table 15.1. *Higher education, enrollment by field, selected years, 1936–1957*

Field	1936–7 No.	%	1944–5 No.	%	1947–8 No.	%	1949–50 No.	%	1957–8 No.	%
Engineering	6,989	16.8	15,047	19.1	27,579	17.8	30,300	26.0	177,600	40.9
Natural science	5,485	13.2	6,177	7.8	10,060	6.5	7,000	6.0	27,100	6.2
Agriculture and forestry	2,590	6.2	6,042	7.7	10,179	6.6	10,400	8.9	37,200	8.6
Health	3,395	8.2	6,343	8.0	11,855	7.6	15,200	13.0	54,800	12.6
Total, science and technology	18,459	44.5	33,609	42.6	59,673	38.5	62,900	54.0	296,700	68.3
Social studies and arts	23,052	55.5	45,300	57.4	95,363	61.5	53,600	46.0	137,900	31.7
Total	41,511	100.0	78,909	100.0	155,036	100.0	116,500	100.0	434,600	100.0

Note: Social studies and arts includes political science and law, education, finance and economics, and literature and arts.
Source: Appendix, Table A.7.

included a department of chemical engineering, modeled on the program at M.I.T., the alma mater of department chairman Ts'ao Pen-hsi. The college of science numbered 400 undergraduate and 72 graduate students. Most faculty who returned to their posts came back to an empty shell. Ninety percent of the Tsinghua campus had been destroyed by Japanese and Kuomintang troops, who were billeted there after the war. The science buildings, although still standing, were stripped of all useful equipment. During the next two years, classes were held, laboratories reopened, and a measure of normalcy restored, but events outside the university overwhelmed efforts within. Civil war and hyperinflation drained funds from education, making it impossible to carry on teaching and research. Students demonstrated, government troops cracked down, and soon the Communists were in Peiping.[1]

The Nationalists faced similar problems restoring industrial production, although they placed greater emphasis on this goal and achieved better results, as indicated by the record of the chemical sector. By the time the Nationalists arrived on the scene, most former Japanese enterprises in Manchuria had been stripped or destroyed by Russian forces that occupied this area after the War, factories in Taiwan had been damaged by Allied bombing, and industrial plants in China proper were worn thin by misuse and neglect. The National Resources Commission, which was responsible for retaking control of these enterprises, seized more than two thousand units worth an estimated $1.8 billion: $1.2 billion in Manchuria, including $96 million in chemical plants and $81 million in liquid fuels; $212 million in Taiwan, with $19 million in chemicals and $7 million in liquid fuels; and the rest in China proper, for which we have no breakdown by sector.[2]

The Nationalists derived little economic benefit from the reoccupation of Manchuria. The industries of this region, which included several large plants for the manufacture of acids, sodas, chemical fertilizer, petroleum products, and other chemicals, had been built with Japanese capital, engineered by Japanese technicians, and run by Japanese managers. Chinese had participated only as low-level laborers and consumers of the simplest end products. Whatever survived the war was removed by the Russians. Edwin Pauley, who headed an American inspection team that toured this region in the summer of 1946, estimated losses caused by the Soviet occupation at $2 billion and reduction in productive capacity for most basic industries at 50 to 75 percent. In some cases, Communist forces moved in behind the departing Russians, denying the Nationalists

---

[1] *Ch'ing-hua ta-hsueh hsiao-shih kao*, 431–8, 448–53, 457–8.
[2] Inflation: Cheng Yu-kwei, *Foreign Trade and Industrial Development of China*, 156, notes that from December 1945 to August 1948, commodity prices in Shanghai increased by more than 5,000 times and the cost of foreign exchange by more than 7,000 times. NRC: ibid., 163, 266.

access to the ruins. Even if the Chinese had been able to restore this industry, the results were difficult to foresee, for the factories had been force-fed with external capital and operated on a "non-profit" basis. Most were unlikely to stand the test of the marketplace.[3]

Taiwan, the other part of China that experienced the mixed blessing of Japanese colonization, was also in shambles. The Japanese built several modern enterprises on the island, including factories for the manufacture of sulfuric acid, caustic soda, alcohol, chemical fertilizer, and refined petroleum products. Many were destroyed in whole or in part by Allied bombing. A few survived the war and formed the basis of industrial corporations, set up and run by the Nationalists after 1945.[4]

Despite the many problems, by 1949 the Chinese chemical industry had made a substantial recovery. Twelve sulfuric acid plants – ten in China proper, one each in Manchuria and Taiwan – produced 100 thousand tons annually, roughly balancing current consumption. The Yungli factory at Tangku turned out 40 thousand tons of soda ash in 1948, although demand remained high and imports of expensive foreign soda continued. The output of caustic soda returned to the level achieved in China proper before the war, around 7 thousand tons per year. Annual production of ethyl alcohol exceeded 31 million gallons, three times the peak rate of Nationalist China during the war. The output of refined petroleum products approached 20 million gallons, a 50 percent increase over the wartime totals of the Manchurian and Kansu refineries combined. The chemical fertilizer industry fared less well: Production of ammonium sulfate in 1949 was only 27 thousand tons, all made at Hsiehchiatien, versus the 1941 peak of 227 thousand tons, most of which had come from Japanese plants in Manchuria.[5]

Whatever the achievements, this was the last act on a stage that was itself collapsing. Amidst the spreading civil war, Chinese factories, schools, academies, and institutes operated sporadically, if at all. In the absence of a stable, secure environment, no new technologies were introduced, no new enterprises launched, no scientific work of any significance performed. By the middle of 1949, most parts of China were under Communist control, or soon would be, and the Nationalists were in flight. The fate of China passed to its new rulers.

[3] Edwin W. Pauley, *Report on Japanese Assets in Manchuria to the President of the United States, July 1946* (Washington, DC: GPO, 1946), 37. For details on Manchurian chemical industries, see: James Reardon-Anderson, "China's Modern Chemical Industry, 1860–1949," 219–20.

[4] Ibid., 220–1.

[5] Sulfuric acid: Shen Tseng-tso, "Liu-suan kung-yeh chi ch'i hsin ch'ü-shih," 11. Soda ash: Ch'en Chen and Yao Lo, eds., *Chung-kuo chin-tai kung-yeh-shih tsu-liao*, 1:519. Caustic soda: Yü Jen-chün, "Wu-kuo shih-yen tien-chieh kung-yeh," 4. Fertilizer: Jung-chao Liu, *China's Fertilizer Economy*, 7, 129; and Ch'en Chen and Yao Lo, eds., *Chung-kuo chin-tai kung-yeh-shih tsu-liao*, 1:519. Other figures: Cheng Yu-kwei, *Foreign Trade and Industrial Development of China*, 267.

Chemistry and the chemical industry after 1949

During their first decade in power, the Communists tried and failed to balance the conflicting forces that had shaped the development of Chinese science during the previous century. At first, Peking sought to build on the achievements of the past. Factories were reopened, rebuilt, and expanded. Scientists and engineers were urged to stay or return from abroad to serve China under new management. All of the major prewar institutions – the colleges and universities, academies and institutes, associations, societies, and journals – were taken over, reorganized, and given a place in the new order. The Communists tried to accommodate the needs of the scientific establishment, while pushing science and scientists to produce more practical results. Like China's previous rulers, they found relations with the scientists difficult but manageable. In the end, however, Mao and the party were provoked, or chose, to make science, along with everything else in China, conform to the harsh vision that had come to dominate the Communist movement in Yenan. With this, the progress of science as the work of autonomous social forces was halted, a narrow orthodoxy imposed, and the healthy development of Chinese science undone.

The advance of the chemical industry after 1949 shows how the new regime tried to carry on where the Nanking government had left off. Under China's First Five-Year Plan (1952–7), Peking expanded heavy industry with the help of Soviet economic and technical assistance. Russian engineers oversaw the completion of 130 industrial plants, including several in the chemical sector that were based on factories previously erected by the Japanese in Manchuria or by the Chinese themselves in China proper. Kirin, the site of several large enterprises, became the center of this industry. The ammonium sulfate plants at Hsiehchiatien and Dairen were expanded, doubling China's capacity to produce chemical fertilizer. A large refinery was built in Lanchow to treat crude petroleum from the Yumen fields. Other chemical factories were erected at sites throughout the country.[6]

The goal was to spur rapid growth of basic industries through the application of capital investment and technical expertise, and it worked. The portion of fixed investment committed to industry rose from 36 percent in 1952 to 52 percent in 1957, and producer goods received between 76 and 88 percent of the industrial share. Production in the modern manufacturing sector doubled in the years 1952–7, and the output by value of seventeen leading producer goods increased by an average of 165 percent. By comparison, the increase in production of

[6] Ammonium sulfate plants: Jung-chao Liu, *China's Fertilizer Economy*, 10–14, 128–30. Other details: Nai-ruenn Chen and Walter Galenson, *The Chinese Economy under Communism* (Chicago: Aldine, 1969), 114–16.

Table 15.2. *Chemical production in Manchuria and China proper,*
selected years (1,000 tons)

Product	Pre-1949 peak	1947–9	1952	1953	1954	1955	1956	1957
Soda ash	110	40	192	223	309	405	476	506
Caustic soda	20	7	79	88	115	137	156	198
Sulfuric acid	—	100	190	260	344	375	517	632
Ammonium sulfate	227	27	181	226	298	332	532	631

Sources: For period through 1949: Pauley, *Japanese Assets*, 201. For 1952–7
for ammonium sulfate: Jung-chao Liu, *China's Fertilizer Economy*, 12–13; for
other chemicals: Liu Ta-chung and Yeh Kung-chia, *The Economy of the Chinese
Mainland* (Princeton: Princeton University Press, 1965), 454.

caustic soda (125 percent) was below, soda ash (164 percent) equal to,
and sulfuric acid (233 percent) and ammonium sulfate (249 percent)
above the average. Table 15.2 shows the annual output of each of these
chemicals at their pre-1949 peak, for the years just before 1949, and
during the First Five-Year Plan.[7]

Peking followed a similar strategy toward scientific and technical personnel, recruiting the leaders of prewar China and promoting them to
positions of responsibility in the People's Republic. After taking power,
the Communists encouraged Chinese scientists and engineers to remain
in China or return from abroad. On the whole, the response was positive:
By the end of 1952, over fifteen hundred students, most of whom had
been studying in Europe or the United States, came home. In chemistry,
only two notable figures, Wu Hsien and Sah Pen-t'ieh, took refuge in the
West, while the rest of those who led the development of this field in
China during the Republican period agreed to serve the new regime. As
the data in Table 15.3 show, these men were well rewarded with high
positions in teaching, research, and administration.[8]

While absorbing the pre-1949 scientific establishment, Peking shifted
the emphasis of Chinese science toward more practical applications. One

[7] Ibid., 50, 56, 62.
[8] Fifteen hundred students: John M. H. Lindbeck, "Organization and Development of
Science," *Scientists in Communist China*, ed. Sidney H. Gould (Washington, DC:
American Association for the Advancement of Science, 1961), 18. The following
discussion of PRC science policy is based on this work and on Theodore Hsi-en Chen,
"Science, Scientists, and Politics," *Scientists in Communist China*, 59–80; Leo A.
Orleans, *Professional Manpower and Education in Communist China* (Washington DC:
GPO, 1961), 101–23; and Richard Suttmeier, *Research and Revolution: Science Policy
and Societal Change in China* (Lexington, MA: Lexington Books, 1974), 29–78.

Table 15.3. *Leading chemists of pre-1949 China in the People's Republic of China*

Name	Position in the PRC
Chang Ta-yü (1910–)	Researcher, Chinese Academy of Sciences
Chang Tsing-lien (1908–)	Department chairman, Peking University
Chang Tzu-kao (1886–1976)	Professor, Tsinghua University
Chi Yuoh-fong (1899–1982)	Researcher, Chinese Academy of Sciences
Chou Tsan-quo (1885–1966)	Researcher, Chinese Academy of Sciences
Chuang Chang-kong (1894–1962)	Director, Institute of Organic Chemistry, CAS
Fu Ying (1902–79)	Professor, Peking and Tsinghua Universities
Hou Te-pang (1890–1974)	Vice minister, Ministry of Chemical Industry
Huang Tzu-ch'ing (1900–82)	Professor, Peking and Tsinghua Universities
Jen Hung-chün (1886–1961)	Standing Committee, All-China Federation of Natural Science Societies
Kao Tsi-yu (1902–)	Professor and vice president, Nanking University
Sun Ch'eng-o (1911–)	Department chairman, Peking University
Tai An-pang (1901–)	Professor and dean, Nanking University
Ts'ai Liu-sheng (1902–)	Professor, Kirin University
Tseng Chao-lun (1899–1967)	Department chairman, Peking University Director, Institute of Chemistry, CAS Vice minister, Ministry of Education
Wang Chin (1888–1966)	Professor, Hangchow University
Wu Hsüeh-chou (1902–)	Director, Institute of Applied Chemistry, CAS
Yang Shih-hsien (1896–)	President, Nankai University
Yüan Han-ch'ing (1905–)	Researcher, Chinese Science and Technology Information Institute

Sources: "Chung-kuo k'o-hsüeh-yüan hua-hsüeh-pu hsüeh-pu wei-yüan chien-chieh," 552–64; and *Chung-kuo k'o-hsüeh-chia ts'u-tien* [Dictionary of Chinese scientists], 4 vols., (Shan-tung k'o-hsüeh chi-shu ch'u-pan-she, 1982). See also individual biographies of these men in *HHTP*, 1979–83.

measure of their success is the redistribution of enrollments in higher education. Despite earlier efforts to promote the study of applied science, the Nationalists had succeeded in raising enrollments in engineering to over 20 percent and in all fields of science and technology to over 50 percent only briefly, during the early years of the war. By contrast, in the first year of the People's Republic, 1949–50, a majority of all students in higher education were enrolled in pure and applied sciences, and by 1957–8 more than 40 percent were in engineering and more than 60 percent in engineering, agriculture, and medicine combined. After 1949, the percentage of students in the *pure* sciences dropped to below the lowest level recorded under the Nationalists, and enrollments in social studies and humanities, other than education, were reduced to a tiny fraction of the pre-1949 standard (Table 15.1 and Appendix, Table A.7).

The allocation of resources between pure and applied science was one issue that divided Communist authorities and the scientific community, just as it had divided scientists and the state in the pre-1949 era. There were other problems as well. In the "thought reform" movement of 1951–2, scientists were attacked for their bourgeois class backgrounds and pro-American attitudes, for failing to appreciate the aid and example of the Soviet Union, for adopting a "purely technical viewpoint" that ignored the social and political implications of science, and for pursuing their own research instead of following the dictates of the party. On their side, many scientists were equally unhappy with the new order, which placed them under control of uneducated Communist cadres and foreign advisers, burdened them with political and administrative duties, provided too few facilities, and left them with little freedom to pursue projects outside the narrow confines of the state plan.

Despite these tensions, some of which were unavoidable, Communist authorities tried to correct the abuses and improve the lot of the scientists, whose contributions were valued by the new regime. One indication of this attitude was the dismissal in 1952 of Lo T'ien-yü, director of the Institute of Genetics of the Chinese Academy of Sciences. As head of the biology department of the Natural Science Institute in Yenan, Lo had criticized the academic bias of Hsü T'e-li and defended the pragmatic and populist approach of *Cheng feng*. Lo's stock rose accordingly, and he emerged in the early 1950s as a prominent spokesman for the PRC on scientific affairs. His problems began when he made increasingly bitter attacks on scientists who questioned the theories of Soviet geneticist T.D. Lysenko, which Lo embraced. In some instances, these broadsides against otherwise repectable Chinese scholars made it difficult for Peking to attract scientists back to China and keep the confidence of those who had returned. In the end, Lo was dismissed, *not* for his Lysenkoist views, which retained their currency in China until the mid 1950s, but because he made life unbearable for so many of his colleagues. The treatment of scientific and technical personnel improved under the 1954 reforms of the Chinese Academy of Sciences, the twelve-year plan for scientific development adopted in 1956, and the ideologically relaxed atmosphere of the mid-1950s. The bond between the scientists and the state was uneasy, as it had been since the 1930s, but unbroken.[9]

All this changed in the summer of 1957, when in reaction to criticisms made in the Hundreds Flowers Campaign, the system snapped back, and the spirit of *Cheng feng* returned to haunt Chinese science. Tseng Chao-lun, one of the few second-generation scientists whose career under both the Nationalists and Communists combined a commitment to scholarship with the willingness to speak out on matters of public policy, played a prominent role in this affair. In June 1957, when elements of the nonparty elite were urged to "let a hundred flowers bloom," Tseng, representing the Science Planning Group of the Democratic League, published an article attacking the CCP for abusing the scientists in several of the ways already described. In the reaction that followed, he was branded a "Rightist," dismissed from his job, and eventually forced to leave Peking.[10]

From this time on, Chinese science was increasingly twisted to fit a mold reminiscent of *Cheng feng*. Rigid orthodoxy, backed by central political authority, crushed academic and intellectual freedom. Theoretical education and pure research — "science for its own sake" — yielded to simple utilitarianism. The sense of participation in a global quest for knowledge gave way to a preoccupation with China. Populism, the call to learn from and "serve the people," particularly the peasantry, replaced book learning and devotion to expertise. Political pressure was relaxed for a time in the early 1960s, only to return with a vengeance in the Cultural Revolution (1966–76). Among the chemists mentioned in this story, Tseng Chao-lun, Chang Tzu-kao, Huang Ming-lung, Ting Hsü-hsien, Wang Chin, and Yang Shih-hsien are just a few of the senior scientists whom Communist sources admit were tortured or subjected to other forms of abuse during this period. Some, including Tseng, Huang, Ting, and Wang, are said to have died as a result of injuries suffered in the Cultural Revolution.[11]

Why have the Chinese been unable to find and keep the balance between authority and freedom that has, when present, advanced the development of science in that country? And why has this problem persisted, despite changes in both politics and society, throughout the past century and more? In part, the trends we have observed arise from a conflict in the views of Chinese statesmen and scientists toward what

---

[9] Lo T'ien-yü: Laurence Schneider, ed., *Lysenkoism in China: Proceedings of the 1956 Qingdao Genetics Symposium*, a special issue of *Chinese Law and Government* 19, no. 2 (Summer 1986): vii–ix.

[10] Tseng Chao-lun: Liu Kuang-ting, "Ts'ung Tseng Chao-lun ti tsao-yü t'an-ch'i," 15–16.

[11] Biographies of these men appear in recent issues of *HHTP*. Tseng: 9 (1980):559–67. Chang: 10 (1980):622–6. Huang: 1 (1980):54–7. Ting: 6 (1979):547–52. Wang: 9 (1982):553–8. Yang: 12 (1983):753–7.

might be called the "crisis of modernization." On one side, China's political elite has been anxious, even desperate about the need to reunify the great empire, defend it against foreign threats, compete successfully with other nations, and perhaps most important of all ascend to and keep their place at the top of the slippery pole of power. The pressure on these men has been enormous, even by twentieth-century standards, and they have responded by demanding obedience to a program that enriches and strengthens the state and does so in short order. On the other side, the scientific elite has been taught, largely in the West or by its example, to be more patient, invest in knowledge that is not and cannot be immediately useful, and leave to others the problems of social and economic development. For the most part, Chinese scientists have been good nationalists, willing to accept the hardships of living and working in China for the satisfaction of serving their country. But they have not shared the feeling of desperation characteristic of China's political leaders, and this has been one source of the unstable relationship between these two parties.

A second factor, closely related to the first, has been the persistent urge toward unity that marks Chinese culture and politics. The dominant role of ideology – Confucian, "scientistic," Marxist, or other – in Chinese society reveals an underlying faith in a single unified truth about man, nature, and the cosmos. In this view, reality is not or should not be composed of disconnected pieces – political theories, artistic styles, scientific laws. Instead, all belong or should belong to one system. Projected onto politics, this ideological imperative yields the ideal of the universal state. China should not be a congeries of autonomous groups – labor unions, professional associations, political parties – that is drawn together, if at all, by the interaction and deals struck among these islands of authority. Rather, state and society should share the same goals and be governed by omnicompetent generalists, steeped in approved ethical ideology, and serving the common good. From the point of view of the ruling elite, technical and professional experts are a danger because they threaten the power of the governors, and because they undermine the principle of national unity and singleness of purpose. It is better for all Chinese to sail on one ship, under the command of the same "great helmsman."

In the story just told, neither the preferences of the statesmen nor the urge toward unity has remained consistently in the ascendance. During the early years of this century, the state was too weak to enforce policies of any type. After the reunification of the country under the Nationalists, a generation of well-trained scientists was able to fend off the proponents of state power, maintain their own sphere of professional activity, and achieve a balance that was unusual for China and unusually good for Chinese science. Even during the war, this balance held in the areas under Nationalist control. The devouring of the scientists by a voracious political elite was neither inevitable nor preordained. History might have taken a different turn.

Yet Chinese proponents of state power, of ideological and organizational unity, were persistent and in the end victorious. They dominated the late Ch'ing, returned, after a temporary eclipse, in the Nanking Decade, and swept aside all vestiges of scientific and other forms of professional autonomy in the Rectification Movement of the 1940s. The reemergence of central authority set the stage for science under the Communists.

The pressure of politics on science in Communist China has proved unbearable, in part because of the absence of party cadres with both political experience and scientific training. These were the men and women who might have emerged from the Natural Science Institute and other schools in Yenan, but were discredited, purged, or deflected to other careers as a result of *Cheng feng*. In chemistry, only those identified with industrial practice – Li Ch'iang, Ch'ien Chih-tao, Lin Hua, and Hua Shou-chün – survived the rigors of Rectification to play a role in the post-1949 era. According to Yü Kuang-yüan, a member of the original NSI staff and a high-ranking science administrator in the PRC, Yenan's "scientists" were on the whole a poorly educated lot who made few contributions to the advance of knowledge, but formed a vital link between the party and the leaders of the Chinese scientific community, most of whom had no connection to the Communists before 1949. As classmates and colleagues of China's true sciences – men and women who had been trained abroad, served in Nationalist China in the 1930s and for the most part during the war – the Yenan veterans were able to identify and recruit technical personnel needed by the new regime and to understand the choices posed by China's scientific elite. Yü himself played this role. A few others – Wu Heng, Ch'en K'ang-pai, and Hua Shou-chün – also helped direct the development of science in the PRC. But the careers of many more were destroyed by *Cheng feng*, which, as Hsü T'e-li warned, stifled the development of people and institutions that would be needed by China long after the war and the revolution were over. When in the mid-1950s the Communists recognized the need for better coordination of scientific affairs, they had to recruit literary figures, like Kuo Mo-jo and Ch'en Po-ta, or former military commanders, Ch'en I and Nieh Jung-chen, none of whom had previous experience in the field of science. And when, in the late 1950s and 1960s, they turned against the scientists, there were few in the party to speak for "the study of change" or any of the other disciplines that have helped to make a modern China.[12]

---

[12] Yü: Yü Kuang-yüan, "K'o-hsüeh ti kuang-hui," *KJCC*, 1:125.

Appendix

Table A.1. *Chinese terms for chemical elements*

Atomic number	Element	Traditional	Hobson 1855	Martin 1868	Kerr/Ho 1871	Fryer/Hsü 1872	Billequin 1873	MOE/NICT 1933
1	Hydrogen		輕氣	輕氣	輕	輕氣	輕氣	氫
2	Helium							氦
3	Lithium				鋰	鋰	鋰	鋰
4	Glucinum				鈍	鉻	鉗	
5	Beryllium							鈹
5	Boron			硼精	硼	硶	硼精	硼
6	Carbon	石炭	炭	炭精	炭	炭	炭精	碳
7	Nitrogen		淡氣	硝氣	淡	淡氣	硝氣	氮
8	Oxygen		養氣	養氣	養	養氣	養氣	氧
9	Fluorine			弗	弗	弗氣	弗	氟
10	Neon							氖
11	Sodium	鹻		鹻精	鈉	鈉	鹻	鈉
12	Magnesium			鎂精	鎂	鎂	鎂	鎂
13	Aluminum	礬石		礬精	釩	鋁	鋁	鋁
14	Silicon	玻璃		玻精	玻	矽	矽精	矽
15	Phosphorus			光藥	燐	燐	燐	磷
16	Sulfur	硫黃	硫黃	硫黃	磺	硫	硫黃	硫
17	Chlorine			鹽氣	綠	綠氣	綠氣	氯
18	Argon							氬
19	Potassium			灰精	鉀	鉀	鉀	鉀
20	Calcium	石灰		炭精	鈣	鈣	碳	鈣
21	Scandium							鈧
22	Titanium				鈦	鉗	銻	鈦
23	Vanadium				鑕	釩	a	釩
24	Chromium				鉻	鉻	鉻	鉻

377

Table A.1. (cont.)

Atomic number	Element	Traditional	Hobson 1855	Martin 1868	Kerr/Ho 1871	Fryer/Hsü 1872	Billequin 1873	MOE/NICT 1933
25	Manganese	無名異		無名異	錳	錳	鏋	錳
26	Iron	鐵	鐵	鐵	鐵	鐵	鐵	鐵
27	Cobalt	扁青		鉆	鴿	鈷	錆	鈷
28	Nickel			鑷	鎘	鎳	鐸	鎳
29	Copper	銅	銅	銅	銅	銅	銅	銅
30	Zinc	白鉛	精鏑	白鉛	鋰	鋅	鎳	鎵
31	Gallium							鎵
32	Germanium							鍺
33	Arsenic			信石	磇	砷	砒	砷
34	Selenium				硒	硒	玥	硒
35	Bromine				溴	溴	溴	溴
36	Krypton							氪
37	Rubidium				鑪	鉚		銣
38	Strontium				鎴	鎴	鍢	鍶
39	Yttrium				鎴	釱		釔
40	Zirconium				鎚	鋯		鋯
41	Columbium				鉫			鈳[b]
	Niobium					鈮		鈮
42	Molybdenum				鉬	鉬		鉬
43	Masurium							鎷[b]
	Technetium							鍀
44	Ruthenium				鉝	釕		釕
45	Rhodium				鎑	鎑	鉚	銠
46	Palladium				鈀	鈀		鈀
47	Silver	銀		白銀	銀	銀	銀	銀

No.	Element					
48	Cadmium	鎘		鎘	鎘	
49	Indium	銦	*a*	銦	銦	
50	Tin	錫	錫	錫	錫	錫
51	Antimony	銻	銻	銻	銻	銻
52	Tellurium	碲	碲	碲	碲	
53	Iodine	碘		碘	碘	海蘭
54	Xenon	氙				
55	Cesium	銫		銫	銫	
56	Barium	鋇		鋇	鋇	
57	Lanthanum	鑭	*a*	鑭	鑭	
58	Cerium	鈰		鈰	鑭	
59	Praseodymium					
60	Neodymium					
61	Illinium *b*					
	Promethium					
62	Samarium					
63	Europium					
64	Gadolinium	釓				
65	Terbium	鋱		鋱	鉱	
66	Dysprosium	鏑		鏑	鏑	
67	Holmium	鈥				
68	Erbium	鉺		鉺	鉺	
69	Thulium	銩				
70	Ytterbium	鐿				
71	Lutetium *c*					
72	Hafnium	鉿		鉿	鉿	
73	Tantalum	鉭		鉭	鉭	
74	Tungsten	鎢		鎢	鎢	
75	Rhenium					
76	Osmium	鋨	鋨	鋨	鋨	

Table A.1. (cont.)

Atomic number	Element	Traditional	Hobson 1855	Martin 1868	Kerr/Ho 1871	Fryer/Hsü 1872	Billequin 1873	MOE/NICT 1933
77	Iridium				鏭	鏔	鉣	銥
78	Platinum	金銀		白金	鉑	鉑	鉑	鉑
79	Gold		金	黃金	金	金	金	金
80	Mercury	水銀	水銀	水銀	汞	汞	水銀	汞
81	Thallium				鉈	鉿		鉈
82	Lead	鉛	鉛	黑鉛	鉛	鉛	黑鉛	鉛
83	Bismuth			鐵	鉍	鉍	鑅	鉍
84	Polonium							釙 [b]
85	Alabamine							硪 [b]
86	Radon							氡
87	Virginium							鈁 [b]
88	Radium							鐳
89	Actinium							錒
90	Thorium				釖	鋀		釷
91	Protoactinium							鏷
92	Uranium				鈾	鈾		鈾

[a] Characters for elements 24, 48, and 56 on Billequin's list are illegible.

[b] Since 1933 the characters for five elements (41, 43, 61, 85, and 87) have been changed to correspond to new Western names.

[c] In one case (element 71), the character has been changed, although the Western name remains the same.

Sources: Traditional: Joseph Needham, Science and Civilisation in China, 5:2, Table 95, 162–84. Terms for 1855, 1868, and 1973: Chang Tzu-kao, "Ho Liao-jan ti 'Hua-hsüeh ch'u-chieh,'" 45–6; and Chang Tzu-kao and Yang Ken, "Ts'ung 'Hua-hsüeh ch'u-chieh' ho 'Hua-hsüeh chien-yüan' k'an," 352. Terms for 1871: John Kerr and Ho Liao-jan, Hua-hsüeh ch'u-chieh, 1:6. Terms for 1872: John Fryer and Hsü Shou, trans., Hua-hsüeh chien-yüan, 6a–6b. Terms for 1933: Hua-hsüeh ming-ming yüan-tse.

Table A.2. *Enrollments in Chinese schools, 1912–1946*

School	1912	1916	1922	1929	1932	1936	1941	1946
Primary								
Lower			5,965,957					
Higher			615,378					
Total	2,795,475	3,843,454	6,601,802	8,882,077	12,223,066	18,364,956	15,058,051	23,813,705
Secondary								
Middle								
Junior					348,412	393,691	586,985	
Senior					61,174	88,831	116,771	
Total	59,971	75,595	118,658		409,586	432,522	703,756	1,495,874
Normal	28,525	24,959	43,846		99,606	37,902	91,239	245,609
Vocational	9,469	10,524	20,300		38,015	56,822	51,557	137,040
Total	97,965	111,078	182,568		547,207	627,246	846,522	1,878,523
Higher								
University	481	1,446		21,320	35,640	37,330	51,861	110,438
Jr. college	39,633	15,795		7,803	7,070	4,592	7,596	18,898
Total	40,114	17,241	34,380	29,123	42,710	41,922	59,457	129,336

Sources: Primary: CYNC (1948), 1,455. Figures for 1922: George Twiss, *Science and Education in China*, 120. Secondary: CYNC (1934), 3:194; CYNC (1948), 1,428 and 1,433–4. Higher: CYNC (1948), 1,400.

Table A.3. *Students accorded permission to study abroad, by course, 1929–1946*

Year	Lit.	Law	Comm.	Ed.	Arts & Hum.	Nat. Sci.	Eng.	Med.	Agri.	Sci. & Tech.	Misc.	Total
1929	266	568	62	75	971	129	249	104	66	548	9	1,528
%	17.4	37.2	4.1	4.9	63.5	8.4	16.3	6.8	4.3	35.9	0.6	100.0
1930	166	307	43	56	572	77	165	109	49	400	0	972
%	17.1	31.6	4.4	5.8	58.8	7.9	17.0	11.2	5.0	41.2	0	100.0
1931	57	108	11	45	221	64	79	60	17	220	0	441
%	12.9	24.5	2.5	10.2	50.1	14.5	17.9	13.6	3.9	49.9	0	100.0
1932	98	179	25	40	342	49	76	53	35	213	21	576
%	17.0	31.1	4.3	6.9	59.4	8.5	13.2	9.2	6.1	37.0	3.6	100.0
1933	77	143	31	49	300	62	131	82	44	319	2	621
%	12.3	23.0	5.0	7.9	48.3	10.0	21.1	13.2	7.1	51.4	0.3	100.0
1934	99	234	43	52	428	116	164	79	72	431	0	859
%	11.5	27.2	5.0	6.0	49.8	13.5	19.1	9.2	8.4	50.2	0	100.0
1935	117	246	70	73	506	135	174	104	113	526	1	1,033
%	11.3	23.8	6.8	7.1	49.0	13.1	16.8	10.1	10.9	50.9	0.1	100.0
1936	108	227	64	64	463	97	183	127	119	526	13	1,002
%	10.8	22.7	6.4	6.4	46.2	9.7	18.3	12.7	11.9	52.5	1.3	100.0
1937	20	61	33	24	138	46	107	34	41	228	0	336
%	6.0	18.2	9.8	7.1	41.1	13.7	31.8	10.1	12.2	67.9	0	100.0

1938	2	7	1	3	13	18	34	20	7	79	0	92
%	2.2	7.6	1.1	3.3	14.1	19.6	37.0	21.7	7.6	85.9	0	100.0
1939	1	9	1	9	20	20	13	8	4	45	0	65
%	1.5	13.8	1.5	13.8	30.8	30.8	20.0	12.3	6.2	69.2	0	100.0
1940	8	10	7	7	32	8	25	11	10	54	0	86
%	9.3	11.6	8.1	8.1	37.2	9.3	29.1	12.8	11.6	62.8	0	100.0
1941	3	11	4	2	20	8	19	4	6	37	0	57
%	5.2	19.3	7.0	3.5	35.1	14.0	33.3	7.0	10.5	54.9	0	100.0
1942	15	39	13	6	73	32	103	7	13	155	0	228
%	6.6	17.1	5.7	2.6	32.0	14.0	45.2	3.1	5.7	58.0	0	100.0
1943	37	53	84	7	181	28	124	9	17	178	0	359
%	10.3	14.8	23.4	1.9	50.4	7.8	34.5	2.5	4.7	49.5	0	100.0
1944	8	11	10	5	34	27	164	23	57	271	0	305
%	2.6	3.6	3.3	1.6	11.1	8.8	53.8	7.5	18.7	38.9	0	100.0
1945	0	0	0	0	0	5	0	0	3	8	1	8
%	0	0	0	0	0	62.5	0	0	37.5	100	0	100.0
1946	94	145	57	25	321	92	205	49	63	409	0	730
%	12.9	19.9	7.8	3.4	44.0	12.6	28.1	6.7	8.6	56.0	0	100.0

N.B.: Totals for 1929–31 exclude students whose fields were unknown.
Source: CYNC (1948), 1,416.

Table A.4. *Average annual Chinese students in America by field,*
1905–1950

Subject	1905–9	1914–18	1920–7	1931–8	1941–50	Total
Humanities	10	71	112	250	187	
%	9.3	9.3	10.0	16.5	9.5	11.5
Social Science	17	103	167	197	285	
%	15.9	13.5	14.9	13.0	14.5	14.0
Business	13	71	146	133	147	
%	12.1	9.3	13.0	8.8	7.5	9.3
Education	5	65	94	236	226	
%	4.7	8.5	8.4	15.6	11.5	11.4
Arts, total	45	310	519	816	845	
%	42.1	40.7	46.3	53.9	42.9	46.3
Engineering	46	264	359	369	652	
%	43.0	34.7	32.1	24.4	33.1	30.9
Science	8	77	115	120	271	
%	7.5	10.1	10.3	7.9	13.7	10.8
Medicine	3	60	77	147	141	
%	2.8	7.9	6.9	9.7	7.2	7.8
Agriculture	5	46	43	60	54	
%	4.7	6.0	3.8	4.0	2.7	3.8
Military Science	0	4	7	3	8	
%	0	0.5	0.6	0.2	0.4	0.4
Sciences, total	62	451	601	699	1,126	
%	57.9	59.3	53.7	46.1	57.1	53.6
Subtotal	107	761	1,120	1,515	1,971	
%	100	100	100	100	100	100
Preparatory/ unknown	70	226	203	313	239	
Total	177	987	1,323	1,828	2,210	

Note: These figures represent the average number of Chinese students for one
year during the period in question.
Source: Y. C. Wang, *Chinese Intellectuals and the West*, 510–11.

Table A.5. Doctorates by Chinese students in Europe and America, 1905–1949

Pure sciences

Year	Chemistry			Physics and math			Biology			Earth science			Total		
	U.S.	Eur	Ttl	U.S.	Eur	Ttl	U.S.	Eur	Ttl	U.S.	Eur	Ttl	U.S.	Eur	Ttl
1905–9	0	1	0	0	0	0	0	0	0	0	0	0	0	1	1
1910–14	0	2	3	0	0	0	2	2	2	0	0	0	2	2	4
1915–19	2	2	4	4	1	5	2	2	4	2	0	2	10	5	15
1920–4	8	4	12	9	2	11	4	4	8	0	1	1	21	11	32
1925–9	19	18	37	19	11	30	16	14	30	1	5	6	55	48	103
1930–4	39	26	65	17	20	37	36	23	59	4	6	10	96	75	171
1935–9	28	33	61	21	37	58	29	27	56	5	16	21	83	113	196
1940–4	20	17	37	32	24	56	24	15	39	10	4	14	86	60	146
1945–9	38	16	54	41	21	62	35	8	43	19	6	25	133	51	184
Total	154	119	273	143	116	259	148	93	241	41	38	79	486	366	852

Applied sciences

Year	Engineering			Chemical engineering			Agriculture and forestry			Medicine			Total		
	U.S.	Eur	Ttl	U.S.	Eur	Ttl	U.S.	Eur	Ttl	U.S.	Eur	Ttl	U.S.	Eur	Ttl
1905–9	0	0	0	0	0	0	0	0	0	1	0	1	1	0	1
1910–14	0	0	0	2	0	2	0	0	0	0	0	0	2	0	2
1915–19	1	0	1	1	0	1	0	0	0	0	5	5	2	5	7
1920–4	4	3	7	4	0	4	1	0	1	4	64	68	13	67	80
1925–9	6	1	7	6	1	7	3	3	6	10	92	102	25	97	122
1930–4	12	6	18	11	3	14	4	1	5	12	71	83	39	81	120
1935–9	20	25	45	4	4	8	5	9	14	11	132	143	40	170	210
1940–4	24	18	42	12	9	21	6	1	7	20	21	41	62	49	111
1945–9	39	31	70	9	2	11	6	4	10	19	23	42	73	60	133
Total	106	84	190	49	19	68	25	18	43	77	408	485	257	529	786

Table A.5. (cont.)

| | Totals | | | | | | | | |
| Year | Pure and applied sciences | | | Humanities and social sciences | | | Total | | |
	U.S.	Eur	Ttl	U.S.	Eur	Ttl	U.S.	Eur	Ttl
1905–9	1	1	2	3	3	6	4	4	8
1910–14	4	2	6	10	3	13	14	5	19
1915–19	12	10	22	22	13	35	34	23	57
1920–4	34	78	112	49	30	79	83	108	191
1925–9	80	145	225	105	79	184	185	224	409
1930–4	135	156	291	104	161	265	239	317	556
1935–9	123	283	406	91	186	277	214	469	683
1940–4	148	109	257	84	91	175	232	200	432
1945–9	206	111	317	99	49	148	305	160	465
Total	743	895	1,638	567	615	1,182	1,310	1,510	2,820

Notes: "U.S." figures include 28 Ph.D.'s by Chinese in Canadian universities. "Europe" includes Germany, France, United Kingdom, Netherlands, Austria, Switzerland, Belgium, and Italy. "Engineering" includes all subfields except chemical, which is listed separately.
Sources: Yüan T'ung-li, *Doctoral Dissertations by Chinese Students in Great Britain and Northern Ireland, 1916–1961; A Guide to Doctoral Dissertations by Chinese Students in America, 1905–1960;* and *A Guide to Doctoral Dissertations by Chinese Students in Continental Europe, 1907–1962.*

Table A.6. *Standard middle school curricula, 1932*

Subject	First year		Second year		Third year	
	1st term	2d term	1st term	2d term	1st term	2d term
Junior middle school						
Civics	2	2	2	2	1	1
Phys. training	3	3	3	3	3	3
National language	6	6	6	6	6	6
English	5	5	5	5	5	5
Mathematics	4	4	5	5	5	5
Botany	4	—	—	—	—	—
Geology	—	4	—	—	—	—
Physics	—	—	3	3	—	—
Chemistry	—	—	—	—	4	3
Health	1	1	1	1	1	1
History	2	2	2	2	2	2
Geography	2	2	2	2	2	2
Manual work	2	2	2	2	4	4
Drawing	2	2	2	2	1	1
Music	2	2	1	1	1	1
Total	35	35	35	34	35	34

Table A.6. (cont.)

Subject	First year		Second year		Third year	
	1st term	2d term	1st term	2d term	1st term	2d term
Senior middle school						
Civics	2	2	2	2	2	2
Physical training	2	2	2	3	3	3
National language	4	4	4	4	4	4
English	5	5	5	5	5	5
Mathematics	4	4	3	3	3	3
Botany	5	—	—	—	—	—
Zoology	—	5	—	—	—	—
Physics	—	—	—	—	6	6
Chemistry	2	2	4	4	—	—
Chinese history	2	2	2	2	2	2
Foreign history	—	—	—	2	—	—
Chinese geography	2	2	2	—	—	—
Foreign geography	—	—	—	2	2	2
Health	—	—	—	—	2	2
Logic	—	—	—	—	—	—
Minerology	—	—	3	—	—	—
Military training	3	3	3	3	—	—
Drawing	1	1	2	2	1	—
Music	1	1	1	1	—	1
Total	33	33	33	33	30	30

Source: The China Year Book (1933), 531–2.

388

Table A.7. Enrollments in Chinese higher education by field, 1928–1957

Year	Arts and humanities										Science and technology										A&H + S&T	Other	Total
	Lit.	%	Law	%	Comm.	%	Ed.	%	Total	%	Sc.	%	Eng.	%	Agr.	%	Med.	%	Total	%			
1928	5,464	21.8	9,466	37.8	1,595	6.8	1,661	6.6	18,286	73.0	1,910	7.5	2,777	11.1	1,085	4.3	977	3.9	6,749	27.0	25,035	163	25,198
1929	6,171	21.2	11,434	39.3	1,567	5.7	2,082	7.1	21,354	73.3	2,291	7.5	3,144	10.8	1,294	4.4	1,138	3.9	7,767	26.7	29,121	0	29,121
1930	7,708	20.5	15,899	42.3	2,025	5.4	2,561	6.8	28,193	75.0	2,872	7.5	3,734	9.9	1,419	3.8	1,350	3.6	9,375	25.0	37,568	0	37,568
1931	10,066	22.8	16,487	37.2	2,156	4.9	4,231	9.6	32,940	74.5	3,930	8.9	4,084	9.3	1,413	3.2	1,800	4.1	11,227	25.5	44,167	0	44,167
1932	9,312	22.1	14,523	34.5	2,867	6.8	3,368	8.0	30,070	71.5	4,359	9.9	4,439	10.5	1,557	3.7	1,852	4.4	12,007	28.5	42,077	633	42,710
1933	8,703	20.3	12,913	30.1	3,167	7.4	4,004	9.3	28,787	67.1	4,722	11.0	5,263	12.3	1,690	3.9	2,458	5.7	14,133	32.9	42,920	16	42,936
1934	7,921	19.0	11,029	26.4	3,033	7.3	4,059	9.7	26,042	62.4	5,324	12.8	5,910	14.2	1,831	4.4	2,633	6.3	15,698	37.6	41,740	28	41,768
1935	9,596	23.4	8,794	21.4	2,951	7.2	2,741	6.7	24,082	58.6	6,272	15.3	5,514	13.4	2,163	5.3	3,041	7.4	16,990	41.4	41,072	56	41,128
1936	8,364	20.1	8,253	19.8	3,243	7.8	3,292	7.9	23,152	55.6	5,485	13.2	6,989	16.8	2,590	6.2	3,395	8.2	18,459	44.4	41,611	311	41,922
1937	4,140	13.4	7,125	23.0	1,846	6.0	2,451	7.9	15,562	50.2	4,458	14.4	5,768	18.6	1,802	5.8	3,386	10.9	15,414	49.8	30,976	212	31,188
1938	4,852	13.6	7,024	19.7	2,809	7.9	3,027	8.5	17,712	49.6	4,802	13.4	7,321	20.5	2,257	6.3	3,623	10.1	18,003	50.4	35,715	465	36,180
1939	5,137	11.7	8,777	19.9	3,690	8.4	3,796	8.6	21,400	48.6	5,828	13.2	9,501	21.6	2,994	6.8	4,322	9.8	22,645	51.4	44,045	377	44,422
1940	5,920	11.3	11,172	21.3	5,199	9.9	4,823	9.2	27,114	51.8	6,090	11.6	11,226	21.4	3,575	7.0	4,271	8.2	25,262	48.2	52,376		52,376
1941	6,156	10.4	12,085	20.3	7,231	12.2	5,919	10.0	31,391	52.8	6,202	10.4	12,584	21.2	4,573	7.9	4,607	7.7	28,060	47.2	59,457		59,457
1942	7,055	11.0	12,598	19.7	7,691	12.0	7,626	11.9	34,970	54.5	5,852	9.1	13,129	20.4	5,038	7.9	5,108	8.0	29,127	45.4	64,057		64,097
1943	8,455	11.5	15,377	20.9	9,039	12.3	8,804	11.9	41,675	56.6	6,099	8.3	14,582	19.8	5,599	7.6	5,714	7.8	31,994	43.4	73,699		73,699
1944	9,102	11.5	15,990	20.3	9,742	12.3	10,466	13.3	45,300	57.4	6,177	7.8	15,047	19.1	6,042	7.7	6,343	8.0	33,609	42.6	78,909		78,909
1945	9,967	11.9	17,774	21.3	9,697	11.6	11,709	14.0	49,147	58.9	6,480	7.8	15,200	18.2	6,380	7.6	6,291	7.5	34,351	41.1	83,498		83,498
1946	14,524	11.2	28,276	21.9	13,851	10.7	18,389	14.2	75,040	58.0	9,091	7.0	24,389	18.9	9,364	7.2	11,452	8.9	54,296	42.0	129,336		129,336
1947	18,446	11.9	37,780	24.4	17,698	11.4	21,439	13.8	95,363	61.5	10,060	6.5	27,579	17.8	10,179	6.6	11,855	7.6	59,673	38.5	155,036		155,036
1948											unavailable												
1949	14,600	12.5	7,300	6.3	19,400	16.7	12,300	10.6	53,600	46.0	7,000	6.0	30,300	26.0	10,400	8.9	15,200	13.0	62,900	54.0	116,500		116,500
1950							13,300	9.6	38,500	27.8			38,500	27.8			17,400	12.5			138,700		138,700
1951							18,200	11.7	48,500	31.2			48,500	31.2			21,400	13.8			155,600		155,600
1952	17,100	8.9	3,800	2.0	22,000	11.5	31,800	16.6	74,700	39.1	9,600	5.0	66,600	34.9	15,500	8.1	24,700	12.9	116,400	60.9	191,100		191,100
1953	16,900	8.0	3,900	1.8	13,500	6.4	41,100	19.4	75,400	35.5	12,400	5.8	80,000	37.7	15,400	7.3	29,000	13.7	136,800	64.5	212,200		212,200
1954	20,900	8.3	4,000	1.6	11,200	4.4	55,000	21.7	91,100	36.0	17,100	6.8	95,000	37.5	15,900	6.3	33,900	13.4	161,900	64.0	253,000		253,000
1955	21,100	7.3	4,800	1.7	11,400	4.0	63,000	21.9	100,300	34.8	20,000	6.9	109,600	38.1	21,600	7.5	36,500	12.7	187,700	65.5	288,000		288,000
1956							99,000	24.3			25,000	6.1	150,000	36.8							408,000		408,000
1957	23,300	5.4	9,300	2.1	12,700	2.9	92,600	21.3	137,900	31.7	27,100	6.2	177,600	40.9	37,200	8.6	54,800	12.6	296,700	68.3	434,600		434,600

Sources: Figures for 1928–47: CYNC (1948), 525–6. Figures for 1949–57: Leo Orleans, *Professional Manpower and Education in Communist China*, 68–9.

Table A.8. *Sex ratios of men (n) to women (1.0) in Chinese higher education by course, 1932–1946*

Year	Lit.	Law	Comm.	Ed.	Sci.	Eng.	Med.	Agr.	Total
1932									7.3
1933									6.3
1934	2.9	11.9	5.9	2.1	4.4	58.7	4.6	21.3	5.7
1935	3.6	10.5	5.0	1.3	4.6	41.7	4.8	26.3	5.4
1936	3.3	10.3	5.5	1.9	4.1	49.6	3.9	22.1	5.6
1937	2.4	7.8	4.1	1.9	3.3	40.8	3.5	12.5	4.8
1938	1.8	6.4	4.8	1.5	3.2	24.4	2.3	8.7	4.4
1939	2.8	5.9	3.6	1.4	3.3	33.5	3.0	8.2	4.7
1940	2.0	5.9	3.0	1.2	3.1	27.9	2.5	6.9	4.1
1941	2.2	5.4	3.2	1.5	3.0	19.7	2.2	5.9	4.0
1942	2.6	6.4	3.6	1.6	3.1	20.3	2.1	5.9	4.2
1943	2.9	6.9	3.5	1.5	3.0	23.6	2.2	5.7	4.4
1944	2.7	7.0	3.9	1.4	3.0	23.6	2.1	5.6	4.3
1945	2.7	7.9	4.0	1.3	2.8	26.0	1.8	4.8	4.3
1946	2.5	7.7	3.5	1.0	2.8	37.8	2.6	6.4	4.5

Source: CYNC (1948), 1,413.

Glossary

a-mo-ni-a	阿摩尼亞
Academia Sinica (Chinese Academy of Sciences)	國立中央研究院
Ai Szu-ch'i	艾思奇
An-sai County	安塞縣
Anti-Japanese University	抗日大學
Beipei	北碚
Border Region Match Factory	邊區火柴廠
Bureau of Industrial Research	工業試驗所
Ch'a-fang Arsenal	茶坊兵工廠
Chang Ch'ao-chün	張朝俊
Chang Chün-mai	張君勱
Chang-erh-ts'un	張二村
Chang, H. T.	張鴻釗
Chang I-tsun	張儀尊
Chang Ta-yü	張大煜
Chang Tsing-lien	張青蓮
Chang Tzu-kao	張子高
Chao Yüan-jen	趙元任
Che-hsüeh tao ho-ch'u ch'ü?	哲學到何處去？
Ch'en Chen-hsia	陳振夏
Chen-Hua Paper Factory	振華造紙廠
Ch'en K'ang-pai	陳康白
Ch'en K'o-chung	陳可忠
Chen Ko-kuei	陳克恢
Ch'en Kuang-fu	陳光甫
Ch'en Kuo-fu	陳果夫
Ch'en Li-fu	陳立夫
Ch'en Po-ta	陳伯達
Ch'en Tu-hsiu	陳獨秀
Ch'en Yu-gwan	陳裕光
Cheng Chen-wen	鄭貞文

Cheng feng	整風
Ch'eng Shu-jen	程叔仁
ch'i (force)	氣
ch'i (gas radical)	气
Ch'i-chi fa-jen	汽機發軔
Ch'i Ju-shan	齊如山
Ch'i-li-pu	七裏舖
Ch'i-li-ts'un	七里村
ch'i-yü	祈雨
Chi Yuoh-fong	紀育灃
Ch'iang hsüeh-hui	強學會
Chiang Monlin	蔣夢麟
Chiang T'ung-yin	蔣同寅
Chi'ao-erh-kou	橋兒溝
Chiao-hui hsin-pao	教會新報
Ch'ieh-yin-tzu	切音字
Ch'ien Chih-tao	錢志道
Ch'ien Tuan-sheng	錢端升
chin	金
Chin Hsin-yüan	金沁園
Chin-k'uei County	金匱縣
Chin P'ei-sung	金培松
chin-shih	進士
China Alcohol Factory	中國酒精廠
China Foundation for the Promotion of Education and Culture	中華教育文化基金董事會
China Institute for Economic and Statistical Research	經濟統計研究所
China Society of Chemical Industry	中華化學工業會
Chinese Chemical Society	中國化學會
Chinese Industrial Chemical Research Institute	中華工業化學研究所
Chinese Scientific Book Depot	格致書室
Chinese Society of Chemical Engineers	中國化學工程學會
ch'ing-ch'i	輕氣
Ching-hsing Mining Bureau	井陘礦務局
Ch'ing-hua hsüeh-pao	清華學報
ching-shih	經世
Chiu-ta Refined Salt Company	久大精鹽工廠
Ch'iu Yan-tsz	趙恩賜

Chou P'ei-yüan	周培源
Chou Tsan-quo	趙承嘏
Chou Tso-min	周作民
ch'ü	區
Chu Chia-hua	朱家驊
chu-chiao	助教
chü-jen	舉人
Chu Ju-hwa, Edith	朱汝華
Ch'ü Po-ch'uan	屈伯傳
chüan	卷
Chuang Chang-kong	莊長恭
Chung Hsi wen-chien lu	中西聞見錄
Consultative Council	參議會
fa-t'uan	法團
Fan Ching-sheng	范靜生
Fan Hsü-tung	范旭東
fan-i-kuan	翻譯館
Fan Jui	范銳
Fan Yüan-lien	范源濂
fang-hsiang	芳香
Fei Hsiao-t'ung	費孝通
fen	分
Feng Yün-hao	豐雲鶴
Fu Ssu-nien	傅斯年
Fu Ying	傅鷹
Golden Sea Research Institute of Chemical Industry	黃海化學工業研究社
Hao Ching-sheng	郝景盛
Henry Lester Institute	萊斯特醫藥研究所
Ho-ch'ü County	河曲縣
Ho Liao-jan	何瞭然
Ho-shui County	合水縣
Hou Te-pang	侯德榜
Hsi Han-po	席漢伯
Hsi Kan	席淦
Hsi-nan lien-ho ta-hsüeh	西南聯合大學
hsi-t'ung	系統
hsiao-ch'iang-shui	硝強水
hsieh-sheng	諧聲
Hsiehchiatien	卸甲甸

Hsien-yang Alcohol Factory	咸陽酒精廠
hsin	鋅
Hsin chiao-yü	新教育
Hsin nsüeh pao	新學報
Hsin shih-chi	新世紀
Hsü-chia-hui	徐家匯
Hsü Chien-yin	徐建寅
hsü-chih	須知
Hsü Hsüeh-ts'un	徐雪村
Hsü Hua-feng	徐華封
Hsü Shou	徐壽
Hsü T'e-li	徐特立
hsüeh-hui	學會
hsüeh-she	學社
Hsüeh-sheng-hui pi-pao	學生會壁報
Hu Hsien-su	胡先驌
Hu Mei	胡美
Hu Ming-fu	胡明復
Hu Shih	胡適
Hua Heng-fang	華蘅芳
hua-hsüeh	化學
Hua-hsüeh t'ung-hsün	化學通訊
Hua Shou-chün	華壽俊
huang-ch'iang-shui	磺強水
huang-chiu	黃酒
Huang Ming-lung	黃鳴龍
Huang Tzu-ch'ing	黃子卿
Huang Wen-wei	黃文煒
hui-i	會意
i-chih	易知
i-t'o	以脫
International Dispensary	五洲固本造藥廠
Jan-liao chuan-pao	燃料專報
Jen	仁
jen Hung-chün	任鴻雋
Jeu Kia-khwe	裘家奎
Kagaku shinsho	化學新書
K'ai-ch'eng Acid Company	開城製酸公司
kan-chiu	乾酒

K'ang Ti	康迪
Kansu Petroleum Bureau	廿肅油礦局
k'ao-cheng	考證
Kao Ch'ung-hsi	高崇熙
kao-liang	高粱
Kao Tsi-yu	高濟宇
Kiangnan Arsenal	江南製造局
Kiangsu Chemical Works (Major Brothers), Ltd.	江蘇藥水廠
Kienwei	犍爲
Kincheng Bank	金城銀行
ko-chih	格致
Ko-chih hsin-pao	格致新報
k'o-hsüeh kung-li	科學公理
k'o-hsüeh yüan-ti	科學園地
Ko Tao-yin	葛道殷
ko-wu	格物
ko-wu chih-chih	格物致知
kou-wen	鉤吻
Ku Chieh-kang	顧頡剛
Ku Shih Pien	古史辨
Ku Yü-chen	顧毓珍
Ku Yü-hsiu	顧毓琇
Ku Yü-tsuan	顧毓瑔
kuan-tu shang-pan	官督商辦
Kuang-hsüeh hui	廣學會
Kuang-Hua Farm	光華農場
K'un-yü	崑崙
Kuo Mo-jo	郭沫若
Kuo Ping-wen	郭秉文
kuo-ts'ui	國粹
kuo-yü	國語
Kwangtung Provincial Electro-chemical Factory	廣東省營苛性鈉廠
Kwangtung Provincial Fertilizer Factory	廣東省營肥田料廠
Kwangtung Provincial Sugar Factory	廣東省營糖廠
Kwangtung Provincial Sulfuric Acid Factory	廣東省營硫酸廠
Lai-yang	萊陽
Lee Fang-hsuin	李方訓

Lee, T. D.	李政道
Leo Shoo-tze	劉樹杞
li	理
Li Chi	李濟
Li Ch'iang	李強
Li-chung Acid Works	利中製酸公司
Li Erh-k'ang	李爾康
Li Fu-ch'un	李富春
Li Shan-lan	李善蘭
Li Sheo-hen	李壽恆
Li Shih-tseng	李石曾
Li Su	李蘇
Li Ta-chang	李大璋
Li Tuan-fen	李端棻
Li Yün-hua	李運華
Liang-Kwang Provincial Sulfuric Acid Factory	兩廣省辦硫酸廠
Lien-ta	聯大
Lih Kun-hou	酈堃厚
Lin Hua	林華
Lin Shan	林山
Liu Hsien-i	劉咸一
Liu Jang	劉讓
Liu Ta-chün	劉大鈞
Lo Feng-lu	羅豐祿
Lo T'ien-yü	樂天宇
Lu Yü-tao	盧于道
Lung-hua Powder Plant	龍華火藥廠
Lung Yün	龍雲
ma-huang	麻黃
ma-lan	馬蘭
Ma Tsu-sheng	馬祖聖
Mei-hua shu-kuan	美華書館
mi-i-t'o-li-ni	米以脫里尼
mi-meng-shui	迷蒙水
Military Industrial Bureau	軍事工業局
Mo-hai shu-kuan	墨海書館
Nankai University Institute of Applied Chemistry	南開大學應用化學研究所
National Academic Work Advisory Board	全國學術工作諮詢處

National Academy of Peiping	國立北平研究院
National Bureau of Industrial Research	中央工業試驗所
National Institute for Compilation and Translation	國立編譯館
National Research Council	評議會
National Resources Commission	資源委員會
Natural Science Institute	自然科學院
Natural Science Research Society	自然科學研究會
New China Chemical Factory	新華化學廠
New China Drug Factory	新華藥廠
Nung-hsüeh pao	農學報
Ou-yu tsa-lu	歐游雜錄
pai-ch'ien	白鉛
pei-mu	貝母
Pei-ta	北大
Pei-yang hsi-hsüeh	北洋西學
Pen Ts'ao Kang Mu	本草綱目
P'eng Jui-hsi	彭瑞熙
P'eng-shan	彭山
pi-li	比例
Ping Chih	秉志
po-ching	玻精
Pohai Chemical Works	渤海化學工業公司
pu-hsi-pan	補習班
P'u-i Sugar Factory	溥益糖廠
Sah Pen-t'ieh	薩本鐵
Sah Pen-t'ung	薩本棟
San-pien	三邊
sanso	酸素
Science Society of China	中國科學社
Seimi kaiso	舍密開宗
Shanghai Commercial and Savings Bank	上海商業銀行
Shanghai Polytechnic Institute	格致書院
shao-chiu	燒酒
Shen Hung	沈鴻
shen-lou-tzu	神樓子
sheng-ch'i	生氣
sheng-yüan	生員

shih	石
shui	水，氵
Shui Ching Chu	水經注
Sinyuan Fuels Research Laboratory	沁園燃料研究室
suiso	水素
Sun Ch'eng-o	孫承諤
Sun Hsüeh-wu	孫學悟
Sun Hung-ju	孫洪儒
Ta-chung che-hsüeh	大眾哲學
Ta-hsüeh	大學
Tai An-pang	戴安邦
T'ang Yüeh	唐鉞
t'i	體
tien	電
T'ien-ch'u Monosodium Glutamate Factory	天廚味精廠
t'ien-jan ching-hsiang	天然景象
T'ien kung k'ai wu	天工開物
T'ien-li Nitrogen Gas Company	天利淡氣製品公司
t'ien-lung ching-hsiang	天龍景象
T'ien-yüan Electrochemical Factory	天原電化廠
Ting Hsü-hsien	丁緒賢
Ting-pien County	定邊縣
Ting Wen-chiang	丁文江
Ts'ai Liu-sheng	蔡鎦生
Ts'ai Yüan-p'ei	蔡元培
Ts'ao Pen-hsi	曹本熹
Tseng Chao-lun	曾昭掄
Tsiang T'ing-fu	蔣廷黼
Tsungli yamen	總理衙門
Tu-fu ch'uan-k'ou	杜甫川口
Tung-li Oil Works	動力油料廠
T'ung Wen Kuan	同文館
Tung Wen-li	董文立
Tzu-fang-kou	紫坊溝
tzu-jan chih kung-li	自然之公理
Tzuliuching	自流井
Udagawa Yōan	宇田川榕庵

Wan-kuo kung-pao	萬國公報
Wang Chen-sheng	汪振聲
Wang Chin	王璡
Wang Feng-tsao	汪鳳藻
Wang Pao-jen	王葆仁
Wang Shih-chen	王士珍
Wang Shih-chieh	王世杰
West China Academy of Sciences	中國西部科學院
Wei sheng lun	衛生論
Wong Wen-hao	翁文灝
Woo Yui-hsun	吳有訓
Wu Ch'eng-lo	吳承洛
Wu-ch'i-chen	吳旭鑌
Wu Chih-hui	吳稚暉
Wu Heng	武衡
Wu Hsien (T'ao-min)	吳憲(陶民)
wu-hsing	五行
Wu Hsüeh-chou	吳學周
wu-ming-i	無名異
wu-shen	巫神
Wu, T. Y.	吳大猷
Wu Tsao-hsi	吳藻溪
Wu Yü-chang	吳玉章
Wu Yün-ch'u	吳蘊初
Wutungchiao	五通橋
Ya-ch'üan tsa-chih	亞泉雜誌
yang	陽
Yang, C. N.	楊振甯
yang-ch'i	養氣
Yang Ch'üan	楊銓
yang-ko	秧歌
Yang Shih-hsien	楊石先
Yeh Ch'ing	葉青
yen-ch'iang shui	鹽強水
Yen Fu	嚴復
yen-hu-so	延胡索
Yen P'ei-lin	閻沛霖
Yen-wu-kou	煙霧溝
Yenchang County	延長縣
Yentai	煙台
yin	陰
Yü Ho-ch'in	虞和欽

Yü Kuang-yüan	于光遠
Yü Kung	禹貢
Yü Ta-wei	俞大維
yüan	元
Yüan Han-ch'ing	袁翰青
Yümen County	玉門縣
yung	用
Yung-p'ing	永坪
Yung Wing	容閎
Yungli Chemical Industry Company	永利化學工業公司
Yungli Soda Manufacturing Company	永利製鹼公司

Bibliography

PERIODICALS

Acta Brevia Sinensia. Natural Science Society of China. Chungking, 1943–5.

Chiao-yü tsa-chih 教育雜誌 [Chinese educational review]. Shang-hai shang-wu yin-shu-kuan 上海商務印書館 [Shanghai Commercial Press]. Shanghai, Changsha, Hong Kong, 1909–48.

Chieh-fang jih-pao 解放日報 [Liberation daily]. Yenan, 1941–7.

The China Medical Missionary Journal, 1887–1909; *The China Medical Journal,* 1909–31. The China Medical Missionary Association. Peking, Shanghai.

Chinese Journal of Physiology [Chung-kuo sheng-li-hsüeh tsa-chih]. 中國生理學雜誌 Chinese Society of Physiology [Chung-kuo sheng-li-hsüeh hui]. 中國生理學會 Peking, 1927–   .

The Chinese Recorder and Missionary Journal. Foochow, Shanghai, 1868–72, 1874–1941.

Ch'ün chung 群眾 [The masses]. Ch'ün-chung chou k'an she 群眾周刊社 [*The masses* weekly publication society]. Hankow, Chungking, Shanghai, 1937–47.

Chung-hua chiao-yü chieh 中華教育界 [Chung-hua educational review]. Chung-hua chiao-yü-chieh tsa-chih she 中華教育界雜誌社 [Chung-hua Educational Review Magazine Society]. Shanghai, 1912–37.

Chung-kuo k'o-chi shih-liao 中國科技史料 [China historical materials of science and technology]. "Chung-kuo k'o-chi shih-liao" pien-chi-pu 「中國科技史料」編輯部 [Editorial department of "China historical..."]. Peking, 1980–   .

Hsin chiao-yü 新教育 [New education]. Shang-hai hsin chiao-yü kung-chin-she 上海新教育共進社 [Shanghai Common Progress Society for New Education]. Shanghai, 1919–25.

Hsin ch'ing-nien 新青年 [New youth]. Shanghai, Peking, Canton, 1915–22.

Hua-hsüeh kung-ch'eng 化學工程 [Journal of chemical engineering, China]. Chung-kuo hua-hsüeh kung-ch'eng hsüeh-hui 中國化學工程學會 [Chinese Institute of Chemical Engineers]. Tientsin, 1934–49.

Hua-hsüeh kung-yeh 化學工業 [Chemical industry]. Chung-hua hua-hsüeh kung-yeh hui 中華化學工業會 [Chinese Society of Chemical Industry]. Peking, Shanghai, Chungking, 1923–49. Title varies, vols. 1–3: *Chung-hua hua-hsüeh kung-yeh-hui hui chih* 中華化學工業會會誌 [The journal and proceedings of the China Society of Chemical Industry].

Hua-hsüeh shih-chieh 化學世界 [Chemical world]. Shang-hai-shih hua-hsüeh hua-kung hsüeh-hui 上海市化學化工學會 [Shanghai Chemical Industry Society]. Shanghai, 1946–66.

Hua-hsüeh t'ung-pao 化學通報 [Chemistry]. K'o-hsüeh ch'u-pan-she 科學出版社 [Science Publishers]. Peking, 1951– .

Japanese Studies in the History of Science. Nippon Kagakusi Gakkai 日本科學 史學會 [The History of Science Society of Japan]. Tokyo, 1962–

Journal of the American Chemical Society. The American Chemical Society. Washington. DC, 1979– .

Journal of Biological Chemistry. Rockefeller Institute for Medical Research. Baltimore, 1905– .

Journal of Chemical Physics. American Institute of Physics. Lancaster, PA, 1933– .

Journal of the Chinese Chemical Society [Chung-kuo hua-hsüeh-hui hui-chih]. 中國化學會會誌 Chinese Chemical Society [Chung-kuo hua-hsüeh hui]. 中國化學會 Nanking, Chungking, 1933–66.

Ko-chih hui-pien 格致彙編 [The Chinese scientific magazine]. John Fryer. Shanghai, 1876–7, 1880–1, 1890–2.

K'o-hsüeh 科學 [Science]. Chung-kuo k'o-hsüeh-she 中國科學社 [Science Society of China]. Shanghai, 1915–58.

Kung-yeh chung-hsin 工業中心 [Industrial center]. Chung-yang kung-yeh shih-yen-so 中央工業試驗所 [National Bureau for Industrial Research]. Nanking, Chungking, 1932–49.

Kuo-li Pei-p'ing yen-chiu-yüan yüan-wu hui-pao 國立北平研究院院務滙報 [Bulletin of the National Academy of Peiping]. Peking: Peiping Academy of Sciences, 1930.

National Central University Science Reports, Series A, Physical Sciences [Kuo-li chung-yang ta-hsüeh k'o-hsüeh yen-chiu-lu, chia tsu, wu-chih k'o-hsüeh]. 國立中央大學科學研究錄，甲組物質科學 National Central University [Kuo-li chung-yang ta-hsüeh]. 國立中央大學 Nanking, 1930–3.

Proceedings of the Society for Experimental Biology and Medicine. Society for Experimental Biology and Medicine. New York, 1903– .

Science Quarterly of the National University of Peking [Kuo-li Pei-ching ta-hsüeh, tzu-jan k'o-hsüeh chi-k'an]. 國立北京大學自然科學季刊 National University of Peking, College of Science [Kuo-li Pei-ching ta-hsüeh li-hsüeh-yüan]. 國立北京大學理學院 Peiping, 1929–35.

Science Reports of National Tsing Hua University, Series A: Mathematical and Physical Science [Kuo-li Ch'ing-hua ta-hsüeh li-k'o pao-kao, ti-i-chung.] 國立清華大學理科報告，第一種 National Tsing Hua University [Kuo-li Ch'ing-hua ta-hsüeh]. 國立清華大學 Peiping, 1931–3.

Tu-li p'ing-lun 獨立評論 [Independent critic]. Tu-li p'ing-lun she 獨立評論社 [Independent Critic Society]. Peiping, 1932–7.

Tung-fang tsa-chih 東方雜誌 [Eastern miscellany]. Tung-fang tsa-chih she 東方雜誌社 [Eastern Miscellany Society]. Shanghai, Changsha, Hongkong, Chungking, 1904–48.

Tzu-jan k'o-hsüeh shih yen-chiu 自然科學史研究 [Studies in the history of natural sciences]. "Tzu-jan k'o-hsüeh shih yen-chiu" pien-chi wei-yüan-hui 「自然 科學史研究」編輯委員會. [Editorial Committee for the *TJKHSYC*]. Peking, 1981– .

Wen-hua hsien-feng 文化先鋒 [Cultural pioneer]. Wen-hua hsien-feng she 文化 先鋒社 [Cultural pioneer society]. Chungking, 1942–8.

Yenching Natural Science News. Yenching University, College of Sciences. Peking, 1933–6.

CHINESE AND JAPANESE

Ch'ai Ching-hsü. 柴景旭 "Chin Ku T'ang-shan nan-Man kung-yeh ts'an-kuan pao-kao" 津沽唐山南滿工業參觀報告 [Report on an inspection of the industry of Tientsin, Tangku, Tangshan, and southern Manchuria]. *HHKY* 5, no. 2 (May 1930): 92–145.

Chang Chen. 張珍 "Chin-Ch'a-Chi ken-chü-ti kung-yeh chi-shu yen-chiu-shih ti kai-mao chi ch'eng-chiu" 晉察冀根據地工業技術研究室的概貌及成就 [Description and accomplishments of the Industrial Technology Research Laboratory of the Shansi-Chahar-Hopeh Base Area]. *KJCC*, 5:252–62.

Chang Ch'in. "Pu-shu Lien-ta ti yü-le-i-shih-chu-hsing" 補述聯大的育樂衣食住行 [More on education, recreation, clothing, food, housing, and activities at Associated University]. *Ch'ing-hua hsiao-yu t'ung-hsün* 清華校友通訊 [Tsinghua alumni bulletin] 77 (31 October 1981).

Chang Ch'ing-lien. 張青蓮 "Hsü Shou yü 'Hua-hsüeh chien-yüan'" 徐壽與「化學鑑原」 [Hsü Shou and the "Hua-hsüeh chien yüan"]. *CKKCSL* 6, no. 4 (1985):54–6.

Chang Chün-mai. 張君勱 "Jen-sheng-kuan" 人生觀 [The philosophy of life]. *KHYJSK*.

Chang Hung-yüan. 張洪沅 "San-shih-nien-lai Chung-kuo chih hua-hsüeh kung-ch'eng" 三十年來中國之化學工程 [Chinese chemical engineering during the past thirty years]. *San-shih-nien-lai chih Chung-kuo kung-ch'eng*. 三十年來之中國工程 Vol. 1. Ed. Chou K'ai-ch'ing. 周開慶 Taipei: Wen-hua shu-chü 文化書局, 1967.

Chang I-tsun. 張儀尊 "Chung-kuo ti hua-hsüeh" 中國的化學 [Chinese chemistry]. *Chung-hua min-kuo k'o-hsüeh-chih*. 中華民國科學誌 Ed. Li Hsi-mou. 李熙謀

Chang Po. 張博 "Wo so chih-tao ti Yen-an Hsing-Hua p'i-ke-ch'ang" 我所知道的延安興華皮革廠 [The Yenan Hsing-Hua Leather Factory that I knew]. *KJCC*, 3:301–10.

Chang Te-yao. 張德耀 "Chin-Ch'a-Chi k'ang Jih ken-chü-ti chün-shih kung-yeh chien-she kung tso hui-i" 晉察冀抗日根據地軍事工業建設工作回憶 [Recollections of military industry construction work in the Chin-Ch'a-Chi Anti-Japanese Base]. *KJCC*, 1:182–97.

Chang Te-yao. "K'ang-Jih chan-cheng shih-ch'i Chin-Ch'a-Chi ken-chü-ti ti cha-yao sheng-ch'an" 抗日戰爭時期晉察冀根據地的炸藥生產 [Production of explosives in the Shansi-Chahar-Hopeh Base Area during the War of Resistance]. *KJCC*, 4:296–304.

Chang Tzu-kao, 張子高 ed. *Chung-kuo ku-tai hua-hsüeh-shih* 中國古代化學史 [History of ancient Chinese chemistry]. Hong Kong: Shang-wu yin-shu-kuan, 商務印書館 1977.

Chang Tzu-kao. "Ho Liao-jan ti 'Hua-hsüeh ch'u-chieh' tsai hua-hsüeh yüan-su i-ming shang ti li-shih i-i" 何瞭然的「化學初階」在化學元素譯名上的歷史意義 [Historical significance of Ho Liao-jan's "Hua-hsüeh ch'u-chieh" for the translation of names of chemical elements]. *Ch'ing-hua ta-hsüeh hsüeh-pao* 清華大學學報 [Tsinghua University Journal] 9, no. 6 (December 1962): 41–7.

Chang Tzu-kao and Yang Ken. 楊根 "Chieh-shao yu-kuan Chung-kuo chin-tai hua-hsüeh shih ti i-hsiang ts'an-k'ao tsu-liao – 'Ya-ch'üan tsa-chih'" 介紹有關中國近代化學史的一項參考資料「亞泉雜誌」 [Introducing reference material related to the history of modern chemistry in China – the "Ya-ch'üan

Magazine"]. *HHTP* 1 (1965):55–9.

Chang Tzu-kao and Yang Ken. "Hsü Shou fu-tzu nien-p'u" 徐壽父子年譜 [Annual record of Hsü Shou and son]. *CKKCSL* 2, no. 4 (1981):55–62.

Chang Tzu-kao and Yang Ken. "Ts'ung 'Hua-hsüeh ch'u-chieh' ho 'Hua-hsüeh chien-yüan' k'an wo-kuo tsao-ch'i fan-i ti hua-hsüeh shu-chi ho hua-hsüeh ming-ts'u" 從「化學初階」和「化學鑑原」看我國早期翻譯的化學書籍和化學名詞 [Viewing our country's early translations of chemistry books and chemical terminologies through *First step to chemistry* and *Mirror of chemical science*]. *TJKHSYC* 1, no. 4 (1982):349–55.

Chang Tzu-kao and Yang Ken. "Ya-p'ien chan-cheng i'-ch'ien hsi-fang hua-hsüeh ch'uan-ju Chung-kuo ti ch'ing-k'uang" 鴉片戰爭以前西方化學傳入中國的情況 [Introduction of Western chemistry into China before the Opium War]. *Chung-kuo ku-tai hua-hsüeh-shih*. Ed. Chang Tzu-kao.

Chao Erh-hsün, 趙爾巽 et al. *Ch'ing-shih kao* 清史稿 [A draft history of the Ch'ing Dynasty]. Vol. 46. Peking: Chung-hua shu-chü, 中華書局 1977.

Chao I-feng. 趙一峯 "Fa-chan pien-ch'ü k'o-hsüeh shih-yeh" 發展邊區科學事業 [Develop science in the border regions]. *CFJP*, 29 November 1941, p. 4.

Ch'en Chen 陳真 and Yao Lo, 姚洛 eds. *Chung-kuo chin-tai kung-yeh-shih tsu-liao* 中國近代工業史資料 [Materials on China's modern industrial history]. 4 vols. Peking: San-lien shu-tien, 三聯書店 1957–61.

Ch'en Chen-hsia 陳振夏 and Hu Hua-ch'in. 胡華欽 "Yen-ch'ang shih-yu-ch'ang mu-ch'ien ti kai-liang ho chin-hou ti fa-chan chi-hua" 延長石油廠目前的改良和今後的發展計劃 [Present improvements and plans for future development of the Yenchang Petroleum Factory] (Report of 1939). *KJCC*, 3:152–6.

Ch'en Chien-shan. 陳兼善 "Chung-hsüeh-hsiao chih po-wu-hsüeh chiao-shou" 中學校之博物學教授 [Natural history teaching in middle schools]. *CYTC* 14, no. 6 (1922), bk 35:19,927–42.

Ch'en Hsin-wen. 陳歆文 "Chi Yung-li hua-hsüeh kung-yeh kung-szu p'ei-yang jen-ts'ai ti tso-fa" 記永利化學工業公司培養人才的作法 [Record of the methods of developing human resources of the Yungli Chemical Industry Company]. *CKKCSL* 2, no. 3 (1981):28–34.

Ch'en Hsin-wen. "Chung-kuo hua-kung hsüeh-hui" 中國化工學會 [China Society of Chemical Industry]. *CKKCSL* 3, no. 3 (1982):57–62.

Ch'en Hsün-tz'u. 陳訓慈 "So-wang yü Chung-kuo k'o-hsüeh-chia-che" 所望於中國科學家者 [Expectations of Chinese scientists]. *KH* 20, no. 10 (October 1936): 884–90.

Ch'en Kuo-fu, 陳果夫 "Kai-ke chiao-yü ch'u-pu fang-an" 改革教育初步方案 [Preliminary plan for educational reform]. *Ch'en Kuo-fu hsien-sheng ch'üan-chi* 陳果夫先生全集 [Complete works of Ch'en Kuo-fu]. Vol. 1. Taipei: Cheng-chung shu-chü, 正中書局 1952.

Ch'en Tai-sun. 陳岱孫 "Kuan-yü ta-hsüeh pi-yeh-sheng chih-yeh wen-t'i i-ke chien-i" 關于大學畢業生職業問題一個建議 [A proposal on the problem of employment of university graduates). *TLPL* 211 (26 July 1936):8–12.

Ch'en T'ao-sheng. 陳駒聲 "Chi-nan hua-hsüeh kung-yeh chih kai-k'uang" 濟南化學工業之概況 [Status of the Tsinan chemical industry]. *HHKY* 3, no. 1 (January 1925): 213–58.

Ch'en T'ao-sheng. "Hu Han hua-hsüeh kung-yeh k'ao-ch'a-chi" 滬漢化學工業考察記 [Report on an investigation of the chemical industries of Shanghai and Hankow]. *HHKY* 2, no. 1 (January 1924): 211–52.

Ch'en Te-yüan. 陳德元 "Liu-suan yü hsiao-suan chih-tsao fang-fa chih ko-ming"

硫酸與硝酸製造方法之革命 [Revolution in the method of manufacture of sulfuric and nitric acids]. *HHKY* 12, no. 1 (January 1937):2–10.

Ch'en T'i-jung. 陳體榮 "San-shih-nien-lai Chung-kuo chih lien-yu kung-yeh" 二十年來中國之煉油工業 [China's oil refining industry during the past thirty years]. *San-shih-nien-lai chih Chung-kuo kung-ch'eng.* Vol. 2. Ed. Chou K'ai-ch'ing.

Ch'en Tu-hsiu. 陳獨秀 "Tsai lun K'ung-chiao wen-t'i" 再論孔教問題 [Again discussing the problem of Confucianism]. *HCN* 2, no. 5 (1 January 1917): 1st article.

Ch'eng Fang-p'ing. 程方平 "Hsü T'e-li t'ung-chih yü tzu-jan k'o-hsüeh chiao-yü" 徐特立同志與自然科學教育 [Comrade Hsü T'e-li and natural science education]. *CKKCSL* 3, no. 4 (1982):4–9.

Ch'eng Ping-hua. 程炳華 "Chiu-chi shih-yeh ta-hsüeh-sheng chung ying chu-i ti chi-tien" 救濟失業大學生中應注意的幾點 [A few noteworthy points on the relief of unemployment among university students]. *TLPL* 219 (20 September 1936), 17–19.

Chi Hung-k'un. 季鴻崑 "Hsü Chien-yin yü Chung-kuo wu-yen huo-yao ti yen-chih" 徐建寅與中國無煙火藥的研製 [Hsü Chien-yin and research and development on Chinese smokeless gunpowder]. *TJKHSYC* 4, no. 1 (1985): 90–8.

Chi Hung-k'un and Wang Chih-hao. 王治浩 "Wo-kuo Ch'ing-mo ai-kuo k'o-hsüeh-chia Hsü Chien-yin" 我國清末愛國科學家徐建寅 [Patriotic scientist in our country during the late Ch'ing, Hsü Chien-yin]. *TJKHSYC* 4, no. 3 (1985): 184–94.

Ch'i Ju-shan. 齊如山 "Ch'i Ju-shan tzu-chuan".齊如山自傳 [Autobiography of Ch'i Ju-shan]. *Chung-kuo i-chou* 中國一周 [China newsweek] 240 (November 1954):18–19.

Chi Tse. 李澤 "Chih-ke-ch'ang tzu-chi chih-tsao tan-ning" 製革廠自己製造丹寧 [The leather factories themselves make tannin]. *CFJP*, 15 December 1943. Reprinted in *KJCC*, 4:103–4.

Chia Kuan-jen. 賈觀仁 "Chung-teng hsüeh-hsiao li-hua-hsüeh chiao-chou fa kai liang i-chien shu" 中等學校理化學教授法改良意見書 [Proposals for improving science teaching methods in middle schools]. *CYTC* 10, no. 10 (1918), bk 26:14,309–12.

Chiang Hsiang. 江湘 "Chen-Hua tsao-chih-ch'ang ts'an-kuan chi" 振華造紙廠參觀記 [Notes on visit to the Chen Hua Paper Factory]. *KJCC*, 2:178–81.

Chiang Hsiang. "Hsin Hua hua-hsüeh-ch'ang ch'an-p'in tseng-chia, ping tsai shih-yen chih-tsao hsin ch'an-p'in" 新華化學廠產品增加，并在試驗製造新產品 [Production increases and new research and development products of the New China Chemical Factory]. *KJCC*. 2:188–90.

Chiang Ping-jan. 蔣丙然 "Erh-shih-nien-lai Chung-kuo ch'i-hsiang shih-yeh kai-k'uang" 二十年來中國氣象事業概況 [Chinese meteorology during the past twenty years]. *KH* 20, no. 8 (1936): 623–42.

Chiang Shu-yüan. 蔣樹源 "Chi-nien wo-kuo shih-chiu shih-chi cho-yüeh ti hua-hsüeh-chia Hsü Shou" 紀念我國十九世紀卓越的化學家徐壽 [In memory of our country's outstanding chemist of the nineteenth century, Hsü Shou]. *Hsü Shou ho Chung-kuo chin-tai hua-hsüeh shih*, 67–72.

Chiang T'ing-fu. 蔣廷黻 "Ch'en Kuo-fu hsien-sheng ti chiao-yü cheng-ts'e" 陳果夫先生的教育政策 [Educational policy of Mr. Ch'en Kuo-fu]. *TLPL* 4 (12 June 1932):6–8.

Chiang T'ing-fu. "Tui ta-hsüeh hsin-sheng kung-hsien chi-tien i-chien" 對大學新

生貢獻幾點意見 [A few ideas on the contributions of new university students]. *TLPL* 69 (24 September 1933):5–10.

Chiao-yü-pu hua-hsüeh t'ao-lun-hui chuan-k'an 教育部化學討論會專刊 [Report of the Chemistry Forum of the Ministry of Education]. Nanking: Kuo-li pien-i-kuan 國立編譯館 [National Institute for Compilation and Translation], 1932.

Chih Feng. 知峰 "Fan Hsü-tung: wo-kuo hua-hsüeh kung-yeh ti t'o-huang-che" 范旭東：我國化學工業的拓荒者 [Fan Hsü-tung: trailblazer of our country's chemical industry]. *CKKCSL* 1, no. 3 (1980):2–9, 20.

Ch'ing-hua ta-hsüeh hsiao-shih pien-hsieh-tsu 清華大學校史編寫組 [Committee for Editing and Writing the History of Tsinghua University]. *Ch'ing-hua ta-hsüeh hsiao-shih kao* 清華大學校史稿 [Draft history of Tsinghua University]. Peking: Chung-hua shu-chü, 中華書局 1981.

Chou Ch'ang-shou. 周昌壽 "I-k'an k'o-hsüeh shu-chi k'ao-lüeh" 譯刊科學書籍考略 [Study of translation and publication of science books]. *Chang Chü-sheng hsien-sheng ch'i-shih sheng-jih chi-nien lun-wen chi* 張菊生先生七十生日紀念論文集 [Essays in honor of the seventieth birthday of Mr. Chang Chü-sheng]. Ed. Hu Shih, 胡適 et al. Shanghai: Shang-wu yin-shu-kuan, 商務印書館 1936.

Chou Ch'ang-yün. "T'u-jang fei-liao-hsüeh" 土壤肥料學 [Soil fertilization studies]. *Chung-hua min-kuo k'o-hsüeh-chih*. 中華民國科學誌 Ed. Li Hsi-mou. 李熙謀

Chou Fa-ch'i. 周發岐 "Hua-hsüeh yen-chiu yü hua-hsüeh kung-yeh" 化學研究與化學工業 [Chemical research and chemical industry]. *HHKC* 6, nos. 3–4 (December 1939):66–7.

Chou Hsin, 周昕 Yang Ken, 楊根 and Pai Kuang-mei. 白廣美 "Yu-hsiu ti chiao-yü-chia hua-hsüeh-shih-chia – Chang Tzu-kao chiao-shou" 優秀的教育家，化學史家–張子高教授 [Outstanding educator and historian of chemistry – Professor Chang Tzu-kao]. *HHTP* 10 (1980):622–6.

Chou K'ai-ch'ing, 周開慶 ed. *San-shih-nien-lai chih Chung-kuo kung-ch'eng* 三十年來之中國工程 [Chinese engineering during the past thirty years]. 2 vols. Taipei: Wen-hua shu-chü, 文化書局 1967.

Chou T'ien-tu. 周天度 *Ts'ai Yüan-p'ei chuan* 蔡元培傳 [Biography of Ts'ai Yüan-p'ei]. Peking: Jen-min ch'u-pan-she, 人民出版社 1984.

Chu Chia-hua. 朱家驊 "Ch'ing-nien yü k'o-hsüeh" 青年與科學 [Youth and science]. *Chu Chia-hua hsien-sheng yen-lun-chi* 朱家驊先生言論集 [Dissertations of Dr. Chu Chia-hua]. Ed. Wang Yee-chun 王聿均 and Sun Pin. 孫斌 Taipei: Institute of Modern History, Academia Sinica, 1977.

Chu Chia-hua. "Ting Wen-chiang yü chung-yang yen-chiu-yüan" 丁文江與中央研究院 [Ting Wen-chiang and the Academia Sinica]. *Chu Chia-hua hsien-sheng yen-lun-chi* [Dissertations of Dr. Chu Chia-hua]. Ed. Wang Yee-chun and Sun Pin. Taipei: Institute of Modern History, Academia Sinica, 1977.

Ch'ü Po-ch'uan 屈伯傳 and Wei Chih. 衛之 "Yen-an tzu-jan k'o-hsüeh-yüan" 延安自然科學院 [The Yenan Natural Science Institute]. *KJCC.* 4:242–60.

Chu Yüan-shan. 朱元善 "Sheng-ch'an chu-i chih li-k'o chiao-shou" 生產主義之理科教授 [Science teaching for production]. *CYTC* 9, no. 1 (1917), bk 21: 11,403–9.

Chung-kung jen-ming-lu 中共人名錄 [Biographies of Chinese Communist personalities]. Rev. ed. Taipei: Institute of International Relations, 1978.

Chung-kuo chiao-yü nien-chien, ti-erh-tz'u 中國教育年鑑，第二次 [The second China education year book]. Ed. Chiao-yü-pu 教育部 [Ministry of Education]. Shanghai: Shang-wu yin-shu-kuan, 商務印書館 1948.

Chung-kuo chiao-yü nien-chien, ti-i-tz'u 第一次 [The first China education year book]. 4 vols. Ed. Chiao-yü-pu [Ministry of Education]. Shanghai: K'ai-ming shu-tien, 開明書店 1934.

"Chung-kuo hua-hsüeh-hui ti nien-hui huo-tung" 中國化學會的年會活動 [Annual meetings of the Chinese Chemical Society]. *HHTP* 9 (1982):552.

Chung-kuo k'o-hsüeh-chia ts'u-tien 中國科學家辭典 [Dictionary of Chinese scientists]. 4 vols. Tsinan: Shan-tung k'o-hsüeh chi-shu ch'u-pan-she, 山東科學技術出版社 1982.

"Chung-kuo k'o-hsüeh-yüan hua-hsüeh-pu hsüeh-pu wei-yüan chien-chieh" 中國科學院化學部學部委員簡介 [Brief introduction to the members of the Department of Chemistry of the Chinese Academy of Sciences]. *HHTP* 9 (1981):552–64.

Ch'ung Shih. 崇實 "Tzu-jan k'o-hsüeh ti yen-chiu kung-tso" 自然科學的研究工作 [Natural science research work]. *PCF*, 100–5.

"Chung-shih chung-yang yen-chiu-yüan p'ing-i-hui chih chien-i" 重視中央研究院評議會之建議 [Emphasizing the proposal of the National Research Council of the Academic Sinica]. *Wen-hua hsien-feng* 文化先鋒 [Cultural pioneer] 3, no. 12 (21 March 1944):2.

Fan Ching-p'ing. 范敬平 "Kuo-ch'an liu-suan chih fen-hsi" 國產硫酸之分析 [Analysis of sulfuric acid produced in China]. *KYCH* 5, no. 12 (December 1936):578–82.

Fan Jui 范銳 [Fan Hsü-tung]. 范旭東 "Yung-li chih-chien kung-szu ta-shih-chi" 永利製鹼公司大事紀 [Description of the Yungli Soda Manufacturing Company]. *HHKY* 2, no. 1 (January 1924):253–60.

Fan Mu-han. 范慕韓 "Shen-Kan-Ning pien-ch'ü ti kung-yeh chien-she" 陝甘寧邊區的工業建設 [Industrial development in the Shen-Kan-Ning Border Region]. *KJCC*, 5:167–82.

Fang Ai-chi. 方藹吉 "Wo-kuo tsui-tsao ti tsao-ch'uan chuan-k'o hsüeh-hsiao – Fu-chou ch'uan-cheng-chü ch'ien-hsüeh-t'ang" 我國最早的造船專科學校 – 福州船政局前學堂 [Our country's earliest shipyard technical school – The Foochow Naval Yard First School]. *CKKCSL* 6, no. 5 (1985):57–62.

Fang Hsin-fang, 方心芳 Wei Wen-te, 魏文德 and Chao Po-ch'üan. 趙博泉 "Huang-hai hua-hsüeh kung-yeh yen-chiu-she kung-tso kai-yao" 黃海化學工業研究社工作概要 [Summary of the work of the Golden Sea Research Insitute for Chemical Industry]. *HHTP* 9 (1982):559–64.

Fu Lan-ya [John Fryer]. 傅蘭雅 *Hua-hsüeh hsü-chih* 化學須知 [Chemistry, outline series]. Shanghai: n.p., 1886.

Fu Lan-ya [John Fryer]. *Hua-hsüeh i-chih* 化學易知 [Chemistry, handbook series]. Shanghai: I-chih shu-hui, 易知書會 1881.

Han Tsu-k'ang 韓組康 and Chang Yüan-lang. 章元琅 "Chung-kuo hua-hsüeh-chia tui-yü fen-hsi hua-hsüeh ti kung-hsien" 中國化學家對於分析化學的貢獻 [Contributions of Chinese chemists to analytical chemistry]. *HHSC* 8, no. 10 (1953):340–5.

Hao Ching-sheng. 郝景盛 "K'ang-chan ch'i-nien lai chih k'o-hsüeh" 抗戰七年來之科學 [Science during the last seven years of the War of Resistance]. *Chung-kuo chan-shih hsüeh-shu* 中國戰時學術 [Chinese wartime scholarship]. Ed. Sun Pen-wen. 孫本文 Shanghai: Cheng-chung shu-chü, 正中書局 1946.

Ho Ch'un-po. 何純渤 "Wu Yü-chang t'ung-chih yü tzu-jan k'o-shüeh yen-chiu-hui" 吳玉章同志與自然科學研究會 [Comrade Wu Yü-chang and the Natural Science Research Society]. *KJCC* 2:64–71.

Ho Hsin 合信 [Benjamin Hobson]. *Po-wu hsin-pien* 博物新編 [Natural philosophy and natural history]. Shanghai: Mo-hai shu-kuan, 墨海書館 1855.

Ho Shang-p'ing. 何尚平 "Tan-chih hua-hsüeh fei-liao yü wo-kuo nung-yeh" 氮質化學肥料與我國農業 [Nitrogenous chemical fertilizer and our country's agriculture]. *HHKY* 8, no. 1 (January 1933):21–30.

Hou Te-feng 侯德封 and Ts'ao Kuo-ch'üan. 曹國權 "San-shih-nien-lai Chung-kuo chih mei-k'uang shih-yeh" 三十年來中國之煤礦事業 [China's coal mining industry during the past thirty years]. *San-shih-nien-lai chih Chung-kuo kung-ch'eng.* Vol. 2. Ed. Chou K'ai-ch'ing.

Hsi-hsüeh fu-ch'iang ts'ung-shu 西學富強叢書 [Collection on wealth and power through Western studies]. Comp. Chang Yin-huan 張蔭桓. 48 vols. Hung-wen shu-chü 鴻文書局,1896.

Hsi Yeh. 西野 "T'ien-li tan-ch'i-ch'ang kai-k'uang" 天利淡氣廠概況 [Status of the T'ien-li Nitrogen Gas Factory]. *HHSC* 6, no. 2 (February 1951): 10–16.

Hsieh Chia-jung 謝家榮 and Chin K'ai-ying. 金開英 "Jan-liao yen-chiu yü Chung-kuo ti jan-liao wen-t'i" 燃料研究與中國的燃料問題 [Fuels research and China's fuel problem]. *KH* 17, no. 10 (October 1933):1717–29.

Hsieh Kuo-chen. 謝國楨 *Ming-Ch'ing chih chi tang-she yün-tung k'ao* 明清之際黨社運動考 [Examination of the movement of associations during the Ming-Ch'ing period]. Shanghai, 1934. Reprinted in Taipei: Commercial Press, 1967.

Hsü Ch'ang-yü. 徐昌裕 "Yen-ch'ang shih-yu-ch'ang hui-i p'ien-tuan" 延長石油廠回憶片斷 [Recollections of the Yenchang Petroleum Factory]. *KJCC*, 5:213–29.

Hsü Chen-ya 徐振亞 and Juan Chen-k'ang. 阮愼康 "Hsü Shou fu tzu, tsu-sun i-chu chien-chieh" 徐壽父子,祖孫譯著簡介 [Brief introduction to the translations of Hsü Shou, his sons, and descendants]. *CKKCSL* 7, no. 1 (1986): 48–55.

Hsü Chih-fang. 許植方 "Chung-kuo hua-hsüeh-chia tui-yü yao-wu hua-hsüeh ti kung-hsien" 中國化學家對於藥物化學的貢獻 [Contributions of Chinese chemists to pharmacological chemistry]. *HHSC* 10, no. 7 (July 1955):296–306.

Hsü Hsien-kung. 徐賢恭 "Chung-kuo hua-hsüeh-chia tui-yü yu-chi hua-hsüeh ti kung-hsien" 中國化學家對於有機化學的貢獻 [Contributions of Chinese chemists to organic chemistry]. *HHSC* 9, nos. 6–8 (1954):230–3, 326–32.

Hsü I-sun, 徐以愻 comp., *Tung hsi-hsüeh shu-lu* 東西學書錄 [Eastern bibliography of Western studies]. Shanghai: Chiang-nan chih-tsao-chü, 江南製造局 1899.

Hsü Ming-hsiu. 許明修 "Chien-k'u ch'uang-yeh, wei chiao-hsüeh ho pien-ch'ü chien-she fu-wu" 艱苦創業,爲教學和邊區建設服務 [The difficult enterprise, service to education and the construction of the border regions]. *KJCC*, 1:126–9.

Hsü Shan-hsiang. 徐善祥 "Chung-kuo jen-tsao fei-liao chih hsien-tsai chi chiang-lai" 中國人造肥料之現在及將來 [The present and future of China's artificial fertilizer]. *HHKY* 8, no. 1 (January 1933):1–20.

Hsü Shou ho Chung-kuo chin-tai hua-hsüeh shih 徐壽和中國近代化學史 [Hsü Shou and the history of chemistry in modern China]. Ed. Yang Ken. 楊根 Peking: K'o-hsüeh chi-shu wen-hsien ch'u-pan-she, 科學技術文獻出版社 1986.

Hsü T'e-li. 徐特立 "Tsen-ma-yang fa-chan wo-men ti tzu-jan k'o-hsüeh" 怎麼樣發展我們的自然科學 [How to develop our natural sciences]. *CFJP*, 24 and 25 September 1941, p. 3.

Hsü T'e-li. "Tui Niu-tun ying-yu ti jen-shih" 對牛頓應有的認識 [What we should know about Newton]. *CFJP*, 14 January 1943, p. 4.

Hu Chi-ch'üan. 胡吉金 "Hui-i Yen-an tzu-jan k'o-hsüeh-yüan ti hsüeh-hsi sheng-huo" 回憶延安自然科學院的學習生活 [Remembering the study life at the Yenan Natural Science Institute]. *KJCC* 1:137–43.

Hu Heng-ch'en. 胡衡臣 "Ch'u-chi chung-hsüeh ti li-hua chiao-hsüeh- fa" 初級中學的理化教學法 [Physics and chemistry teaching methods in lower middle schools]. *CYTC* 17, no. 6 (1925), bk 44:26, 171–82.

Hu Hsien-su. 胡先驌 "Erh-shih-nien-lai Chung kuo chih wu-hsüeh chih chin-pu" 二十年來中國植物學之進步 [Progress in Chinese botany during the past twenty years]. *KH* 19, no. 10 (1935): 1555–9.

Hu Hsien-su. "K'o-hsüeh yü chien-kuo" 科學與建國 [Science and national reconstruction]. *Wen-hua hsien-feng* 文化先鋒 [Cultural pioneer] 2, no. 21 (10 October 1943):5–7.

Hu Po-yüan. 胡博淵 "San-shih-nien-lai Chung-kuo chih kang-t'ieh shih-yeh" 三十年來中國之鋼鐵事業 [China's iron and steel industry during the past thirty years]. *San-shih-nien-lai chih Chung-kuo kung-ch'eng*. Vol. 2. Ed. Chou K'ai-ch'ing.

Hua-hsüeh chien-yüan 化學鑑原 [Mirror of chemical science]. Trans. John Fryer 傅蘭雅 and Hsü Shou 徐壽 of David A. Wells, *Wells's Principles and Applications of Chemistry* (New York, 1858). Shanghai: Kiangnan Arsenal, 1872. *Hsi-hsüeh fu-ch'iang ts'ung-shu*. Vols. 9–10. 1902.

Hua-hsüeh chien-yüan hsü-pien 化學鑑原續編 [Continuation of mirror of chemical science]. Trans. John Fryer and Hsü Shou of "Organic Chemistry," a section in Charles L. Bloxam, *Bloxam's Chemistry: Inorganic and Organic, with Experiments* (London, 1867). Shanghai: Kiangnan Arsenal, 1875.

Hua-hsüeh chien-yüan pu-pien 化學鑑原補編 [Supplement to mirror of chemical science]. Trans. John Fryer and Hsü Shou of "Chemistry of Non-metallic Elements" and "Chemistry of Metals," sections in Charles L. Bloxam, *Bloxam's Chemistry: Inorganic and Organic, with Experiments* (London, 1867). Shanghai: Kiangnan Arsenal, 1883.

Hua-hsüeh chih-nan 化學指南 [Guide to chemistry]. Trans. Anatole Billequin 畢利干 and Lien Tzu-chen 聯子振 of Faustino Malaguti, *Leçons élémentaires de chimie* (Paris, 1853–68). Peking: T'ung Wen Kuan, 同文館 1873.

Hua-hsüeh ch'iu-shu 化學求數 [Quantitative chemical analysis]. Trans. John Fryer and Hsü Shou of Karl R. Fresenius, *Quantitative Chemical Analysis* (London, 1876). Shanghai: Kiangnan Arsenal, 1883.

Hua-hsüeh ch'u-chieh 化學初階 [First step to chemistry]. Trans. John Kerr [Chia Yüeh-han] 嘉約翰 and Ho Liao-jan 何瞭然 of David A. Wells, *Well's Principles and Applications of Chemistry* (New York, 1858). Canton: Po-chi i-yüan, 博濟醫院 1871–5.

Hua-hsüeh fen-yüan 化學分原 [Chemical analysis]. Trans. John Fryer and Hsü Chien-yin 徐建寅 of John E. Bowman, *An Introduction to Practical Chemistry* (London, 1866). Shanghai: Kiangnan Arsenal, 1872.

Hua-hsüeh k'ao-chih 化學考質 [Chemical analysis]. Trans. John Fryer and Hsü Shou of Karl R. Fresenius, *Manual of Qualitative Chemical Analysis* (London, 1875). Shanghai: Kiangnan Arsenal 1883.

Hua-hsüeh kung-i 化學工藝 [Chemical manufacturing]. Trans. John Fryer and Wang Chen-sheng 汪振聲 of Georg Lunge, *A Theoretical and Practical Treatise on the Manufacture of Sulphuric Acid and Alkali* (London, 1880). Shanghai: Kiangnan Arsenal, 1898.

Hua-hsüeh ming-ming yüan-tse (tseng-ting-pen) 化學命名原則(增訂本)

[Principles of chemical nomenclature (revised and enlarged edition)]. Ed. Kuo-li pien-i-kuan 國立編譯館 [National Institute for Compilation and Translation], promulgated by the Ministry of Education, November 1932. Taipei: Cheng-chung shu-chü, 正中書局 1965.

Hua-hsüeh shan-yüan 化學闡原 [Explanation of the principles of chemistry]. Trans. Anatole Billequin, Ch'eng Lin, 承霖 and Wang Chung-hsiang 王鍾祥 of Karl R. Fresenius, *Chemical Analysis* (ed. unknown). Peking: T'ung Wen Kuan, 1882.

Hua-hsüeh ts'ai-liao Chung-Hsi ming-mu-piao 化學材料中西名目表 [Chinese-Western glossary of chemical substances]. Shanghai: Kiangnan Arsenal, 1885.

Hua Shou-chün. 華壽俊 "K'ang-Jih fei-tsao" 抗日肥皂 [Resist Japan soap]. *KJCC*, 4:323–6.

Hua Shou-chün. Tzu-fang-kou hua-hsüeh-ch'ang" 紫坊溝化學廠 [Tzu-fang-kou Chemical Factory]. *KJCC*, 5:230–3.

Hua Shou-chün. "Yung tzu-jan pien-cheng-fa chih-tao k'o-hsüeh chi-shu ti yen-chiu" 用自然辯證法指導科學技術的研究 [Use the dialectics of nature to direct scientific and technical research]. *PCF*, 303–7.

Huang Hai-lin, 黃海霖 "Tzu-li keng-sheng hsieh-tso fu-wu – i Shen-Kan-Ning ch'a-fang ping-kung-ch'ang chi-ch'i-pu" 自力更生協作服務…憶陝甘寧茶坊兵工廠機器部 [Self-reliance, cooperation, and service – remembering the machine department of the Shen-Kan-Ning Ch'a-fang Arsenal]. *KJCC*, 2:223–31.

I Jui-chen. 宜瑞珍 "Ma-lan-ts'ao tsao-chih shih-yen ti ch'eng-kung chieh-chüeh-le pien-ch'ü ti chih-chang k'un-nan" 馬蘭草造紙試驗的成功，解決了邊區的紙張困難 [Success of the experiments in making paper from *ma-lan* grass solved the paper problem of the border regions]. *KJCC*, 1:163–6.

Jen Hung-chün. 任鴻雋 "Chung-chi-hui yü Chung-kuo k'o-hsüeh" 中基會與中國科學 [The China Foundation and Chinese science]. *KH* 17, no. 9 (1933): 1521–4.

Jen Hung-chün. "Chung-kuo k'o-hsüeh-she chih kuo-ch'ü chi chiang-lai" 中國科學社之過去及將來 [The past and future of the Science Society of China]. *KH* 8, no. 1 (1923):1–9.

Jen Hung-chün. "Chung-kuo k'o-hsüeh-she she-shih chien-shu" 中國科學社社史簡述 [Brief account of the history of the Science Society of China]. *CKKCSL* 4, no. 1 (1983):2–13.

Jen Hung-chün. "I-ke kuan-yü li-k'o chiao-k'o-shu ti tiao-ch'a" 一個關於理科教科書的調查 [An investigation of science textbooks]. *KH* 17, no. 12 (1933): 2029–34.

Jen Hung-chün. "Jen Shu-yung hsien-sheng chih chiang-yen" 任叔永先生之講演 [Speech of Mr. Jen Shu-yung]. "Fu Ch'uan k'ao-ch'a-t'uan tsai Ch'eng-tu ta-hsüeh yen-shuo lu" 赴川考察團在成都大學演說錄 [Record of speeches made at Chengtu University by the Investigating Committee to Szechwan]. *KH* 15, no. 7 (1931): 1168–9.

Jen Hung-chün. "K'ang-chan hou ti k'o-hsüeh" 抗戰後的科學 [Science after the War of Resistance]. *TFTC* 37, no. 13 (1940):21–3.

Jen Hung-chün. "K'o-hsüeh-she kai-tsu shih-mo" 科學社改組始末 [Complete account of the reorganization of the Science Society]. *KH* 2, no. 1 (1916): 127–35.

Jen Hung-chün. "Shih-nien-lai Chung-chi-hui shih-yeh ti hui-ku" 十年來中基會事業的回顧 [Look back at ten years of work of the China Foundation]. *TFTC* 32, no. 7 (1935):19–25.

Jen Hung-chün. "Tao Hu Ming-fu" 悼胡明復 [Eulogy to Hu Ming-fu]. *KH* 13, no. 6 (1928): 822–6.

Jen Hung-chün. "Wai-kuo k'o-hsüeh-she chi pen-she chih li-shih" 外國科學社及本社之歷史 [The history of foreign science societies and our society]. *KH* 3, no. 1 (1917):2–18.

Jen Hung-chün. "Wu-kuo k'o-hsüeh yen-chiu chuang-k'uang chih i-pan" 吾國科學研究狀況之一斑 [Note on scientific research in our country]. *KH* 13, no. 8 (1929):1063–9.

Jen Hung-chün. "Wu-shih-nien-lai ti k'o-hsüeh" 五十年來的科學 [Science during the past fifty years]. *Wu-shih-nien-lai ti Chung-kuo* 五十年來的中國 [China during the past fifty years]. Ed. P'an Kung-chan. 潘公展 Chungking: Sheng-li ch'u-pan-she, 勝利出版社 1945.

K'ang-Jih chan-cheng shih-ch'i chieh-fang-ch'ü k'o-hsüeh chi-shu fa-chan shih tsu-liao 抗日戰爭時期解放區科學技術發展史資料 [Historical materials on the development of science and technology in the liberated areas during the War of Resistance]. 5 vols. Ed. Wu Heng. 武衡 Peking: Chung-kuo hsüeh-shu ch'u-pan-she, 中國學術出版社, 1983–5.

K'ang Ti. 康迪 "Tui Lo T'ien-yü t'ung-chih 'T'an kuan-yü Yen-an kan-pu hsüeh-hsiao ti chüeh-ting' chih shang-ch'üeh" 對樂天宇同志「談關於延安幹部學校的決定」之商榷 [Discussion of comrade Lo T'ien-yü's "On the resolution on the Yenan Cadre School"]. *CFJP*, 25 September 1942, p. 4. Reprinted in *PCF*, 141–6.

Kao Ai-t'ing. 高藹亭 "Hui-i Chin-Ch'a-Chi pien-ch'ü kung k'uang-chü ti ch'ou-chien ho fa-chan" 回憶晉察冀邊區工礦局的籌建和發展 [Remembering the establishment and development of the Industrial and Mining Bureau of the Chin-Ch'a-Chi Border Region]. *KJCC*, 2:244–9.

Kao I-sheng, 高怡生 Chu Jen-hung, 朱任宏 and Hsieh Yü-yüan. 謝毓元 "Wo-kuo Chung-ts'ao-yao hua-hsüeh yen-chiu ti hsien-ch'ü-che – Chao Ch'eng-ku chiao-shou" 我國中草藥化學研究的先驅者... 趙承嘏教授 [Pioneer in the study of Chinese medicinal herbs – Professor Chou Tsan-quo]. *HHTP* 3 (1980):178–81.

Kao Su. 高蘇 "Chung-kuo chih-chien kung-yeh ti hsien-ch'ü – Hou Te-pang po-shih" 中國製鹼工業的先驅－侯德榜博士 [Pioneer of China's soda industry – Dr. Hou Te-pang]. *HHTP* 5 (1979):461–8.

Kao Tzu-li. 高自立 "Wei kung-yeh-p'in ti ch'üan-mien tzu-chi erh fen-tou" 爲工業品的全面自給而奮鬥 [Struggle for complete self-sufficiency in industrial products]. *KJCC* 2:5–15.

K'o-hsüeh yü jen-sheng-kuan 科學與人生觀 [Science and the philosophy of life]. Ed. Ch'en Tu-hsiu 陳獨秀 and Hu Shih. 胡適 Shanghai: Ya-tung, 亞東 1923. Reprinted as *K'o-hsüeh yü jen-sheng-kuan chih lun-chan* 科學與人生觀之論戰 [Debate on science and the philosophy of life]. Ed. Wang Meng-tsou. 江孟鄒 Hong Kong: Chinese University of Hong Kong, 1973.

"K'o-hsüeh-yüan chan-k'ai t'ao-lun 'Ju-ho yü shih-chi lien-hsi?'" 科學院展開討論「如何與實際聯繫」 [Science Institute opens discussion of "How to link up with reality?"]. *CFJP*, 11 October 1942, p. 2.

"K'o-hsüeh-yüan chiao-yü fang-chen ti t'ao-lun tsung-chieh" 科學院教育方針的討論總結 [Summary of discussion on the educational policy of the Science Institute]. *CFJP*, 9 November 1942. Reprinted in *PCF*, 147–51.

Ku Hsing. 顧型 "Li-k'o chiao-shou ko-hsin chih yen-chiu" 理科教授革新之研究 [Research on the reform of science teaching]. *CYTC* 10, no. 1 (1918), bk 24: 13,076–84.

Ku Yü-chen. 顧毓珍 "Chung-kuo hua-hsüeh-chia tui-yü hua-hsüeh kung-ch'eng ti kung-hsien" 中國化學家對於化學工程的貢獻 [Contributions of Chinese chemists to chemical engineering]. *HHSC* 8, no. 10 (1953):333–9.

Ku Yü-chen and Fan Ching-p'ing. 范敬平 "Hua-hsüeh fen-hsi-shih kung-tso kai-k'uang" 化學分析試工作概況 [Survey of the work of the chemical analysis laboratory]. *KYCH* 4, no. 1 (1935):20–3.

Ku Yü-hsiu. 顧毓琇 "Ch'i k'o-hsüeh t'uan-t'i lien-ho nien-hui ti i-i ho shih-ming" 七科學團體聯合年會的意義和使命 [Significance and mission of the joint annual meeting of seven science organizations]. *TLPL* 215 (23 August 1936): 8–10.

Ku Yü-hsiu. "Chuan-men jen-ts'ai ti p'ei-yang" 專門人才的培養 [Cultivating technical personnel]. *TLPL* 71 (12 November 1933):8–11.

Ku Yü-hsiu. "Chün-shih ti chi-hsüeh-hua yü k'o-hsüeh-hua" 軍事的機械化與科學化 [The mechanization and scientification of warfare]. *TLPL* 201 (17 May 1936):25–8.

Ku Yü-hsiu "K'o-hsüeh yen-chiu yü kuo-chia hsü-yao" 科學研究與國家需要 [Scientific research and the nation's needs]. *TLPL* 210 (19 July 1936):5–8.

Ku Yü-hsiu. "Wo-men hsü-yao tsen-yang ti k'o-hsüeh" 我們需要怎樣的科學 [What kind of science do we need?]. *TLPL* 33 (1 January 1933):12–15.

Ku Yü-tsuan. 顧毓瑔 "I-nien-pan i-lai chih Chung-yang kung-yeh shih-yen-so" 一年半以來之中央工業試驗所 [National Bureau of Industrial Research during the past year and a half]. *KYCH* 5, no. 1 (1936):1–50.

K'un-yü ko-chih 坤輿格致 [Investigation of the earth]. Trans. Johann Schreck 鄧玉函 [Terrentius] of Georgius Agricola, *De Re Metallica* (1556). Peking, 1639.

K'ung-chi ko-chih 空際格致 [Treatise on the material composition of the universe]. Trans. Alfonso Vagnoni 高一志 of Aristotle, *In Libros Meteorum* (Portugal, 1593). Shanghai, 1633.

Kuo-li chung-yang yen-chiu-yüan chih-yüan-lu 國立中央研究院職員錄 [Record of personnel of the Academia Sinica]. Academia Sinica, 1934.

Kuo-li chung-yang yen-chiu-yüan yüan-shih-lu 國立中央研究院院士錄 [Roster of members of the Academia Sinica]. Vol. I. Ed. Chu Chia-hua. 朱家驊 Academia Sinica, 1948.

Kuo-li Pei-p'ing yen-chiu-yüan kung-tso pao-kao 國立北平研究院工作報告 [Work report of the National Academy of Peiping]. Kunming: Pei-p'ing yen-chiu-yüan, 北平研究院 1939, 1942.

"Kuo-li Pei-p'ing yen-chiu-yüan shih-chou-nien chi-nien" 國立北平研究院十週年紀念 [Commemoration of the tenth anniversary of the Peiping Academy]. *KH* 24, no. 2 (1940):144–6.

Kuo Mo-jo. 郭沫若 *Wo-ti yu-nien* 我的幼年 [My youth]. Shanghai: Ch'üan-ch'iu shu-tien, 全球書店 1947.

"Kuo-nei kung-yeh" 國內工業 [Chinese industry]. *HHKC* 3, no. 1 (March 1936): 74–7.

Kuo Pen-la. 郭本瀾 "Chang-chia-k'ou chih liang ta hua-hsüeh kung-yeh" 張家口之兩大化學工業 [Two great chemical industries of Changchiakou]. *HHKY* 1, no. 2 (July 1923):137–9.

Leng Tzu-sheng. 冷自生 "Hsin-Hua yao-ch'ang ti yu-nien shih-tai hui-i p'ien-tuan" 新華藥廠的幼年時代回憶片斷 [Memories of my youth in the New China Drug Factory]. *KJCC*, 3:321–8.

Li Ch'iang, 李強 et al. "Tzu-jan k'o-hsüeh chiao-yü yü kung-yeh chien-she" 自然科學教育與工業建設 [Natural science education and industrial reconstruction]. *CFJP*, 9 November 1942. Reprinted in *PCF*, 132–6.

Li Ch'iao-p'ing. 李喬平 *Chung-kuo hua-hsüeh shih* 中國化學史 [History of Chinese chemistry]. 2 vols. Taipei: T'ai-wan shang-wu yin-shu-kuan, 臺灣商務印書館 1976.

Li Ch'iao-p'ing. "Kuan-shui yü hua-hsüeh kung-yeh" 關稅與化學工業 [Customs duties and the chemical industry]. *HHKY* 6, no. 1 (January 1931):8–18.

Li Chih-ch'uan 李祉川 and Ch'en Hsin-wen. 陳歆文 "'Hou-shih chien-fa' ti tan-sheng ho fa-chan" 「侯氏鹼法」的誕生和發展 [Birth and development of the "Hou soda method"]. *HHTP* 8 (1982):475–9.

Li Erh-k'ang. 李爾康 "Erh-shih-san nien-tu chung-yang kung-yeh shih-yen-so hua-hsüeh-tsu kung-tso chih hui-ku" 二十三年度中央工業試驗所化學組工作之回顧 [Look back at the work of the chemical department of the National Bureau of Industrial Research during 1934]. *KYCH* 4, no. 1 (1935):6–9.

Li Erh-k'ang. "Wo-kuo hua-hsüeh kung-yeh kai-k'uang chi ch'i fa-chan t'u-ching" 我國化學工業概況及其發展途徑 [The situation and path of development of our country's chemical industry]. *KYCH* 6, nos. 7–8 (1937): 268–80, nos. 9–12:339–57.

Li Erh-k'ang. "Yen-chiu hua-hsüeh-che ying-yu chih jen-shih" 研究化學者應有之認識 [What knowledge chemical researchers ought to have]. *KYCH* 4, no. 8 (1935):405–7.

Li Fu-ch'un. 李富春 "Li Fu-ch'un t'ung-chih kei tzu-jan k'o-hsüeh-hui ti i-feng hsin" 李富春同志給自然科學會的一封信 [Letter from comrade Li Fu-ch'un to the Natural Science Society]. *CFJP* 30 January 1943, p. 4.

Li Hsi-mou, 李熙謀 ed. *Chung-hua min-kuo k'o-hsüeh-chih* 中華民國科學誌 [Record of science in the Republic of China]. Taipei: Chung-hua wen-hua ch'u-pan shih-yeh wei-yüan-hui, 中華文化出版事業委員會 1955.

Li K'un-hou. 酈堃厚 "Chiang-nan chih-tsao-chü yü Chung-kuo hsien-tai hua-hsüeh" 江南製造局與中國現代化學 [The Kiangnan Arsenal and contemporary Chinese chemistry]. *Chung-kuo k'o-hsüeh-shih lun-chi.* 中國科學史論集 Ed. Lin Chih p'ing. 林致平

Li P'u. 李普 "Chieh-fang-ch'u ti hua-hsueh kung-ch'ang" 解放區的化學工廠 [Chemical factories in the liberated areas]. *Ch'ün-chung* 群衆 [The masses], 13, no. 2 (27 October 1946):26–8.

Li Wei-chen. 李維禎 "K'ang-Jih chan-cheng shih-ch'i wo-chün yao-hsüeh shih-yeh ti fa-chan" 抗日戰爭時期我軍藥學事業的發展 [Development of our army's pharmaceutical industry during the War of Resistance]. *KJCC*, 5: 318–38.

Li Ya-tung. 李亞東 "Hsü Kuang-ch'i ti hua-hsüeh ch'eng-chiu" 徐光啓的化學成就 [The chemical achievements of Hsü Kuang-ch'i]. *Literature and History in China* (Shanghai) 3 (1984):29–39.

Li Ya-tung. "Hsü Shou so-i hua-hsüeh chu-tso ti yüan-pen" 徐壽所譯化學著作的原本 [Original chemical works translated by Hsü Shou]. *HHTP* 3 (1985): 52–5.

Liang Ch'i-ch'ao. 梁啓超 "Ou-yu chung chih i-pan kuan-ch'a chi i-pan kan-hsiang" 歐遊中之一般觀察及一般感想 [General observations and impressions on a European journey]. *Liang Jen-kung wen-chi* 梁任公文集 [Collected works of Liang Jen-kung]. Hong Kong: San-ta ch'u-pan-she, 三達出版社 n.d.

L'ien-ta pa-nien 聯大八年 [Eight years at Associated University]. Kunming: Hsi-nan lien-ta hsüeh-sheng ch'u-pan-she, 西南聯大學生出版社 1946.

Lin Chih-p'ing 林致平 [Ling Chih-Bing], ed. *Chung-kuo k'o-hsüeh-shih lun-chi*

中國科學史論集 [Essays on the history of science in China]. Taipei: Chung-hua wen-hua ch'u-pan shih-yeh wei-yüan-hui, 中華文化出版事業委員會 1958.

Lin Hua. 林華 "Shen-Kan-Ning pien-ch'ü po-li t'ao-ts'u ho nai-huo ts'ai-liao kung-yeh ti ch'uang-chien" 陝甘寧邊區玻璃陶瓷和耐火材料工業的創建 [Creation of the glass, porcelain, and fire-resistant materials industry in the Shensi-Kansu-Ninghsia Border Region]. *KJCC*, 4:305–22.

Lin Shan. 林山 "Kuan-yü fa-chan wo-men ti tzu-jan k'o-hsüeh chiao-yü yü kung-tso ti wo-chien" 關於發展我們的自然科學教育與工作的我見 [My views on developing our natural science education and work]. *CFJP*, 25 September 1942, p. 4.

Lin Wen-chao. 林文照 "Chung-yang yen-chiu-yüan kai-shu" 中央研究院概述 [Summary of the Academia Sinica]. *CKKCSL* 6, no. 2 (1985):21–8.

Lin Ying. 林英 "Lien-ta pa-nien" 聯大八年 [Eight years at Associated University]. *Ta Kung Pao* 大公報 (Shanghai) (26 November 1946):7.

Liu Hui. 劉惠 "Chung-kuo hua-hsüeh hui" 中國化學會 [Chinese Chemical Society]. *CKKCSL* 3, no. 3 (1982):53–61.

Liu Kuang-ting. 劉廣定 "Chu-ming yu-chi hua-hsüeh-chia Chuang Ch'ang-kung chien-chuan" 著名有機化學家莊長恭簡傳 [Brief biography of the famous organic chemist, Chuang Chang-kong]. *Chuan-chi wen-hsüeh* 傳記文學 [Biographical Essays]. Vol. 39, no. 4, 31–3.

Liu Kuang-ting. "Chung-kuo hua-hsüeh chiao-yü fa-chan chien-shih" 中國化學教育發展簡史 [Brief history of the development of Chinese chemical education]. *Hua-hsüeh* 化學 [Chemistry] (The Chinese Chemical Society, Taiwan, China) 43, no. 4 (December 1985):A152–A163.

Liu Kuang-ting. "Ts'ung Tseng Chao-lun ti tsao-yü t'an-ch'i" 從曾招掄的遭遇談起 [On the vicissitudes of Tseng Chao-lun's life]. *Chung-hua tsa-chih* 中華雜誌 [China magazine], 19, no. 218 (September 1981):15–16.

Liu Shu-k'ai. 劉樹楷 "Wo-kuo chin-tai hua-hsüeh ti ch'i-meng-che Hsü Shou" 我國近代化學的啓蒙者徐壽 [Pioneer in our country's modern chemistry, Hsü Shou]. *Hsü Shou ho Chung-kuo chin-tai hua-hsüeh shih*, 44–66.

Liu Ta. 劉達 "Wo-kuo chih-chien kung-yeh yü Ying-kuo Pu-nei-men kung-szu ti tou-cheng" 我國製鹼工業與英國卜內門公司的鬥爭 [Struggle between our country's soda industry and the Brunner-Mond Company of England]. *CKKCSL* 1, no. 2 (1980):101–3.

Liu Ta-chün 劉大鈞 [D. K. Lieu]. *Chung-kuo kung-yeh tiao-ch'a pao-kao* 中國工業調查報告 [Report on a survey of Chinese industry]. 3 vols. Shanghai: Ching-chi t'ung-chi yen-chiu-so, 經濟統計研究所 1937.

Liu Ting, 劉鼎 Ch'ien Wei-jen, 錢維人 and Fu Chiang. 傳江 "Yen-ch'ang Yung-p'ing yu-k'uang tiao-ch'a pao-kao chi ch'u-pu i-chien-shu" 延長永坪油礦調查報告及初步意見書 [Investigation report and preliminary ideas on the Yenchang and Yung-p'ing oil fields]. *KJCC*, 2:121–6.

Lo T'ien-yü. 樂天宇 "Tu 'Kuan-yü Yen-an kan-pu hsüeh-hsiao ti chüeh-ting'" 讀「關於延安幹部學校的決定」 [On reading the "Resolution on the Yenan Cadre School"]. *CFJP*, 23 July 1942, p. 4.

Lu Pao-yü. 陸寶愈 "San-shih-nien-lai Chung-kuo chih chiu-chīng kung-yeh" 三十年來中國之酒精工業 [China's alcohol industry during the past thirty years]. *San-shih-nien-lai chih Chung-kuo kung-ch'eng*. Vol. 2. Ed. Chou K'ai-ch'ing.

Lu Pin. 盧斌 "Huang-hai hua-hsüeh kung-yeh yen-chiu-she" 黃海化學工業研究社 [Golden Sea Research Institute for Chemical Industry]. *CKKCSL* 2, no. 1 (1981):56–60.

Lu Yü-tao. 盧于道 "Erh-shih-nien-lai chih Chung-kuo tung-wu-hsüeh" 二十年來

之中國動物學 [Twenty years of Chinese zoology]. *KH* 20, no. 1 (1936):41–48.

Lu Yü-tao. "Hsien-shih-hsing ti k'o-hsüeh yen-chiu" 現實性的科學研究 [Practical scientific research]. *Wen-hua hsien-feng* 文化先鋒 [Cultural pioneer] 1, no. 9 (27 October 1942):11–14.

Lu Yü-tao. "K'ang-chan ch'i-nien lai chih k'o-hsüeh chieh" 抗戰七年來之科學界 [The scientific community during the last seven years of the War of Resistance]. *Chung-kuo chan-shih hsüeh-shu* 中國戰時學術 [Chinese wartime scholarship]. Ed. Sun Pen-wen. 孫本文 Shanghai: Cheng-chung shu-chü, 正中書局 1946.

Luan Hsüeh-ch'ien. 欒學謙 "Ko-chih shu-yüan chiao-yen hua-hsüeh chi" 格致書院教演化學記 [Record of studying chemistry at the Shanghai Polytechnic Institute]. Quoted in "San-shih-nien-ch'ien wu-kuo k'o-hsüeh chiao-yü chih i-pan" 三十年前吾國科學教育之一斑 [Note on science education in our country thirty years ago]. *KH* 8, no. 4 (1924):430–2.

"Lun ching-chi yü chi-shu kung-tso" 論經濟與技術工作 [On economic and technical work]. *CFJP*, 2 June 1941. Reprinted in *KJCC*, 1:50–3.

Ma Hai-p'ing. 馬海平 "Shen-Kan-Ning pien-ch'ü k'o-hsüeh chi-shu ho tzu-jan pien-cheng-fa yen-chiu kai-k'uang" 陝甘寧邊區科學技術和自然辯證法研究概況 [Situation of research on science, technology, and the dialectics of nature in the Shen-Kan-Ning Border Region]. *KJCC*, 1:103–19.

Mao Yüan-yao. 毛遠耀 "Hui-i tsai Yen-an ts'an-chia chün-shih kung-yeh ti p'ien-tuan" 回憶在延安參加軍事工業的片斷 [Note on remembering joining military industry in Yenan]. *KJCC*, 1:176–81.

Meng Chen 孟真 [Fu Ssu-nien]. 傅斯年 "Chiao-yü kai-ke chung chi-ko chü-t'i shih-chien" 教育改革中幾個具體事件 [Some concrete issues in educational reform]. *TLPL* 10 (24 July 1932):6–9.

Meng Chen [Fu Ssu-nien]. "Chiao-yü peng-k'uei chih yüan yin" 教育崩潰之原因 [The reason for the collapse of education]. *TLPL* 9 (17 July 1932):2–6.

Meng Chen [Fu Ssu-nien]. "Kai-ke kao-teng chiao-yü chung chi ko wen-t'i" 改革高等教育中幾個問題 [Some questions on the reform of higher education]. *TLPL* 14 (21 August 1932):2–6.

Min Erh-ch'ang, 閔爾昌 comp. *Pei chuan chi pu* 碑傳集補 [Supplement to the collection of epitaphs], chüan 43. Vol. 998 of *Chin-tai Chung-kuo shih-liao ts'ung-k'an* 近代中國史料叢刊 [Collected materials on modern Chinese history]. Ed. Shen Yün-lung. 沈雲龍 Taipei: Wen-hai ch'u-pan-she, 文海出版社 1973.

Minami Manshū tetsudō kabushinki gaisha, Tenshin jimusho chōsaka 南滿洲鐵道株式會社，天津事務所調查課 [South Manchurian Railway Company, Tientsin Work Investigation Section]. *Shina ni okeru san sōda oyobi chisso kōgyō* 支那に於ける酸曹達及室素工業 [China's acid, soda, and nitrogen industries]. Tientsin, 1937. (Cited as *Shina Kōgyō*.)

"Mo-fan ch'ang-chang Ch'en Chen-hsia t'ung-chih ch'uang-chien mei-yu kung-yeh ch'eng-chi hao" 模範廠長陳振夏同志創建煤油工業成績好 [The good achievements of model factory manager comrade Ch'en Chen-hsia in establishing the kerosene industry]. *CFJP*, 18 May 1944. Reprinted in *KJCC*, 3:130–2.

"Mo-fan kung-ch'eng-shih Ch'ien Chih-tao t'ung-chih ch'uang-li pien-ch'ü chi-pen hua-hsüeh kung-yeh" 模範工程師錢志道同志創立邊區基本化學工業 [Model engineer comrade Ch'ien Chih-tao establishes the border region's basic chemical industry]. *CFJP*, 16 May 1944. Reprinted in *KJCC*, 2:61–3.

"Mo-fan kung-ch'eng-shih Shen Hung t'ung-chih" 模範工程師沈鴻同志 [Model engineer, comrade Shen Hung]. *CFJP*, 10 May 1944. Reprinted in *KJCC*, 2: 57–60.

Mo Ssu. 摩斯 "Lun kai-liang li-hua chiao-shou-fa" 論改良理化敎授法 [On improving physics and chemistry teaching methods]. *CHCYC* 1, no. 2 (1913): 23–4.

Ou-yang I. 歐陽詣 "Pei-p'ing hua-hsüeh kung-yeh k'ao-ch'a-chi" 北平化學工業考察記 [Notes on an investigation of the Peiping chemical industries]. *HHKY* 4, no. 2 (October 1929):75–100.

Ou-yang I. "P'ing-tung hua-hsüeh kung-yeh k'ao-ch'a-chi" 平東化學工業考察記 [Record of an investigation of the chemical industries east of Peiping]. *HHKY* 5, no. 1 (February 1930):77–110.

"Pa-lu-chün chih-yao-ch'ang chih-ch'eng yao-p'in ssu-shih chung" 八路軍製藥廠製成藥品40種 [Forty types of drugs manufactured by the Eighth Route Army Drug Factory]. *KJCC*, 2:195.

Pai Kuang-mei 白廣美 and Yang Ken. 楊根 "Hsü Shou yü 'Huang Hu' hao lun-ch'uan" 徐壽與「黃鵠」號輪船 [Hsü Shou and the "Yellow Swan" steamship]. *TJKHSYC* 3, no. 3 (1984):284–90.

Pai T'ao. 白桃 "She-shih erh-t'ung k'o-hsüeh chiao-yü ti hsin lu-hsien" 設施兒童科學教育的新路線 [Constructing a new road of science education for children]. *CHCYC* 20, no. 12 (1933):23–32.

P'an Chi-hsing. 潘吉星 "A-ke-li-k'o-la ti *K'uang-yeh ch'üan-shu* chi ch'i tsai Ming-tai Chung-kuo ti liu-ch'uan" 阿格里柯拉的「礦冶全書」及其在明代中國的流傳 [Agricola's *De Re Metallica* and its transmission to China during Ming times]. *TJKHSYC* 2, no. 1 (1983):32–44.

P'an Chi-hsing. "Ming-Ch'ing shih-ch'i (1640–1910) hua-hsüeh i-tso shu-mu k'ao" 明清時期(1640–1910)化學譯作書目考 [Study of catalogs of chemical translations of the Ming-Ch'ing period]. *CKKCSL* 5, no. 1 (1984):23–38.

P'an Chi-hsing. "T'an 'hua-hsüeh' i-ts'u tsai Chung-kuo ho Jih-pen ti yu-lai" 談「化學」一詞在中國和日本的由來 [Discussing the origin of the term "hua-hsüeh" in China and Japan]. *Ch'ing-pao hsüeh-k'an* 情報學刊 1 (1981):62–5.

P'an Chi-hsing. "Wo-kuo Ming-Ch'ing shih-ch'i kuan-yü wu-chi-suan ti chi-tsai" 我國明清時期關於無機酸的記載 [Chinese records of inorganic acids from the Ming and Ch'ing periods]. *Ta tzu-jan t'an-so* 大自然探索 [Explorations of nature] 3 (1983).

P'an Chi-hsing and K'o Kuei-hua. 柯桂華 "Hua-hsüeh-chia hua-hsüeh-shih-chia Ting Hsü-hsien chiao-shou ti i-sheng" 化學家，化學史家丁緒賢教授的一生 [The life of Professor Ting Hsü-hsien, chemist and chemical historian]. *HHTP* 6 (1979):547–52.

P'an Chün-hsiang. 潘君祥 "Wu-hsü shih-ch'i ti wo-kuo tzu-jan k'o-hsüeh hsüeh-hui" 戊戌時期的我國自然科學學會 [Scientific societies in our country at the time of the 1898 reforms]. *CKKCSL* 4, no. 1 (1983):28–30.

Pao Tsun-p'eng. 包遵彭 *Chung-kuo hai-chün shih* 中國海軍史 [History of the Chinese Navy]. Taipei: Hai-chün ch'u-pan-she, 海軍出版社 1951.

Pei-ching ta-hsüeh hsiao-shih 北京大學校史 [History of Peking University]. Ed. Su Ch'ao-jan, 蕭超然 Sha Chien-sun, 沙健孫 Chou Ch'eng-en 周承恩 and Liang Kuei. 梁桂 Shanghai: Shang-hai chiao-yü ch'u-pan-she, 上海教育出版社 1981.

Pei-p'ing yen-chiu-yüan 北平研究院 [The National Academy of Peiping]. Nanking(?): Hsing-cheng-yüan, hsin-wen-chü, 行政院，新聞局 1948.

P'eng Erh-ning. 彭爾寧 "Yen-an tzu-jan k'o-hsüeh-yüan – wo-ti mu-hsiao" 延安自然科學院—我的母校 [The Yenan Natural Science Institute – my alma mater]. *KJCC*, 3:332–42.

P'eng Kuang-ch'in. 彭光欽 "K'o-hsüeh ti ying-yung" 科學的應用 [Application of science], *TLPL* 199 (3 May 1936): 11–13.

P'eng Kuang-ch'in. "Lun k'o-hsüeh yen-chiu chih t'ung-chih" 論科學研究之統制 [Discussing control of scientific research]. *TLPL* 214 (16 August 1936):7–9.

"Pien-ch'ü huo-ch'ai-ch'ang shih-chih lü-suan-chia ch'eng-kung" 邊區火柴廠試製氯酸鉀成功 [Success in experimental production of potassium chlorate by the Border Region Match Factory]. *CFJP*, 26 June 1945. Reprinted in *KJCC*, 5:94.

Ping Chih. 秉志 "K'o-hsüeh tsai Chung-kuo chih chiang-lai" 科學在中國之將來 [The future of science in China]. *KH* 18, no. 3 (1934):301–4.

Ping Chih. "Kuo-nei sheng-wu k'o-hsüeh chin-nien-lai chih chin-chan" 國內生物科學近年來之進展 [Recent development of biological sciences in China]. *TFTC* 28, no. 13 (1931):99–110.

Ping Chih. "Kuo-nei sheng-wu k'o-hsüeh (fen-lei-hsüeh) chin-nien-lai chih chin-chan" 國內生物科學(分類學)近年來之進展 [Recent developments in Chinese biological sciences (taxonomy)]. *KH* 28, no. 3 (1934):414–35.

Sah Pen-t'ung. 薩本棟 "Ch'un-ts'ui k'o-hsüeh yü shih-yung k'o-hsüeh" 純粹科學與實用科學 [Pure science and applied science]. *TLPL* 236 (30 May 1937): 14–17.

"Shan-tung Tzu-po Hsin-Hua yao-ch'ang ch'ang-shih" 山東淄博新華藥廠廠史 [History of the Shantung Tzu-po New China Drug Factory]. *KJCC*, 4: 367–77.

Shen Chin-t'ai. 沈覲泰 "Chung-kuo hua-hsüeh-shih kai-shu" 中國化學史概述 [Summary of the history of chemistry in China]. *Chung-kuo k'o-hsüeh-shih lun-chi*. 中國科學史論集 Ed. Lin Chih-p'ing. 林致平

Shen Chin-t'ai. "Hua-hsüeh kung-ch'eng-hsüeh" 化學工程學 [Chemical engineering]. *Chung-hua min-kuo k'o-hsüeh-chih*. 中華民國科學誌 Ed. Li Hsi-mou. 李熙謀

"Shen-Kan-Ning pien-ch'ü ping-kung fa-chan chien-shih" 陝甘寧邊區兵工發展簡史 [Brief history of the Shen-Kan-Ning Border Region arsenals]. *KJCC*, 1:167–75.

Shen-Kan-Ning pien-ch'ü tzu-jan pien-cheng-fa yen-chiu tsu-liao 陝甘寧邊區自然辯證法研究資料 [Research materials on the dialectics of nature in the Shen-Kan-Ning Border Region]. Ed. Shen-hsi sheng kao-teng yüan-hsiao tzu-jan pien-cheng-fa yen-chiu hui, Yen-an ta-hsüeh fen-hui 陝西省高等院校自然辯證法研究會延安大學分會 [Shensi Provincial Higher Schools Dialectics of Nature Research Society, Yenan University Branch]. Sian: Shen-hsi jen-min ch'u-pan-she. 陝西人民出版社 1984.

Shen Kuo-ch'iang. 沈國強 "'Chung-kuo hua-hsüeh ts'o-yao' ti chu-yao ch'eng-chiu ho ying-hsiang" 「中國化學撮要」的主要成就和影響 [Essential accomplishments and influence of "Chinese Chemical Abstracts"] *CKKCSL* 4, no. 2 (1983):89–92.

Shen Tseng-tso. 沈增祚 "Liu-suan kung-yeh chi ch'i hsin ch'ü-shih" 硫酸工業及其新趨勢 [The sulfuric acid industry and its new direction]. *KYCH* 12, no. 1 (October 1948):10–13.

Shu Yung 叔永 [Jen Hung-chün]. "Tang-hua chiao-yü shih k'o-neng ti ma?" 黨化教育是可能的嗎 [Is a party-ized education possible?]. *TLPL* 3 (5 June 1932):12–15.

Shu Yung [Jen Hung-chün]. "Tsai-lun tang-hua chiao-yü" 再論黨化教育 [More on party-ized education]. *TLPL* 8 (10 July 1932):10–13.

Ssu Hao. 思浩 "Mu-ch'ien Chung-kuo kung-ch'eng chiao-yü chu wen-t'i" 目前中

國工程教育諸問題 [Various questions on current Chinese engineering education]. *Hsüeh-hsi sheng-huo* 學習生活 [Study life] 3, no. 2 (20 July 1942): 92–5.

Sugawara, Kunika. 菅原國香 "Kagaku toiu yōgo no honpō de no shutsugen shiyō ni kansuru ikkōsatsu" 「化學」という用語の本邦での出現・使用に關する一考察 [A historical study of the use of the term *kagaku* for chemistry in Japan]. *Kagakushi* 化學史研究 (Tokyo: Journal of the Japanese Society for the History of Chemistry, 1987), 29–40.

Sun Hsüeh-wu. 孫學悟 "T'i-ch'ang tzu-jan k'o-hsüeh chiao-yü ti chi-chien chi-wu" 提倡自然科學教育的幾件急務 [Urgent tasks for promoting natural science education]. *TLPL* 34 (8 January 1933):9–12.

Sun I. 孫逸 "Tu Ku Yü-hsiu 'Wo-men hsü-yao tsen-yang ti k'o-hsüeh' hou" 讀顧毓琇 "我們需要怎樣的科學" 後 [After reading Ku Yü-hsiu's "What kind of science do we need?"] *TLPL* 36 (22 January 1933):14 –18.

Sun Yü-t'ang, 孫毓棠 comp. *Chung-kuo chin-tai kung-yeh shih tsu-liao, ti-i chi, 1840–1895 nien* 中國近代工業史資料，第一輯，1840–1895年 [Source materials on the history of modern industry in China, first collection, 1840–1895.] 2 vols. Peking: K'o-hsüeh ch'u-pan-she, 科學出版社 1957.

Sung Tzu-ch'eng. 宋子成 "Hou Te-pang ch'eng-kung chih lu" 侯德榜成功之路 [Hou Te-pang's road of success]. *CKKCSL* 1, no. 1 (May 1980):26–39.

Tai An-pang. 戴安邦 "Chung-kuo hua-hsüeh chiao-yü ti hsien-chuang" 中國化學教育的現狀 [Present situation of Chinese chemical education]. *KH* 24, no. 2 (February 1940):89–109.

T'ang Chih-chün. 湯志鈞 *Wu-hsü pien-fa shih lun-ts'ung* 戊戌變法史論叢 [Collected essays on the history of the 1898 reforms]. Wuhan: Hu-pei jen-min ch'u-pan-she, 湖北人民出版社 1957.

T'ang Jo-wang 湯若望 [Adam Schall von Bell] and Chiao Hsü. 焦勗 *Huo-kung ch'ieh-yao* 火攻挈要 [Essentials of gunnery]. Peking, 1643.

T'ang Yüeh. 唐鉞 "I-ke ch'ih-jen ti shuo-meng" 一個癡人的説夢 [Gibberish of an idiot]. *KHYJSK.*

T'ao Ying-hui. 陶英惠 "Ts'ai Yüan-p'ei yü chung-yang yen-chiu-yüan (i-chiu-erh-ch'i – i-chiu-ssu-ling)" 蔡元培與中央研究院（一九二七～ 一九四〇）[Ts'ai Yüan-p'ei and the Academia Sinica, 1927–1940]. *Chung-yang yen-chiu-yüan chin-tai-shih yen-chiu-so chi-k'an* 中央研究院近代史研究所期刊 [Bulletin of the Institute of Modern History, Academia Sinica] 7 (June 1978): 1–50.

"T'ien-ch'u wei-ching chih-tsao kai-k'uang" 天廚味精製造概況 [Summary of T'ien-ch'u monosodium glutamate manufacturing]. *HHSC* 3, no. 6 (1948): 9–11.

T'ien Yü-lin, 田遇霖 Kao I-sheng, 高怡生 and Huang Yao-tseng. 黃耀曾 "Wo-kuo yu-chi hua-hsüeh ti hsien-ch'ü – Chuang Ch'ang-kung chiao-shou" 我國有機化學的先驅···莊長恭教授 [Pioneer of our country's organic chemistry – Professor Chuang Chang-kong]. *HHTP* 4 (1979):365–72.

Ting Ssu. 定思 "K'o-hsüeh ching-shen, k'o-hsüeh t'ai-tü" 科學精神，科學態度 [Scientific spirit, scientific attitude]. *Ch'ün-chung* 羣衆 [The masses] 9, no. 12 (June 1944):498–504.

Ting Szu-hsien. 丁嗣賢 "Chien-she kuo-fang hua-hsüeh kung-yeh chih wo-chien" 建設國防化學工業之我見 [My views on establishing a chemical industry of national defense]. *HHKC* 2, no. 1 (June 1935):74–83.

Ting Wei-liang 丁韙良 [W. A. P. Martin]. *Ko-wu ju-men* 格物入門 [Natural philosophy]. 7 vols. Peking: T'ung Wen Kuan. 同文館 1868.

Ting Wen-chiang. 丁文江 "Chung-yang yen-chiu-yüan ti shih-ming" 中央研究院

的使命 [Historic mission of the Academia Sinica]. *TFTC* 32, no. 2 (1935): 5–8.

Ting Wen-chiang. "Hsüan hsüeh yü k'o-hsüeh – ta Chang Chün-mai" 玄學與科學—答張君勱 [Metaphysics and Science – a reply to Chang Chün-mai]. *KHYJSK.*

Ts'ai Yüan-p'ei. 蔡元培 "Lun ta-hsüeh ying-she ko k'o yen-chiu-so chih li-yu" 論大學應設各科研究所之理由 [Reasons that universities should establish research institutes]. *TFTC* 32, no. 1 (January 1935):13–14.

Ts'ao Hui-ch'ün. 曹惠群 "Ta-hsüeh-sheng yü hua-hsüeh kung-yeh" 大學生與化學工業 [University students and chemical industry]. *HHKY* 8, no. 2 (July 1933):1–7.

Tseng Chao-lun. 曾昭掄 "Chiang-nan chih-tsao-chü shih-tai pien-i chih hua-hsüeh shu-chi chi ch'i so-yung chih hua-hsüeh ming ts'u" 江南製造局時代編譯之化學書籍及其所用之化學名詞 [Books on chemistry edited during the era of the Kiangnan Arsenal and the chemical nomenclature used in them]. *Hua-hsüeh* 化學 [Chemistry] 3, no. 5 (1936):746–62.

Tseng Chao-lun. "Chih- te chu-i ti ta-hsüeh-sheng ch'u-lu wen-t'i" 值得注意的大學生出路問題 [The problem of the future of university students deserves attention]. *Hsin min-tsu* 新民族 [The new nation] 2, no. 1 (17 July 1938): 5–7.

Tseng Chao-lun. "Chung-kuo hsüeh-shu ti chin-chan" 中國學術的進展 [Development of Chinese scholarship]. *TFTC* 38, no. 1 (1941):56–9.

Tseng Chao-lun. "Chung-kuo k'o-hsüeh hui-she kai-shu" 中國科學會社概述 [Summary of Chinese scientific societies]. *KH* 20, no. 10 (October 1936):798–830.

Tseng Chao-lun. "Erh-shih-nien-lai Chung-kuo hua-hsüeh chih chin-chan" 二十年來中國化學之進展 [Development of chemistry in China during the past twenty years]. *KH* 19, no. 10 (October 1935):1514–59.

Tseng Chao-lun. "K'o-hsüeh ming-ts'u chung ti tsao-tzu wen-t'i" 科學名詞中的造字問題 [Problem of the creation of characters in scientific terminology]. *Chung-kuo yu-wen* 中國語文 [Chinese language] 14 (October 1953):3–4.

Tseng Chao-lun. "Kuo-nan ch'i-chien k'o-hsüeh-chieh t'ung-jen ying fu ti tse-jen" 國難期間科學界同人應負的責任 [Responsibilities that must be borne by our scientific colleagues during the period of national crisis]. *KH* 20, no. 4 (1936):255–6.

Tseng Chao-lun. *Ta-liang shan I-ch'ü k'ao-ch'a-chi* 大涼山夷區考察記 [Notes on an investigation of the Yi region of Taliangshan]. Shanghai: Tu-shu ch'u-pan-she, 讀書出版社 1945.

Tseng Chao-lun. *Tien-K'ang tao-shang* 滇康道上 [On the Yunnan-Sikang Road]. Kweilin: Wen-yü shu-tien, 文有書店 1943.

Tso Yün. 左雲 "Chi k'o-hsüeh-yüan ti ch'ün-chung t'uan-t'i" 記科學院的羣衆團體 [Record of the mass organizations in the Science Institute]. *CFJP*, 1 February 1943, p. 4.

Tung Wen-li. 董文立 "Chieh-fang-ch'ü kung-ch'ang chi-shu kung-tso tien-ti" 解放區工廠技術工作點滴 [A note on the technical work of the factories in the liberated areas]. *KJCC*, 1:158–62.

"Tzu-jan k'o-hsüeh-yüan chi-hsü t'ao-lun chiao-hsüeh fang-chen" 自然科學院繼續討論教學方針 [The Natural Science Institute continues discussion of education policy]. *CFJP*, 23 October 1942, p. 2.

"Tzu-jan k'o-hsüeh yen-chiu-hui hsüan-yen" 自然科學研究會宣言 [Pronouncement of the Natural Science Research Society]. *PCF*, 87–9.

Tzu Nien. 梓年 "Fa-hui 'Wu-ssu' yün-tung so t'i-chang ti k'o-hsüeh ching-shen"

發揮 "五四" 運動所提倡的科學精神 [Develop the scientific spirit advocated by the "May Fourth" movement]. *Ch'ün-chung* 羣眾 [The masses] 2, no. 24/25 (15 May 1939), 797–8.

Wang Chih-chia. 王志稼 "Wo-kuo k'o-hsüeh chiao-yü chin-hou ying chü chih fang-chen" 我國科學教育今後應具之方針 [Plans for the future of our science education]. *KH* 24, no. 5 (May 1940):347–51.

Wang Chih-hao 王治浩 and Hsing Jun-ch'uan. 邢潤川 "Chih-ming hsüeh-che, hua-hsüeh-chia Tseng Chao-lun chiao-shou" 知名學者，化學家曾昭掄教授 [Famous scholar and chemist, Professor Tseng Chao-lun]. *HHTP* 9 (1980): 559–67.

Wang Chih-hao and Yang Ken. 楊根 "Ko-chih shu-yüan yü 'Ko-chih hui-pien'" 格致書院與「格致彙編」 [The Shanghai Polytechnic Institute and the Chinese Scientific Magazine]. *CKKCSL* 5, no. 2 (1984):59–64.

Wang Chin. 王璡 "Ch'u-chi chung-hsüeh chih hun-ho tzu-jan k'o-hsüeh chiao-hsüeh wen-t'i" 初級中學之混合自然科學教學問題 [The problem of teaching general science in lower middle schools]. *KH* 13, no. 8 (1929):1092–101.

Wang Chin. "Wu-chu-ch'ien hua-hsüeh ch'eng-fen chi ku-tai ying-yung ch'ien, hsi, hsin, la k'ao" 五銖錢化學成份及古代應用鉛錫鋅鑞考 [Chemical composition of Chinese coins and investigation of the use of lead, tin, and zinc in ancient (coins)]. *KH* 8, no. 8 (1923):839–54.

Wang Ching-hsi. 汪敬熙 "T'i-ch'ang k'o-hsüeh yen-chiu tsui ying chu-i ti i-chien-shih – jen-ts'ai ti p'ei-yang" 提倡科學研究最應注意的一件事—人材的培養 [In support of the most important aspect of scientific research – cultivation of human talent]. *TLPL* 26 (13 November 1932):10–14.

Wang Ching-hsi. "Tsai t'an-t'an tsen-yang t'i-ch'ang k'o-hsüeh yen-chiu" 再談談怎樣提倡科學研究 [Another discussion of how to promote scientific research]. *TLPL* 76 (12 November 1933):5–8.

Wang Ching-yü, 汪敬虞 comp. *Chung-kuo chin-tai kung-yeh shih tsu-liao, ti-erh-chi, 1895–1914 nien* 中國近代工業史資料，第二輯，1895–1914年 [Source materials on the history of modern industry in China, 2d collection, 1895–1914]. 2 vols. Peking: K'o-hsüeh ch'u-pan-she, 科學出版社 1957.

Wang Chung-t'ien. 汪忠天 "Kuo-nei ta-hsüeh chi chuan-men hsüeh-hsiao pi-yeh-sheng chiu-yeh ch'ing-k'uang ti i-ke tiao-ch'a" 國內大學及專門學校畢業生就業情況的一個調查 [Investigation into the employment situation for graduates of Chinese universities and higher technical schools]. *CHCYC* 22, no. 6 (1934):49–60.

Wang Erh-min. 王爾敏 *Ch'ing-chi ping-kung-yeh ti hsing-ch'i* 清季兵工業的興起 [The rise of the armament industry during the late Ch'ing]. Taipei: Institute of Modern History, Academia Sinica, 1963.

Wang Hsin-min. 王新民 "Hua-tung chieh-fang-ch'ü ti chün-shih kung-yeh" 華東解放區的軍事工業 [Military industry in the East China Liberated Area]. *KJCC*, 5:286–307.

Wang Hsiu-lu. 王岫廬 "Chung-hsüeh chih k'o-hsüeh chiao-yü" 中學之科學教育 [Middle school science education]. *KH* 7, no. 11 (1922):1121–30.

Wang Hsü-chiu. 王旭九 "K'ang-Jih chan-cheng ho chieh-fang chan-cheng shih-ch'i Chiao-tung kung-yeh yen-chiu-shih ti k'o-yen kung-tso" 抗日戰爭和解放戰爭時期膠東工業研究室的科研工作 [The scientific research work of the Chiao-tung Industrial Research Laboratory during the War of Resistance and the War of Liberation]. *KJCC* 4:268–80.

Wang Shih-mo. 王世模 "Tan-pai-chih chung chih ku-suan yü ku-suan-na t'iao-wei-fen" 蛋白質中之穀酸與穀酸鈉調味粉 [Glutamic acid and monosodium glutamate seasoning in proteins]. *KH* 17, no. 7 (July 1933):1018–48.

Wang T'ao, 王韜 ed. *Ko-chih shu-yüan k'e-i* 格致書院課藝 [Themes of the Shanghai Polytechnic Institute]. Shanghai: Ko-chih shu-yüan, 格致書院 1886–93. Microfilm. East Asian Library, University of California, Berkeley.

Wang Tsu-t'ao. 王祖陶 "Chung-kuo chin-tai ti hua-hsüeh chiao-yü" 中國近代的化學教育 [Modern Chinese chemical education]. *CKKCSL* 5, no. 2 (1984):93–4.

Wang Yüan-te 王元德 and Liu Yü-feng, 劉玉峰 eds. *Wen-hui-kuan chih* 文會館志 [History of Tengchow College]. Weihsien, 1913.

Wang Yü-sheng, 王玉生 ed. *Chung-kuo k'o-hsüeh chi-shu shih-kao* 中國科學技術史稿 [Draft history of Chinese science and technology]. 2 vols. Peking: K'o-hsüeh ch'u-pan-she, 科學出版社 1982.

Wei T'ing-ying. 魏廷英 "Chi-nan, Ch'ing-tao, Te-chou, T'ang-shan, T'ang-ku hua-hsüeh kung-yeh k'ao-ch'a-chi" 濟南青島德州唐山塘沽化學工業考察記 [Report on an investigation of the chemical industries of Tsinan, Tsingtao, Tehchow, Tangshan, and Tangku]. *HHKY* 2, no. 2 (July 1924):201–20.

Wei Yün-kung, 魏允恭 ed. *Chiang-nan chih-tsao-chü chi*. 江南製造局記 [Record of the Kiangnan Arsenal]. Shanghai: Shang-hai wen-pao shu-chü, 上海文寶書局 1905.

Wong Wen-hao. 翁文灝 "I-ke ta-p'o fan-men ti fang-fa" 一個打破煩悶的方法 [A method for overcoming depression]. *TLPL* 10 (24 July 1932):2–5.

Wu Ch'eng-lo. 吳承洛 "Ch'üan-kuo k'o-hsüeh chiao-yü she-pei kai-yao" 全國科學教育設備概要 [Survey of science teaching facilities in the entire country]. *KH* 9, no. 8 (1924):950–77.

Wu Ch'eng-lo. "Chung-kuo hua-hsüeh kung-yeh she-chi chi yüan-liao wen-t'i" 中國化學工業設計及原料問題 [Question of plans and materials for the Chinese chemical industry]. *HHKY* 4, no. 2 (October 1929):5–21.

Wu Ch'eng-lo. "San-shih-nien-lai Chung-kuo chih suan-chien hua-hsüeh kung-ch'eng" 三十年來中國之酸鹼化學工程 [China's acid and soda chemical engineering during the past thirty years]. *San-shih-nien-lai chih Chung-kuo kung-ch'eng*. Vol. 1. Ed. Chou K'ai-ch'ing.

Wu Ch'eng-lo. "San-shih-nien-lai Chung-kuo chih hua-hsüeh kung-yeh" 三十年來中國之化學工業 [China's chemical industry during the past thirty years]. *San-shih-nien-lai chih Chung-kuo kung-ch'eng*. Vol. 2. Ed. Chou K'ai-ch'ing.

Wu Ch'eng-lo. "Ts'ung kuo-chi mao-i kuan-ch'a Chung-kuo hua-hsüeh kung-yeh chih hsien-chuang" 從國際貿易觀察中國化學工業之現狀 [Observing the present situation of China's chemical industry on the basis of international trade]. *HHKY* 3, no. 2 (July 1925):1–72.

Wu Ch'eng-lo. "Ts'ung nung-shang-pu kung-szu chu-ts'e kuan-ch'a Chung-kuo hua-hsüeh kung-yeh chih hsien-chuang" 從農商部公司註冊觀察中國化學工業之現狀 [Observing the present situation of China's chemical industry on the basis of the corporation registration of the Ministry of Agriculture and Commerce]. *HHKY* 2, no. 2 (July 1924):11–76.

Wu Ch'eng-lo. "Ts'ung Shang-hai hua-hsüeh kung-i chan-lan-hui kuan-ch'a Chung-kuo hua-hsüeh kung-yeh chih hsien-chuang" 從上海化學工藝展覽會觀察中國化學工業之現狀 [Observing the present situation of the Chinese chemical industry on the basis of the Shanghai Chemical Handicraft Exhibition]. *HHKY* 2, no. 1 (January 1924):7–56.

Wu Chih-hui. 吳稚暉 "I-ke hsin hsin-yang ti yü-chou-kuan chi jen-sheng-kuan" 一個新信仰的宇宙觀及人生觀 [The cosmology and philosophy of life of a new belief]. *KHYJSK*.

Wu Chih-hui. "Ta jen shu" 答人書 [Answer to a letter]. *Wu Chih-hui hsüeh-shu lun-chu* 吳稚暉學術論著 [Selections of Wu Chih-hui's academic writings]. Shanghai: Ch'u-pan ho-tso-she,' 出版合作社 1926.

Wu Chin, 武進 et al., eds *Hsüeh-sheng tzu-tien* 學生字典 [Student dictionary]. Shanghai: Shang-wu yin-shu-kuan, 商務印書館 1915.

Wu Heng. 武衡 "Shen-Kan-Ning pien-ch'ü ti k'o-hsüeh p'u-chi kung-tso" 陝甘寧邊區的科學普及工作 [Scientific popularization work in the Shen-Kan-Ning Border Region]. *KJCC*, 5:156–66.

Wu Heng and Yen P'ei-lin. 閻沛霖 "Huai-nien Hsü-lao, wo-men ti hao lao-shih" 懷念徐老，我們的好老師 [Remembering old Hsü, our great teacher]. *KJCC*, 1:90–7.

Wu Hsien 吳憲 [T'ao-ming]. 濤鳴 "Chung-kuo ti ping ying-kai tsen-ma chih" 中國的病應該怎麼治 [How can (we) cure China's illness]. *TLPL* 51 (21 May 1933):32–6.

Wu Hsien. "Kuan-yü k'o-hsüeh yen-chiu chih wo-chien" 關於科學研究之我見 [My views on scientific research]. *TLPL* 101 (20 May 1934):15–17.

Wu Hsien [T'ao-ming]. "Ting Hsien chien-wen tsa-lu" 定縣見聞雜錄 [Notes on a visit to Ting Hsien]. *TLPL* 4 (12 June 1932):13–18.

Wu Hsien. "Tsai lun ch'ih-fan wen-t'i" 再論吃飯問題 [More on the food problem]. *TLPL* 205 (14 June 1936):14–16.

Wu Hsien [T'ao-ming]. "Ts'ung hsiao-hai shuo tao ta-jen" 從小孩說到大人 [Discussing adults from the point of view of childhood]. *TLPL* 57 (2 July 1933): 19–21.

Wu Hsien. "Wu kuo-jen chih ch'ih-fan wen-t'i" 吾國人之吃飯問題 [The food problem of our people]. *TLPL* 2 (29 May 1932):15–19.

Wu Hsien. *Ying-yang kai-lun* 營養概論 [Principles of nutrition]. Shanghai: Commercial Press, 1929.

Wu Tsao-hsi. 吳藻溪 "Erh-ch'i k'ang-chan ti k'o-hsüeh yün-tung" 二期抗戰的科學運動 [Scientific movement in the second stage of the War of Resistance]. *Ch'ün-chung* 羣眾 [The masses] 3, nos. 13–14 (27 August 1939):343–7, 372–6.

Wu Yüan-ku. 伍源古 "Chin-Ch'a-Chi k'ang-Jih ken-chü-ti ti ti-lei yü ti-lei-chan" 晉察冀抗日根據地的地雷與地雷戰 [Mines and mine warfare of the Shansi-Chahar-Hopeh Anti-Japanese Base Area]. *KJCC*, 5:273–85.

Yang Ken. 楊根 "Wo-kuo chin-tai hua-hsüeh hsien-ch'ü-che Hsü Shou ti sheng-p'ing chi chu-yao kung-hsien" 我國近代化學先驅者徐壽的生平及主要貢獻 [The life and important contributions of the vanguard of our country's modern chemistry, Hsü Shou]. *HHTP* 4 (1984):71–7.

Yang Kuo-liang 楊國樑 and Cheng T'ang. 正棠 "Wo-kuo hua-hsüeh-shih ho fen-hsi hua-hsüeh yen-chiu ti k'ai-t'a-che – Wang Chin chiao-shou" 我國化學史和分析化學研究的開拓者— 王璡教授 [Pioneer in our country's chemical history and analytical chemical research – Professor Wang Chin]. *HHTP* 9 (1982):553–8.

Yang Mo. 楊模 ed. *Hsi-Chin ssu-che shih-shih hui-ts'un* 錫金四哲事實彙存 [Records of the four scholars of Wuhsi and Chin-k'uei].Wuhsi, 1910.

Yeh Hsiao-ch'ing. 葉曉青 "Chin-tai hsi-fang k'o-chi ti yin-chin chi ch'i ying-hsiang" 近代西方科技的引進及其影響 [Introduction and influence of modern Western science and technology]. *Li-shih yen-chiu* 歷史研究 [Historical research] 1 (1982):3–17.

Yen Chi-tz'u. 嚴濟慈 "Chin-shu-nien-lai kuo-nei chih wu-li-hsüeh yen-chiu" 近數年來國內之物理學研究 [Physics research in China in recent years]. *TFTC*

32, no. 1 (1935):15–20.

Yen Chi-tz'u. "Erh-shih-nien-lai Chung-kuo wu-li-hsüeh chih chin-chan" 二十年來中國物理學之進展 [Development of Chinese physics during the past twenty years]. *KH* 19, no. 11 (1935):1705–16.

Yen Chih-hsien. 嚴志弦 "Chung-kuo hua-hsüeh-chia tui-yü wu-chi hua-hsüeh ti kung-hsien" 中國化學家對於無機化學的貢獻 [Contributions of Chinese chemists to inorganic chemistry]. *HHSC* 8, nos. 8–9 (1953):260–3, 296–300.

Yen Fu. 嚴復 *Ming-hsüeh* 名學 [Logic]. Trans. of John Stuart Mill, *System of Logic*. In vol. 2 of *Yen i ming-chu ts'ung-k'an* 嚴譯名著叢刊 [A collection of Yen Fu's translated works]. Shanghai: Shang-wu yin-shu-kuan, 商務印書館 1931.

Yen Lin-shan 閻林山 and Ma Tzung-liang. 馬宗良 "Hsü-chia-hui t'ien-wen-t'ai ti chien-li ho fa-chan" 徐家滙天文台的建立和發展 [Establishment and development of the Zikawei Observatory]. *CKKCSL* 5, no. 2 (1984):65–72.

Yen Yen-ts'un. 嚴演存 "Ching-hsing lien-chiao-ch'ang fu-ch'an chih ch'i-yu" 井陘煉焦廠副產之汽油 [Byproduct gasoline of the Ching-hsing Coking Plant]. *HHKY* 10, no. 1 (January 1935):69–80.

Yü Jen-chün. 俞人駿 "Wo-kuo shih-yen tien-chieh kung-yeh chih ching-chi kuan" 我國食鹽電解工業之經濟觀 [Economic outlook for our country's salt electrolysis industry]. *HHKY* 19, nos. 1–2 (April 1947):4–9.

Yü Kuang-yüan. 于光達 "K'o-hsüeh ti kuang-hui tsai Yen-an shan-yao" 科學的光輝在延安閃耀 [The brilliance of science sparkles in Yenan]. *KJCC*, 1:120–4.

Yüan Han-ch'ing. 袁翰青 *Chung-kuo hua-hsüeh-shih lun-wen chi* 中國化學史論文集 [Collected essays on the history of Chinese chemistry]. Peking: Hsin-chih san-lien shu-tien, 新知三聯書店 1956.

Yüan Han-ch'ing. "Lun t'i-kao k'o-hsüeh ti yen-chiu" 論提高科學的研究 [On raising up scientific research]. *Hsin min-tsu* 新民族 [The new nation] 1, no. 12 (15 May 1938):4–7.

Yüan Han-ch'ing. "Tsui-tsao i-ch'eng Chung-wen ti liang-pu fen-hsi hua-hsüeh shu chi yüan-tso-che hsiao-chuan" 最早譯成中文的兩部分析化學書及原作者小傳 [The first two books on analytical chemistry translated into Chinese and a short biography of the author]. *HHTP* 2 (1982):115–17.

Yüan Han-ch'ing and Meng Nai-ch'ang. 孟乃昌 "Hsü i 'Hua-hsüeh k'ao-chih' ho 'Hua-hsüeh ch'iu-shu'" 徐譯「化學考質」和「化學求數」 [Hsü's translations, "Qualitative analytical chemistry" and "Quantitative analytical chemistry"]. *Hsü Shou ho Chung-kuo chin-tai hua-hsüeh shih*, 119–41.

"Yung-li hua-hsüeh kung-yeh kung-szu liu-suan-ya-ch'ang ch'eng-li ching-kuo chi ch'i kai-k'uang" 永利化學工業公司硫酸錏廠成立經過及其概況 [The process of establishment and current status of the Yungli Chemical Industry Company Ammonium Sulfate Factory]. *HHKC* 4, no. 2 (June 1937):183–95.

"Yung-li liu-suan-ya-ch'ang ch'eng-kung chih i-i" 永利硫酸錏廠成功之意義 [Significance of the success of the Yungli Ammonium Sulfate Factory]. *HHKC* 4, no. 2 (June 1937):111–2.

WESTERN LANGUAGE

Academia Sinica (1928–1948). N.p., n.d.

The Academia Sinica and Its National Research Institutes. Nanking: Academia Sinica, 1931.

Adolph, William H. "Chemical Industry in China." *Journal of Industrial and Engineering Chemistry* 13, no. 12 (December 1921): 1099.

Adolph, William H. "Science in the Middle School." *Educational Review* 10, no. 2 (April 1918):120–4.

Alleton, Viviane and Jean-Claude Alleton. *Terminologie de la Chimie en Chinois Moderne* [Modern Chinese chemical terminology]. Maison des Sciences de l'Homme, Materiaux pour l'Etude de l'Extreme-Orient Moderne et Contemporain, Etudes Linguistique, 1. Paris: Mouton, 1966.

Allison, Andrew. "Middle School Science." *Educational Review* 12, no. 2 (April 1920):135–43.

American Men of Science. 9th ed. Ed. Jaques Cattell. Lancaster, PA: Science Press, 1955.

Atwell, William. "From Education to Politics: The Fu She." *The Unfolding of Neo-Confucianism.* Ed. William T. deBary. New York. Columbia University Press, 1975.

Badger, W. L., and E. M. Baker. *Inorganic Chemical Technology.* New York: McGraw-Hill, 1941.

Balme, Harold. *China and Modern Medicine: A Study in Medical Missionary Development.* London: United Council for Missionary Education, 1921.

Band, Claire, and William Band. *Two Years with The Chinese Communists.* New Haven: Yale University Press, 1948.

Band, William. *Science in the Christian Universities at Chengtu, China.* New York: Associated Boards for Christian Colleges in China, 1945.

Bennett, Adrian A. *John Fryer: The Introduction of Western Science and Technology into Nineteenth-Century China.* Cambridge: Harvard University Press, 1967.

Bennett, Adrian A. *Missionary Journalist in China: Young J. Allen and His Magazines. 1860–1883.* Athens: University of Georgia Press, 1983.

Bennett, Adrian A., comp. *Research Guide to the Chiao-hui hsin-pao (The Church News), 1868–1874.* San Francisco: Chinese Materials Center, 1975.

Bennett, Adrian A., comp. *Research Guide to the Wan-kuo kung-pao (The Globe Magazine), 1874–1883.* San Francisco: Chinese Materials Center, 1976.

Biggerstaff, Knight. "Shanghai Polytechnic Institution and Reading Room: An Attempt to Introduce Western Science and Technology to the Chinese." *Pacific Historical Review* 25 (1956):127–49.

Biggerstaff, Knight. *The Earliest Modern Government Schools in China.* Ithaca: Cornell University Press, 1961.

Boorman, Howard L., and Richard C. Howard, eds. *Biographical Dictionary of Republican China.* 4 vols. New York: Columbia University Press, 1967–71.

Borg, Dorothy. *The United States and the Far Eastern Crisis of 1933–1938.* Cambridge: Harvard University Press, 1964.

Bowers, John Z. *Western Medicine in a Chinese Palace: Peking Union Medical College, 1917–1951.* New York: Macy Foundation, 1971.

Briere, O. "L'effort de la Philosophie Marxiste en Chine." *Bulletin de l'Université l'Aurore*, serie 3, tome 8, no. 3 (1947):309–47.

Briere, O. "Les Courants Philosophiques en Chine depuis 50 ans (1898–1950)." *Bulletin de l'Université l'Aurore*, serie 3, tome 10, no. 40 (1949):561–654.

Britton, Roswell. *The Chinese Periodical Press, 1800–1912.* Shanghai: Kelly and Walsh, 1933.

Brown, Shannon R. "The Transfer of Technology to China in the Nineteenth Century: The Role of Direct Foreign Investment." *Journal of Economic History* 39, no. 1 (March 1979):181–97.

Buck, Peter. *American Science and Modern China, 1876–1936.* Cambridge,

Eng.: Cambridge University Press, 1980.

Bullock, Mary Brown. *An American Transplant: The Rockefeller Foundation and Peking Union Medical College.* Berkeley: University of California Press, 1980.

Burton, Ernest Dewitt. "Journal and Record of Interviews and Observations, University of Chicago, Oriental Education Investigation, 1909." Unpub. ms., Missionary Research Library, Union Theological Seminary, New York.

Burton, Ernest DeWitt, and Thomas Chrowder Chamberlin. "Report of the Oriental Educational Commission of the University of Chicago, Part VI, China (December 1909)." Unpub. ms., Missionary Research Library, Union Theological Seminary, New York.

The Cambridge History of China. Vols. 10–11, *Late Ch'ing, 1800–1911.* Vols. 12–13, *Republican China, 1912–1949.* Ed. John K. Fairbank et al. Cambridge, Eng.: Cambridge University Press, 1978–86.

Chan, Anthony B. *Arming the Chinese: The Western Armaments Trade in Warlord China, 1920–28.* Vancouver: University of British Columbia Press, 1982.

Chang, C. Y. "Botanical Work in China during the War, 1937–43." *Acta Brevia Sinensia* 6 (April 1944):3–6.

Chang, H. T. "On the History of the Geological Science in China." *Bulletin of the Geological Society of China* 1, no. 1 (1922):4–7.

Chang, Hao. *Liang Ch'i-ch'ao and Intellectual Transition in China, 1890–1907.* Cambridge: Harvard University Press, 1971.

Chang, John. *Industrial Development in Pre-Communist China: A Quantitative Analysis.* Chicago: Aldine, 1969.

Chang Kwang-chih. *Shang Civilization.* New Haven: Yale University Press, 1980.

Chang Kwang-chih. *The Archaeology of Ancient China.* Rev. and enl. ed. New Haven: Yale University Press, 1968.

Chang, Tsing lien. "Recent Researches on Heavy Water." *Science* 100 (14 July 1944):29–30.

Chang, Y. C. "The Research Activity of the National Institute of Astronomy." *Acta Brevia Sinensia* 2 (March 1943):7–8.

Ch'en Ch'i-t'ien, Gideon. *Tseng Kuo-fan: Pioneer Promoter of the Steamship in China.* Peiping: Yenching University, 1935. Reprint. New York: Paragon Book Reprint Corporation, 1961.

Ch'en Li-fu. *Chinese Education during the War (1937–42).* Chungking: Ministry of Education, 1942.

Ch'en Li-fu. *Philosophy of Life.* Trans. Jen Tai. New York: Philosophical Library, 1948.

Chen, Nai-ruenn, and Walter Galenson. *The Chinese Economy under Communism.* Chicago: Aldine, 1969.

Chen, Theodore Hsi-en. "Science, Scientists, and Politics." *Sciences in Communist China.* Ed. Sidney H. Gould. Washington, DC: American Association for the Advancement of Science, 1961.

Cheng Tsung-hai. "Elementary Education in China." Chinese National Association for the Advancement of Education. *Bulletin* 2, no. 14 (1923).

Cheng Yu-kwei. *Foreign Trade and Industrial Development of China.* Washington, DC: University Press of Washington, DC, 1956.

Ch'i Hsi-sheng. *Warlord Politics in China, 1916–1928.* Stanford: Stanford

University Press, 1976.

The China Handbook. Ed. Chinese Ministry of Information, Chungking. New York: Macmillan, 1937–45.

China Industrial Handbooks: Kiangsu. Shanghai: Ministry of Industry, 1933.

The China Year Book. Ed. H. G. W. Woodhead. Shanghai: North China Daily News and Herald, 1912–1939.

Christian Education in China: A Study Made by an Educational Commission Representing the Mission Boards and Societies Conducting Work in China. New York: Committee of Reference and Counsel of the Foreign Missions Conference of North America, 1922.

Coble, Parks M., Jr. *The Shanghai Capitalists and the Nationalist Government, 1927–1937.* Cambridge: Harvard University Press, 1980.

Cole, James H. *Shaohsing: Competition and Cooperation in Nineteenth-Century China.* Tucson: University of Arizona Press, 1986.

Condliffe, J. B. "The Nankai Institute of Economics." *There is Another China.* New York: King's Crown Press, 1948.

Corbett, Charles Hodge. *Lingnan University.* New York: Trustees of Lingnan University, 1963.

Corbett, Charles Hodge. *Shantung Christian University (Cheeloo).* New York: United Board for Christian Colleges in China, 1955.

Covell, Ralph. *W. A. P. Martin: Pioneer of Progress in China.* Washington, DC: Christian University Press, 1978.

Craig, Albert. "Science and Confucianism in Tokugawa Japan." *Changing Japanese Attitudes Toward Modernization.* Ed. Marius Jansen. Princeton: Princeton University Press, 1965.

Cressy, Earl H. "Christian Colleges in China: Statistics." China Christian Educational Association. *Bulletin,* nos. 30, 33, 35 (1933–5).

Cressy, Earl H. "Christian Higher Education in China, A Study for the Year, 1925–26." China Christian Educational Association. *Bulletin* 28 (1928).

Croizier, Ralph C. *Traditional Medicine in Modern China: Science, Nationalism, and Tensions of Cultural Change.* Cambridge: Harvard University Press, 1968.

Crosland, Maurice P. *Historical Studies in the Language of Chemistry.* New York: Dover, 1978.

Data Papers on China, 1931. Shanghai: China Institute of Pacific Relations, 1931.

Dernberger, Robert. "The Role of the Foreigner in China's Economic Development, 1840–1949." *China's Modern Economy in Historical Perspective.* Ed. Dwight H. Perkins. Stanford: Stanford University Press, 1975.

Doke, Tatsumasa. "Yōan Udagawa: A Pioneer Scientist of Early Nineteenth Century Feudalistic Japan." *JSHS* 12 (1973):99–120.

Doolittle, Justus. *Social Life of the Chinese.* London: Sampson, Low, Son and Marston, 1868.

Duus, Peter. "Science and Salvation in China: The Life and Work of W. A. P. Martin (1827–1916)." *American Missionaries in China: Papers from Harvard Seminars.* Ed. Kwang-ching Liu. Cambridge: Harvard University Press, 1966.

Eastman, Lloyd. *The Abortive Revolution: China under Nationalist Rule, 1927–37.* Cambridge: Harvard University Press, 1974.

Edmunds, Charles K. "Modern Education in China." Department of the Interior, Bureau of Education, *Bulletin* 44. Washington, DC: Government Printing

Office, 1919.

Edwards, Dwight W. *Yenching University.* New York: United Board for Christian Higher Education, 1959.

Elman, Benjamin. *From Philosophy to Philology: Intellectual and Social Aspects of Change in Late Imperial China.* Cambridge: Harvard University Press, 1984.

Faber, Knud. *Report on Medical Schools in China.* Geneva: League of Nations Health Organization, 1931.

Fairbank, Wilma. *America's Cultural Experiment in China, 1942–1949.* Washington, DC: Government Printing Office, 1976.

Fan, C. Y. "Advance of Physics in War-time China." *Acta Brevia Sinensia* 8 (December 1944):3–5.

Fenn, William Pruviance. *Christian Higher Education in Changing China, 1880–1950.* Grand Rapids, MI: W. Eerdmans, 1976.

Feuchtwang, Stephan. "School-Temple and City God." *Studies in Chinese Society.* Ed. Arthur P. Wolf. Stanford: Stanford University Press, 1978.

Feuerwerker, Albert. "Economic Trends in the Late Ch'ing Empire, 1870–1911." *The Cambridge History of China,* vol. 11.

Feuerwerker, Albert. *Economic Trends in the Republic of China, 1912–1949.* Ann Arbor: Center for Chinese Studies, University of Michigan, 1977.

Fewsmith, Joseph. *Party, State, and Local Elites in Republican China.* Honolulu: University of Hawaii Press, 1985.

Fisher, Daniel W. *Calvin Wilson Mateer: Forty-Five Years a Missionary in Shantung Province, China.* Philadelphia: Westminster Press, 1911.

Folsom, Kenneth E. *Friends, Guests, and Colleagues: The Mu-fu System in the Late Ch'ing Period.* Berkeley: University of California Press, 1968.

Fong, H. D. "China's Industrialization: A Statistical Survey." *Data Papers on China.* Shanghai: China Institute of Pacific Relations, 1931.

Forman, Harrison. *Report from Red China.* New York: Holt, 1945.

Freedman, Maurice. "Sociology in and of China." *British Journal of Sociology* 13 (1962):106–16.

Freyn, Hubert. *Chinese Education in the War.* Shanghai: Kelly and Walsh, 1940.

Fryer, John. "Science in China." *Nature* (May 1881):9–11, 54–57.

Fung, Yu-lan. *A History of Chinese Philosophy.* 2 vols. Trans. Derk Bodde. Princeton: Princeton University Press, 1953.

Furth, Charlotte. "Intellectual Change: From the Reform Movement to the May Fourth Movement, 1895–1920." *The Cambridge History of China,* vol. 12.

Furth, Charlotte. *Ting Wen-chiang: Science and China's New Culture.* Cambridge: Harvard University Press, 1970.

Furton, Joseph S. *Molecules and Life: Historical Essays on the Interplay of Chemistry and Biology.* New York: John Wiley and Sons, 1972.

Giles, Herbert. *A Chinese–English Dictionary.* Shanghai, 1912. Reprint. Taipei: Ch'eng-Wen Publishing Company, 1967.

Giquel, Prosper. *The Foochow Arsenal and Its Results.* Trans. H. Lang. Shanghai: *Shanghai Evening Courier,* 1874.

Glasstone, Samuel, Keith J. Laidler, and Henry Eyring. *The Theory of Rate Processes.* New York: McGraw-Hill, 1941.

Grabau, A. W. "Palaeontology." *Symposium on Chinese Culture.* Ed. Sophia H. Chen Zen.

Graham, Loren R. *Science and Philosophy in the Soviet Union.* New York: Knopf, 1972.

Greene, John C. "Science, Learning and Utility: Patterns of Organization in the Early American Republic." *The Pursuit of Knowledge in the Early American Republic*. Ed. Alexandra Oleson and Sanborn C. Brown. Baltimore: Johns Hopkins University Press, 1976.

Greene, Roger. "Aspects of Science Education." *There is Another China*. New York: King's Crown Press, 1948.

Gregg, Alice H. *China and Educational Autonomy*. Syracuse: Syracuse University Press, 1946.

Gubbins, John Harrington. *A Dictionary of Chinese–Japanese Words in the Japanese Language*. 2d ed. Tokyo: Maruya and Co., 1908.

Gulick, Edward V. *Peter Parker and the Opening of China*. Cambridge: Harvard University Press, 1973.

Haber, L. F. *The Chemical Industry during the Nineteenth Century: A Study of the Economic Aspect of Applied Chemistry in Europe and North America*. Oxford: Clarendon Press, 1958.

Hardie, D. W. F. *Electrolytic Manufacture of Chemicals from Salt*. London: Oxford University Press, 1959.

Haurowitz, Felix. *Chemistry and Biology of Proteins*. New York: Academic Press, 1950.

Hayhoe, Ruth. "Towards the Forging of a Chinese University Ethos: Zhendan and Fudan, 1903–1919." *China Quarterly* 94 (June 1983):323–41.

Hemeling, Karl. *English–Chinese Dictionary of the Standard Chinese Spoken Language*. 1916. Reprint. Freeport, NY: Books for Libraries Press, 1973.

Henderson, Lawrence J. *Blood: A Study in General Physiology*. New Haven: Yale University Press, 1928.

Hepburn, James Curtis. *A Japanese–English and English–Japanese Dictionary*. 3d ed. Tokyo: Maruya and Co., 1886.

Hoh Yam Tong. "The Boxer Indemnity Remissions and Education in China." Ph.D. diss., Columbia Teachers College, 1933.

Holubnychy, Vsevolod. "Mao Tse-tung's Materialistic Dialectics." *China Quarterly* 19 (July–September 1964):3–37.

Hou Chi-ming. *Foreign Investment and Economic Development in China, 1840–1937*. Cambridge: Harvard University Press, 1965.

Hou Te-pang. *Manufacture of Soda with Special Reference to the Ammonia Process*. New York: Hafner, 1969.

Howard, Richard C. "K'ang Yu-wei (1858–1927): His Intellectual Background and Early Thought." *Confucian Personalities*. Ed. Arthur F. Wright and Denis Twitchett. Stanford: Stanford University Press, 1962.

Hsiao Kung-chuan. *Rural China: Imperial Control in the Nineteenth Century*. Seattle: University of Washington Press, 1960.

Hsü, Yung Ying. *A Survey of the Shensi-Kansu-Ninghsia Border Region*, 2 vols. New York: Institute of Pacific Relations, 1945.

Hummel, Arthur W. *Eminent Chinese of the Ch'ing Period (1644–1912)*. 2 vols. Washington. DC: Government Printing Office, 1943.

Huntress, Ernest H., and Samuel P. Mulliken. *Identification of Pure Organic Compounds*. New York: Wiley and Sons, 1941.

Hyatt, Irwin T., Jr. *Our Ordered Lives Confess: Three Nineteenth-Century American Missionaries in East Shantung*. Cambridge: Harvard University Press, 1976.

Ihde, Aaron J. *The Development of Modern Chemistry*. New York: Harper and Row, 1964.

Jackson, Rev. James J. "Objects, Methods, and Results of Higher Education in our Mission Schools." *CRMJ* 24, no. 1 (January 1893):7–12.

Jansen, Marius. "Japan and the Chinese Revolution of 1911." *The Cambridge History of China*, vol. 11.

Joravsky, David. *Soviet Marxism and Natural Science, 1917–1932*. New York: Columbia University Press, 1961.

Judd, Ellen R. "Prelude to the 'Yan'an Talks': Problems in Transforming a Literary Intelligentsia." *Modern China* 11, no. 3 (July 1985):377–408.

Kennedy, Thomas L. "Chang Chih-tung and the Struggle for Strategic Industrialization: The Establishment of the Hanyang Arsenal, 1884–95." *Harvard Journal of Asian Studies* 33 (1973):154–82.

Kennedy, Thomas L. *The Arms of Kiangnan: Modernization in the Chinese Ordnance Industry, 1860–1895*. Boulder: Westview Press, 1978.

Kimball, Alice H., comp. *Bibliography of Research on Heavy Hydrogen Compounds*. New York: McGraw-Hill, 1949.

Kirby, William C. "Technocratic Organization and Technological Development in China: The Nationalist Experience and Leagacy, 1928–1953." *Science and Technology in Post-Mao China*. Ed. Denis Fred Simon and Merle Goldman. Cambridge: Harvard University Press, 1989.

Klein, Donald W., and Anne B. Clark. *A Biographical Dictionary of Chinese Communism, 1921–1965*. 2 vols. Cambridge: Harvard University Press, 1971.

Knight, Edgar W. "Christian Education." *Laymen's Foreign Missions Inquiry Fact-Finders' Reports: China*. Vol. 5, pt 2. Ed. Orville A. Petty. New York: Harper and Bros., 1933.

Kuhn, Philip. "The Development of Local Government." *The Cambridge History of China*, vol. 13.

Kuno, Yoshi S. *Educational Institutions in the Orient with Special Reference to Colleges and Universities in the United States*. Part 2, *Chinese Educational Institutions*. Berkeley: University of California Press, 1928.

Kuo Ping Wen. *The Chinese System of Public Education*. New York: Teachers College, Columbia University, 1915.

Kuo Tze-hsiung. "Higher Education in China." *Information Bulletin* (Nanking: Council of International Affairs) 3, no. 2 (21 January 1937):29–50.

Kwei, Chi-ting. "The Status of Physics in China." *American Journal of Physics* 12, no. 1 (February 1944):13–18.

Kwok, D. W. Y. *Scientism in Chinese Thought, 1900–1950*. New York: Biblo and Tannen, 1971.

Lamberton, Mary. *St. John's University, Shanghai, 1876–1951*. New York: United Board for Christian Colleges in China, 1955.

League of Nations' Mission of Educational Experts. *The Reorganization of Education in China*. Paris: League of Nations' Institute of Intellectual Co-operation, 1932.

Lee, Tsui-hua Yang. "Geological Sciences in Republican China, 1912–1937." Ph.D. diss., State University of New York at Buffalo, 1985.

Lewis, Ida Belle. "A Study of Primary Schools." *Laymen's Foreign Missions Inquiry Fact-Finders' Reports: China*. Vol. 5, pt 2. Ed. Orville A. Petty. New York: Harper and Bros., 1933.

Li Chi. "Archaeology." *Symposium on Chinese Culture*. Ed. Sophia H. Chen Zen.

Li Ch'iao-p'ing. *The Chemical Arts of Old China*. Easton, PA: Journal of

Chemical Education, 1948.

Li Jung. "An Account of the Salt Industry at Tzu-liu-ching." *ISIS* 39, no. 118 (November 1948):228–34.

Li Lo-yüan. "Low Temperature Carbonization of Szechwan Coal." *HHKC* 6, nos. 3–4 (December 1939):33–7.

Li San-pao. "Letters to the Editor in John Fryer's *Chinese Scientific Magazine*, 1876–1892: An Analysis." *Chin-tai-shih yen-chiu-so chi-k'an* [Quarterly of the Research Institute on Modern History] 4 (1974):729–77.

Liao T'ai-ch'u. "Rural Education in Transition: A Study of the Old-fashioned Chinese Schools (Szu-shu) in Shantung and Szechuan." *Yenching Journal of Social Studies* 4, no. 2 (February 1949):19–69.

Lieu, D. K. *The Growth and Industrialization of Shanghai.* Shanghai: China Insititute of Pacific Relations, 1936.

Lindbeck, John M. H. "Organization and Development of Science." *Sciences in Communist China.* Ed. Sidney H. Gould. Washington, DC: American Association for the Advancement of Science. 1961.

Linden, Allen Bernard. "Politics and Higher Education in China: The Kuomintang and the University Community, 1927–37." Ph.D. diss., Columbia University, 1969.

Lissak, Ormond. *Ordnance and Gunnery.* New York: Wiley, 1915.

Liu Jung-chao. *China's Fertilizer Economy.* Chicago: Aldine, 1970.

Liu Kwang-ching. "Early Christian Colleges in China." *Journal of Asian Studies* 20, no. 1 (1960):71–8.

Liu Ta-chung and Yeh Kung-chia. *The Economy of the Chinese Mainland: National Income and Economic Development, 1933–1959.* Princeton: Princeton University Press, 1965.

Lo, Jung-pang. *K'ang Yu-wei: A Biography and a Symposium.* Tucson: University of Arizona Press, 1967.

Lockhart, William. *The Medical Missionary in China.* London: Hurst and Blackett, 1861.

Lubot, Eugene. *Liberalism in an Illiberal Age: New Cultural Liberals in Republican China, 1919–1937.* Westport, CT: Greenwood Press, 1982.

Lund, Renville C. "The Imperial University of Peking." Ph.D. diss., University of Washington, 1956.

Lutz, Jessie G. *China and the Christian Colleges, 1850–1950.* Ithaca: Cornell University Press, 1971.

McIntosh, Gilbert. *The Mission Press in China.* Shanghai: American Presbyterian Mission Press, 1895.

Mao Tse-tung. *Selected Works of Mao Tse-tung.* 4 vols. Peking: Foreign Languages Press, 1967.

Mao's China: Party Reform Documents, 1942–1944. Trans. Boyd Compton. Seattle: University of Washington Press, 1966.

Marble, Alexander. "Otto Folin: Benefactor of Diabetics through Biochemistry." *Diabetes* 2 (1953):503–5.

Maritime Customs of China. *Returns of Trade and Trade Reports* (1906–1919). *Foreign Trade of China* (1920–1931). *The Trade of China* (1932–1938). Shanghai: Inspector General of Customs.

Martin, W. A. P. *A Cycle of Cathay.* New York: Fleming H. Revell, 1900.

Martin, W. A. P. *The Awakening of China.* New York: Doubleday, 1907.

Medicine in China. New York: China Medical Commission of the Rockefeller Foundation, 1914.

Meskill, John. "Academies and Politics in the Ming Dynasty." *Chinese Government in Ming Times: Seven Studies.* Ed. Charles O. Hucker. New York: Columbia University Press, 1969.

Mill, John Stuart. *System of Logic.* Vol. 1. London: Longmans, Green, Reader, and Dyer, 1872.

Monroe, Paul. "A Report on Education in China." *The Institute of International Education Bulletin* 4 (20 October 1922).

Morgan, L. G. *The Teaching of Science to the Chinese.* Hong Kong: Kelly and Walsh, 1933.

Nakayama Shigeru. *Academic and Scientific Traditions in China, Japan, and the West.* Trans. Jerry Dusenbury. Tokyo: University of Tokyo Press, 1974.

Nathan, Andrew J. *Peking Politics, 1918–1923: Factionalism and the Failure of Constitutionalism.* Berkeley: University of California Press, 1985.

The National Geological Survey of China, 1916–1931. Peiping, 1931.

National Sun Yat-sen University: A Short History. N.p., 1936.

Needham, Joseph. *Chinese Science.* London: Pilot Press, 1945.

Needham, Joseph. *The Grand Titration: Science and Society in East and West.* London: Allen and Unwin, 1972.

Needham, Joseph. *Science and Civilisation in China.* 5 vols. Cambridge, Eng.: The University Press, 1954–

Needham, Joseph, and Dorothy Needham. *Science Outpost: Papers of the Sino–British Science Co-operation Office, 1942–46.* London: Pilot Press, 1948.

Orleans, Leo A. *Professional Manpower and Education in Communist China.* Washington, DC: Government Printing Office, 1961.

Oya, Shin'ichi. "Reflections on the History of Science in Japan." *Science and Society in Modern Japan: Selected Historical Sources.* Ed. Nakayama Shigeru, David I. Swain, and Yagi Eri. Cambridge: M.I.T. Press, 1974.

Paterno, Roberto. "Devello Z. Sheffield and the Founding of the North China College." *American Missionaries in China: Papers from Harvard Seminars.* Ed. Liu Kwang-ching. Cambridge: Harvard University Press, 1966.

Pauley, Edwin W. *Report on Japanese Assets in Manchuria to the President of the United States, July 1946.* Washington, DC: Government Printing Office, 1946.

Price, Jane L. *Cadres, Commanders, and the Commissars: The Training of the Chinese Communist Leadership, 1920–45.* Boulder: Westview Press, 1976.

Pusey, James Reeve. *China and Charles Darwin.* Cambridge: Harvard University Press, 1983.

Rankin, Mary Backus. *Elite Activism and Political Transformation in China: Zhejiang Province, 1865–1911.* Stanford: Stanford University Press, 1986.

Rawlinson, John L. *China's Struggle for Naval Development, 1839–1895.* Cambridge: Harvard University Press, 1967.

Rawski, Thomas G. *China's Transition to Industrialism.* Ann Arbor: University of Michigan Press, 1980.

Reardon-Anderson, James. "China's Modern Chemical Industry, 1860–1949." *OSIRIS,* 2d ser., 2 (1986), 177–224.

Remer, C. F. *Foreign Investments in China.* New York: Macmillan, 1933.

Report of the Board of Trustees for Administration of Boxer Indemnity Funds Remitted by the British Government. Nanking, 1931–6.

Report of the China Foundation for the Promotion of Education and Culture. Peking and Shanghai, 1926–48.

Reynolds, David C. "The Advancement of Knowledge and the Enrichment of Life: The Science Society of China and the Understanding of Science in the Republic, 1914–1930." Ph.D. diss., University of Wisconsin, 1983.

Richard, Timothy. *Forty-five Years in China*. New York: Frederick A. Stokes, 1916.

Richthofen, Ferdinand von. *China: Ergebnisse eigener Reisen und darauf gegründeter Studien* [China, the results of my travels and the studies based thereon]. 5 vols. Berlin: D. Reimer, 1877–1912.

Rockefeller Foundation, China Medical Board, Annual Reports. New York, 1914–26.

Rolland, Stephane. *Le Langage Chimique Chinois* [The Chinese chemical language]. Taipei: European Languages Edition, 1985.

Salaff, Stephen. "A Biography of Hua Lo-keng." *Science and Technology in East Asia*. Ed. Nathan Sivin. New York: Science History Publications, 1977.

Schneider, Laurence A. "Genetics in Republican China." Paper presented to conference sponsored by the Rockefeller Foundation, Pocantico Hills, NY, May 1984.

Schneider, Laurence A. *Ku Chieh-kang and China's New History*. Berkeley: University of California Press, 1971.

Schneider, Laurence, ed. *Lysenkoism in China: Proceedings of the 1956 Qingdao Genetics Symposium*. A special issue of *Chinese Law and Government* 19, no. 2 (Summer 1986).

Schoppa, R. Keith. *Chinese Elites and Political Change: Zhejiang Province in the Early Twentieth Century*. Cambridge: Harvard University Press, 1982.

Schram, Stuart R., trans. *The Political Thought of Mao Tse-tung*, rev. and enl. ed. New York: Praeger, 1969.

Schran, Peter. *Guerrilla Economy: The Development of the Shensi-Kansu-Ninghsia Border Region, 1937–1945*. Albany: State University of New York Press, 1976.

Schwartz, Benjamin. *In Search of Wealth and Power: Yen Fu and the West*. New York: Harper and Row, 1964.

Science Society of China, ed. *The Science Society of China: Its History, Organization, and Activities*. Shanghai: The Science Press, 1931.

Selden, Mark. *The Yenan Way in Revolutionary China*. Cambridge: Harvard University Press, 1971.

Seybolt, Peter. "Terror and Conformity: Counterespionage Campaigns, Rectification, and Mass Movements, 1942–1943." *Modern China* 12, no. 1 (January 1986):39–73.

Shaffer, Philip. "Otto Folin (1867–1934)." *Journal of Nutrition* 52 (1954):3–11.

Shen, T. H. "First Attempts to Transform Chinese Agriculture." *The Strenuous Decade: China's Nation-Building Efforts, 1927–37*. Ed. Paul K. T. Sih. New York: St. John's University Press, 1970.

Shimao, Eikoh. "The Reception of Lavoisier's Chemistry in Japan." *ISIS* 63 (1972):309–20.

Shriner, Ralph L., and Reynold C. Fuson. *The Systematic Identification of Organic Compounds: A Laboratory Manual*. New York: Wiley, 1935, 1940, 1948.

Sivin, Nathan. "Chinese Alchemy and the Manipulation of Time." *Science and Technology in East Asia*. New York: Science History Publications, 1977.

Sivin, Nathan. "Wang Hsi-shan." *Dictionary of Scientific Biography*. 15 vols.

New York: Scribner's Sons, 1970–8, 14:159–68.

Sivin, Nathan. "Why the Scientific Revolution did not take place in China – or didn't it?" *Transformation and Tradition in the Sciences: Essays in Honor of I. Bernard Cohen.* Ed. Everett Mendelsohn. Cambridge, Eng.: Cambridge University Press, 1984.

Snow, Edgar. *Red Star Over China.* New York: Grove Press, 1961.

"Statistical Summaries of Chinese Education." Chinese National Association for the Advancement of Education. *Bulletin* 2, no. 16 (1923).

Stein, Gunther. *The Challenge of Red China.* New York: McGraw-Hill, 1945.

Su, Frank Kai-ming, and Alvin Barber. "China's Tariff Autonomy, Fact or Myth." *Far Eastern Survey* 5, no. 12 (June 1936):115–22.

Sugimoto, Masayoshi, and David L. Swain, eds. *Science and Culture in Traditional Japan, A.D. 600–1854.* Cambridge: M.I.T. Press, 1978.

Summary Report of the Activities of the China Foundation for the Promotion of Education and Culture, from 1925 to 1945. N.p., December 1946.

Sun, E-tu Zen. "The Growth of the Academic Community, 1912–1949." *The Cambridge History of China,* vol. 12.

Sun Hsüeh-wu. "The Hwang-hai (Golden Sea) Research Institute of Chemical Industry." *Science and Technology in China* 1, no. 4 (August 1948): 69–70.

Sung Ying-hsing. *T'ien-kung k'ai-wu: Chinese Technology in the Seventeenth Century.* Trans. E-tu Zen Sun and Shiou-chuan Sun. University Park: Pennsylvania State University Press, 1966.

Suttmeier, Richard P. *Research and Revolution: Science Policy and Societal Change in China.* Lexington, MA: Lexington Books, 1974.

Tanaka, Minoru. "A Note on the Development of Chemistry in Japan." *JSHS* 7 (1968):61–70.

Taylor, F. Sherwood. *A History of Industrial Chemistry.* New York: Abelard-Schuman, 1957.

Technical Terms: English and Chinese. Ed. the Educational Association of China. Shanghai: Presbyterian Mission Press, 1904.

Tisdale, W. E. "Report of Visit to Scientific Institutions in China (September–December 1933)." RG 1, ser. 601, box 40. Rockefeller Archive Center, Pocantico Hills, NY.

Tseng Chao-lun. "Progress of Chemical Research in China." *Acta Brevia Sinensia* 3 (May 1943):15–18.

Tseng Chao-lun and Hu Mei. "Gluten Hydrolysis and Preparation of d-Glutamic Acid Hydrochloride." *JCCS* 3, no. 2 (June 1935):154–72.

Tsien, Tsuen-hsuin. "Western Impact on China through Translation." *Far Eastern Quarterly* 13 (1954):305–27.

Twiss, George Ransom. *Science and Education in China.* Shanghai: Commercial Press, 1925.

Van Slyke, Lyman P., ed. *The Chinese Communist Movement: A Report of the United States War Department, July 1945.* Stanford: Stanford University Press, 1968.

Wakeman, Frederic, Jr. *History and Will: Philosophical Perspectives of Mao Tse-tung's Thought.* Berkeley: University of California Press, 1973.

Wang Shih-chieh. "Education." *The Chinese Yearbook, 1937.* Shanghai: Commercial Press, 1937.

Wang, Y. C. *Chinese Intellectuals and the West, 1872–1949.* Chapel Hill: University of North Carolina Press, 1966.

Wang Yu-ch'uan. "The Development of Modern Social Science in China." *Pacific*

Affairs 11, no. 3 (September 1938):345–62.

Watson, Andrew. *Mao Zedong and the Political Economy of the Border Region.* Cambridge, Eng.: Cambridge University Press, 1980.

Wells, David Ames. *Wells's Principles and Applications of Chemistry.* New York: Ivison, Phinney and Co., 1864.

Whitmore, Ralph D. "Engineering Education in China." *The Tsing Hua Journal* 2, no. 5 (March 1917):1–25.

Who's Who in China. 4th ed. Shanghai: China Weekly Review, 1931.

Wong, K. Chimin. *Lancet and Cross: Biographical Sketches of Fifty Pioneer Medical Missionaries in China.* Shanghai: Council on Christian Medical Work, 1950.

Wong, K. Chimin, and Wu Lien-teh. *History of Chinese Medicine.* Shanghai: National Quarantine Service, 1936.

Wong, Siu-lun. *Sociology and Socialism in Contemporary China.* London: Routledge and Kegan Paul, 1979.

Wong, W. H. "Chinese Geology." *Symposium on Chinese Culture.* Ed. Sophia H. Chen Zen.

Wu, Daisy Yen. *Hsien Wu, 1893–1959: In Loving Memory.* Boston: Author, 1959.

Wu Hsien. "Chinese Diet in the Light of Modern Knowledge of Nutrition." *Chinese Social and Political Science Review* 2 (1927):56–81.

Wu Hsien. *A Guide to Scientific Living.* Taipei: Academia Sinica, 1963.

Wu Hsien. "Nutritional Deficiencies in China and Southeast Asia." *Fourth International Congress on Tropical Medicine and Malaria.* Washington, DC: Department of State, 1948.

Wu Hsien. *Principles of Physical Biochemistry.* Peiping: Peiping Union Medical College, 1934.

Wu Yun-to. *Son of the Working Class: An Autobiography of Wu Yun-to.* Peking: Foreign Languages Press, 1956.

Yang Tsun-yi. "Development of Geology in China since 1911." *China Institute Bulletin* 3, no. 5 (February 1939):131–7.

Yeh Ch'i-sun. "Work of the Academia Sinica, 1937–42." *Quarterly Bulletin of China Bibliography* 3, no. 1 (1943):7–20.

Yen Wei-ching. *An English and Chinese Standard Dictionary.* 2d ed. Shanghai: Commercial Press, 1908.

Young, Arthur N. *China's Nation-Building Effort, 1927–1937.* Stanford: Hoover Institution Press, 1971.

Yuan T'ung-li. *Doctoral Dissertations by Chinese Students in Great Britain and Northern Ireland, 1916–1921.* Reprinted from *Chinese Culture* 4, no. 4 (March 1963).

Yuan T'ung-li. *A Guide to Doctoral Dissertations by Chinese Students in America, 1905–1960.* Washington, DC: Sino–American Cultural Society, 1961.

Yuan T'ung-li. *A Guide to Doctoral Dissertations by Chinese Students in Continental Europe, 1907–1962.* Reprinted from *Chinese Culture Quarterly* 5:3, 4 and 6:1.

Zen, Sophia H. Chen, ed. *Symposium on Chinese Culture.* New York: Paragon, 1969.

Zhang Yunming. "Ancient Chinese Sulfur Manufacturing Processes." *ISIS* 77 (1986):487–97.

Recent Studies of the East Asian Institute

CHINA'S POLITICAL ECONOMY: THE QUEST FOR DEVELOPMENT SINCE 1949, by Carl Riskin. Oxford: Oxford University Press, 1987.

ANVIL OF VICTORY: THE COMMUNIST REVOLUTION IN MANCHURIA, by Steven I. Levine. New York: Columbia University Press, 1987.

SINGLE SPARKS: CHINA'S RURAL REVOLUTIONS, edited by Kathleen Hartford and Steven M. Goldstein. Armonk, NY: M. E. Sharpe, 1987.

URBAN JAPANESE HOUSEWIVES: AT HOME AND IN THE COMMUNITY, by Anne E. Imamura. Honolulu: University of Hawaii press, 1987.

CHINA'S SATELLITE PARTIES, by James D. Seymour. Armonk, NY: M. E. Sharpe, 1987.

THE JAPANESE WAY OF POLITICS, by Gerald L. Curtis. New York: Columbia University Press, 1988.

BORDER CROSSINGS: STUDIES IN INTERNATIONAL HISTORY, by Christopher Thorne. Oxford and New York: Basil Blackwell, 1988.

THE INDOCHINA TANGLE: CHINA'S VIETNAM POLICY, 1975–1979, by Robert S. Ross. New York: Columbia University Press, 1988.

REMAKING JAPAN: THE AMERICAN OCCUPATION AS NEW DEAL, by Theodore Cohen, Herbert Passin, ed. New York: The Free Press, 1987.

KIM IL SUNG: THE NORTH KOREAN LEADER, by Dae-Sook Suh. New York: Columbia University Press, 1988.

JAPAN AND THE WORLD, 1853–1952: A BIBLIOGRAPHIC GUIDE TO RECENT SCHOLARSHIP IN JAPANESE FOREIGN RELATIONS, by Sadao Asada. New York: Columbia University Press, 1988.

CONTENDING APPROACHES TO THE POLITICAL ECONOMY OF TAIWAN, edited by Edwin A. Winckler and Susan Greenhalgh. Armonk, NY: M. E. Sharpe, 1988.

AFTERMATH OF WAR: AMERICANS AND THE REMAKING OF JAPAN, 1945–1952, by Howard B. Schonberger. Kent, OH: Kent State University Press, forthcoming.

NEIGHBORHOOD TOKYO, by Theodore C. Bestor. Stanford: Stanford University Press, 1989.

MISSIONARIES OF THE REVOLUTION: SOVIET ADVISERS AND CHINESE NATIONALISM, by C. Martin Wilbur and Julie Lien-ying How. Cambridge, MA: Harvard University Press, 1989.

EDUCATION IN JAPAN, by Richard Rubinger and Beauchamp. Honolulu: University of Hawaii Press, 1989.

FINANCIAL POLITICS IN CONTEMPORARY JAPAN, by Frances Rosenbluth. Ithaca: Cornell University Press, 1989.

THAILAND AND THE UNITED STATES: DEVELOPMENT, SECURITY AND FOREIGN AID, by Robert Muscat. New York: Columbia University Press, 1990.

ANARCHISM AND CHINESE POLITICAL CULTURE, by Peter Zarrow. New York: Columbia University Press, 1990.

CHINA'S CRISIS: DILEMMAS OF REFORM AND PROSPECTS FOR DEMOCRACY, by Andrew J. Nathan. New York: Columbia University Press, 1990.

STATE POWER, FINANCE AND INDUSTRIALIZATION OF KOREA, by Jung-Eun Woo. New York: Columbia University Press, forthcoming.

SUICIDAL NARRATIVE IN MODERN JAPAN: THE CASE OF DAZAI OSAMU, by Alan Wolfe. Princeton: Princeton University Press, forthcoming.

COMPETITIVE TIES: SUBCONTRACTING IN THE JAPANESE AUTOMOTIVE INDUSTRY, by Michael Smitka. New York: Columbia University Press, forthcoming.

Index

References to illustrations are printed in boldface type.

Academia Sinica, 82, 135, 178, 181, 189, 211, 214, 230–1, 237; and National Research Council, 250–5, 306; during war, 311, 315
Academy of Current Affairs, 161
acids, 4, 34, 38, 153, 159, 226, 261–4, 274, 278–9, 281, import and production figures, 283; during war, 294, 297, 300–1, 311; postwar 367
activation energy, 235
Adolph, William, 121, 125, 139
Agricola, 31
agricultural research and education, 77, 96, 116, 135–6; 1927–37, 183, 190, 215–17, 229, 264; chemistry, 227, 256; Nationalist areas during war, 306–7; Communist areas during war, 348, 354, 358; enrollments in higher education, 366, 389; Chinese students abroad, 382–6; women in, 390
Ai Szu-ch'i, 321, 323
alchemy, 2, 20, 31–2
alcohol, ethyl (ethanol), 169, 244, 258, 269; in Nationalist areas during war, 294, 300–2, 306; alcohol factory, 300; in Communist areas during war, 311, 328, 330, 332, 335, 342; postwar, 368
All-China Federation of Natural Science Societies, 371
American Cyanamid Company, 267
American War Production Mission, 308
ammonia, 3, 45, 264, 268
ammonium chloride (fertilizer), 295
ammonium sulfate, 3, 160, 167; manufacture in, 1930s, 264–9, 266, 278; in Communist areas during war, 330; post-1949, 368–70
Amoy, 64
An-sai County, 326, 335
Andersson, J. G., 134, 137
Anking arsenal, 23, 24, 154
Anti-Japanese University, 341
Anyang, 137
archaeology, 137
Aristotle, 2, 31
arsenals: built by "self-strengtheners," 8, 15, 73–4, 54, 57, 157; by Li Hung-chang, 61–2; Kiangnan, 153–4; manufacture of sulfuric acid in, 170, 262, 264; in Communist areas, 331, 350; see also individual arsenals by location
astronomy, 5–6; traditional, 20; Jesuit translations, 6, 28, 30–1; Protestant translations, 34, 36, 56–7; education in, 1930s, 72, 83, 111, 183, 218; in wartime, 309

Beipei, 311
Billequin, Anatole, 35–6, 56, 110; names of chemical elements, 40, 377–80
Biological Research Institute, Science Society of China, 99
biology, 5, 58, 82–3; in universities, 1895–1927, 113–16, 119–20, 122–3; research, 135; in universities, 1927–37, 181, 183, 214, 217–18, 224–7; in Nationalist areas during war, 304, 309; in Communist areas during war, 346–7, 354, 358; Chinese Ph.D.'s in, 385
bleaching powder, 3, 156, 302, 335; manufacture by electrolysis, 259–60, 284, 294
Book of Odes, 20
Border Region Match Factory, 338
botany, 57, 200, 211, 225; in middle school curricula, 387–8
Boussingault, Jean, 3, 56
Boxer Indemnity, 9, 77, 114; fellows, 140–1, 150, 165, 187, 224; U.S. and U.K. funds, 179, 198–202, 237–8
Boxer Rebellion, 109–10, 155
Brunner-Mond and Company, 163, 165, 279, 294

Buddhism, 21, 45, 65
Bukharin, Nikolai, 320
Bureau of Industrial Research (Peking), 100, 151

Calmette, Leon, 246
Canton, xvii, 64, 107, 227, 326; missionaries in, 29, 34, 60; Arsenal, 157; chemical manufacturing in, 243, 260–3, 268, 272
caustic soda, *see* sodium hydroxide
Central Observatory, 135
Central Research Institute (Yenan), 346
Ch'a-fang Arsenal, 326–7, 329
Chang Ch'ao-chün, 346
Chang Chih-tung, 155
Chang Chün-mai, 92, 204
Chang-erh-ts'un, 336
Chang, H. T., 100, 134
Chang I-tsun, 225
Chang Kwang-chih, 137
Chang, T. S., 309
Chang Ta-yü, 224, 371
Chang Tsing-lien, 222, 235, 310, 371
Chang Tzu-kao, 115, 224, 371, 373
Changchiakou, 163
Changlu salt fields, 161
Changsha, 70, 161, 344
Chao Yüan-jen, 97
Chekiang, 61, 80, 106, 259, 262, 326; society in, 177–81
Chekiang University, 220, 222, 226, 327
Chemical Bulletin (journal), 197
Chemical Engineering (journal), 196
Chemical Industry (journal), 101
chemistry, history of in Europe, 2–5
Chemistry (journal), 197
Chemistry Forum, 181, 191–2, 196
chemistry of natural defense, 188
Ch'en Chen-hsia, 333–4
Ch'en Hua Paper Factory, 335
Ch'en I, 375
Ch'en K'ang-pai, 345, 358, 375
Ch'en K'o-chung, 306
Chen Ko-kuei, 149–50, 152, 180, 200
Ch'en Kuang-fu, 164
Ch'en Kuo-fu, 205; in debate on educational policy, 1932–34, 208–13, 229; and educational reforms, 254
Ch'en Li-fu, 205; minister of education during war, 306–9, 316
Ch'en Po-ta, 321, 375
Chen-tan (L'Aurore) University, 123
Ch'en Tu-hsiu, 90
Ch'en Yu-gwan, 125
Cheng Chen-wen, 81
Cheng feng Campaign, 347, 351–60, 364–5, 372–5
Ch'eng Shu-jen, 338
Chengtu, 264, 289, 299, 303
ch'i [vapors], 6, 21
Ch'i Ju-shan, 56
Ch'i-li-pu, 342
Ch'i-li-ts'un, 334
ch'i-yü [praying for rain], 362
Chi Yuoh-fong, 371
Ch'iang hsüeh-hui [Society for the Study of (Self) strengthening] 94
Chiang Monlin, 202, 227
Chiang T'ung-yin, 70–1
Ch'iao-erh-kou, 332
Chiao-hui hsin pao [The church news], 45
Chiao-tung Industrial Research Laboratory, 330–1, 337
Chiao-tung Liberated Area, 330–1, 337

437

Chiao-t'ung University, 220, 223, 226
ch'ieh-yin-tzu (phonetic term), 43
Ch'ien Chih-tao, 327, 356, 375
Ch'ien Tuan-sheng, 314–15
Chile saltpeter (sodium nitrate), 3
Chin-Ch'a-Chi Arsenal, 330
Chin Hsin-yüan, 240
Chin-k'uei County, 21
Chin P'ei-sung, 311
China Alcohol Factory, 269, 300
China Christian Educational Association, 122
China Foundation for the Promotion of Education and Culture, 82, 149, 199–200, 227
China Institute for Economic and Statistical Research, 273
China Medical Board, 120
China Medical Missionary Association, 50
China Merchant Navigation Company, 333
China Society of Chemical Industry, 100
Chinese Academy of Sciences (PRC), 186, 239, 371–2
Chinese Botanical Society, 197
Chinese Chemical Society, 187, 196–7
Chinese Geographical Society, 197
Chinese Industrial Chemical Research Institute, 261
Chinese Journal of Physiology, 141
Chinese Mathematical Society, 197
Chinese Meteorological Society, 100, 197
Chinese National Association for the Advancement of Education, 128
Chinese Physical Society, 197
Chinese Physiological Society, 141
Chinese Science and Technology Information Institute, 371
Chinese Scientific Book Depot, 48
Chinese Society of Chemical Engineers (CSCE), 196
Chinese Society of Chemical Industry (CSCI), 196, 261
Chinese Sociological Society, 138
Chinese-Western Glossary of Chemical Substances, 41
Chinese Zoological Society, 197
Ch'ing dynasty, policies affecting science, 8, 9, 14–15, 65, 73
Ching-hsing Mining Bureau, 233, 240–1, 241, 271, 279
chiu-li [science], 85, 88
Chiu-ta Refined Salt Company (Tangku), 151, 162; history, 162–5; during war, 296
Ch'iu Yan-tsz, 125
ch'iung-li [science], 85–6
chlorine, liquid, 259–60, 284
Chou Jen, 97
Chou P'ei-yüan, 309
Chou Tsan-quo, 81, 149–50, 200, 237, 252, 371
Chou Tso-min, 164
Chou Yang, 358
Chu Chia-hua, 306, 315
Chu Hsi, 66–7
Chu Ju-hwa, Edith, 225, 310; biography, 234–5; research, 310
Ch'ü Po-ch'uan, 338, 345, 348, 350
Chuang Chang-kong, 180, 252, 371; biography, 186–7; research, 237, 249
Ch'ün-chung [The masses], 314, 316
Chung-Hsi wen-chien-lu [The Peking magazine], 46
Chungking, 11, 141, 243; during war, 289, 293, 297–9, 303, 311, 313–17, 326, 334, 343
coal tar, 3, 233; distillation of, 1927–37, 240–2, 270–2, 279; during war, 298
Coble, Parks, 284–5
Coltman, Dr. Robert, 109
Columbia Teachers College, 112, 128
Columbia University, 82, 166
Commercial Press, 191
Confucianism, 2, 9, 28, 85, 89, 91, 93, 96, 101, 374; of K'ang Yu-wei, 65–8; revival under Kuomintang, 204–5

contact method, for manufacture of sulfuric acid, 263–4, 267–8
Cordyalis [yen-hu-so], 150, 200
Cornell University, 82, 97, 124, 191
Cressy, Earl H., 122–3, 130
Cultural Revolution, 373
Curie, Marie, 349

Dairen, 369
Dalton, John, 2, 37
Darwin, Charles, and darwinism, 89, 115
De Re Metallica, 31
Deborin, Abram, 320–2
Democratic League, 188, 373
denaturation of proteins, 145–7
Dernberger, Robert, 278
Dewey, John, 93
dialectical materialism, 320, 322–3
dialectics of nature, 320
dipole moment, 236
Discourses on Salt and Iron, 161
drugs, manufacture of, 3–4, 162, 274, 294, 311, 325; in Communist areas during war, 336–7
Du Pont Corporation, 268
Dutch learning [Rangaku], 51
dyes, manufacture of, 3–4, 162, 260, 262, 270, 274, 294, 306, 311, 336

Eastman, Lloyd, 284–5
economics, 138
education: 19th century: missionary, 24–7, 57–60; government, 53–7; 1895–1927: primary and secondary, 104–9; higher government, 109–16; higher missionary, 116–26; scientific, 126–31; 1927–37: debate on policy, 208–13; reforms, 213–18; scientific, 218–29; in Nationalist areas during war, 303–9; in Communist areas during war, 342–7, 343–5; postwar, 365–7; enrollments, 366, 381, 389; Chinese students abroad, 382–6; middle school curricula, 1932, 387–8; women in, 390
Educational Association of China, 48
Eighth Route Army, 324, 336–7
Einstein, Albert, 250
electrolysis of salt, 259–61, 279, 284
elements, Chinese names for, 40–3, 377–80
Eli Lilly Company, 150
empiricism [k'ao-cheng], 23
Engels, Frederick, 320, 322
engineering, 116, 183, 190; personnel, 165; in higher education, 1927–37, 215–18, 224, 226, 229; chemical, 256; in Nationalist areas during war, 303–9; in Communist areas during war, 348, 350, 354; postwar, 365–7; enrollments in higher education, 366, 389; Chinese students abroad, 382–6; women in, 390
Ephedra vulgaris [ma-huang], 149–50, 200
ephedrine, 149–50, 337
Epstein, Israel, 364
ethanol, see alcohol, ethyl
examinations, civil service, 14–15, 19–20, 81, 94, 104
Executive Yuan, 254
explosives, manufacture of, 3, 31; 19th century, 62–3; 1895–1927, 153–7; 1927–37, 262–4; in Communist areas during war, 325–31
Eyring, Henry, 235

fa-t'uan [associations established by law], 96
Faber, Knud, 225
Fan Ching-sheng, see Fan Yüan-lien
Fan Hsü-tung, 101, 151, 239, 261, 278–9; biography and career in north China, 160–7, 170; and Yungli company, 265; during war, 294–5
Fan Jui, see Fan Hsü-tung
Fan Yüan-lien, 161–2

Fei Hsiao-t'ung, 222
Feng Yün-hao, 125
fermentation, 239, 244–8, 256, 286
fertilizer, chemical, 3, 239, 258; manufacture, 1927–37, 264–9, 278- 9; in Nationalist areas during war, 295, 300, 311; in Communist areas during war, 330; postwar, 367–8
Feuerwerker, Albert, 278
First Five-Year Plan, 369
five elements [wu-hsing], 6, 31
Folin, Otto, 141, 143, 145
Folin-Wu method of blood analysis, 142
Fong, H. D., 138
Foochow, 61, 140, 165
Foochow Naval Academy, 89, 140
Foochow Navy Yard, 54; schools, 54–5
Foochow University, 299
Forman, Harrison, 334, 361
Frank, Henry, 121, 139
Fresenius, Karl, 39
Fritillaria [pei-mu], 150
Fryer, John, 17, 22, 26, 33, 154; biography, 25–8; at Kiangnan Arsenal, 29–30; translations of, 36–48, 52, 64, 191–2; names of chemical elements, 40–3, 377–80; and Shanghai Polytechnic Institute, 68; and Japanese translations, 84–5
Fu-jen University, 187
Fu Ssu-nien, 202, 211–2, 222
Fu Ying, 371
Fukien, 69, 80, 110, 113, 140, 165, 178, 186, 300
Fukien Anhwei Railroad School, 165
Furth, Charlotte, 93
Futan University, 123, 221

Galileo, 349
gasoline: research on, 1927–37, 240–4, 247–8; refining of, 272–3, 294; production in Nationalist areas during war, 297–300, 302, 311; in Communist areas during war, 334–5
Gay-Lussac towers, 261–2
Gelsemium [kou-wen], 150
General Missionary Conference, 48
Geological Society of China, 100, 135, 197
geology, 5, 36, 57, 84, 115, 183, 219, 225, 309; research in, 134–5; Chinese Ph.D.'s in, 385; in middle school curricula, 387–8
Giles, Herbert, 88
Giquel, Prosper, 54
glass, 158–9, 162–3, 278, 281, 294, 310, 342, 345, 348; chemical glassware, 305; manufacture in Communist areas, 335–6;
Glover towers, 261–2
Golden Sea Research Institute of Chemical Industry, 151, 170, 239, 246
Gottingen, 186, 237
Grabau, A. W., 115, 134–5
Grand Canal, 10
Great Rear, 288, 303, 314, 318
Great Wall, 163, 360
Green, Roger, 116
gunpowder, manufacture of: 19th century, 62–3; smokeless, 63, 153-7, 279; in Communist areas during war, 326–31, 356

Haber (synthetic ammonia) process, 267, 295
Haiphong, 259, 304
Hangchow, 64, 69, 226
Hangchow University, 371
Hankow, 243, 279, 326
Hanyang Arsenal, 63, 261, 326; manufacture of gunpowder, 155–7, 168
Hanyang Iron Works, 271
Hao Ching-sheng, 315
Harvard Medical School, 141
heavy water (deuterium oxide), 235, 310

Heilongtan, 310
Henry Lester Institute of Medical Research, 136, 139
herbal drugs, research on, 149–50, 200, 349
Himalayan Hump, 301, 314, 318
Histories of the Former and Latter Han Dynasties, 20
Ho-ch'ü County, 328
Ho, Franklin, 138
Ho Liao-jan, 35; names of chemical elements, 40, 42, 377–80
Ho-shui County, 362
Hobson, Benjamin, 22–3, 34–6, 51, 60; names of chemical elements, 40, 377–80
Honan, 137, 262
Hong Kong, 277, 295, 317, 343
Hopeh, 240, 271, 329
Hou Chi-ming, 273
Hou Te-pang, 180, 252; early career, 165–7, 170; and Yungli Chemical plant, 267; "united soda method," 294-5; in PRC, 371
Hsi Han-po, see Hsi Kan
Hsi Kan, 56–7
hsieh-sheng (type of Chinese ideograph), 194
Hsiehchiatien, 265–8, 266–7, 368–9
Hsien-yang Alcohol Factory, 269
Hsin-hsüeh pao [New study journal], 95
Hsü Chien-yin, 23, 30, 38, 48; introduction of smokeless gunpowder 154–6, 161
Hsü Hsüeh-ts'un, see Hsü Shou
Hsü Hua-feng, 168
Hsü Shou, 18, 66, 72–3, 154, 168; early years, 17–25, 28; at Kiangnan Arsenal, 29–30, 32, 34; translations, 36–50, 52, 64, 191–2; names of chemical elements, 40–3, 377–80; Shanghai Polytechnic Institute, 68; and Japanese translations, 84–5
Hsü T'e-li, 344–58, 372, 375
hsüeh-hui [study society], 94–5, 98
hsüeh-she [scholarly society], 98
Hsüeh-sheng tzu-tien [Student dictionary], 88
Hu Hsien-su, 136, 152, 315
Hu Mei, 235
Hu Ming-fu, 97
Hu Shih, 93, 97, 137, 202
Hua Heng-fang, 21–5, 30
hua-hsüeh [chemistry], 1, 35, 51, 71
Hua-hsüeh chien-yüan [Authentic mirror of chemical science], 37, 40, 49
Hua-hsüeh chih-nan [Guide to chemistry], 35
Hua-hsüeh ch'iu-shu [Quantitative analytical chemistry], 38
Hua-hsüeh ch'u-chieh [First step to chemistry], 35, 40
Hua-hsüeh kung-yeh [Chemical industry], 101
Hua-hsüeh shan yüan [Explanations of the principles of chemistry], 35
Hua-hsüeh shih-chieh [Chemical world], 101
Hua Shou-chün, 332, 335, 345–6, 375
huang chiu [yellow wine], 243
Huang Ming-lung, 373
Huang Tzu-ch'ing, 224, 371
Huang Wen-wei, 125
hui-i (type of Chinese ideograph), 194
Hunan, 69–70, 161, 187, 281, 344
Hundred Days Reforms, 109
Hundred Flowers Movement, 189, 373
Huo-kung ch'ieh-yao [Essentials of gunnery], 31
Hupeh, 155
hydrochloric acid, 32, 34, 169, 262; electrolysis, 259–60; manufacturers, 276–7, 279; import and production figures, 283–4; during war, 294, 302; illustration of factory, 301; in Communist areas, 327

Ikeda, Kikunae, 168
immunology, 147
Imperial Maritime Customs Service, 135
Imperial University of Peking, 109–10, 124

India, 250, 301, 318
industry, chemical, 11, 62–3; military, early 20th century, 153–7; civilian, early 20th century, 157–69; electrolysis, 259–61; sulfuric acid, 261–4; fertilizer, 264–9; alcohol, 269; organic, 270–3; overview and analysis, 273–86; in Nationalist areas during war, 293–302; in Communist areas during war, 323–5, 331–8; in PRC, 369–70
Inkstone Book Store [Mo-hai shu-kuan], 21, 22, 46
Inner Mongolia, 163, 332, 337
Inoue, Kowashi, 86
Institute of Applied Chemistry, 371
Institute of Chemistry, National Academy of Peiping, 238, 310
Institute of Chemistry, Academia Sinica, 186, 237, 311; in PRC, 371
Institute of Genetics, 372
Institute of Geology, 135
Institute of History and Philology, 137, 211
Institute of Materia Medica, 150
Institute of Organic Chemistry, 186, 371
instruments, chemical, 47
International Dispensary, 270
"investigation of things" [ko-wu or ko-chih], 17, 104, 109, 111; Hsü Shou and, 20-1, 23; K'ang Yu-wei and, 66–7; literati and, 68–71; science and, 83–8
ionic potentials, 310

J. Llewellyn and Company, 278
Jan-liao chuan-pao [Fuels report], 240
Japan: chemistry in, 34, 36, 39, 42, 50–1; scientific language in 81–8; influence on Chinese education, 110, 116, 124; chemical research, 149; chemical industry, 242, 265
Jen Hung-chün, 81–2, 90, 103, 371; and Science Society of China, 97, 97–9
Jeu Kia-khwe, 125
Johns Hopkins University, 117, 119
Journal of Sociology, 138
Journal of the Chinese Chemical Society, 187, 197, 231

kagaku [chemistry], 51
kagaku [science], 86–7
Kagaku Shinso [New book on chemistry], 51
K'ai-ch'eng Acid Company, 262, 276, 281
Kaiping Coal Mines, 151
kakubutsu [science or physics], 85–7
kakuchi [science or physics], 85–6
K'ang Hsi Dictionary, 43
K'ang Ti, 354
K'ang Yu-wei, 64–8, 71, 89, 94–5, 204
Kansu, 297, 360, 362
Kansu Petroleum Bureau, 297
Kao Ch'ung-hsi, 224, 235
kao-liang, 245
Kao Tsi-yu, 225, 371
kerosene, 240, 258, 272; in Nationalist areas during war, 297–9; in Communist areas during war, 334–5
Kerr, John Glasgow, 35, 37, 52, 60; names of chemical elements, 40, 42, 377-80
Kiangnan (region), 18–19
Kiangnan Arsenal, 17, 24–5, 168; translation bureau, 28–30; translations of, 36–45, 50–2, 84, 192; manufacture of gunpowder, 62–3; influence of translations, 64–5, 73; smokeless gunpowder, 154–6
Kiangsu, 18, 80; and scientific education, 106, 108, 110, 112; and second-generation scientists, 177–9, 181
Kiangsu Chemical Works, see Major Brothers
Kienwei, 298
Kincheng Bank (Tientsin), 164, 169
Kirin, 369
Kirin University, 371
ko-chih, see "investigation of things"
Ko-chih hsin-pao [Journal of science], 95

Ko-chih hui-pien [Chinese scientific magazine], 45–7, 64, 66, 155
K'o-hsüeh [Science magazine], 90, 97, 99, 313
k'o-hsüeh [science], 86–8, 90
Ko Tao-yin, 71
ko-wu, see "investigation of things"
Ko-wu ju-men [Natural philosophy], 34
Kryer, Carl, 30
Ku Chieh-kang, 137
Ku shih pien [Critical reviews of ancient history], 137
Ku Yü-chen, 311
Ku Yü-hsiu, 212, 254
Ku Yü-tsuan, 239, 311
Kuang-Hua Drug Factory, 336
Kuang-Hua Farm, 342, 356–8
K'un-yü ko-chih [Investigation of the earth], 31
K'un-yü Mountains, 331
K'ung-chi ko-chih [Treatise on the material composition of the universe], 31
Kung-hsien Arsenal, 157, 264
Kung-yeh chung-hsin [Industrial Center], 239
Kunming, 11, 186, 188, 234–5, 238, 330; wartime universities in, 288, 303–5, 309–11, 314–17
Kuno, Yoshi, 130, 227
Kuo Mo-jo, 126, 375
Kuo Ping-wen, 112
Kuomintang, in debates on science policy, 1927–37, 7, 10–11, 203–6; on education, 208–13; on research, 250–5
Kuomintang Central Political Council, 209, 251
Kwangsi, xvii, 262–3, 279, 344
Kwangtung, xvii, 65, 69, 80, 125, 178, 181, 242; manufacture of chemicals in, 262–3, 272, 279
Kwangtung Provincial Electrochemical Factory, 260, 275, 277
Kwangtung Provincial Fertilizer Factory, 268
Kwangtung Provincial Sugar Factory, 269
Kwangtung Provincial Sulfuric Acid Factory, 263–4, 276, 281
Kweichow, 288
Kwok, D. W. Y., 93
kyuri [natural philosophy], 85–6, 88

Lai-yang, 337
Laminaria japonica, 168
Lanchow, 314, 369
Lavoisier, Antoine, 2, 4, 32, 51
law of periodicity, 39
lead chamber, method for manufacture of sulfuric acid, 156, 263; described in China, 160; in Communist areas, 328, 330
League of Nations Health Organization, 225
Leblanc, Nicolas, 3; Leblanc method of soda manufacture, 165, 294–5
Lee Fang-hsuin, 125
Lee, T. D., 309
Lehigh University, 113
Lenin, 320, 322–3
Lenin Normal School, 344
Leo Shoo-tze, 165, 227, 236
Lew, Timothy Ting-fang, 130
li [principle], 6, 21
Li Chi, 137
Li Ch'iang, 350, 356, 358, 375
Li-chung Acid Company, 262, 276, 280
Li Erh-kang, Richard, 250, 311
Li Fu-ch'un, 342, 344, 348, 357
Li Hung-chang, 61, 73, 155
Li Shan-lan, 21–3, 25, 56–7
Li Sheo-hen, 227
Li Shih-tseng, 82
Li Su, 345
Li Ta-chang, 356
Li Tuan-fen, 104

Li Yün hua, 224
Liang Ch'i-ch'ao, 64, 91–2, 161
Liang-Kwang Provincial Sulfuric Acid Factory, 263, 276
Liaoning, 264
Liberation Daily [Chieh-fang jih-pao], 335, 348, 353, 358, 361–2
Lien-ta, see National Southwest Associated University
Lieu, D. K. (Liu Ta-chün), 273–4, 277
Lih Kun-hou, 226
Lin Hua, 335, 345, 375
Lin Shan, 354–5
Lin Tse-hsü, xvii–xviii
Lingnan University, 124–5, 139, 221, 223
literati, 15, 27, 53; and science, 63, 68–71, 73
Liu Hsien-i, 335, 345, 356
Liu Jang, 337
Liu Ta-chung, 277
Lo Feng-lu, 55, 61
Lo T'ien-yü, 353–6, 358, 372
London Missionary Society, 21
Long March, 289, 319, 326, 334, 339, 344
Lou, H. O., 227
Lu Yü tao, 313, 316
Lung-hua Powder Plant, 154–5
Lung Yun, 314
Lysenko, T. D., 372
Lysol, 270

ma-huang, see Ephedra vulgaris
ma-lan grass, 335
Ma, S. T., 309
Ma Tsu-sheng, 222
MacGowan, John, 30
Major Brothers, 261–2, 278
Malaguti, Faustino, 35
Manchuria, 163, 186, 337; Japanese manufacture in, 242–3, 273, 277–8; postwar development in, 367–70
Mao Tse-tung, 102, 291; philosophical writings, 321–3; and Hsü T'e-li, 344–5; and Cheng feng, 353, 356, 358–9, 363, 369
Martin, W. A. P.: scientific writings, 34–7; names of chemical elements, 40, 42, 377–80; and T'ung Wen Kuan, 56; and Imperial University, 109–10
Marxism, 7, 93, 290, 314, 374; interpretation of history, 284–5; and science, 319–23; in Yenan, 341, 352, 354
Massachusetts Institute of Technology (MIT), 141, 165–6, 187
Mateer, Calvin, 27, 57–9, 109
matches, manufacture of, 158, 168, 274, 278; in Communist areas during war, 325, 337–8
mathematics, 6, 36, 83, 89, 183; Hsü Shou and, 20–1, 28; Jesuits and, 30–1, 72; in modern schools, 19th century, 56–8; in schools, 1895–1927, 107, 110–12, 120, 122; in schools, 1927–37, 214–15, 217–18, 224–5; in Nationalist areas during war, 304, 309; in Communist areas during war, 345, 347; Chinese Ph.D.'s in, 385; in middle school curricula, 387–8
McLean, Franklin, 145
medicine and medical education: in 19th century, 20, 59–60; 1895–1927, 116–20, 135; 1927–37, 183, 190, 215–16, 225, 228–9, 256; during war, 306–7, 347; enrollments in higher education, 366, 389; Chinese students abroad, 382–6; women in, 390
meteorology, 5, 34, 83, 135, 183
methylene, 45
Military Industrial Bureau, 350, 358
Mill, John Stuart, 87, 89
Ming dynasty, 18, 67, 89, 94
Mining and Engineering College of the Hupeh Province Board of Mines, 54
Ministry of Agriculture and Commerce, 100, 151
Ministry of Chemical Industry (PRC), 371
Ministry of Economic Affairs (MEA), 293–4

Ministry of Education: 1912, 106; 1927–37, 190–1, 211; educational reforms led by, 213–15, 227–9; during war, 309; in PRC, 371
Ministry of Heavy Industry (PRC), 239
Ministry of Industry, 189, 263
Mirsky, Alfred, 146
missionaries, Jesuit, 6; translations, 20, 22, 28, 30–2, 36, 42
missionaries, Protestant, 1, 8–9, 14–17; John Fryer and, 22, 25–8; translations by, 32–6, 46; schools maintained by, 53, 57–60, 105, 116–26, 190, 218–19, 223, 228, 230; chemical research by, 139
monosodium glutamate (MSG), 168–9, 188, 234, 278–9
Monroe, Paul, 128
Moscow, 344
Mucor, 246
Mukden, 157, 326

Nagayoshi, Nagai, 149
Nakayama, Shigeru, 86
Nankai Economic and Social Quarterly, 138
Nankai University, 121, 221; Institute of Economics, 138; Institute of Applied Chemistry, 262; during war, 303–6; in PRC, 371
Nanking, 69, 105, 107, 139, 181, 237, 239, 243, 265, 290, 293, 298, 327, 352, 365
Nanking Arsenal, 24, 63, 154
Nanking Higher Normal College, 112–13, 226
Nanking University, 121, 123, 124–6, 130, 217, 221, 223, 227; college of agriculture, 124; in PRC, 371
National Academic Work Advisory Board (NAWAB), 183
National Academy of Peiping, 150, 181, 189, 238, 310, 311
National Agricultural Research Bureau, 189, 256
National Bureau of Industrial Research (NBIR), 189; research in, 1927–37, 239–48, 242, 250; during war, 298–9, 299, 311
National Central University, 112, 187, 220, 223; during 1930s, 225–6, 231; publication, 225; during war, 306
National Chekiang University, 226–7
National Chungshan University, 263
national essence [kuo-ts'ui], 204
National Geological Survey, 82, 100, 134–5, 151, 189, 240
National Health Administration, 289
National Institute for Compilation and Translation (NICT), 190–1; names for chemical elements, 377–80
National Institute of Health, 141
national language [kuo-yü], 106; in middle school curricula, 387–8
National Peking University, 100, 115, 130, 137; research in, 1927–37, 187–8, 209, 231, 234–6; education in, 1927–37, 220, 227; during war, 303–4; in PRC, 371
National Research Council, 248, 55, 298, 315
National Resources Commission (NRC), 293–4, 297, 308, 367
National Southeastern University, 112–15, 121, 130, 139, 225
National Southwest Associated University, 303–6, 310–11, 312, 330
National Sun Yat-sen University, 227
National Taiwan University, 186, 222
Natural Science Institute (NSI), 327, 348; educational program in, 342–7; Cheng feng and, 350–9, 372, 375; Student Association Bulletin of, 356
Natural Science Research Society (NSRS), 342, 347–8, 350, 352, 358
Needham, Dorothy, 295, 308
Needham, Joseph, 5, 31, 302–3, 317
New Century [Hsin shih-chi], 87
New China Chemical Factory, 332, 336–7, 342
New China Drug Factory, 337

New Culture, 7, 9, 79, 89–90, 103, 115, 319; debate
with Kuomintang right, 202–7
New Fourth Army, 331
New Life Movement, 205
New Text School, 67
New Youth [*Hsin ch'ing-nien*], 90
Newton, Issac, 357
Nieh Jung-chen, 375
nitric acid, 3, 32, 34, 63, 262; and manufacture of
explosives, 153–6, 170; factory, 267; manufacturers,
276, 279–80; import and production figures, 283; in
Nationalist areas during war, 302; in Communist areas
during war, 327–31
Nobel Prize, 233, 250, 309
North China Chemical Research Institute, 139
North China College, 58
North China Herald, 107
Northern Wei dynasty (A.D. 424–532), 20
Nung-hsüeh pao [Journal of agriculture], 95
Nystrom, E. T., 110

Oriental Education Commission of the University of
Chicago, 111
Ostwald process, manufacture of nitric acid, 268

P'an, C. C., 226
paper, manufacture of: 1895–1927, 158–9, 162–3;
1927–37, 260, 278, 281; in Nationalist areas during
war, 294, 306, 310–11; in Communist areas during
war, 325, 328, 333, 346, 356; from *ma-lan* grass, 335
Parker, Peter, 60
Pauley, Edwin, 367
Pauling, Linus, 146
Pei-yang hsi-hsüeh, 111
Peiping, 210, 253, 365,
Peiping Academy, *see* National Academy of Peiping
Peiyang University, 111
Peking, xvii, 34, 53, 69, 92, 94, 100, 105, 133, 135,
150, 210, 213, 236, 290, 326, 329, 352
Peking Imperial University, 113
Peking Union Medical College (PUMC), 77; education
in, 118–20, 118–21; research in, 132, 136, 139,
140–50, 142–4, 152, 249
Peking University, *see* National Peking University
Peking University (Methodist), 58
Pen ts'ao kang mu [Great pharmacopoeia], 149
P'eng Jui-hsi, 70
P'eng-shan, 294
People's Republic of China (PRC), 186, 331, 369–72,
375
Perkins, William, 3
petroleum, 258; early exploration, 272–3; in Nationalist
areas during war, 294, 297–8; in Communist areas
during war, 325, 333–5; postwar, 367–8
physics, 5; introduced to China, 19th century, 34, 36,
57–8, 83–4; education, 1895–1927, 107, 109–11,
113–15, 119–20, 122–3; research in China, 136–7;
education, 1927–37, 181, 183, 200, 211, 214,
217–18, 224–7; in Nationalist areas during war, 304,
309; in Communist areas during war, 320, 345, 347,
350, 354; Chinese Ph.D.'s in, 385; in middle school
curricula, 387–8
Pierle, C.A., 115
Ping Chih, 97, 110, 136, 152, 185
Po-wu hsin-pien [Natural philosophy and natural
history], 22, 34
Pohai Chemical Works, 275–6, 280
potassium nitrate, *see* saltpeter
Presbyterian Mission Press, 46
Principles of Chemical Nomenclature, 191–4
Principles of Physical Biochemistry, 141
P'u-i Sugar Factory, 269

Rangaku [Dutch studies], 85
Rankin, Mary, 61

Rawski, Thomas, 284–5
Read, Bernard, 139
Rectification Movement, *see Cheng feng* Campaign
Red Guards, 189
Remer, C. F., 273
research, chemical: early efforts, 49, 138–40; by Wu
Hsien, 142–8; on herbals, 148–50; applied, 151,
238–48; in universities and academies, 1927–37,
230–8, 255; on gasoline, 240–4, 247–8; on
fermentation, 244–8; during war, 310–11
research, scientific (other than chemical): beginnings in
China 132–8; debate over in 1930s, 248–55; during
war, 309–11
Ricci, Matteo, 20
Richard, Timothy, 110
Richthofen, Ferdinand von, 134
Rockefeller Foundation, 9, 77, 114, 225, 228; and
PUMC, 116–20, 124, 132, 140; contributions in
China, 198, 202, 218–19
Rockefeller Institute, 143

Sah Pen-t'ieh, Peter, 179–80, 224; biography, 186–7,
304, 370; research, 231–3, 249
Sah Pen-t'ung, Adam, 186, 255
salt, refined: industry established in China, 160–7, 170;
electrolysis of, 259–61; 1927–37, 278–80; in
Nationalist areas during war, 294–6; salt wells, 296; in
Communist areas during war, 324, 336
saltpeter (potassium nitrate), 63, 154, 156
San-pien subregion, 360
Schmidt, Carl, 150
School and Textbook Committee, 48
School and Textbook Series, 45
Science Society of China (SSC), 79, 82; early history,
96–101, 132; board of directors, 97; membership,
1930s, 182–3
scientism, 7, 67, 77; in early 20th century, 88–93, 101–2;
and Chinese Marxism, 319–23, 374
seimi [chemistry], 51
Seimi Kaiso [Foundation of chemistry], 51
self-strengthening, 8, 50, 61, 73, 204, 210; and
introduction of science, 14–16, 23–4
Shanghai: introduction of science in 19th century, 18,
20–2, 24, 26, 29, 34, 45, 48, 64, 69–70; during
1895–1927, 88, 101, 105, 135, 150–1, 165, 168–9;
during 1927–37, 186, 237, 243–4; chemical industries
in, 259–63, 268–70, 274, 279, 290; during war, 293,
295, 304, 320–1, 323, 326, 331, 333, 341, 352, 365
Shanghai Commercial and Savings Bank, 164
Shanghai Gas Company, 270
Shanghai Polytechnic Institute, 47, 49, 65, 68
Shansi, 110, 280, 328, 330
Shansi-Chahar-Hopeh (Chin-Ch'a-Chi) Military Region,
329–30, 335
Shansi Imperial University, 110
Shansi-Suiyüan Border Region, 328
Shantung, 330–1, 337
Shantung Christian University, 121, 221
Shaohsing wine, 245–6
Shen Hung, 326–7, 350
Shen-Kan-Ning Border Region Consultative Council,
350–1
shen lou tzu [spirit tower], 362
Shensi, 262; petroleum in, 240, 242, 262, 272, 297;
Communists in, 323, 326, 332, 334, 336–7, 360–2
Shensi-Kansu-Ninghsia (Shen-Kan-Ning) Border Region:
industry in, 323–6, 330–8; papermaking in, 333;
education in, 340, 343–5, 364; antisuperstition
campaign in, 360–2
Sherman, Albert, 235
Shihchiachuang, 240, 271
shipyards, 8, 15, 24, 57, 61
Shui-ching-chu, 20
Sian, 303, 327, 336–7

Sinyuan Fuels Research Laboratory, 240–1, 243, 298
Sivin, Nathan, 72
Snow, Edgar, 326
soap, manufacture of: 1895–1927, 158–9, 162–3;
  1927–37, 260, 262, 274, 278, 281; in Communist
  areas during war, 325, 332–3, 342, 348
social studies, 122, 183, 215, 307; research in, 138;
  enrollments in higher education, 366, 389; Chinese
  students abroad, 382–6
Société Astronomique de Chine, 100, 197
Society for the Diffusion of Useful Knowledge, 46
sociology, 138
Socony Oil Company, 335
soda, imports of, 159, 281–4
soda ash, see sodium carbonate
sodium bicarbonate (baking soda), 162, 337
sodium carbonate (soda ash), natural, 163, 332, 337
sodium carbonate, synthetic, 3–4, 38, 156; 1895–1927,
  160, 162–7, 170, 226; 1927–37, 239, 275–84; in
  Nationalist areas during war, 294–7, 300–2, 311; in
  Communist areas, 342; post-1949, 367–70
sodium chloride (salt), 162, 259, 295
sodium hydroxide (caustic soda), 162; manufacture of by
  electrolysis, 259–60; manufacturers, 275; assessment,
  1927–37, 281–4; in Nationalist areas during war, 294,
  302; in Communist areas during war, 332, 337;
  post-1949, 368–70
sodium nitrate (Chile saltpeter), 3
sodium silicate, 162, 275–6, 282
sodium sulfate, 294, 336
sodium sulfide, 162, 275, 282, 302
Solvay, Ernest, 165, 167
Solvay (ammonia) process, 165–7, 295
South Manchurian Railway Company, 273–4
Soviet Union, 252, 265, 315, 369, 372
Spring and Autumn Annals, 20
St. John's University, 58, 151, 221, 225
Standard Oil Company, 272
statecraft [ching shih], 23
Stein, Gunther, 327
Stuhlmann, Carl, 56
sugar, manufacture of, 239, 294
sulfanilamides, 311
sulfonamides, 337
sulfur, 63, 154, 156, 280, 328–30
sulfuric acid, 3, 32, 34, 62–3; in manufacture of
  explosives, 153–7, 170; lead chamber method, 160;
  manufacture in 1930s, 258–9, 261–4; contact method,
  263–4, 267–8, 280–1; manufacturers, 276; import and
  production figures, 283; in Nationalist areas during
  war, 294, 302; in Communist areas during war,
  327–30, 336, 338; postwar, 368, 370
Sun Ch'eng-o, 235, 310, 371
Sun Hsüeh-wu, 151
Sun Hung-ju, 345
Sun Yat-sen, 203, 205
Sun Yat-sen University, 220
Sung dynasty, 89, 280
Sung Fa-hsiang, 110
superstition, campaign against in Communist areas,
  359–63
Szechwan, 18, 82, 126, 239, 241–2, 244; in Nationalist
  areas during war, 288, 293–6, 298–300
Szent-Gyorgyi, Albert, 233

Ta-hsüeh [Great learning], 66, 69
Tagore, 250
Tai An-pang, 125, 371
Taiping Rebellion, 22–4, 61
Taiwan, 140, 367–8
Taiyuan, 110–11, 157, 264, 279, 327
T'an Ssu-t'ung, 64
T'ang Yüeh, 91
Tangku, 151, 162, 165–6, 260, 268, 295

tanning, 158, 226–7, 259, 286, 306; research on, 236; in
  Communist areas, 325, 336
Tao, Y. C., 125
Taoism, 65, 93, 204
Tarle, M., 139
Tehchow Arsenal, 261, 264
Tengchow, 57–9, 109
textiles, manufacture of, 158–9, 162, 260, 262, 274,
  281, 310; in Communist areas, 324–5
Thomson, James C., 121, 125, 139
T'ien-ch'u Monosodium Glutamate Factory, 169, 259,
  280
T'ien-kung k'ai-wu, 246, 330
T'ien-li Nitrogen Gas Company, 268, 276
T'ien-yüan Electrochemical Factory, 259–60, 275–6
Tientsin, 61, 64, 101, 290, 326; academic center, 111,
  138–9; industry in, 154, 162, 236, 262–3, 279
Tientsin Arsenal, 24, 62–3
Ting Hsü-hsien, 81, 373
Ting-pien County, 360
Ting Wen-chiang: proponent of science, 82, 92, 134;
  secretary-general, Academia Sinica, 251–3
Tisdale, W. E., 181; survey of Chinese universities,
  218–23, 226–7
translation, of scientific texts, 15; at Kiangnan Arsenal,
  29–30, 36–40, 65; before 1870, 30–6; of chemical
  terms, 40–5, 190–5; in popular journals, 45–8; in
  Japan, 42, 50–1; of the term science 82–8; after 1900,
  83–4; Japanese influence on, 83–8; names of chemical
  elements, 377–80
translation bureau [fan-i-kuan], Kiangnan Arsenal,
  29–30, 36
Ts'ai Liu-sheng, 125, 236, 371
Ts'ai Yüan-p'ei, 202, 209, 213; president, Academia
  Sinica, 251, 253, 315
Ts'ao Pen-hsi, 367
Tseng Chao-lun, 52, 133; biography, 179, 187–8;
  research, 225, 233–6, 255, 310; in PRC, 371, 373
Tseng Kuo-fan, 23–5, 29–30, 73, 187
Tsiang T'ing-fu, 202, 210–12
Tsinan, 279; Incident (1927), 269
Tsinghua University, 77, 92; early history, 113–15, 137;
  1927–37, 187, 198, 210, 212, 222–6; publications,
  224–5; research at, 231–2, 235; during war, 303–4,
  332; in PRC, 365, 367, 371
Tsingtao, 337
Tsungli Yamen, 24, 53
Tu-fu-ch'uan-k'ou, 342
Tu-li p'ing-lun [Independent critic], 141, 210
Tung-fang tsa-chih [Eastern miscellany], 252
Tung Li Oil Works, 243, 298–9
tung oil, 125, 243, 298, 312
T'ung Wen Kuan, 24, 34–6, 53, 56, 109–10
Tung Wen-li, 332, 336–7, 345–6
Tungpei University, 186
Twiss, George Ransom, 128
Tzu-fang-kou, 327
Tzuliuching, 295, 296

Udagawa, Yōan, 51, 84
United Chemical Works, 260
United Front, 253, 290, 323–4, 359
University of California, Davis, 187
University of Chicago, 186
University of Michigan, 234
University of Wisconsin, 150, 186, 224, 235

Van Slyke, Donald, 143, 145
vegetable oil, synthesis of gasoline from, 243–4, 335
vitamin C, 232–3
vitamin K, 232–4, 238, 311

Wan-kuo kung-pao [The globe magazine], 45
Wang Chen-sheng, 38

Wang Chin, 113, 139, 237, 371, 373
Wang Feng-tsao, 110
Wang Pao-jen, 225
Wang Shih-chen, 327, 332, 345
Wang Shih-chieh, 213
Wang, Y. C., 178–9, 182
Wang Yang-ming, 67
Welch, William, 117
Wells, David Ames, 35, 44
Wen-hua hsien-feng [Cultural pioneer], 316
West China Academy of Sciences, 244, 298
Whang, S. L., 227
Whang Siar-hong, 228
Williamson, Alexander, 51
Wilson, Earl, 121, 125, 139
Wilson, Stanley D., 113–15, 119, 125, 139
Windaus, Adolph, 186, 187, 237
women in science, 217–18, 390
Wong, B. F., 263
Wong Wen-hao, 134, 152, 182, 203
Woo Sho-chow (Wu Hsüeh-chou), 237, 311, 371
Woo Yui-hsun, 137
Wu Ch'eng-lo, 165
Wu-ch'i-chen, 326
Wu Chih-hui, 82, 87, 90
Wu Ching, 296
Wu Heng, 346, 348, 375
Wu Hsien, 144, 152, 165, 180, 187–8, 252, 370;
   biography, 140–2; research, 142–8, 249; political
   views, 213
Wu Hsüeh-chou, see Woo Sho-chow
wu-shen [wizards], 360–1
Wu, T. Y., 309
Wu Tsao-hsi, 317
Wu Yü-chang, 348, 358
Wu Yün-ch'u, 160, 268, 278–9; biography and early
   career, 167–70; electrolysis of salt, 259–61
Wuchow, 263
Wuhan, 293, 332
Wuhsi, 18–25, 30

Wutungchiao, 295
Wylie, Alexander, 30, 51

Ya-ch'üan tsa-chih [Ya-ch'üan magazine], 84
Yale University, 232
Yang, C. N., 309
Yang Ch'üan, 97
yang-ko [folk dance], 361
Yang Shih-hsien, 371, 373
Yeh Ch'ing, 320–2
Yeh Kung-chia, 277
Yen, Daisy, 145, 148
Yen Fu, 87–9, 93
Yen P'ei-lin, 345
Yen-wu-kou, 334
Yenan Forum on Literature and Art, 353
Yenan University, 358
Yenchang County, 272–3, 333–4
Yenching University, 121, 124, 125, 130, 139; 1927–37,
   217, 221–3, 231, 236
Yentai, 337
Ying-yang kai-lun [Principles of nutrition], 148
Yü Ho-ch'in, 192
Yü Kuang-yüan, 348
Yü-kung, 20
Yü Ta-wei, 187
Yüan Han-ch'ing, 225, 314, 371
Yüan Shih-kai, 76, 161
Yümen County, 297, 369
Yung-p'ing, 334
Yung Wing, 30
Yungli Chemical Industry Company, 264–9, 266–7, 274
Yungli Soda Manufacturing Company, 164; early
   history, 164–7, 170; during 1927–37, 260, 275,
   279–83; during war, 294–5
Yunnan, 288, 304

Zikawei [Hsü-chia-hui] Observatory, 135
zoology, 34, 57, 110, 200, 225; in middle school
   curricula, 388